Advanced Materials for Societal Implementation

Advanced Materials for Societal Implementation

Editors

Hideyuki Kanematsu
Yoshikazu Todaka
Takaya Sato

MDPI • Basel • Beijing • Wuhan • Barcelona • Belgrade • Manchester • Tokyo • Cluj • Tianjin

Editors

Hideyuki Kanematsu
Collaborative Research
Promotion Center
National Institute of
Technology (KOSEN), Suzuka
College
Suzuka
Japan

Yoshikazu Todaka
Department of Mechanical
Engineering
Toyohashi University of
Technology
Toyohashi, Aichi
Japan

Takaya Sato
President
National Institute of
Technology (KOSEN),
Okinawa College
Nago, Okinawa
Japan

Editorial Office
MDPI
St. Alban-Anlage 66
4052 Basel, Switzerland

This is a reprint of articles from the Special Issue published online in the open access journal *Materials* (ISSN 1996-1944) (available at: www.mdpi.com/journal/materials/special_issues/ KOSEN_ISATE2021_materials).

For citation purposes, cite each article independently as indicated on the article page online and as indicated below:

LastName, A.A.; LastName, B.B.; LastName, C.C. Article Title. *Journal Name* **Year**, *Volume Number*, Page Range.

ISBN 978-3-0365-6132-5 (Hbk)
ISBN 978-3-0365-6131-8 (PDF)

© 2023 by the authors. Articles in this book are Open Access and distributed under the Creative Commons Attribution (CC BY) license, which allows users to download, copy and build upon published articles, as long as the author and publisher are properly credited, which ensures maximum dissemination and a wider impact of our publications.

The book as a whole is distributed by MDPI under the terms and conditions of the Creative Commons license CC BY-NC-ND.

Contents

About the Editors vii

Preface to "Advanced Materials for Societal Implementation" ix

Ryuto Kamimura, Hideyuki Kanematsu, Akiko Ogawa, Takeshi Kogo, Hidekazu Miura and Risa Kawai et al.
Quantitative Analyses of Biofilm by Using Crystal Violet Staining and Optical Reflection
Reprinted from: *Materials* **2022**, *15*, 6727, doi:10.3390/ma15196727 1

Hikonaru Kudara, Hideyuki Kanematsu, Dana M. Barry, Akiko Ogawa, Takeshi Kogo and Hidekazu Miura et al.
Proposal for Some Affordable Laboratory Biofilm Reactors and Their Critical Evaluations from Practical Viewpoints
Reprinted from: *Materials* **2022**, *15*, 4691, doi:10.3390/ma15134691 11

Nozomu Adachi, Haruki Ueno, Satoshi Morooka, Pingguang Xu and Yoshikazu Todaka
Deformation Texture of Bulk Cementite Investigated by Neutron Diffraction
Reprinted from: *Materials* **2022**, *15*, 4485, doi:10.3390/ma15134485 23

Ryo Satoh, Takashi Morinaga and Takaya Sato
Novel Dry Spinning Process of Natural Macromolecules for Sustainable Fiber Material -1- Proof of the Concept Using Silk Fibroin
Reprinted from: *Materials* **2022**, *15*, 4195, doi:10.3390/ma15124195 31

Masamoto Tafu, Juna Nakamura, Momoka Tanii, Saori Takamatsu and Atsushi Manaka
Improved On-Site Characterization of Arsenic in Gypsum from Waste Plasterboards Using Smart Devices
Reprinted from: *Materials* **2022**, *15*, 2446, doi:10.3390/ma15072446 41

Akihiro Takahashi, Naoyuki Yamamoto, Yu Ooka and Toshinobu Toyohiro
Tensile Examination and Strength Evaluation of Latewood in Japanese Cedar
Reprinted from: *Materials* **2022**, *15*, 2347, doi:10.3390/ma15072347 49

Mitsuharu Todai, Keisuke Fukunaga and Takayoshi Nakano
Athermal Phase and Lattice Modulation in Binary Zr-Nb Alloys
Reprinted from: *Materials* **2022**, *15*, 2318, doi:10.3390/ma15062318 65

Kazuyuki Furuya, Shiro Jitsukawa and Takayuki Saito
Application of the Sinter-HIP Method to Manufacture Cr–Mo–W–V–Co High-Speed Steel via Powder Metallurgy
Reprinted from: *Materials* **2022**, *15*, 2300, doi:10.3390/ma15062300 79

Nobumitsu Hirai, Masaya Horii, Takeshi Kogo, Akiko Ogawa, Daisuke Kuroda and Hideyuki Kanematsu et al.
Simple Methods for Evaluating Acid Permeation and Biofilm Formation Behaviors on Polysiloxane Films
Reprinted from: *Materials* **2022**, *15*, 2272, doi:10.3390/ma15062272 89

Koichiro Ogata, Tsutomu Harada, Hideo Kawahara, Kazuki Tokumaru, Riho Abe and Eiji Mitani et al.
Characteristics of Vibrating Fluidization and Transportation for Al_2O_3 Powder
Reprinted from: *Materials* **2022**, *15*, 2191, doi:10.3390/ma15062191 97

Hiroyuki Arafune, Yuma Watarai, Toshio Kamijo, Saika Honma and Takaya Sato
Mechanical and Lubrication Properties of Double Network Ion Gels Obtained by a One-Step Process
Reprinted from: *Materials* **2022**, *15*, 2113, doi:10.3390/ma15062113 115

Koichi Shigeno, Takuma Yano and Hirotaka Fujimori
Solid-State-Activated Sintering of $ZnAl_2O_4$ Ceramics Containing $Cu_3Nb_2O_8$ with Superior Dielectric and Thermal Properties
Reprinted from: *Materials* **2022**, *15*, 1770, doi:10.3390/ma15051770 125

Ryosuke Hashimoto, Toshiya Itaya, Hironaga Uchida, Yuya Funaki and Syunsuke Fukuchi
Properties of Magnetic Garnet Films for Flexible Magneto-Optical Indicators Fabricated by Spin-Coating Method
Reprinted from: *Materials* **2022**, *15*, 1241, doi:10.3390/ma15031241 145

Yasuyuki Ueda, Yuki Kurokawa, Kei Nishii, Hideyuki Kanematsu, Tadashi Fukumoto and Takehito Kato
Morphology Control of Monomer–Polymer Hybrid Electron Acceptor for Bulk-Heterojunction Solar Cell Based on P3HT and Ti-Alkoxide with Ladder Polymer
Reprinted from: *Materials* **2022**, *15*, 1195, doi:10.3390/ma15031195 155

Vanpaseuth Phouthavong, Ruixin Yan, Supinya Nijpanich, Takeshi Hagio, Ryoichi Ichino and Long Kong et al.
Magnetic Adsorbents for Wastewater Treatment: Advancements in Their Synthesis Methods
Reprinted from: *Materials* **2022**, *15*, 1053, doi:10.3390/ma15031053 165

Satoshi Achira, Yohei Abe and Ken-ichiro Mori
Self-Pierce Riveting of Three Thin Sheets of Aluminum Alloy A5052 and 980 MPa Steel
Reprinted from: *Materials* **2022**, *15*, 1010, doi:10.3390/ma15031010 209

Akira Yamauchi and Masashi Kurose
Effect of Sb and Zn Addition on the Microstructures and Tensile Properties of Sn–Bi-Based Alloys
Reprinted from: *Materials* **2022**, *15*, 884, doi:10.3390/ma15030884 227

Beomdeok Seo, Hideyuki Kanematsu, Masashi Nakamoto, Yoshitsugu Miyabayashi and Toshihiro Tanaka
Copper Surface Treatment Method with Antibacterial Performance Using "Super-Spread Wetting" Properties
Reprinted from: *Materials* **2022**, *15*, 392, doi:10.3390/ma15010392 239

Ryo Shomura, Ryota Tamate and Shoichi Matsuda
Lithium-Ion-Conducting Ceramics-Coated Separator for Stable Operation of Lithium Metal-Based Rechargeable Batteries
Reprinted from: *Materials* **2022**, *15*, 322, doi:10.3390/ma15010322 253

Michiko Yoshitake
Tool for Designing Breakthrough Discovery in Materials Science
Reprinted from: *Materials* **2021**, *14*, 6946, doi:10.3390/ma14226946 265

Masakazu Kobayashi, Masanobu Izaki, Pei Loon Khoo, Tsutomu Shinagawa, Akihisa Takeuchi and Kentaro Uesugi
High-Resolution Mapping of Local Photoluminescence Properties in CuO/Cu_2O Semiconductor Bi-Layers by Using Synchrotron Radiation
Reprinted from: *Materials* **2021**, *14*, 5570, doi:10.3390/ma14195570 281

About the Editors

Hideyuki Kanematsu

Professor Kanematsu, a Fellow of the ASM International (the USA) and a Fellow of IMF (United Kingdom), is a Professor Emeritus at the National Institute of Technology at Suzuka College and a Visiting Professor at Nagoya University. He holds an M. Eng. degree and a Ph.D. in Materials Science and Engineering from Nagoya University in Japan. He has been an active researcher in Surface Science and Engineering Materials for many years and has over 600 academic publications. He has numerous honors to his name, including Japan's Minister of Education, Culture, Sports, Science, and Technology Award (considered the number one prize from the Minister of Education in the Japanese government).

Yoshikazu Todaka

Prof. Yoshikazu Todaka is a Professor at the Toyohashi University of Technology. He holds a doctorate of engineering from the Toyohashi University of Technology in Japan. He has been an active researcher in physical metallurgy for many years and has around 200 academic publications.

Takaya Sato

Dr. Takaya Sato is currently the president of the National Institute of Technology (NIT), Okinawa College, JAPAN, and a professor of polymer chemistry. He studied polymer and fiber chemistry at Shinshu University and Kyoto University. He received his Ph.D. in polymer chemistry from Kyoto University in 1992 under the direction of Professor Takeaki Miyamoto. After acquiring a degree, he has been engaged in developing functional polymer materials such as micro-particles for cosmetics, gel materials for microbial carriers, solid polymer electrolytes for lithium-ion batteries, and some novel ionic liquids in Nisshinbo Holdings. Inc. In 2003, it was the first time in the world to commercialize an electric double-layer capacitor using an ionic liquid as an electrolyte after moving to the NIT. Tsuruoka College as a professor in 2004, is continuing the development of ionic liquids and related polymer materials and is vigorously researching to apply them to electrochemical devices and low friction materials.

Preface to "Advanced Materials for Societal Implementation"

Materials science is the fundamental basis for all engineering disciplines. From mechanical engineering to medical science, electrical engineering, and chemical engineering to information technology, every field requires knowledge and information about materials science and engineering. We can say that everything is a material. In Japan, the National Institute of Technology (KOSEN), one of the most prominent higher education organizations, started an extensive research project throughout the country. NIT KOSEN was the first established network in the islands of Japan. The 51 colleges of KOSEN are established across Japan, and they consolidated into one big organization almost 17 years ago. Now, it is actively pursuing an industrial–academia partnership to carry out joint experiments, develop collaboration products, and also to cultivate young engineers for the future through these joint projects. KOSEN started its new nationwide research project, called GEAR 5.0, in May 2020. This is basically a research and education project to train young students heading into industrial fields in KOSEN, so that they can adjust themselves to a new innovative society (Society 5.0 in Japan or Industry 4.0 in Germany) and can actively make significant contributions to these societies. In this large national project for KOSEN, certain main engineering disciplines were chosen, and their research hubs were established in particular areas. Currently, research activities are taking place to boost industrial and economic situations. Materials science and engineering are some of the critical disciplines selected for the GEAR 5.0 project.

In this Special Issue, we broadly called for papers relating to materials science and engineering, according to the philosophy of GEAR 5.0 project and KOSEN research. As mentioned above, topics of interest were not restricted to basic or conventional topics for materials science and engineering. We instead accepted papers dealing with versatile applications of materials science and engineering in various industrial fields. The call for papers was initially based on talks given at ISATE 2021, an international conference with a secretariat in Finland. However, this reprint is a compilation of papers published as a Special Issue of MDPI *Materials*, including, but not limited to, other papers that are oriented toward social implementation as the ultimate goal. The editors hope that this reprint will be used as a good example, a legend, and a textbook for social implementation of materials science.

Hideyuki Kanematsu, Yoshikazu Todaka, and Takaya Sato
Editors

Article

Quantitative Analyses of Biofilm by Using Crystal Violet Staining and Optical Reflection

Ryuto Kamimura [1], Hideyuki Kanematsu [1,*], Akiko Ogawa [1], Takeshi Kogo [1], Hidekazu Miura [2], Risa Kawai [1], Nobumitsu Hirai [1], Takehito Kato [3], Michiko Yoshitake [4] and Dana M. Barry [5,6]

1. National Institute of Technology (KOSEN), Suzuka College, Suzuka 510-0294, Mie, Japan
2. Faculty of Medical Engineering, Suzuka University of Medical Science, Suzuka 510-0293, Mie, Japan
3. National Institute of Technology (KOSEN), Oyama College, Oyama 323-0806, Tochigi, Japan
4. National Institute for Materials Science (NIMS), Tsukuba 305-0044, Ibaraki, Japan
5. Department of Electrical & Computer Engineering, Clarkson University, Potsdam, NY 13699, USA
6. STEM Laboratory, State University of New York, Canton, NY 13617, USA
* Correspondence: kanemats@mse.suzuka-ct.ac.jp; Tel.: +81-59-368-1848

Abstract: Biofilms have caused many problems, not only in the industrial fields, but also in our daily lives. Therefore, it is important for us to control them by evaluating them properly. There are many instrumental analytical methods available for evaluating formed biofilm qualitatively. These methods include the use of Raman spectroscopy and various microscopes (optical microscopes, confocal laser microscopes, scanning electron microscopes, transmission electron microscopes, atomic force microscopes, etc.). On the other hand, there are some biological methods, such as staining, gene analyses, etc. From the practical viewpoint, staining methods seem to be the best due to various reasons. Therefore, we focused on the staining method that used a crystal violet solution. In the previous study, we devised an evaluation process for biofilms using a color meter to analyze the various staining situations. However, this method was complicated and expensive for practical engineers. For this experiment, we investigated the process of using regular photos that were quantified without any instruments except for digitized cameras. Digitized cameras were used to compare the results. As a result, we confirmed that the absolute values were different for both cases, respectively. However, the tendency of changes was the same. Therefore, we plan to utilize the changes before and after biofilm formation as indicators for the future.

Keywords: biofilms; crystal violet; optical reflection; color analyses; XYZ color plane; L*a*b* color plane

1. Introduction

A biofilm (BF) is a thin film of material formed by bacterial activity on the surface of a material or other interface. The fundamental concepts of biofilms have been clarified and explained by many researchers and summarized in some books, reviews, etc. [1–5]. In most cases, they form on materials' surfaces. Therefore, biofilm formation must be affected by materials' surfaces. From this viewpoint, we have tried to show how materials affect biofilm formation and growth [6,7]. BF is composed of about 80% water, EPS (extracellular polymeric substances), and bacteria. It has a characteristic sliminess. This sliminess is said to be caused by quorum sensing, a phenomenon in which bacteria adhere to the surface of material, multiply, and expel polysaccharides outside of the colony. In addition to polysaccharides, proteins, lipids, and nucleic acids (DNA and RNA) are produced in BF, which collectively is called EPS (Figure 1).

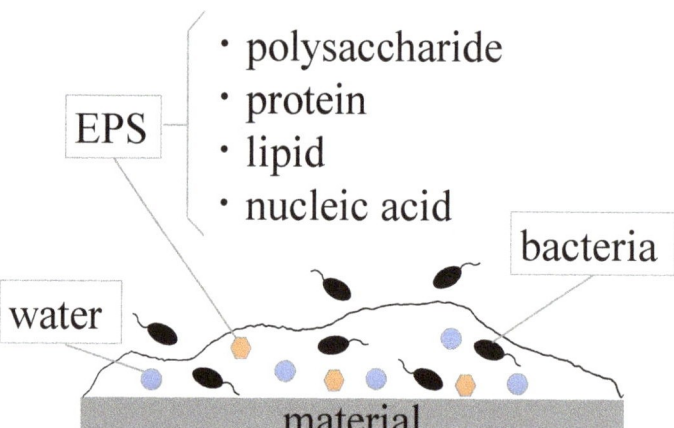

Figure 1. Schematic diagram of biofilms on materials.

Bacteria in BFs have different properties from those of ordinary airborne bacteria. This affects various fields, such as medicine, environmental science, architecture, mechanics, chemical and pharmaceutical engineering, pharmaceuticals, and materials science. To control these effects, BF must be accurately evaluated by taking appropriate measures.

As described above, the appropriate evaluation method for BF is very important and the main premise for the following development of anti-biofilm materials. The evaluation methods are mainly classified into two types. One of them is the evaluation group composed of many versatile analytical instruments. These include various microscopes, such as optical microscopes, electron microscopes, confocal laser microscopes, etc., or various analytical facilities, such as Raman spectroscopy, FT-IR spectroscopy, etc. The other type is the biological evaluation group that is composed of gene analyses, staining methods, etc. These two types are combined appropriately to produce new advanced analytical methods. Examples include electron microscopes [8–12], confocal laser microscopes [13–17], IR measurements [18–21], and Raman spectroscopy [22–29]. Some proposed methods have made great contributions for clarifying the biological essence of biofilms and their relationship with materials and environments. These methods provided us with qualitive, semi-quantitative, and quantitative analyses for our research projects. However, we still need other new evaluation methods for practical applications. Practical applications mean that researchers, engineers, and general users (facing practical industrial or daily life problems) could use them to check biofilms quantitatively as well as qualitatively, and above all, products that have relatively large and unsteady shapes should be analyzed directly. In such a case, the evaluation method requires swiftness and simplicity. From the practical viewpoint, the measurement condition should be close to satisfy those requirements as much as possible. To satisfy the purpose, the SIAA (Society of International Sustaining Growth for Antimicrobial Articles, Japan, Tokyo), composed of more than 1000 Japanese companies in the antimicrobial materials field, are going to establish an ISO and we expect that it would be valid until March 2023. In this method, crystal violet staining [30–32] biofilms are extracted into sodium dodecyl sulfate (SDS) solution and the absorbance by 590 nm light is defined as the quantity of biofilms. However, if the stained colors of specimens could be evaluated directly, the process would be simpler. Therefore, we carried out some experiments as trials to determine the biofilm quantity by measuring surface color at the stained biofilms, so that the newly proposed method would lead to the modified quantification method in the future.

2. Experimental Section

2.1. Substrate Specimens

In this experiment, commercially available PE (polyethylene sheet), and pure Titanium specimens were used as substrates. Thin sheets (0.5 to 1.0 mm thick) of each material were cut into 10×10 mm^2 pieces using metal shears and they were cleaned with alcohol. We used two specimens because we wanted to confirm the applicability of the proposed method in this experiment to both metallic materials and polymeric substances.

2.2. Bacteria

Escherichia coli (*E. coli*, K12 G6) were used as model bacteria in this study. The bacteria were selected due to the following two reasons. First, the model bacteria for this study should have low risk and should be easy to deal with. Next, we often used these bacteria in previous studies and have accumulated versatile data and experiences. Therefore, we used *E. coli* as our model bacteria.

2.3. Biofilm Formation

Biofilm formation was carried out by a static method. Luria–Bertani (LB) liquid medium (2068-75, M9T2881, Nacalai Co., Kyoto, Japan) was autoclaved at 121 °C for 15 min and *E.coli* were added in LB medium, so that the colony formation unit (CFU) per milli liter (mL) was around 1×10^9 after a shaking incubation at 37 °C for 24 h. Next, the bacterial solution was put into 12 plastic wells, so that each well was filled with 1.2 mL of solution. Then, the specimens were immersed into wells for 0, 1, and 3 days at 25 °C in an incubator.

2.4. Raman Spectroscopy

We used Raman spectroscopy as a confirmation method to verify that biofilms were really formed. Specimens with BFs were pretreated by freeze dehydration in advance to carry out Raman spectroscopy. The freeze dehydration process is composed of two steps. One of them is the substitution of water in BFs with alcohol, and the other is vacuuming. The concrete steps are described as follows.

The aqueous solutions were adjusted so that the ratios of distilled water: ethanol (Ethanol, C_2H_5OH, 99.5%, Reagent Special Grade, 057-00451, APQ8101, Wako Pure Chemical Industries, Ltd., Osaka, Japan) were 7:3, 5:5, 3:7, 2:8, 1:9, 0.5:9.5, 0.2:9.8, and 0:1. Solutions of ethanol and t-butyl alcohol (tert-Butyl alcohol, 2-Methyl-2-propanol, special grade, 000-10915, G72121J, Kishida Chemical Co., Osaka, Japan) in the wells were aspirated with a dropper; the adjusted solution was added with a dropper and replaced in turn, and the wells were allowed to stand for 15 min. After alcohol displacement, the samples were frozen in a freezer and vacuumed using a vacuum pump.

Raman spectroscopy was carried out, using a Raman spectrometer (LabRAM HR Evolution, Horiba, Kyoto, Japan). A laser beam (532 nm) was irradiated onto the sample's surface (diameter: approximately 1 μm), and the Raman shift was measured three times (N = 3) under the following conditions: -50% attenuation filter, 3 s exposure time, 5 integration times, grating: 300 gr/mm, and measurement wavelength range: 500 cm^{-1}–3500 cm^{-1}.

2.5. Color Analyses

An aqueous solution containing 0.1% crystal violet (CV) was prepared to stain specimens. The solution was used as a standard solution for ISO. This is because we have investigated some cases using the solution in the past. After bacterial solutions were removed from the wells, the CV solution was put into the well containing the sample and the immersion continued for 30 min. Then, the CV solution was removed, and pure water was poured into the wells to remove non-special absorbed CV, which was washed away from the specimens' surfaces. Then the water was immediately removed. This washing process was repeated three times. As a result, we obtained stained specimens that correspond to the amounts of biofilm present.

To evaluate the extent of biofilm formation on specimens, the staining must be analyzed quantitatively. In usual cases, the stained parts are extracted into a proper solution and the absorbances are measured [33,34]. On the contrary, we measured the stained violet color on specimens by optical reflection, using color meters. Then, by combining three color parameters, L*, a* and b* were obtained [35]. In this study, we analyzed the color reflection of stained parts using photos and image analyses. A digital camera (1066C004, PSG7X Mark II, Canon Inc., Tokyo, Japan), a black box, and a ring light source using a white LED were set up for photographing the specimens. The camera parameters used in the shooting were aperture f = 9.0, shutter speed SS = 1/40, and ISO sensitivity 125. The photographed samples were analyzed using ImageJ, and histograms of each RGB color within the measurement range on the image were obtained. The histograms were converted into the XYZ color system (Equation (1)).

$$\begin{pmatrix} X \\ Y \\ Z \end{pmatrix} = \begin{pmatrix} 0.4124 & 0.3576 & 0.1805 \\ 0.2126 & 0.7152 & 0.0722 \\ 0.0193 & 0.1192 & 0.9505 \end{pmatrix} \quad (1)$$

Then, they were converted into the Yxy color system (Equation (2)), and plotted on the xy chromaticity diagram, according to the following equations:

$$x = \frac{X}{X+Y+Z}, \quad y = \frac{Y}{X+Y+Z} \quad (2)$$

To compare the results by this new method with those by using a color meter in the past, we measured the specimens' stained surfaces, using a color measurement device (Color Reader, CR-13, Konica Minolta Sensing, Inc., Tokyo, Japan). The results were then plotted on an xy color diagram.

3. Results and Discussions

3.1. Confirmation of Biofilm Formation on Both Specimens

The results of the Raman spectrometer measurements of pure titanium specimens immersed in E. coli culture solution (for different periods of time) are shown in Figure 2. The wavenumber (cm^{-1}) is on the horizontal axis and the intensity is on the vertical axis.

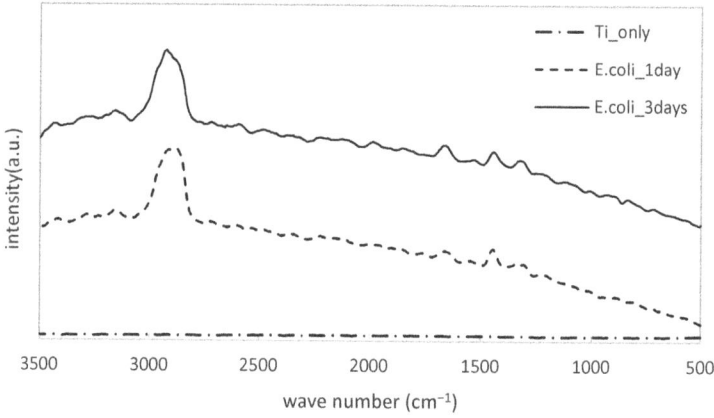

Figure 2. Change of Raman shifts for titanium specimens immersed in LB media filled with E. coli.

The results for the Ti substrate alone showed no specific peaks. On the contrary, specimens immersed in the bacterial solution of LB showed peaks at 2930 cm^{-1}, 1660 cm^{-1}, 1440 cm^{-1}, and 1320 cm^{-1}. These are typical peaks for biofilms as compared to those we previously confirmed for specimens where biofilms formed on them.

Figure 3 shows the results of the Raman spectrometer measurements of the samples immersed in *E. coli* culture solution (for different periods of time) on the PE substrate.

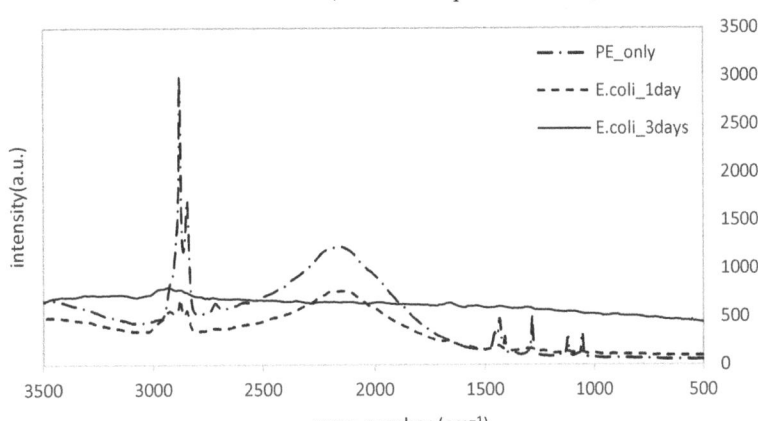

Figure 3. Raman shifts of PE specimens with immersion time.

In the measurement for only the PE substrate, sharp peaks were detected at 2880 cm^{-1}, 1440 cm^{-1}, 1290 cm^{-1}, 1130 cm^{-1}, and 1060 cm^{-1}, and a broad peak was found at around 2160 cm^{-1}, respectively. Obviously, these peaks were derived from PE itself. However, we could observe that these original PE-derived peaks were clearly reduced. Furthermore, the extent of the reduction increased with the immersion time. Figure 3 shows that it was hard for us to analyze biofilm peaks because the PE-derived peaks were relatively strong. We enlarged the results for the specimen immersed in *E. coli* for 3 days. They are displayed in Figure 4.

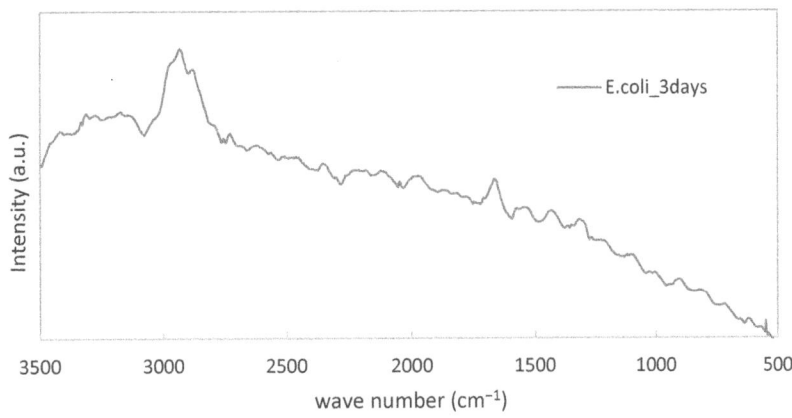

Figure 4. Enlarged results for PE immersed in LB bacterial solution.

In Figure 4, peaks at 2930 cm^{-1}, 1660 cm^{-1}, 1440 cm^{-1}, and 1320 cm^{-1} were also observed, even though they were not so remarkable. Since they were typical peaks for specimens with biofilms, we could confirm biofilms also on the surface of PE.

3.2. Results of Staining Evaluations

After staining samples on Ti substrates with different immersion periods in the *E. coli* culture medium, measurements were made using a colorimeter. The results were plotted on an xy chromaticity diagram, as shown in Figure 5 below.

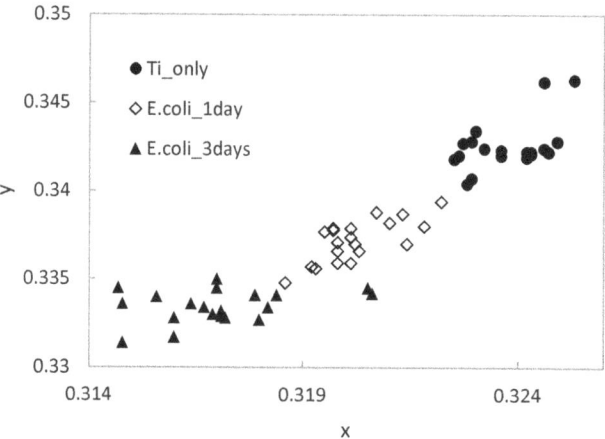

Figure 5. Color changes of stained pure titanium specimens with immersion times.

The dots in the upper right of the figure for the Ti substrate only shift to the lower left as the immersion period in the *E. coli* culture medium increases. The mean values and standard deviations of the colorimetric measurements of Ti substrate and Ti specimen immersed in *E. coli* culture medium for 3 days are shown in Table 1 below.

Table 1. Average values and their standard deviations for stained titanium specimens.

	Average	Standard Deviation
Ti only (x,y)	(0.3237, 0.3425)	(0.0009, 0.0014)
E. coli on Ti for 3 days (x,y)	(0.3171, 0.3335)	(0.0016, 0.0009)

After staining the PE substrate, the samples were immersed in the *E. coli* culture solution for different periods of time. Then, they were measured by using a colorimeter and plotted on an xy chromaticity diagram (Figure 6). In the PE samples, as in the Ti samples, there is a tendency that the point cloud shifts to the lower left as the immersion period increases.

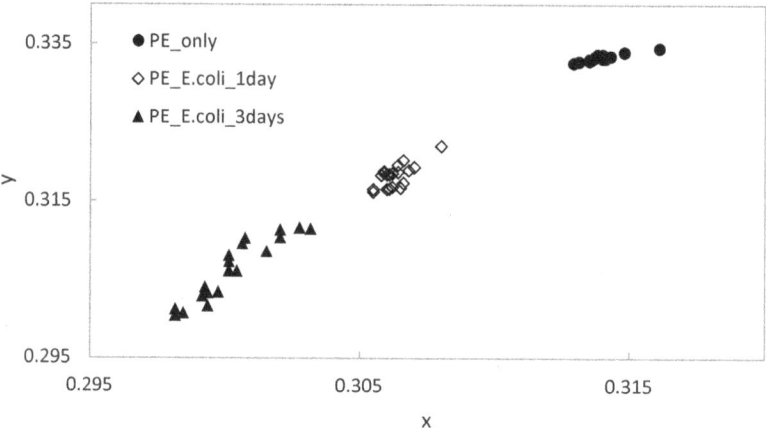

Figure 6. Color changes of stained PE specimens with immersion times.

The color difference between two points (x1, y1) and (x2, y2) on the xy chromaticity diagram is defined as ΔC, according to the following equation:

$$\Delta C = \sqrt{\{(x1-x2)^2 + (y1-y2)^2\}} \quad (3)$$

The color difference between the substrate and samples immersed in the *E. coli* culture medium (for 3 days) in Ti and PE, respectively, was measured using a colorimeter. Table 2 shows the color difference between the two samples.

Table 2. Color difference between titanium and PE specimens.

	Ti	PE
Color Difference ΔC	0.01126	0.03037

After staining samples on Ti substrates with different immersion periods in the *E. coli* culture medium, measurements were made using an image analysis technique. The results were plotted on an xy chromaticity diagram, shown in Figure 7 below.

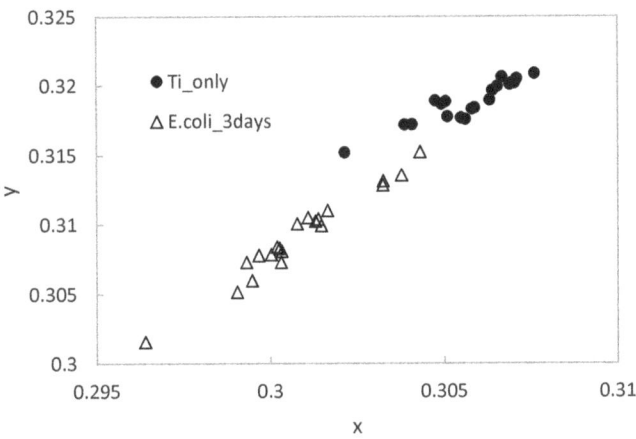

Figure 7. Color changes based on image analyses and calculations.

The mean values and standard deviations of the results of staining the Ti substrate and the samples immersed in the *E. coli* culture medium (for 3 days) are shown in Table 3.

Table 3. Average values and their standard deviations based on image analyses and calculations.

	Average	Standard Deviation
Ti only (x,y)	(0.30670, 0.31886)	(0.00132, 0.00146)
E. coli on Ti for 3 days (x,y)	(0.30087, 0.330926)	(0.00184, 0.00317)

Results for the color evaluations show that the mean values of the two methods (color meter measurements and image analyses) were different, but the standard deviation and the trend of the color change in the immersed samples were similar. The colorimetric method was the same as the image analysis method.

We started this research project to complete the evaluation method as a quantitative one. At this point, we have not completed it. However, we showed that this method has the potential to be a substitution method for the extraction one. To rapidly quantify the biofilm of products or large-scale specimens, the color change should be expressed as concrete figures, such as vector values or more statistical ones. We will continue this project to obtain our final goal.

4. Conclusions

In this experiment, we investigated the process where usual photos were quantified without any instruments, except for the usual digitized cameras. The results were compared by using a digitized camera. We obtained the following results from our experiments:

(1) We confirmed by Raman spectroscopy that biofilms formed both on titanium and PE specimens, respectively.
(2) Although the average of the number of color values obtained by the method using image analysis is different from that by the method using a colorimeter, the accuracy and trend of the shift of the point cloud are almost the same. Therefore, the method using image analysis is effective as an alternative colorimetric method to the method using a colorimeter.
(3) We found that, in the future, it is possible that the image analyses from photos could be applied to the evaluation of biofilms.

Author Contributions: Writing—original draft preparation, R.K. (Ryuto Kamimoto); Data curation, A.O., R.K. (Risa Kawai) and T.K. (Takeshi Kogo); Formal analysis, R.K. (Ryuto Kamimura), A.O. and T.K. (Takeshi Kogo); project administration, H.K., H.M. and D.M.B.; writing—review and editing, D.M.B. and H.K.; funding, H.M. and N.H.; investigation, T.K. (Takeshi Kogo), R.K. (Ryuto Kamimura), A.O. and M.Y.; Methodology, N.H.; Project administration, H.K., H.M., T.K. (Takehito Kato), M.Y. and D.M.B.; supervision, T.K. (Takehito Kato) and N.H. All authors have read and agreed to the published version of the manuscript.

Funding: This work was supported by JSPS KAKENHI (Grants-in-Aid for Scientific Research from the Japan Society for the Promotion of Science, Grant Number 20K05185 and 21K12739). A part of this work was supported by the GEAR 5.0 Project of the National Institute of Technology (KOSEN) in Japan.

Institutional Review Board Statement: Not applicable.

Informed Consent Statement: Not applicable.

Data Availability Statement: Not applicable.

Acknowledgments: We appreciate the Advanced Technology R&D Center of Mitsubishi Electric Co., Japan Food Research Laboratories (JFRL), and The Society of International Sustaining Growth for Antimicrobial Articles (SIAA) for their very useful advice and information about biofilms.

Conflicts of Interest: The authors declare no conflict of interest The funders had no role in the design of the study; in the collection, analyses, or interpretation of data; in the writing of the manuscript; or in the decision to publish the results.

References

1. Characklis, W.G. Fouling biofilm development: A process analysis. *Biotechnol. Bioeng.* **1981**, *23*, 1923–1960. [CrossRef]
2. William, G.C.; Keith, E.C. Biofilms and microbial fouling. In *Advances in Applied Microbiology*; Elsevier: Amsterdam, The Netherlands, 1983; pp. 93–138.
3. Costerton, J.W.; Cheng, K.J.; Geesey, G.G.; Timothy, I.; Ladd, J. Curtis Nckel, Mrinal Dasgupta and Thomas J. Marrie. Bacterial biofilms in nature and disease. *Annu. Rev. Microbiol.* **1987**, *41*, 435–464. [CrossRef] [PubMed]
4. Lappin-Scott, H.M.; William Costerton, J. Bacterial biofilms and surface fouling. *Biofouling* **1989**, *1*, 323–342. [CrossRef]
5. Lappin-Scott, H.M.; Jass, J.; Costerton, J.W. Microbial biofilm formation and characterisation. In *Society for Applied Bacteriology Technical Series, Society for Applied Bacteriology Symposium*; Blackwell Scientific Publications: Oxford, UK, 1993; Volume 30.
6. Kanematsu, H.; Barry, M.D. *Biofilm and Materials Science*; Springer: New York, NY, USA, 2015.
7. Kanematsu, H.; Barry, M.D. *Formation and Control of Biofilm in Various Environments*; Springer Nature: Singapore, 2020; Volume 249.
8. Eighmy, T.T.; Maratea, D.; Bishop, P.L. Electron microscopic examination of wastewater biofilm formation and structural components. *Appl. Environ. Microbiol.* **1983**, *45*, 1921–1931. [CrossRef] [PubMed]
9. Anthony, G.S.; Peter, B.C.; Jurgen, R.; Allan, M.S.; Christopher, R.N.; John, H.; William, C.J. Biliary stent blockage with bacterial biofilm: A light and electron microscopy study. *Ann. Intern. Med.* **1988**, *108*, 546–553.

10. Lawrence, J.R.; Swerhone, G.D.W.; Leppard, G.G.; Araki, T.; Zhang, X.; West, M.M.; Hitchcock, A.P. Scanning transmission X-ray, laser scanning, and transmission electron microscopy mapping of the exopolymeric matrix of microbial biofilms. *Appl. Environ. Microbiol.* **2003**, *69*, 5543–5554. [CrossRef]
11. Priester, J.H.; Horst, A.M.; Van De Werfhorst, L.C.; Saleta, J.L.; Mertes, L.A.; Holden, P.A. Enhanced visualization of microbial biofilms by staining and environmental scanning electron microscopy. *J. Microbiol. Methods* **2007**, *68*, 577–587. [CrossRef] [PubMed]
12. Bossù, M.; Selan, L.; Artini, M.; Relucenti, M.; Familiari, G.; Papa, R.; Vrenna, G.; Spigaglia, P.; Barbanti, F.; Salucci, A.; et al. Characterization of Scardovia wiggsiae Biofilm by Original Scanning Electron Microscopy Protocol. *Microorganisms* **2020**, *8*, 807. [CrossRef] [PubMed]
13. Kuehn, M.; Hausner, M.; Bungartz, H.J.; Wagner, M.; Wilderer, P.A.; Wuertz, S. Automated confocal laser scanning microscopy and semiautomated image processing for analysis of biofilms. *Appl. Environ. Microbiol.* **1998**, *64*, 4115–4127. [CrossRef]
14. Lawrence, J.R.; Neu, T.R. [9] Confocal laser scanning microscopy for analysis of microbial biofilms. In *Methods in Enzymology*; Doyle, R.J., Ed.; Academic Press: San Diego, CA, USA, 1999; Volume 310, pp. 131–144. [CrossRef]
15. Akiyama, H.; Oono, T.; Saito, M.; Iwatsuki, K. Assessment of cadexomer iodine against Staphylococcus aureus biofilm in vivo and in vitro using confocal laser scanning microscopy. *J. Dermatol.* **2004**, *31*, 529–534. [CrossRef]
16. Shukla, S.K.; Rao, T.S. Effect of calcium on Staphylococcus aureus biofilm architecture: A confocal laser scanning microscopic study. *Colloids Surf. B Biointerfaces* **2013**, *103*, 448–454. [CrossRef]
17. Reichhardt, C.; Parsek, M.R. Confocal Laser Scanning Microscopy for Analysis of Pseudomonas aeruginosa Biofilm Architecture and Matrix Localization. *Front. Microbiol.* **2019**, *10*, 677. [CrossRef]
18. Bremer, P.J.; Geesey, G.G. An evaluation of biofilm development utilizing non-destructive attenuated total reflectance Fourier transform infrared spectroscopy. *Biofouling* **1991**, *3*, 89–100. [CrossRef]
19. Jürgen, S.; Hans-Curt, F. Ftir-spectroscopy in microbial and material analysis. *Int. Biodeterior. Biodegrad.* **1998**, *41*, 1–11.
20. Delille, A.; Quilès, F.; Humbert, F. In Situ Monitoring of the Nascent Pseudomonas fluorescens Biofilm Response to Variations in the Dissolved Organic Carbon Level in Low-Nutrient Water by Attenuated Total Reflectance-Fourier Transform Infrared Spectroscopy. *Appl. Environ. Microbiol.* **2007**, *73*, 5782–5788. [CrossRef]
21. Chirman, D.; Pleshko, N. Characterization of bacterial biofilm infections with Fourier transform infrared spectroscopy: A review. *Appl. Spectrosc. Rev.* **2021**, *56*, 673–701. [CrossRef]
22. Samek, O.; Al-Marashi, J.F.M.; Telle, H.H. The potential of raman spectroscopy for the identification of biofilm formation by staphylococcus epidermidis. *Laser Phys. Lett.* **2010**, *7*, 378–383. [CrossRef]
23. Millo, D.; Harnisch, F.; Patil, S.A.; Ly, H.K.; Schröder, U.; Hildebrandt, P. In Situ Spectroelectrochemical Investigation of Electrocatalytic Microbial Biofilms by Surface-Enhanced Resonance Raman Spectroscopy. *Angew. Chem. Int. Ed.* **2011**, *50*, 2625–2627. [CrossRef] [PubMed]
24. Jung, G.B.; Nam, S.W.; Choi, S.; Lee, G.-J.; Park, H.-K. Evaluation of antibiotic effects on Pseudomonas aeruginosa biofilm using Raman spectroscopy and multivariate analysis. *Biomed. Opt. Express* **2014**, *5*, 3238–3251. [CrossRef] [PubMed]
25. Keleştemur, S.; Avci, E.; Çulha, M. Raman and Surface-Enhanced Raman Scattering for Biofilm Characterization. *Chemosensors* **2018**, *6*, 5. [CrossRef]
26. Ogawa, A.; Kanematsu, H.; Sano, K.; Sakai, Y.; Ishida, K.; Beech, I.B.; Suzuki, O.; Tanaka, T. Effect of Silver or Copper Nanoparticles-Dispersed Silane Coatings on Biofilm Formation in Cooling Water Systems. *Materials* **2016**, *9*, 632. [CrossRef]
27. Kanematsu, H.; Kudara, H.; Kanesaki, S.; Kogo, T.; Ikegai, H.; Ogawa, A.; Hirai, N. Application of a Loop-Type Laboratory Biofilm Reactor to the Evaluation of Biofilm for Some Metallic Materials and Polymers such as Urinary Stents and Catheters. *Materials* **2016**, *9*, 824. [CrossRef]
28. Sano, K.; Kanematsu, H.; Hirai, N.; Ogawa, A.; Kougo, T.; Tanaka, T. The development of the anti-biofouling coating agent using metal nanoparticles and analysis by Raman spectroscopy and FIB system. *Surf. Coat. Technol.* **2017**, *325*, 715–721. [CrossRef]
29. Ogawa, A.; Takakura, K.; Hirai, N.; Kanematsu, H.; Kuroda, D.; Kougo, T.; Sano, K.; Terada, S. Biofilm Formation Plays a Crucial Rule in the Initial Step of Carbon Steel Corrosion in Air and Water Environments. *Materials* **2020**, *13*, 923. [CrossRef]
30. Kanematsu, H.; Ikigai, H.; Yoshitake, M. Evaluation of Various Metallic Coatings on Steel to Mitigate Biofilm Formation. *Int. J. Mol. Sci.* **2009**, *10*, 559–571. [CrossRef]
31. Kanematsu, H.; Nakagawa, R.; Sano, K.; Barry, D.M.; Ogawa, A.; Hirai, N.; Kogo, T.; Kuroda, D.; Wada, N.; Lee, S.; et al. Graphene-dispersed silane compound used as a coating to sense immunity from biofilm formation. *Med. Devices Sens.* **2019**, *2*, e10043. [CrossRef]
32. Tanaka, N.; Kogo, T.; Hirai, N.; Ogawa, A.; Kanematsu, H.; Takahara, J.; Awazu, A.; Fujita, N.; Haruzono, Y.; Ichida, S.; et al. In-situ detection based on the biofilm hydrophilicity for environmental biofilm formation. *Sci. Rep.* **2019**, *9*, 8070. [CrossRef]
33. Kanematsu, H.; Barry, D.M. Detection and evaluation of biofilms. In *Formation and Control of Biofilm in Various Environments*; Springer Nature: Singapore, 2020; pp. 111–154.
34. Kanematsu, H.; Barry, D.M.; Ikegai, H.; Mizunoe, Y. Biofilm Control on Metallic Materials in Medical Fields from the Viewpoint of Materials Science–from the Fundamental Aspects to Evaluation. *Int. Mater. Rev.* **2022**, 1–25. [CrossRef]
35. Takayanagi, M.; Kanematsu, H.; Miura, H.; Kogo, T.; Kawai, R.; Ogawa, A.; Hirai, N.; Kato, T.; Yoshitake, M.; Tanaka, T. Biofilms Formed on Metallic Materials by E. Coli and S. Epidermidis and Their Evaluation by Crystal Violet Staining and Its Reflection. *Trans. IMF* **2022**, *100*, 200–207. [CrossRef]

Proposal for Some Affordable Laboratory Biofilm Reactors and Their Critical Evaluations from Practical Viewpoints

Hikonaru Kudara [1], Hideyuki Kanematsu [1,*], Dana M. Barry [2,3], Akiko Ogawa [1], Takeshi Kogo [1], Hidekazu Miura [4], Risa Kawai [1], Nobumitsu Hirai [1], Takehito Kato [5] and Michiko Yoshitake [6]

1. National Institute of Technology (KOSEN), Suzuka College, Suzuka 510-0294, Japan; s18m172@kagawa-u.ac.jp (H.K.); ogawa@chem.suzuka-ct.ac.jp (A.O.); kougo@mse.suzuka-ct.ac.jp (T.K.); kawai-r@mse.suzuka-ct.ac.jp (R.K.); hirai@chem.suzuka-ct.ac.jp (N.H.)
2. Department of Electrical & Computer Engineering, Clarkson University, Potsdam, NY 13699, USA; dmbarry@clarkson.edu
3. STEM Laboratory, State University of New York, Canton, NY 13617, USA
4. Faculty of Medical Engineering, Suzuka University of Medical Science, Suzuka 510-0293, Japan; miura-h@suzuka-u.ac.jp
5. National Institute of Technology (KOSEN), Oyama College, Oyama 323-0806, Japan; kato_t@oyama-ct.ac.jp
6. National Institute for Materials Science (NIMS), Tsukuba 305-0044, Japan; yoshitake.michiko@nims.go.jp
* Correspondence: kanemats@mse.suzuka-ct.ac.jp; Tel.: +81-59-368-1848

Abstract: Biofilms are a result of bacterial activities and are found everywhere. They often form on metal surfaces and on the surfaces of polymeric compounds. Biofilms are sticky and mostly consist of water. They have a strong resistance to antimicrobial agents and can cause serious problems for modern medicine and industry. Biofilms are composed of extracellular polymeric substances (EPS) such as polysaccharides produced from bacterial cells and are dominated by water at the initial stage. In a series of experiments, using Escherichia coli, we developed three types of laboratory biofilm reactors (LBR) to simulate biofilm formation. For the first trial, we used a rotary type of biofilm reactor for stirring. For the next trial, we tried another rotary type of reactor where the circular plate holding specimens was rotated. Finally, a circular laboratory biofilm reactor was used. Biofilms were evaluated by using a crystal violet staining method and by using Raman spectroscopy. Additionally, they were compared to each other from the practical (industrial) viewpoints. The third type was the best to form biofilms in a short period. However, the first and second were better from the viewpoint of "ease of use". All of these have their own advantages and disadvantages, respectively. Therefore, they should be properly selected and used for specific and appropriate purposes in the future.

Keywords: biofilms; laboratory biofilm reactors; LBR

1. Introduction

Problems caused by biofilms are found in various fields. Since biofilms form on materials' surfaces, some interactions between materials and bacterial environments must be related to the formation and growth of biofilms. The concept of biofilms was proposed from the late 1970s to 1980s [1,2] by medical [3] and environmental scientists [4,5]. Biofilms usually form on solid materials. Therefore, it is very important to develop antibiofilm materials [6,7]. To achieve this goal, we need to determine the mechanism for biofilm formation and growth and the factors involved. This requires appropriate evaluation processes (optical microscopes and electron microscopes [8,9], scanning electron microscopes [10–12], confocal laser microscopy [13–17], FTIR [18–21], and Raman spectroscopy [22–25]). When we began our investigations about biofilms from the viewpoint of materials science [26–33], we did not have any appropriate and well-known biofilm evaluation processes to use. Although many methods related to biology, environmental science, and medical science were available, we needed some appropriate processes to evaluate biofilms on materials.

Our evaluation process is composed of two steps. The first step is biofilm formation. Biofilm reactors are used to make biofilms, which might be natural or artificial. Particularly, biofilm reactors used at the laboratory scale are called a laboratory biofilm reactors (LBRs). Then quantitative or qualitative evaluation of biofilms on materials produced in laboratory biofilm reactors (LBRs) is needed. The second step involves The Society of International Sustaining Growth for Antimicrobial Articles (SIAA) in Japan and their current plans to establish an International Standard to evaluate biofilms on materials. However, globally, reactors of the type in the first step have not been considered for standardization yet. Therefore, we must continue our investigations and trials, so that practical engineers can use biofilm reactors as a common evaluation tool. So far, we have developed some laboratory biofilm reactors.

This paper mentions biological researchers and engineers who have used biofilm reactors where the flow factors have not been considered. However, the flow should be incorporated into the artificial production phase of biofilms since biofilms often form on materials' surfaces in fluid environments. For example, consider scale formation in bathtubs, kitchen sinks, and various water and sewage pipes. To simulate these environments, the conditions will differ from case to case. Using common reactors for these items is impossible. On the contrary, we need to devise a biofilm reactor that is simple, intuitive, and practical, as well as applicable to as many cases as possible. We devised three types of biofilm reactors for use at the laboratory scale, where flow factors can be incorporated to some extent and affordable to practical researchers and engineers at the same time. In this paper, we compared them from the practical viewpoint and mentioned problems, citing some concrete examples.

2. Experimental

2.1. Proposals for the New Artificial Laboratory Biofilm Reactors and Their Concepts

We arranged and made a rotary LBR using a magnetic stirrer, so that the culture part and biofilm formation (the specimen part) within the LBR can share a common space. In the past, we developed a flow-type biofilm reactor, where the incubation of bacteria was placed apart from the biofilm formation part. In this case, biofilm formation was accelerated because biofilm growth and bacterial growth were close to each other. A schematic diagram of this is shown in Figure 1. In this reactor, the solution was circulated within the reactor and circulation was caused by the stirrer. Therefore, we tentatively call this the stirrer-driven rotary biofilm reactor (SDRBR). The SDRBR is composed of three-neck flasks (500 mL), a magnetic stirrer, and a fixation jig. The jig is made of metallic materials (stainless steel) and inserted into the flask. The silicon rubber was used at the inlet part to fix the jig where specimens were attached. The jig with a specimen and a stirrer were inserted into the three-neck flask. In the SDRBR, only one specimen can be used in each experiment.

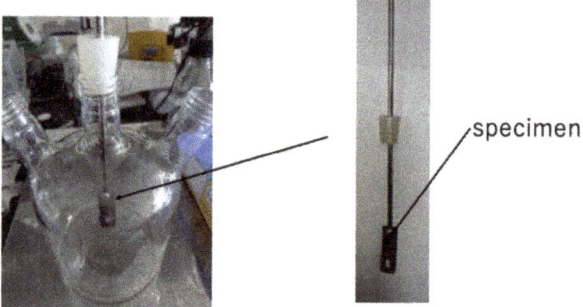

Figure 1. Stirrer-driven rotary biofilm reactor (SDRBR).

On the other hand, we designed and made the other rotation type of LBR. In the rotary LBR with a rotating jig, we developed an LBR in which the area where *E. coli* is cultured and the area where the sample exists are the same as in the rotary LBR with a stirrer. In the rotating jig, a jig to which various samples are fixed, it rotates by coupling with a motor. By this mechanism, the jig has the function of fixing the samples and agitating them. Figure 2 shows the setup. We named this type of LBR: a rotating-platform-driven LBR (RPDLBR).

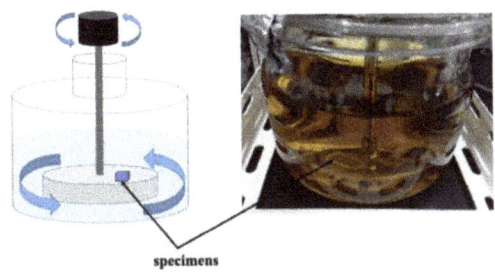

(a) Schematic figure   (b) The appearance

Figure 2. Rotating-platform-driven LBR.

The third type is tentatively called the closed-loop circulation LBR (CLC LBR) by the authors [34]. Polymeric tubes (Suffeed tubes, TERUMO Co., Tokyo, Japan) were connected to a 500 mL 3-neck beaker (SCHOTT DURAN, Jena, Germany), and an acrylic column containing a sample fixed with an acrylic jig (in the center) was incorporated through silicone rubber. A peristaltic pump (tubing pump) was incorporated into the Suffeed connecting tube to create a circulating LBR. The circulating LBR was placed in a table-top clean booth (AS ONE) or in an incubator during the experiment.

The laboratory biofilm reactors, including the CLC LBR used in this study, and their experimental conditions are summarized schematically in Figure 3.

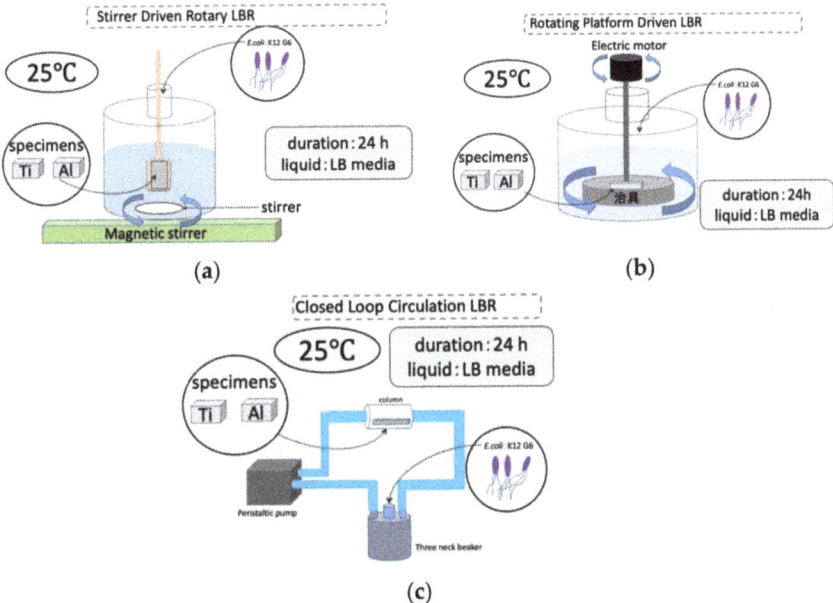

Figure 3. Three kinds of LBRs used in this study and their experimental conditions: (**a**) stirrer-driven rotary biofilm reactor; (**b**) rotating-platform-driven biofilm reactor; (**c**) closed-loop circulation biofilm reactor.

2.2. Biofilm Formation and Evaluations Used as Comparative Examples

2.2.1. Specimens and Bacteria

In this experiment, we concentrated on the differences between the biofilm reactors. We considered their effects on the evaluation results and the various characteristics of each setup (apparatus). Commercial pure metals of titanium and aluminum were used in this experiment. Specimens of these metals do not easily form biofilms and avoid the formation of corrosion products (in our experimental conditions) due to the inherit dense oxide films.

For a source of bacteria, we used *E. coli* (K12 G6). We have often used this kind of bacteria for experiments and can fix biofilm formation. Therefore, data in this experiment can be compared with those from previous experiments.

2.2.2. Preparation of Bacterial Solution and the Biofilm Formation Process

An amount of 12.5 g of LB medium was added to 500 mL of distilled water, stirred for 5 min to dissolve, and autoclaved at 121 °C/15 min. A volume of 500 mL of liquid LB medium was used as a solution for the rotary LBR and circulating LBR stirrer. An amount of 25 g of LB medium was added to 1000 mL of distilled water, stirred for 5 min to dissolve, and autoclaved at 121 °C/15 min. A volume of 1000 mL of liquid LB medium, which was dissolved by stirring and sterilized in an autoclave at 121 °C/15 min, was used as the solution for the rotary LBR with a stirrer. The K12 strain of *E. coli* was used. The bacteria were also cultured successively on LB agar medium. As a pre-culture before the experiment, one colony of *E. coli* (K12) was taken from LB agar medium in a loop and placed in a test tube containing 200 mL of undiluted LB medium and incubated for 18 h. Additionally, then, they were used as bacterial solution.

Various samples were inserted into the jig, which was then bonded to various LBRs. To make the inside of each LBR sterile, the LBR was sealed and autoclaved at 121 °C for 20 min. After undergoing pressure sterilization, 500 μm of pre-cultured *E. coli* was placed in the rotating LBR and circulating LBR using a stirrer and 1000 μm in the rotating LBR (using a rotating jig in a clean bench). This resulted in a concentration of 1 mL/1000 mL in the liquid LB medium of the culture medium in which *E. coli* was cultured. Biofilm formation was performed by operating the various LBRs. The temperature during the experiment was set at 25 °C and the duration of the experiment was 24 h.

2.3. Evaluation of Biofilms

To evaluate biofilms, we used two kinds of evaluation methods. One of them was Raman spectroscopy and the other was crystal violet (CV) staining.

2.3.1. Raman Spectroscopy

Biofilms formed on specimens were freeze-dried to fix the components of biofilms before Raman spectroscopy. Ninety percent of the biofilm's constituents is H_2O. Therefore, if the sample is left in the air, the H_2O evaporates and the biofilm shrinks, changing its structure. To prevent this, freeze-drying was used to solidify the biofilm formed on the sample's surface as a post-experiment sample treatment. This method made it possible for us to avoid cases where planktonic bacteria and polymeric substances derived only from LB media (irrelevant of biofilms) are detected during the evaluation process.

For the freeze-drying procedure, water, ethanol, and t-butyl alcohol were prepared as solutions. First, solutions were prepared by mixing water and ethanol in the following ratios: 7:3, 5:5, 3:7, 2:8, 1:9, 0.5:9.5, 0.2:9.8, and 0:1. The samples were immersed in the solutions of each concentration for 15 min, from left to right, as the concentration of ethanol increased. Next, ethanol: t-butyl alcohol solutions were prepared in the proportions 7:3, 5:5, 3:7, and 0:1, and the samples were immersed in each solution for 15 min, starting from the left, as the concentration of t-butyl alcohol increased. The samples were then placed in a freezer for at least 30 min to freeze. A vacuum pump was used to evacuate the frozen samples.

A laser Raman spectrophotometer (NRS-3100, JASCO Co., Tokyo, Japan) was used. Raman spectroscopy is a method of analyzing molecular structure by irradiating a sample with laser light and analyzing Raman scattering, which is extremely weak compared to Rayleigh scattering. This method is used to analyze and compare samples before and after experiments to analyze the various organic substances in the EPS in biofilms, mainly polysaccharides, nucleic acids, proteins, and lipids. Data analyzed using the Raman spectrophotometer were baseline corrected and smoothed. Peaks were identified, comparing the data obtained in our previous experiments and data from other researchers' studies. Since the apparatus has its own optical microscope, we observed the materials' surfaces by using the function and fixed the place of green laser irradiation (100 mW, 532 nm).

2.3.2. Crystal Staining

Staining with crystal violet stains proteins and polysaccharides contained in the biofilm. First, 0.1% crystal violet is prepared, and the biofilm-formed specimen is immersed in it for 30 min, after which the specimen's surface is rinsed with tap water. The color change in the material's surface is evaluated by using a color meter (CR-13, Konika-Minolta Sensing Co. Ltd., Tokyo, Japan) and their L*, a* and b* values were used for the evaluation of colored surfaces by crystal violet solution. In this paper, we used L, a and b were used as conventional short technical term for L*, a* and b*. Figure 4 shows L-a-b color space and the positions/mutual relations schematically.

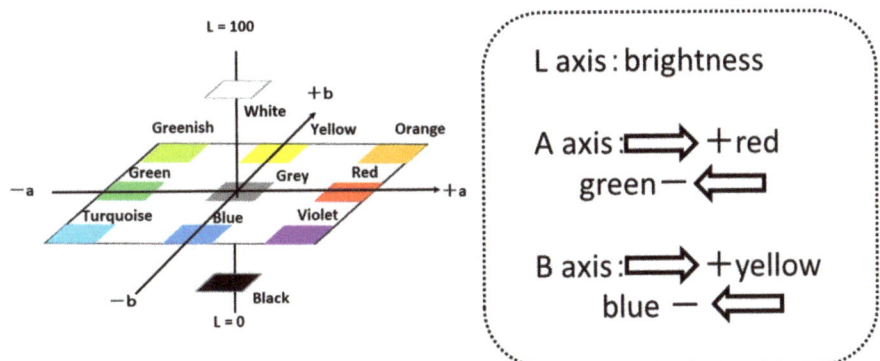

Figure 4. L-a-b color space and mutual relations among various colors.

3. Results

3.1. Results from the Stirrer-Driven Rotary Biofilm Reactor (SDRBR)

The results of the Raman spectroscopic experiments performed in a rotating LBR with a stirrer are shown in Figures 5 and 6. Figure 5 shows the specimens' surfaces observed by the microscope and Figure 6 shows Raman peaks for titanium and aluminum specimens.

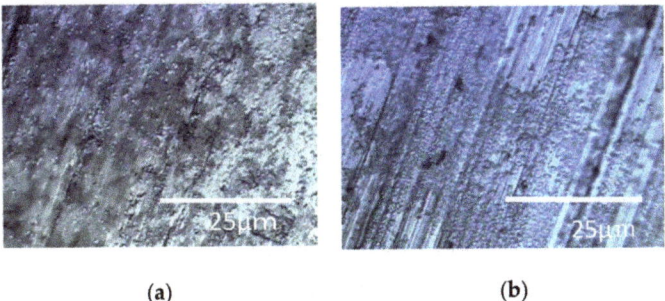

(a) (b)

Figure 5. Specimens' surfaces observed by the optical microscope in the SDRBR: (a) titanium specimen and (b) aluminum specimen.

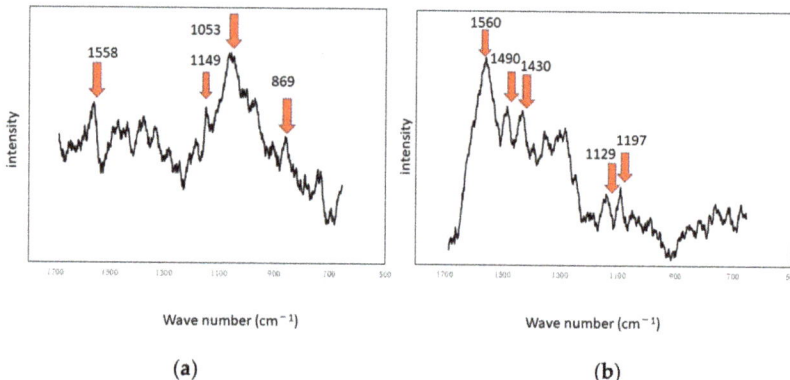

Figure 6. Raman shifts by the SDRBR: (**a**) titanium specimen and (**b**) aluminum specimen.

The shaded areas one observes by using optical microscopes often correspond to biofilms. Therefore, such areas were used as landmarks for observation as shown in Figure 5. The center of each optical microscopic image provides Raman peaks shown in Figure 6, respectively. Since peaks for titanium specimens were small, we presume biofilms in this case were not so remarkable. If we counted four tiny peaks as biofilm components in Figure 6a, the following four peaks can be mentioned: 1558 cm^{-1} (proteins such as amide II) [35], 1149 cm^{-1} (lipids) [36], 1053 cm^{-1} (lipids) [36] and 869 cm^{-1} (polysaccharides) [37]. We presume that these peaks show the existence of biofilms on titanium specimens. However, they can be derived from LB media. Even in such a case, the existence of organic matter shows that the surfaces are sticky and the stickiness obviously shows the existence of biofilms (since stickiness is generally caused by biofilms). From the practical viewpoint, the peaks of organic matter (even after washing and the substitution processes) show the existences of biofilms directly or indirectly. Figure 6b shows Raman peaks observed on the aluminum specimen's surface where the location corresponds to Figure 5b. Figure 6b shows a better view of the Raman peaks at the following locations: 1560 cm^{-1} (amide II) [35], 1490 cm^{-1} (protein) [35], 1430 cm^{-1} (lipids) [36], 1129 cm^{-1} (lipids) [36,38] and 1197 cm^{-1} (lipids) [36,38]. These peaks were derived from biofilms directly or obtained from sticky surfaces caused by biofilms in the same way.

Figure 7 shows the results of staining by using 0.1% crystal violet solutions. L*a*b* values were measured by using the apparatus. To show the extent of staining into violet colors, the values of a* and b* were plotted in the a–b plane. The figures show that the plots of titanium and aluminum specimens moved from the original points to the staining ones. The point corresponding to a violet color is in the fourth quadrant. Both plots of titanium and aluminum tended to move from the first or the second quadrants to the fourth one with staining. The changes show the surfaces were stained by crystal violet due to the existence of biofilms. Additionally, the length of change in the space corresponds to the extent of staining. In this case, the extent of staining for the titanium specimen was smaller than that of the aluminum specimen.

3.2. Results from the Rotating-Platform-Driven LBR (RPDLBR)

The results of the experiments performed on a rotary LBR with a rotating fixture (RPDLBR) are shown in Figures 8 and 9. Figure 8 shows the specimens' surfaces observed by using the optical microscope. Figure 9 shows Raman peaks obtained for titanium and aluminum specimens.

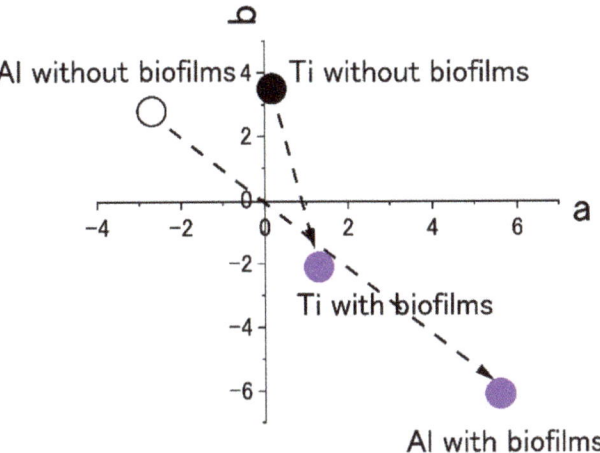

Figure 7. The color changes in specimens' surfaces stained by 0.1% crystal violet in the SDRBR. (a) complementary color dimension between red and green. (b) complementary color dimension between yellow and blue.

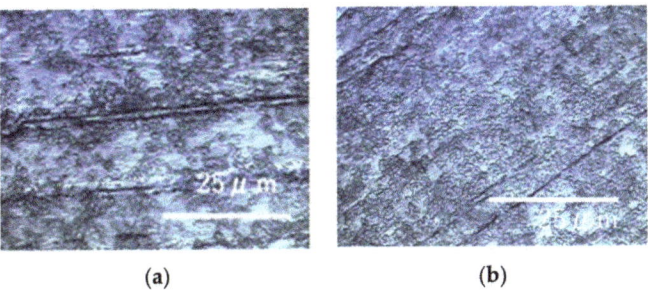

Figure 8. Specimens' surfaces observed by the optical microscope in the RPDLBR: (a) titanium specimen and (b) aluminum specimen.

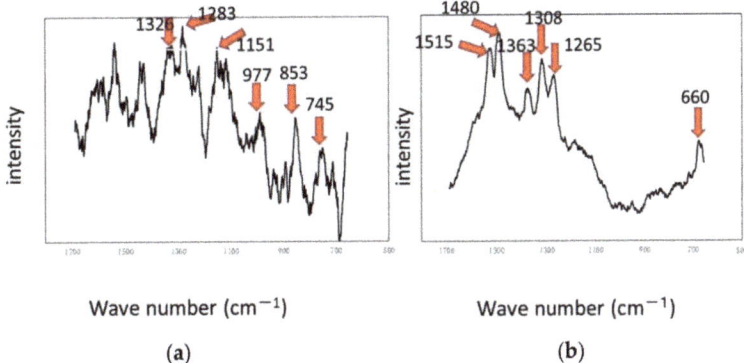

Figure 9. Raman shifts by the RPDLBR: (a) titanium specimen and (b) aluminum specimen.

The results of Raman spectroscopic analysis detected peaks of organic matter that may be of biological origin in the various samples. Therefore, the surface adherends observed by using optical microscopy are considered to mainly be organic materials of biological origin. The following results indicate certain materials for the titanium specimens: 1326 cm^{-1}

(Protein) [38], 1283 cm^{-1} (amide III or protein) [35,36], 1151 cm^{-1} (Lipids) [36], 977 cm^{-1} (nucleic acids or lipid) [35], 853 cm^{-1} (polysaccharides) [37] and 745 cm^{-1} (polysaccharides or lipids) [35]. Most of these materials might be derived from biofilms, even though some of them can come from floating organic compounds in the system. At any rate, the attached organisms show the existence of biofilms, as we already described. Keep in mind that the results displayed in Figure 6 show weak and not clearly defined peaks for the titanium specimens. This information suggests that the extent of biofilm formation is not be very large. On the other hand, the Raman peaks for the aluminum specimens were detected at 1326 cm^{-1} (lipid or protein) [35,38], 1283 cm^{-1} (amide III or lipid) [36,39], 1151 cm^{-1} (lipids) [36], 977 cm^{-1} (Nucleic acids or lipid) [35], 853 cm^{-1} (polysaccharides) [37] and 745 cm^{-1} (protein) [39]. These results suggest that biofilms were formed when the samples were observed locally. The crystal violet staining confirmed that all samples were stained purple, although there were differences in the staining.

Figure 10 shows the color plots of stained surfaces and their changes after biofilm formation. Like the results in Figure 7, the change in surface color shows the formation of biofilms. Even though both specimens showed a color change to blue, the extent was low for the titanium specimen. This was a common pattern in the results of Raman spectroscopy and color measurement. At the same time, this suggests that it is difficult to form biofilms on the specimen and this was similar between the SDRBR and the RPDLBR.

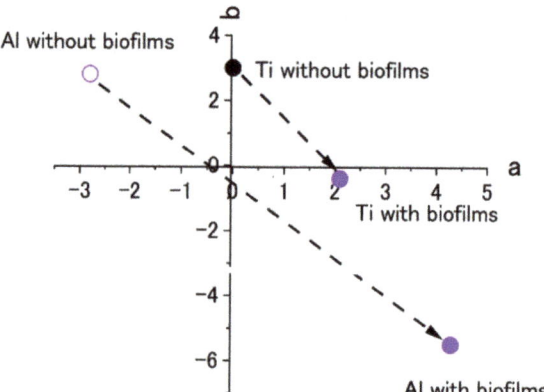

Figure 10. The color changes in specimens' surfaces stained by 0.1% crystal violet in the RPDLBR. (a) complementary color dimension between red and green. (b) complementary color dimension between yellow and blue.

3.3. Results from Using the Closed-Loop Circulation LBR (CLC LBR)

The results of experiments with Ti and Al using a circulating LBR (closed-loop circulation LBR) are shown in Figures 11–13. Figure 11 shows the dark areas which are supposed to biofilms. When the areas were irradiated by a laser beam, peaks were obtained, as shown in Figure 12. The results of Raman spectroscopic analysis (Figure 12) showed that peaks for the titanium specimen were at 1437 cm^{-1} (Lipids) [36,38], 1296 cm^{-1} (amide II or lipids) [35,36], 1125 cm^{-1} (Lipids) [38] and 1058 cm^{-1} (Lipids) [36]. These are considered to indicate biofilm origin. On the other hand, the Raman peaks for the aluminum specimen were at 1430 cm^{-1} (Lipids) [38], 1338 cm^{-1} (protein) [39], 1296 cm^{-1} (lipid or amid III) [36,38] and 1120–1185 cm^{-1} (Lipids and/or proteins) [38]. They also showed the existence of biofilms on the specimen. Compared with the results for the other two types of LBRs, Raman peaks were clearly seen, and their S/N ratios were relatively high, particularly for the titanium specimen. The reason can be attributed to the flow type for this apparatus. Figure 13 shows the color changes before and after biofilm formation. Additionally, in this case, the color change to violet can be confirmed. As for the titanium specimen, the color change to violet can be seen more clearly, as compared to the results for the RPDLBR.

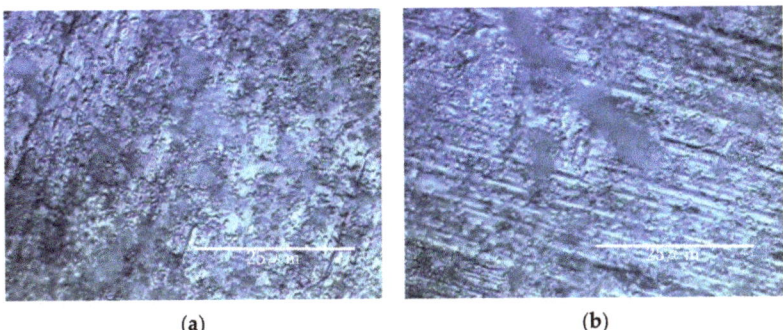

Figure 11. Specimens' surfaces observed by the optical microscope in the CLCLBR: (a) titanium specimen and (b) aluminum specimen.

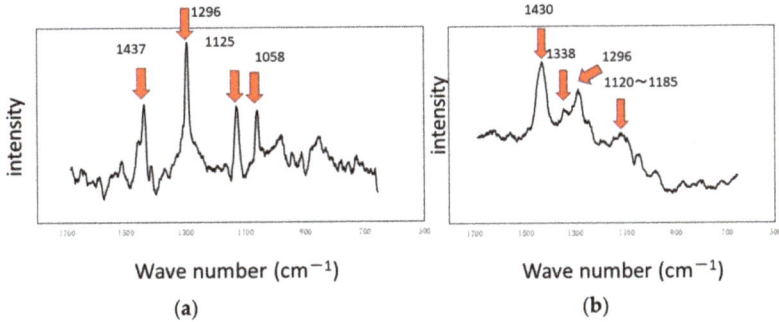

Figure 12. Raman shifts by the CLCLBR: (a) titanium specimen and (b) aluminum specimen.

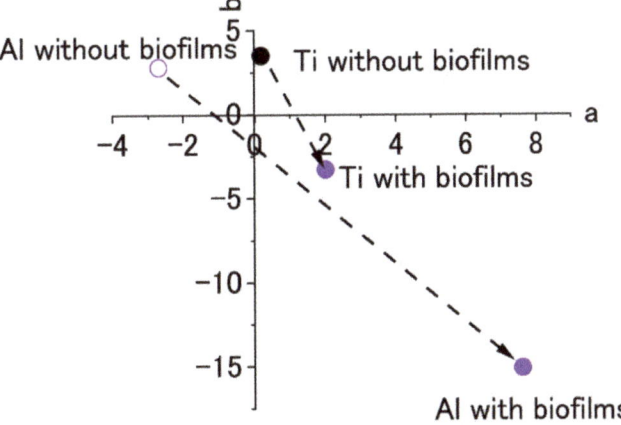

Figure 13. The color changes in specimens' surfaces stained by 0.1% crystal violet in the CLCLBR. (a) complementary color dimension between red and green. (b) complementary color dimension between yellow and blue.

4. Discussion

For all cases described above, biofilms can form on specimens to a greater or lesser extent. Even though generalization of the results might be difficult to determine in this experiment, we can evaluate the characteristics of three kinds of LBRs from the practical viewpoint. The difference also depends on the type of LBR used because each one has its own merits and limitations. Therefore, a certain LBR type should be selected for a specific purpose and application. We compared the characteristics of these LBRs. This information is summarized in Table 1.

Table 1. The comparisons of characteristics among three kinds of LBRs from the practical viewpoint.

LBR	Biofilm Formation	Remarkable Components	Capacity	Ease of Use
SDRBR	medium	proteins	low	simple
RPDLBR	weak	proteins	medium	medium
CLCLBR	strong	Lipids	medium	hard

From the practical viewpoint of social implementation, we analyze and compare the three types of laboratory biofilm reactors, as shown in Table 1. Biofilms are formed by using the three types of reactors. The extent of biofilm formation was a bit weak in the SDRBR and the RPDLBR as compared with the CLCBR. This suggests that parallel flow is effective at forming biofilms. The types of biofilm components depend on the type of LBR used. Proteins are the main component for the SDRBR and the RPDLBR, while lipids occupy biofilms in the CLCLBR. These results may be attributed to the ability of the biofilm components to remain on materials' surfaces against the flow. Liquid flow can remove bacteria and some components of biofilms. Under these experimental conditions (the balance between the adherence force of components and flow strength), the results are shown in Table 1. The structure might limit the capacity of how many specimens can be treated at the same time. The SDRBR can deal with one or two specimens simultaneously. On the other hand, the RPDLBR and the CLCLBR can deal with a couple of specimens. However, more revisions for both types will improve their capacities. As for "ease of use", the SDRBR was the best, followed by the RPDLBR and ending with the CLCLBR. The most difficult problem for this project has been sterilization of the devices. The larger the device, the more difficult sterilization becomes in many ways. This factor should be incorporated into further studies.

5. Conclusions

With pure titanium and aluminum specimens as model metallic materials, we carried out biofilm formation tests, using three different laboratory biofilm reactors that we designed and produced for practical purposes. These LBRs can produce the flow in the systems and can be applied to the practical acceleration tests for industries. They were named the stirrer-driven rotary biofilm reactor (SDRBR), the rotating-platform-driven LBR (RPDLBR) and the closed-loop circulation LBR (CLC LBR). The SDRBR and the RPDLBR belong to the same category because the rotating flow is the driving force to form biofilms. On the other hand, linear parallel flow is added to the specimens' surfaces.

Closed-loop circulation LBRs are a little more difficult to handle, but this type can form biofilms the most effectively.

The rotating flow LBRs can be easily pressurized and sterilized, so they should be easy to handle for biofilm formation using bacteria. They are also considered to be easy to use in real-life, non-living environments because they are not large devices and can provide flow velocity to the sample.

Biofilm formation was observed by using our devised LBRs. However, each one should be selected for different/specific purposes, according to flow types and conditions for practical situations.

To develop anti-biofilm materials and for their societal implementation in the future, these models should be improved further. However, prototypes such as shown in these experiments will be useful and good references, when each is properly selected and used for specific and appropriate purposes.

Author Contributions: Original draft preparation, H.K. (Hikonaru Kudara); project administration, H.K. (Hideyuki Kanematsu), H.M. and D.M.B.; review and editing, D.M.B. and H.K. (Hideyuki Kanematsu); funding, H.M. and N.H.; investigation, T.K. (Takeshi Kogo), R.K., A.O., T.K. (Takehito Kato) and M.Y.; supervision, T.K. (Takehito Kato) and N.H. All authors have read and agreed to the published version of the manuscript.

Funding: This work was supported by JSPS KAKENHI (Grants-in-Aid for Scientific Research from the Japan Society for the Promotion of Science, Grant Number 20K05185 and 21K12739). A part of this work was supported by the GEAR 5.0 Project of the National Institute of Technology (KOSEN) in Japan.

Institutional Review Board Statement: Not applicable.

Informed Consent Statement: Not applicable.

Data Availability Statement: Not applicable.

Acknowledgments: We appreciate the Advanced Technology R&D Center of Mitsubishi Electric Co., Japan Food Research Laboratories (JFRL), and The Society of International Sustaining Growth for Antimicrobial Articles (SIAA) for their very useful advice and information about biofilms.

Conflicts of Interest: The authors declare no conflict of interest.

References

1. Characklis, W.G. Fouling biofilm development: A process analysis. *Biotechnol. Bioeng.* **1981**, *23*, 1923–1960. [CrossRef]
2. William, G.C.; Keith, E.C. Biofilms and Microbial Fouling. In *Advances in Applied Microbiology*; Elsevier: Amsterdam, The Netherlands, 1983; pp. 93–138.
3. Costerton, J.W.; Cheng, K.J.; Geesey, G.G.; Ladd, T.I.; Nickel, J.C.; Dasgupta, M.; Marrie, T.J. Bacterial biofilms in nature and disease. *Annu. Rev. Microbiol.* **1987**, *41*, 435–464. [CrossRef] [PubMed]
4. Lappin-Scott, H.M.; Costerton, J.W. Bacterial biofilms and surface fouling. *Biofouling* **1989**, *1*, 323–342. [CrossRef]
5. Lappin-Scott, H.M.; Jass, J.; Costerton, J.W. Microbial Biofilm Formation and Characterisation. In *Society for Applied Bacteriology Technical Series, Society for Applied Bacteriology Symposium*; Blackwell Scientific Publications: Oxford, UK, 1993; Volume 30.
6. Kanematsu, H.; Barry, M.D. *Biofilm and Materials Science*; Springer: New York, NY, USA, 2015.
7. Kanematsu, H.; Barry, M.D. *Formation and Control of Biofilm in Various Environments*; Springer Nature: Singapore, 2020; Volume 249.
8. Eighmy, T.T.; Maratea, D.; Bishop, P.L. Electron microscopic examination of wastewater biofilm formation and structural components. *Appl. Environ. Microbiol.* **1983**, *45*, 1921–1931. [CrossRef] [PubMed]
9. Anthony, G.S.; Peter, B.C.; Jurgen, R.; Allan, M.S.; Christopher, R.N.; John, H.; William, C.J. Biliary stent blockage with bacterial biofilm: A light and electron microscopy study. *Ann. Intern. Med.* **1988**, *108*, 546–553.
10. Lawrence, J.R.; Swerhone, G.D.W.; Leppard, G.G.; Araki, T.; Zhang, X.; West, M.M.; Hitchcock, A.P. Scanning transmission x-ray, laser scanning, and transmission electron microscopy mapping of the exopolymeric matrix of microbial biofilms. *Appl. Environ. Microbiol.* **2003**, *69*, 5543–5554. [CrossRef] [PubMed]
11. Priester, J.H.; Horst, A.M.; Van De Werfhorst, L.C.; Saleta, J.L.; Mertes, L.A.; Holden, P.A. Enhanced visualization of microbial biofilms by staining and environmental scanning electron microscopy. *J. Microbiol. Methods* **2007**, *68*, 577–587. [CrossRef]
12. Bossù, M.; Selan, L.; Artini, M.; Relucenti, M.; Familiari, G.; Papa, R.; Vrenna, G.; Spigaglia, P.; Barbanti, F.; Salucci, A.; et al. Characterization of Scardovia wiggsiae Biofilm by Original Scanning Electron Microscopy Protocol. *Microorganisms* **2020**, *8*, 807. [CrossRef] [PubMed]
13. Kuehn, M.; Hausner, M.; Bungartz, H.J.; Wagner, M.; Wilderer, P.A.; Wuertz, S. Automated confocal laser scanning microscopy and semiautomated image processing for analysis of biofilms. *Appl. Environ. Microbiol.* **1998**, *64*, 4115–4127. [CrossRef] [PubMed]
14. Lawrence, J.R.; Neu, T.R. Confocal Laser Scanning Microscopy for Analysis of Microbial Biofilms. In *Methods in Enzymology*; Doyle, R.J., Ed.; Academic Press: San Diego, CA, USA, 1999; Volume 310, pp. 131–144. [CrossRef]
15. Akiyama, H.; Oono, T.; Saito, M.; Iwatsuki, K. Assessment of cadexomer iodine against Staphylococcus aureus biofilm in vivo and in vitro using confocal laser scanning microscopy. *J. Dermatol.* **2004**, *31*, 529–534. [CrossRef]
16. Shukla, S.K.; Rao, T.S. Effect of calcium on Staphylococcus aureus biofilm architecture: A confocal laser scanning microscopic study. *Colloids Surf. B Biointerfaces* **2013**, *103*, 448–454. [CrossRef] [PubMed]
17. Reichhardt, C.; Parsek, M.R. Confocal Laser Scanning Microscopy for Analysis of Pseudomonas aeruginosa Biofilm Architecture and Matrix Localization. *Front. Microbiol.* **2019**, *10*, 677. [CrossRef] [PubMed]
18. Bremer, P.J.; Geesey, G.G. An evaluation of biofilm development utilizing non-destructive attenuated total reflectance Fourier transform infrared spectroscopy. *Biofouling* **1991**, *3*, 89–100. [CrossRef]
19. Jürgen, S.; Hans-Curt, F. FTIR-spectroscopy in microbial and material analysis. *Int. Biodeterior. Biodegrad.* **1998**, *41*, 1–11.
20. Delille, A.; Quilès, F.; Humbert, F. In Situ Monitoring of the Nascent Pseudomonas fluorescens Biofilm Response to Variations in the Dissolved Organic Carbon Level in Low-Nutrient Water by Attenuated Total Reflectance-Fourier Transform Infrared Spectroscopy. *Appl. Environ. Microbiol.* **2007**, *73*, 5782–5788. [CrossRef]

21. Chirman, D.; Pleshko, N. Characterization of bacterial biofilm infections with Fourier transform infrared spectroscopy: A review. *Appl. Spectrosc. Rev.* **2021**, *56*, 673–701. [CrossRef]
22. Samek, O.; Al-Marashi, J.F.M.; Telle, H.H. The potential of raman spectroscopy for the identification of biofilm formation by staphylococcus epidermidis. *Laser Phys. Lett.* **2010**, *7*, 378–383. [CrossRef]
23. Millo, D.; Harnisch, F.; Patil, S.A.; Ly, H.K.; Schröder, U.; Hildebrandt, P. In Situ Spectroelectrochemical Investigation of Electrocatalytic Microbial Biofilms by Surface-Enhanced Resonance Raman Spectroscopy. *Angew. Chem. Int. Ed.* **2011**, *50*, 2625–2627. [CrossRef] [PubMed]
24. Jung, G.B.; Nam, S.W.; Choi, S.; Lee, G.-J.; Park, H.-K. Evaluation of antibiotic effects on Pseudomonas aeruginosa biofilm using Raman spectroscopy and multivariate analysis. *Biomed. Opt. Express* **2014**, *5*, 3238–3251. [CrossRef] [PubMed]
25. Keleştemur, S.; Avci, E.; Çulha, M. Raman and Surface-Enhanced Raman Scattering for Biofilm Characterization. *Chemosensors* **2018**, *6*, 5. [CrossRef]
26. Ogawa, A.; Kanematsu, H.; Sano, K.; Sakai, Y.; Ishida, K.; Beech, I.B.; Suzuki, O.; Tanaka, T. Effect of Silver or Copper Nanoparticles-Dispersed Silane Coatings on Biofilm Formation in Cooling Water Systems. *Materials* **2016**, *9*, 632. [CrossRef] [PubMed]
27. Kanematsu, H.; Kanesaki, S.; Kudara, H.; Barry, M.D.; Ogawa, A.; Mizunoe, Y. Biofilm Formation on Titanium Alloy Surfaces in a Laboratory Biofilm Reactor. In *Ceramic Transactions—Advances in Ceramics for Environmental, Functional, Structural, and Energy Applications*; Morsi, M.M., Kumar, S., Colorado, H., Bhalla, A.S., Singh, J.P., Gupta, S., Langhorn, J., Jitianu, A., Jose Manjooran, N., Eds.; John Wiley & Sons Inc.: New York, NY, USA, 2018.
28. Sano, K.; Kanematsu, H.; Hirai, N.; Ogawa, A.; Kougo, T.; Tanaka, T. The development of the anti-biofouling coating agent using metal nanoparticles and analysis by Raman spectroscopy and FIB system. *Surf. Coat. Technol.* **2017**, *325*, 715–721. [CrossRef]
29. Ogawa, A.; Takakura, K.; Hirai, N.; Kanematsu, H.; Kuroda, D.; Kougo, T.; Sano, K.; Terada, S. Biofilm Formation Plays a Crucial Rule in the Initial Step of Carbon Steel Corrosion in Air and Water Environments. *Materials* **2020**, *13*, 923. [CrossRef]
30. Kanematsu, H.; Ikigai, H.; Yoshitake, M. Evaluation of Various Metallic Coatings on Steel to Mitigate Biofilm Formation. *Int. J. Mol. Sci.* **2009**, *10*, 559–571. [CrossRef] [PubMed]
31. Kanematsu, H.; Nakagawa, R.; Sano, K.; Barry, D.M.; Ogawa, A.; Hirai, N.; Kogo, T.; Kuroda, D.; Wada, M.; Lee, S.; et al. Graphene-dispersed silane compound used as a coating to sense immunity from biofilm formation. *Med. Devices Sens.* **2019**, *2*, e10043. [CrossRef]
32. Tanaka, N.; Kogo, T.; Hirai, N.; Ogawa, A.; Kanematsu, H.; Takahara, J.; Awazu, A.; Fujita, N.; Haruzono, Y.; Ichida, S.; et al. In-situ detection based on the biofilm hydrophilicity for environmental biofilm formation. *Sci. Rep.* **2019**, *9*, 8070. [CrossRef]
33. Kanematsu, H.; Barry, M.D. Laboratory Biofilm Reactors. In *Formation and Control of Biofilm in Various Environments*; Springer Nature: Singapore, 2020; pp. 83–110.
34. Kanematsu, H.; Kudara, H.; Kanesaki, S.; Kogo, T.; Ikegai, H.; Ogawa, A.; Hirai, N. Application of a loop-type laboratory biofilm reactor to the evaluation of biofilm for some metallic materials and polymers such as urinary stents and catheters. *Materials* **2016**, *9*, 824. [CrossRef]
35. Larkin, P. General Outline and Strategies for Infrared and Raman Spectral Interpretation. In *Infrared and Raman Spectroscopy: Principles and Spectral Interpretation*; Elsevier: Watham, MA, USA, 2011; pp. 117–133.
36. Czamara, K.; Majzner, K.; Pacia, M.Z.; Kochan, K.; Kaczor, A.; Baranska, M. Raman spectroscopy of lipids: A review. *J. Raman Spectrosc.* **2015**, *46*, 4–20. [CrossRef]
37. Yuen, S.-N.; Choi, S.-M.; Phillips, D.L.; Ma, C.-Y. Raman and FTIR spectroscopic study of carboxymethylated non-starch polysaccharides. *Food Chem.* **2009**, *114*, 1091–1098. [CrossRef]
38. Chao, Y.; Zhang, T. Surface-enhanced Raman scattering (SERS) revealing chemical variation during biofilm formation: From initial attachment to mature biofilm. *Anal. Bioanal. Chem.* **2012**, *404*, 1465–1475. [CrossRef]
39. David, C. *Raman Spectroscopy for Proteins*; Horiba Scientific Co.: Tokyo, Japan, 2012; Available online: www.horiba.com/scientific (accessed on 31 March 2022).

Article

Deformation Texture of Bulk Cementite Investigated by Neutron Diffraction

Nozomu Adachi [1,*], Haruki Ueno [1], Satoshi Morooka [2], Pingguang Xu [2] and Yoshikazu Todaka [1]

1. Department of Mechanical Engineering, Toyohashi University of Technology, 1-1 Hibarigaoka, Tempaku, Toyohashi 441-8580, Aichi, Japan; ueno@martens.me.tut.ac.jp (H.U.); todaka@me.tut.ac.jp (Y.T.)
2. Materials Sciences Research Center, Japan Atomic Energy Agency, 2-4 Shirakata, Tokai, Naka 319-1195, Ibaraki, Japan; morooka.satoshi@jaea.go.jp (S.M.); xu.pingguang@jaea.go.jp (P.X.)
* Correspondence: n-adachi@me.tut.ac.jp; Tel.: +81-532-44-1126

Abstract: Understanding the deformation mechanism of cementite such as on a slip plane is important with regard to revealing and improving the mechanical property of steels. However, the deformation behavior of cementite has not been well investigated because of the difficulty of sample preparation given the single phase structure of cementite. In this study, by fabricating bulk single phase cementite samples using the method developed by the authors, the deformation texture formed by uniaxial compression was investigated using both electron back scatter diffraction and neutron diffraction. The fabricated sample had a random texture before the compression. After applying a compressive strain of 0.5 at 833 K, (010) fiber texture was formed along the compressive axis. It has been suggested from this trend that the primary slip plane of cementite is (010).

Keywords: cementite; deformation texture; neutron diffraction

Citation: Adachi, N.; Ueno, H.; Morooka, S.; Xu, P.; Todaka, Y. Deformation Texture of Bulk Cementite Investigated by Neutron Diffraction. *Materials* 2022, 15, 4485. https://doi.org/10.3390/ma15134485

Academic Editors: Federico Mazzucato and Daniela Kovacheva

Received: 30 April 2022
Accepted: 24 June 2022
Published: 25 June 2022

Publisher's Note: MDPI stays neutral with regard to jurisdictional claims in published maps and institutional affiliations.

Copyright: © 2022 by the authors. Licensee MDPI, Basel, Switzerland. This article is an open access article distributed under the terms and conditions of the Creative Commons Attribution (CC BY) license (https://creativecommons.org/licenses/by/4.0/).

1. Introduction

Pearlite is one of the common phases found in steel which has a lamellar structure with ferrite and cementite. It is well-known that pearlitic steels have superior strength-ductility balance. Pearlitic steels exhibit a high work-hardening rate, and wire-drawn pearlitic steels reach tensile strengths of 6.3 GPa [1]. Pearlitic steels are therefore used as industrial material such as suspension cable for suspension bridges and tire reinforcement steel wire. Although the primary reason of the high work-hardening rate of pearlitic material is thought to be a decrease of lamellar spacing by wire-drawing [2], the origin of such extremely high strength is not yet fully understood since wire-drawing causes the partial chemical decomposition of cementite and the amorphization of cementite. Recent in situ neutron diffraction experiments under tensile loading has revealed that stress partitioning between ferrite and cementite plays an important role in the high strength of pearlitic steels [3]. Since a stress partitioning is generated by a difference of mechanical properties between phases such as strength, deformability, and plastic anisotropy, understanding the deformation mechanism in both phases are quite important to revealing the origin of the strength of pearlite. Whereas the deformation mechanism of ferrite is well-established, the literature reporting the deformation mechanism of cementite is limited. Cementite has an orthorhombic unit cell with a space group of *Pnma* (No. 62) which contains 12 Fe atoms and 4 C atoms [4]. While one type of bonding between Fe atoms is metallic bonding, that between Fe and C atoms is characteristic of covalent bonding, which should affect the slip deformation mechanism of cementite. Recent in situ synchrotron X-ray diffraction experiments have revealed the lattice constants of cementite to be $a = 0.5084$ nm, $b = 0.6747$ nm, and $c = 0.4525$ nm at room temperature [5]. The time-of-flight neutron powder diffraction experiment also reported similar lattice constants [6]. Inoue et al. has investigated the deformation behavior of cementite in cold-rolled carbon steels by observing the dislocation structure using transmission electron microscopy. They have reported that slip planes (100), (010), and (100) were observed and a slip direction of [100] on (010) and [010] on (100) was suggested [7]. However, since cementite is generally formed as one of constitute phases in a multiphase microstructure, the obtainable size of the cementite sample is limited in micrometer scale. Therefore, the deformation behavior of cementite has not been systematically investigated.

Our research group has previously proposed the method to fabricate a bulk single phase cementite sample using the mechanical ball milling and pulse current sintering (PCS) processes [8,9]. This method enabled us to investigate the properties of cementite alone such as hardness [10], elastic properties [11], and the hydrogen permeation property [12]. In this study, by employing the method, we investigated the deformation behavior of cementite by means of a neutron diffraction experiment.

2. Materials and Methods

Pure Fe (particle diameter d_p < 150 μm) and graphite (d_p < 20 μm) powders with 99.9% purity were mixed to be stoichiometric composition of cementite (i.e., Fe-25at%C) and subjected to mechanical ball milling (MM) using the conventional horizontal ball mill. The conditions of MM performed in this study are summarized in Table 1.

Table 1. Conditions of mechanical ball milling (MM) performed in this study.

Milling Receptacle	JIS SUS304 Stainless Steel (Inner Diameter ϕ128 mm, Volume 1.7 L)
Milling media	JIS SUJ2 Bearing Steel ball (Diameter ϕ25 mm)
Milling media weight	3800 g
Powder weight	38 g
Milling time	360 ks
Rotation speed	95 rpm
Atmosphere	Ar

After the MM process, supersaturated Fe + C solid solution is formed. The MM processed powder was subjected to the PCS (Sumitomo Coal Mining Co., Ltd., Tokyo, Japan) at 1173 K for 900 s at a compressive pressure of 50 MPa in a vacuum and then furnace cooled. Cementite is formed during the PCS process, and bulk cementite with a diameter of 10 mm and a height of 8–10 mm was consequently obtained in this study. By using this method, a bulk sample having nearly 95 vol% of cementite can be fabricated [12].

In order to investigate the deformation behavior of cementite, a compressive strain of 0.5 was applied to the bulk cementite sample by compression test using the PCS. The compression tests were performed at 833 K in a vacuum.

The bulk cementite samples before and after the compression test were mechanically polished and finished using a colloidal silica, and then the microstructure was observed using a Schottkey field emission scanning electron microscope (SEM, SU5000 Hitach High-Tech Corporation, Schaumburg, IL, USA) attached with electron back scatter diffraction (EBSD, TexSEM Laboratories, Inc., Draper, UT, USA). EBSD profiles were analyzed using the OIM analysis software.

The as-sintered sample and the compressed sample were subjected to a neutron diffraction (ND) experiment to investigate the deformation texture of cementite. The ND experiment was performed using the diffractometer for residual stress analysis (RESA) at the Japan Research Reactor No.3 (JRR-3) of the Japan Atomic Energy Agency (JAEA). The sample for the ND experiment has a diameter of 10 mm and a height of 10 mm. As the compressed sample has a thickness of 2 mm, five samples were stacked to make a total thickness of 10 mm. In the ND experiment, seven diffraction peaks (002), (201), (211), (102), (112), (221), and (122) were measured while rotating a sample along χ and ϕ with a step of 5° and 15° (see Figure 1 for the definition of χ and ϕ). Here, CD and RD means compressive direction and radial direction, respectively. The obtained ND profiles were analyzed by using MAUD (Materials Analysis Using Diffraction, version 2.97) software to calculate pole figures [13].

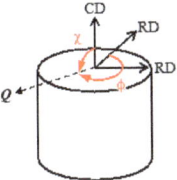

Figure 1. Schematic illustration showing the definition of χ and φ in the ND experiments, where Q is the neutron scattering vector.

3. Results and Discussion
3.1. Texture of Bulk Cementite before Compression Test

Figure 2 shows the phase, inverse pole figure (IPF), and kernel average misorientation (KAM) maps of the as-sintered bulk cementite sample obtained by the EBSD. The phase map indicates that the as-sintered cementite sample has a volume of 96% cementite and contains 4% ferrite, which is in good agreement with our previous investigation using neutron diffraction [12]. We have also investigated the porosity of the sintered sample through the bulk density measurement based on Archimedes' method [12]. It found that the as-sintered sample had a porosity volume of 3.2 %. The cementite phase in the sample has an equiaxed grain shape with an average grain size of 0.6 μm. The KAM map shows that each grain in the sample has relatively low misorientation angle, suggesting that the dislocations introduced by the MM was recovered during the sintering.

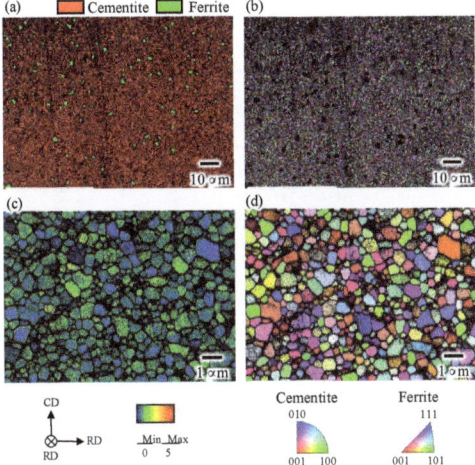

Figure 2. (**a**) Phase, (**b**,**d**) inverse pole figure (IPF), and (**c**) kernel average misorientation maps of the as-sintered bulk cementite sample.

{001}, {010}, and {100} pole figures were also obtained from the EBSD to determine the texture in the as-sintered sample. The results are shown in Figure 3. The maximum texture intensity was only 1.4, showing that the as-sintered sample has a random texture.

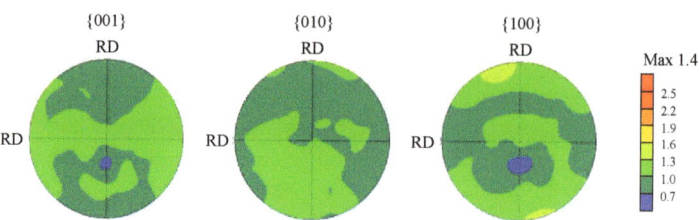

Figure 3. {001}, {010}, and {100} pole figures of as-sintered samples obtained by the EBSD.

As an EBSD method can collect information only from a surface of a sample, the internal texture of the sample can differ from the texture obtained by an EBSD. Considering this problem, the ND experiments were performed in this study. A neutron beam has very high penetration depth against steels over 10 mm, which enables the measuring diffraction from the internal microstructure. Figure 4 shows the examples of the selected seven diffractions of cementite obtained by the ND experiments at χ of 90°.

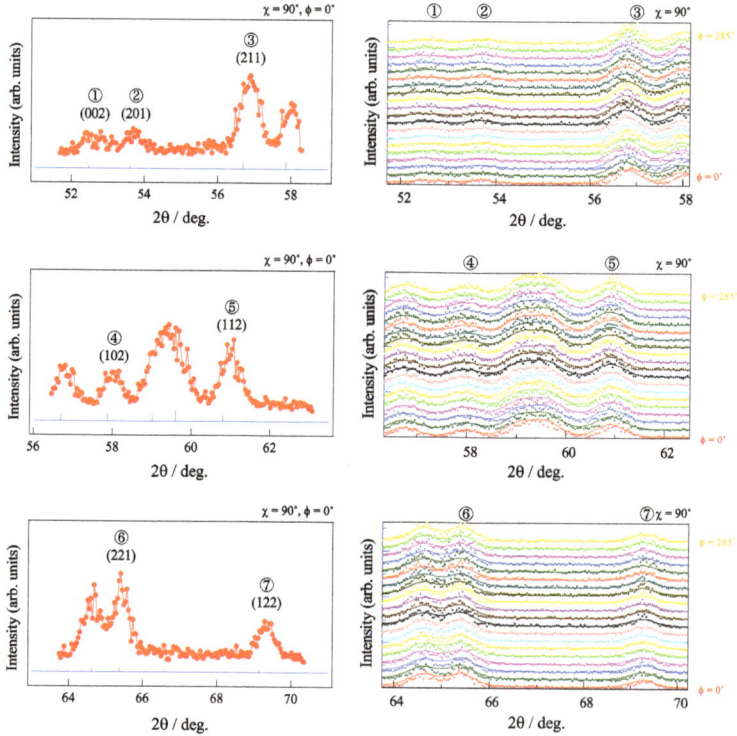

Figure 4. Typical neutron diffraction patterns obtained in the as-sintered sample. The blue lines in the left portion of the figure shows the expected peak positions of cementite. Each solid line in the right portion of the figure is the result of peak fitting.

It can be seen from the profiles that there is no clear difference in peak intensities with varying ϕ, suggesting that the sample has random texture. From the result of peak fitting of each diffraction profile, lattice constants of $a = 0.5085$ nm, $b = 0.6748$ nm, and $c = 0.4521$ nm were obtained, which is a quite reasonable value considering previous studies [4,5,10]. By using diffraction profiles measured along wide range of χ and ϕ, pole figures were calculated. The obtained pole figures are shown in Figure 5. The pole figure shows a maximum texture intensity of 1.3 and indicates that the sample has random texture, which is a result consistent with the result of EBSD, as shown in Figure 3.

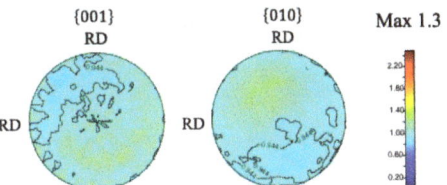

Figure 5. Pole figure maps of as-sintered sample obtained from neutron diffraction profiles.

3.2. Evolution of Deformation Texture in Bulk Cementite by Compression

Figure 6 illustrates the IPF maps observing the CD-RD and RD-RD planes. It can be seen that grains are slightly elongated along RD in the CD-RD plane and are equiaxed in RD-RD planes. By randomly selecting 100 cementite grains from each image, aspect ratios of the cementite grain are calculated to be 0.9 and 1.7 in the RD-RD and CD-RD planes, respectively. This trend is quite reasonable considering that cementite grains were plastically deformed by the introduction of a compressive strain of 0.5. The KAM map is also shown in Figure 6e. It seems that the misorientation in each grain in the compressed sample is smaller compared with that in the as-sintered sample, showing that the atmospheric temperature of 833 K was high enough to cause the dynamic recovery of dislocations introduced by the compression.

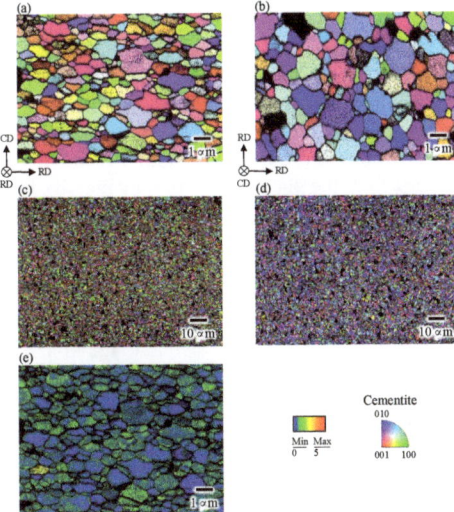

Figure 6. IPF maps of the compressed sample in (**a**,**c**) CD-RD and (**b**,**d**) RD-RD plains. (**e**) KAM maps.

It should be noted that a proportion of grains having an orientation close to (010) is large in the RD-RD plane (Figure 6d), suggesting that the deformation texture was formed. Figure 7 shows the corresponding pole figures. The axisymmetric pole figure can be clearly seen, in other words, a (010) fiber texture was formed along CD. As shown in Figure 8, this trend was also confirmed by the analysis based on the ND profiles performed with the same way as shown in Figures 4 and 5.

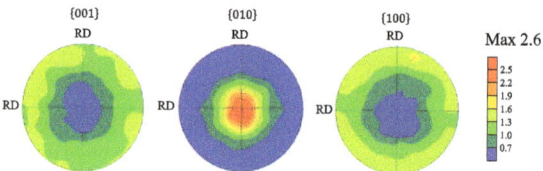

Figure 7. Pole figures of the compressed sample obtained by the EBSD.

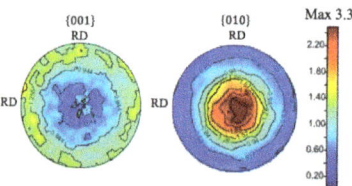

Figure 8. Pole figures of the compressed sample obtained by the neutron diffraction experiments.

In the case of uniaxial compressive deformation, it is generally known that primary slip plain gradually aligns perpendicular to the compression axis (i.e., CD) with increasing compressive strain. The (010) fiber texture found in this study therefore indicates that the primary slip plain of cementite is (010) at least under an elevated temperature of 833 K. As was mentioned in the introduction, Inoue et al. has reported that the possible slip planes of cementite are (100), (010) and (001) [7]. By investigating macroscopic deformation texture by employing the ND experiment, it has been concluded that the primary slip plane is (010). One of the possible reasons why (010) being primary slip plane is chemical bondings across (010). As was mentioned in the introduction section, cementite crystal consists of both metallic and covalent bondings. Figure 9 shows the schematic illustration of the atomic structure and the type of bonding between atoms in the cementite crystal. The trace of (010) was also shown in the figure with a blue line. It can be seen that (010) is the only plane where bondings across the plane consist of only metallic bonding. Since the strength of the metal bonding is much lower compared with covalent bonding, the critical resolved shear stress required to activate slip deformation is thought to be lower compared with other planes. The results obtained in this study showed that the method we used is effective to investigate primary slip plane. However, as shown in von-Mises' criterion [14], at least five independent slip systems are required to accommodate deformation to an arbitrary shape. A detailed investigation is therefore essential to understand the more detailed deformation mechanism of cementite.

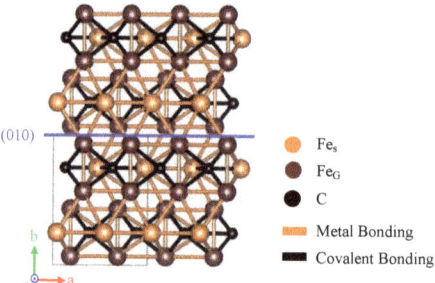

Figure 9. Schematic illustration of cementite observed from [001], where black and orange lines show that a type of bonding between atoms are covalent and metallic bonding, respectively.

4. Conclusions

This study investigated the deformation texture of cementite by preparing nearly 100 vol% of cementite in bulk shape using the mechanical ball milling and pulse current sintering processes. The texture of the samples was measured by both the electron back scattering pattern and the neutron diffraction experiments. The as-sintered bulk cementite sample showed random texture with equiaxed grains having a grain size of 0.6 μm. After the application of a compressive strain of 0.5 at 833 K, the cementite grains were plastically deformed and showed an elongated shape along the radial direction of the sample. The (010) of cementite aligned perpendicularly to the compression direction, showing that the primally slip plane of cementite is (010).

Author Contributions: Conceptualization, N.A. and Y.T.; methodology, N.A., S.M., P.X. and Y.T.; software, H.U., S.M. and P.X.; validation, N.A., H.U. and P.X.; formal analysis, N.A., H.U., S.M. and P.X.; investigation, N.A., H.U., S.M., P.X. and Y.T.; resources, S.M., P.X. and Y.T.; data curation, N.A. and H.U.; writing—original draft preparation, N.A.; writing—review and editing, N.A. and Y.T.; visualization, H.U., S.M. and P.X.; supervision, Y.T.; project administration, Y.T.; funding acquisition, N.A. and Y.T. All authors have read and agreed to the published version of the manuscript.

Funding: This research was funded by the 27th ISIJ Research promotion grant.

Institutional Review Board Statement: Not applicable.

Informed Consent Statement: Not applicable.

Data Availability Statement: The data presented in this study are available on request from the corresponding author.

Acknowledgments: The neutron diffraction experiments were performed at the RESA of JRR-3 with the approval of the Japan Atomic Energy Agency (Proposal No. 21037).

Conflicts of Interest: The authors declare that they have no conflict of interest.

References

1. Li, Y.J.; Choi, P.; Goto, S.; Borchers, C.; Raabe, D.; Kircheim, R. Evolution of strength and microstructure during annealing of heavily cold-drawn 6.3 GPa hypereutectoid pearlitic steel wire. *Acta Mater.* **2012**, *60*, 4005–4016. [CrossRef]
2. Marder, A.R.; Bramfitt, B.L. The effect of morphology on the strength of pearlite. *Metall. Trans. A* **1976**, *7*, 365–372. [CrossRef]
3. Tomota, Y.; Lukas, P.; Neov, D.; Harjo, S.; Abe, Y.R. In situ neutron diffraction during tensile deformation of a ferrite-cementite steel. *Acta Mater.* **2003**, *51*, 805–817. [CrossRef]
4. Fasiska, E.J.; Jeffrey, G.A. On the cementite structure. *Acta Cryst.* **1965**, *19*, 463–471. [CrossRef]
5. Litasov, K.D.; Rashchenko, S.V.; Shmakov, A.N.; Palyanov, Y.N.; Sokol, A.G. Thermal expansion of iron carbides, Fe_7C_3 and Fe_3C, at 297–911 K determined by in situ X-ray diffraction. *J. Alloys Compd.* **2015**, *628*, 102–106. [CrossRef]
6. Wood, I.G.; Vocadlo, L.; Knight, K.S.; Dobson, D.P.; Marshall, W.G.; Price, G.D.; Brodholt, J. Thermal expansion and crystal structure of cementite, Fe_3C, between 4 and 600 K determined by time-of-flight neutron powder diffraction. *J. Appl. Cryst.* **2004**, *37*, 82–90. [CrossRef]
7. Inoue, A.; Ogura, T.; Masumoto, T. Transmission electron microscope study on deformation and fracture of cementite in cold-rolled steels. *Trans. Jpn. Inst. Met.* **1976**, *17*, 149–157. [CrossRef]
8. Umemoto, M.; Liu, Z.G.; Takaoka, H.; Sawakami, M.; Tsuchiya, K.; Masuyama, K. Production of bulk cementite and its characterization. *Metall. Mater. Trans. A* **2001**, *32*, 2127–2131. [CrossRef]
9. Umemoto, M.; Todaka, Y.; Takahashi, T.; Li, P.; Tokumiya, R.; Tsuchiya, K. High temperature deformation behavior of bulk cementite produced by mechanical alloying and spark plasma sintering. *Mater. Sci. Eng. A* **2004**, *375–377*, 894–898. [CrossRef]
10. Umemoto, M.; Liu, Z.G.; Masuyama, K.; Tsuchiya, K. Influence of alloy additions on production and properties of bulk cementite. *Scr. Mater.* **2001**, *45*, 391–397. [CrossRef]
11. Umemoto, M.; Kruger, S.E.; Ohtsuka, H. Ultrasonic study on the change in elastic properties of cementite with temperature and Mn content using nearly full density polycrystalline bulk samples. *Mater. Sci. Eng. A* **2019**, *742*, 162–168. [CrossRef]
12. Adachi, N.; Ueno, H.; Onoe, K.; Morooka, S.; Todaka, Y. Hydrogen permeation property of bulk cementite. *ISIJ Int.* **2021**, *61*, 2320–2322. [CrossRef]
13. Lutterotti, L.; Matthies, S.; Wenk, H.R.; Schultz, A.J.; Richardson, J. Combined texture and structure analysis of deformed limestone from neutron diffraction spectra. *J. Appl. Phys.* **1997**, *81*, 594–600. [CrossRef]
14. Mises, R.V. Mechanik der plastischen Formänderung von Kristallen. *Z. Angew. Math. Mech.* **1928**, *8*, 161–185. [CrossRef]

Article

Novel Dry Spinning Process of Natural Macromolecules for Sustainable Fiber Material -1- Proof of the Concept Using Silk Fibroin

Ryo Satoh, Takashi Morinaga * and Takaya Sato

Department of Creative Engineering, National Institute of Technology (KOSEN), Tsuruoka College, 104 Sawada, Inooka, Tsuruoka 997-8511, Yamagata, Japan; r-satoh@tsuruoka-nct.ac.jp (R.S.); takayasa@tsuruoka-nct.ac.jp (T.S.)
* Correspondence: morinaga@tsuruoka-nct.ac.jp; Tel.: +81-235-25-9121

Abstract: Researchers around the world are developing technologies to minimize carbon dioxide emissions or carbon neutrality in various fields. In this study, the dry spinning of regenerated silk fibroin (RSF) was achieved as a proof of concept for a process using ionic liquids as dissolution aids and plasticizers in developing natural polymeric materials. A dry spinning equipment system combining a stainless-steel syringe and a brushless motor was built to generate fiber compacts from a dope of silk fibroin obtained by degumming silkworm silk cocoons and ionic liquid 1-hexyl-3-methyl-imidazolium chloride ([HMIM][Cl]) according to a general method. The maximum stress and maximum elongation of the RSF fibers were 159.9 MPa and 31.5%, respectively. RSF fibers containing ionic liquids have a homogeneous internal structure according to morphological investigations. Elemental analysis of fiber cross sections revealed the homogeneous distribution of nonvolatile ionic liquid [HMIM][Cl] in RSF fibers. Furthermore, the removal of ionic liquids from RSF fibers through impregnation washing with organic solvents was verified to enhance industrial applications. Tensile testing showed that the fiber strength could be maintained even after removing the ionic liquid. Thermogravimetric analysis results show that the organic solvent 1,1,1,3,3,3-hexafluoro-2-propanol is chemically coordinated to silk fibroin and, as a natural polymer, can withstand heat up to 250 °C.

Keywords: dry spinning; silk fibroin; ionic liquids

1. Introduction

As global warming progresses, various weather abnormalities such as droughts, massive typhoons, and localized heavy snowfalls become more frequent [1]. There are concerns that the greenhouse effect will cause permafrost to thaw, releasing unknown frozen viruses (*Morbillivirus*) and causing new infections [2]. Researchers worldwide are developing technologies to reduce carbon dioxide emissions or carbon neutrality in various fields [3–5].

In this context, as researchers and engineers of textile materials, we are focusing our efforts on developing new energy-effective spinning processes using renewable natural polymer materials, such as textile and polymer materials, which do not depend on petroleum resources [6]. Since the days when naturally occurring polymeric materials such as cellulose, silk, and wool were used directly as fibers, previous textile chemists developed a wet spinning process of dissolving natural polymers in a solvent and spinning the solution. This is referred to as wet spinning [7]. Wet spinning spider silk protein produced by fungi into fibers is now being researched and developed to produce high-strength, high-toughness renewable fibers by Sekiyama et al. [8]. However, these natural polymers are difficult to dissolve in solvents because of their strong intramolecular and intermolecular hydrogen bonds, which often require complicated dissolution processes such as the use of special solvents containing metal ions or minor chemical modifications to break the hydrogen bonds [9,10]. Therefore, in many cases, designing an ecological and cost-effective manufacturing process is challenging. The difficulty in dissolving the polymer is greater for high molecular weight polymers with good fiber properties [11,12]. However, it has been reported that ionic liquids (ILs) can dissolve natural polymers such as cellulose [13] and silk fibroin (SF) [14], and research is ongoing for using ILs as spinning solvents [15–17]. However, due to the high cost of ILs and their nonvolatile nature, the focus of development has been on wet spinning, and little research has been conducted on dry spinning, which is easier, less expensive, and significantly more efficient than wet spinning [18,19]. We have developed a process to obtain natural polymer fibers by dry spinning from a solution of natural polymers dissolved in a solvent containing a small amount of IL as a hydrogen bond cleaver [20]. Typical hydrogen bond cleavers are metal salt compounds dissolved in a solvent that interacts with donor types of substituents such as hydroxyl, carboxylic, and amino groups in the solution to cleave hydrogen bonds [21,22]. During the fiber formation, salt substances crystallize and create voids in the fiber [23]. Using an IL instead of a metal salt compound may prevent the formation of voids [24]. The IL was expected to remain uniformly in the fiber and function as a plasticizer [25]. The nonvolatility of ILs allows them to function as permanent plasticizers and softeners, maintaining the fibers' flexibility even under vacuum conditions. In this study, we describe the development of a process that uses ILs as a dissolution aid for natural-derived macromolecules, including a proof-of-concept study of dry spinning using regenerated SF (RSF).

2. Materials and Methods

2.1. Materials

Ethanol, methanol, and calcium chloride ($CaCl_2$) were purchased from Kanto Chemical (Tokyo, Japan). Sodium carbonate (Na_2CO_3) and lithium chloride were purchased from FUJIFILM Wako Pure Chemical (Osaka, Japan). 1-Hexyl-3-methyl-imidazolium chloride ([HMIM][Cl], ≥95% purity) was purchased from Sigma-Aldrich (St. Louis, MO, USA) as the ionic liquid. 1,1,1,3,3,3-Hexafluoro-2-propanol (HFIP) was purchased from Fluorochem (Derbyshire, UK). All chemical reagents were used without further purification.

Silkworm silk cocoons were kindly donated by Tsuruoka Silk (Yamagata, Japan). Marseilles soap (additive free) was purchased from Miyoshi Soap (Tokyo, Japan). A cellulose dialysis tubing (diameter, 21.4 mm; molecular weight cutoff (MWCO), 12,000–14,000 Da) was purchased from SERVA Electrophoresis (Heidelberg, Germany). Nitrogen (N_2) gas was purchased from Taiyo Nippon Sanso (Tokyo, Japan). Water purification systems (PURELAB flex 3, ELGA LabWater, High Wycombe, UK) were used to obtain ultrapure water.

2.2. Equipment

An SSY-30E stainless steel cylinder (23 mm i.d.) was purchased from Musashi Engineering (Tokyo, Japan). A UNP-23 dispenser nozzle (0.3 mm i.d.) was purchased from Unicontrols (Chiba, Japan). Brushless direct current electric motors (model GFV2G20) and the speed control units (model BMUD30-A2) were purchased from Oriental Motor (Tokyo, Japan). Stainless-steel filter membrane (vertical mesh/horizontal mesh, 500/3600; filtration threshold, 5 µm) was purchased from Yao Kanaami (Osaka, Japan).

2.3. Silk Degumming

Degumming was executed with minor modifications according to the literature [26]. Using a temperature-controlled water bath, 5.0 g of silkworm silk cocoons and 500 mL of alkaline solution (0.25 w/v% Marseilles soap, 0.25 w/v% Na_2CO_3) were boiled at 85 °C for 15 min. After removing the solution by decantation, the residue was rinsed six times with ultrapure water at 80 °C and dried in vacuo overnight. In 100 g of Ajisawa's solution [27] ($CaCl_2$/water/ethanol = 1:8:2, molar ratio), 15 g of degummed silk was dissolved. The degummed silk was thoroughly blended with Ajisawa's solution and heated at 55 °C for 1 h. The silk solution was filtered through a stainless-steel filter membrane. The filtrate was dialyzed against ultrapure water immediately using cellulose dialysis tubing (MWCO: 12,000–14,000 g/mol) at 4 °C for 4 days until the conductivity was lower than 10 mS. The dialyzed solution (diluted with ultrapure water up to 4 w/v%) was then lyophilized, and the resulting SF was desiccated and stored at ambient temperature.

2.4. Preparation of Dope Solution

To 3.0 g of SF, 16.25 g of HFIP was added, and the mixture was heated at 50 °C for 3 h. Subsequently, 0.75 g of [HMIM][Cl] was added dropwise, and the mixture was heated at 50 °C for 1 h. The final dope composition was SF/[HMIM][Cl]/HFIP = 15:3.75:81.25 ($w/w/w$). The resulting IL-containing SF dope solution was brought to ambient temperature and used for dry spinning.

2.5. Dry Spinning and Heat Stretching

A schematic of the dry spinning equipment is shown in Figure 1a. Under N_2 pressure, the dope solution in a stainless-steel cylinder (23 mm i.d.) with a dispenser nozzle (0.3 mm i.d.) was discharged into the air. The distance between the spinneret and the first guide roller was 100 cm. The pressure was controlled between 0.3 and 0.02 MPa to stabilize the spinning operation. Simultaneously, the roller winding speed was reduced from 12.0 to 4.0 m/min. The thread was predried among the guide rollers shortly after discharging and led to the take-up roller (4.2 m/min) as shown in Figure 1a. The resulting solidified thread was dealt as the as-spun fiber. The obtained yarns were dried in vacuo at ambient temperature for >5 h with bobbins.

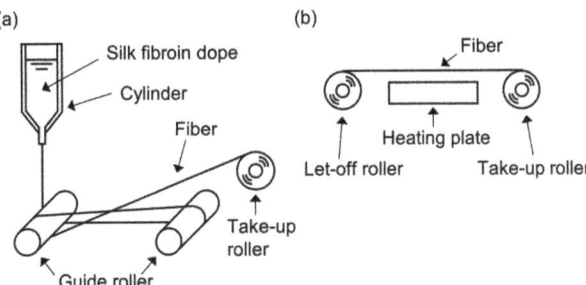

Figure 1. Conceptual images of dry spinning and heat stretching for regenerated silk fibroin. (**a**) Dry spinning system. The extruded dope solution was dried in ambient air to mold into the fiber on the guide rollers equipped with polytetrafluoroethylene tubes to prevent sticking of the dope in process and collected to the take-up roller. (**b**) Heat stretching system. The molded silk fiber was sent from the let-off roller and collected to the take-up roller directly. Spinning fiber was heated for 3 mm tolerance over the heating plate and stretched among the two rollers.

The equipment for heat stretching is shown in Figure 1b. The speeds of the let-off and take-up roller were set to 0.25 and 0.88 m/min, respectively (3.5-fold stretch). A hot plate was located between the two rollers and maintained at 120 °C. The obtained heat-stretched fiber was dried in vacuo for >5 h. To examine the physical properties of the fiber with or without the IL-washing process, the heat-stretched fiber was soaked in methanol to remove [HMIM][Cl].

2.6. Characterization

Scanning electron microscopy (SEM) was performed using a JSM-7100F field-emission scanning electron microscope (JEOL, Tokyo, Japan). The acceleration voltage was 2 kV. Elemental analysis of the fibers was performed using a JED-2300F energy-dispersive X-ray spectroscopy (EDX) instrument (JEOL) equipped with a solid-state detector. An Instron Model 3342 single-column testing system (Instron, Norwood, MA, USA) was equipped with grips with high frictional faces for flats, and a static load cell (maximum load, 500 N) was used for tensile testing. Instron Bluehill Lite Software (version 2.28.832) was used for data collection. Microsoft Excel (version 2202, Microsoft, Redmond, WA, USA) was used to process the data. The fiber specimens (randomly taken) were cut and fixed to paper mounts. The initial gauge length was 20 mm. The elongation rate was 2 mm/min (10% initial length). Using Microsoft Excel, the cross-sectional area of the fiber was determined from the cross-sectional images of the SEM results. A Thermo Plus TG8120 (Rigaku, Tokyo, Japan) was used to perform thermogravimetric (TG)–differential thermal analysis (DTA). Data acquisition was performed at a temperature of 50–400 °C, the rising rate was 10 °C/min, and the flow rate of N_2 gas was 200 mL/min. The sampling time was set at 1.0 s. Al_2O_3 was used as the reference material.

3. Results and Discussion

3.1. Morphological Observation

SEM was used to clarify the surface and cross-sectional images of as-spun (Figure 2a,b) or heat-stretched (Figure 2c,d) RSF fibers. The RSF fibers before/after heat stretching had homogeneous internal structures, contrary to the possibility of forming a heterogeneous sea–island structure. A belt-like flat fiber cross section was observed for both the as-spun and heat-stretched fibers.

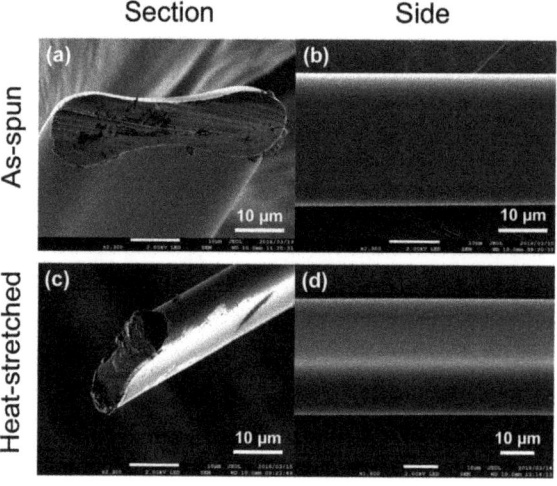

Figure 2. SEM images of regenerated silk fibroin (RSF) fibers: (**a**) sectional and (**b**) side images of RSF as-spun fibers; (**c**) sectional and (**d**) side images of RSF heat-stretched fibers. Each picture has a 10 μm scale bar. The cross-sections were made by direct immersion to liquid nitrogen and cutting with a razor blade.

From the morphology inside the cross-section of the fibers, the void formation could be effectively prevented using ILs instead of metal salts. Because no rapid volatilization of HFIP occurred during the first solidification process, all the RSF fibers had smooth surfaces. The authors considered that belt-shaped RSF fibers were formed and caused by tensile tension from the guide rollers during solidification into fiber bodies or shrinkage tension from drying between the guide rollers. To investigate the more intrinsic mechanical properties, a circular fiber needed to be used as a test specimen for tensile testing. Therefore, extending the vertical distance from the spinneret to the first guide roller and extending the time for HFIP volatilization was effective. The plasticity of nonvolatile ILs is high during the drying process despite the belt-shaped formation. Furthermore, the thermo-melting-like state of the RSF fibers in heat stretching allowed us to achieve further deformed and modified fibers that had star-shaped, Y-shaped, or other odd-shaped cross sections. In the case of the RSF fibers in this research, it may have been possible to develop from dry spinning to melt spinning.

3.2. Elemental Analysis

Figure 3 shows the distribution of ILs in RSF fibers. Elemental mapping revealed the distribution of chlorine atoms specifically in the chemical structure of [HMIM][Cl] using EDX analysis. The uniform distribution of chlorine atoms in the cross-section corresponding to the SEM images was confirmed in the picture.

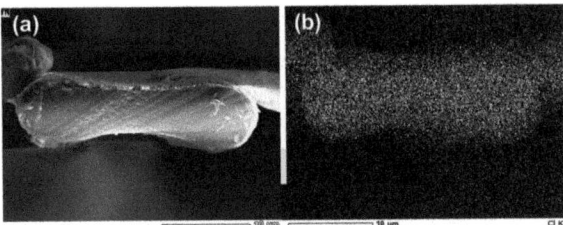

Figure 3. SEM picture (**a**) and an EDX mapping image (**b**) of regenerated silk fibroin fibers. Areas of dense green dots indicate the presence of chlorine atoms. Each picture has a 10 μm scale bar.

3.3. Removal of IL from the RSF Fibers

Figure 4 shows the EDX spectra of heat-stretched RSF fibers before and after methanol immersion. In addition, the extracted EDX spectrum of the RSF fiber after heat stretching is shown in Figure 4a, as well as the 2.6 keV indicated chlorine atoms from [HMIM][Cl]. Figure 4b shows the extracted EDX spectrum of the RSF fiber that was washed with methanol after heat stretching, and chlorine was not confirmed.

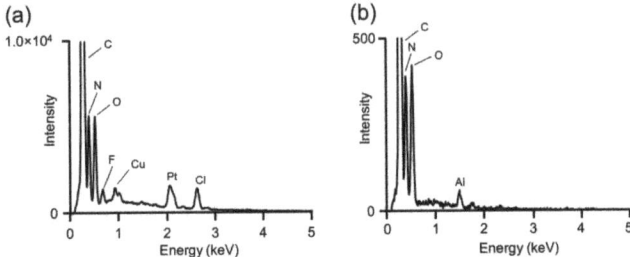

Figure 4. EDX spectra for heat-stretched regenerated silk fibroin (RSF) fibers with/without washing using methanol to confirm the remaining ionic liquid. The RSF fiber (**a**) without washing and (**b**) with washing using methanol. The peaks of copper, platinum, and aluminum in the picture are backgrounds from sample holders or sputter deposition in sample preprocessing.

After heat stretching, IL was easily removed using a washing process that involved immersing the fibers in a suitable organic solvent. This process suggested the possibility of not only removing ILs used specifically as forming aids from RSF fibers as needed but also controlling the amount of IL expected to function as a plasticizer. Furthermore, under solution equilibrium, the immersion process was expected to allow for the exchange and distribution of the ILs eventually remaining in the RSF fibers.

3.4. Mechanical Strength of the RSF Fibers

Figure 5 shows the stress–strain (S–S) plots of heat-stretched RSF fiber (closed circles, ●) and the plots washed with methanol (open squares, □) to compare the mechanical strength of the RSF fiber from a practical viewpoint. Methanol was selected as a typical solvent for ILs. The IL-washing process after heat stretching was designed for industrial application, and methanol was selected as a typical solvent for [HMIM][Cl], which was added as a plasticizer. The main physical values that characterize the RSF fibers in the tensile testing are shown in Table 1. For non-washed (closed circle plots, ●) and washed-off IL (open square plots, □), corresponding to Figure 5, the tensile stress at the point of the maximum load was 159.9 and 105.4 MPa (tensile strain: 3.6% and 2.3%, respectively), the fracture strain was 31.5% and 35.8%, Young's modulus was 4.86 and 4.83 GPa (to the top yield), and toughness was 43.2 and 39.4 MJ/m^3 (where calculated from the area under the curve integration with a trapezoidal approximation of the plots), respectively.

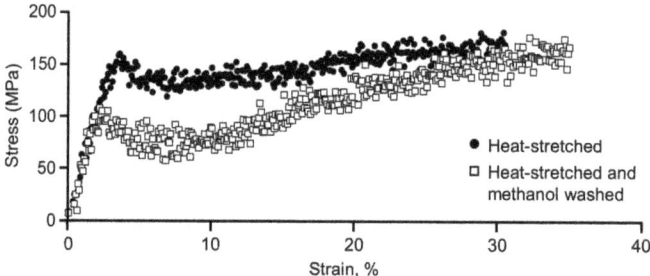

Figure 5. Stress–strain (S–S) plots for heat-stretched regenerated silk fibroin (RSF) fibers. Closed circles (•) represent S–S plots of heat-stretched RSF without wash using methanol. Open squares (□) represent S–S plots of heat-stretched RSF after wash using methanol.

Table 1. Mechanical property values of heat-stretched regenerated silk fibroin (RSF) fibers on the tensile testing.

Heat-Stretched RSF	Top Yield (MPa)	Fracture Strain (%)	Young's Modulus (GPa)	Toughness (MJ/m^3)
Non-washed (containing IL)	159.9	31.5	4.86	43.2
Washed off IL	105.4	35.8	4.83	39.4

The process of washing off the IL did not cause the significant degradation of fiber properties according to the S–S plots after heat stretching. The tensile strain at the yield point was reduced by 1.3% points when methanol was used. This is reasonably explained by the loss of IL, which allowed it to function as a plasticizer. Conversely, the fracture strain increased by 4.3% points in heat-stretched RSF fibers (closed circles (•) in Figure 5) compared with that in the RSF without IL washing (open squares (□) in Figure 5). These results were considered a trade-off between the increased secondary structure of SF protein from an alcohol-induced β-coil structure [28] in SF molecules and the loss of flexibility due to the removal of IL as a plasticizer.

The authors hypothesized that fibers containing ILs were momentarily fractured because of an imbalance between the relaxing fiber's body caused by the plasticizing function of ILs and the fracture of microscopic defects caused by elongation, in which the tearing force prevailed. The methanol-washed fibers lost ILs; however, the secondary structure was increased overall, as described in the literature [28]. The SF molecules themselves exhibit more elongation, as shown in the data. There would be no relaxation in the elongation mechanism as there would be in the presence of the plasticizer, and an instantaneous fracture of the microscopic defects in the fiber was occurring.

3.5. Thermophysical Properties

Figure 6 shows the TG–DTA curves of as-spun RSF fiber. Data showed a 2.0 wt% loss relative to the initial weight from 50 °C to 100 °C, with a further 5.1 wt% loss in steps from 110 °C to 185 °C. Desorption of equilibrated moisture from the atmosphere would account for the initial 2.0 wt%. Because the fiber already contains ILs that interact with hydrogen bonds among SF molecular chains, it could be assumed that the moisture was volatilized by heating below 100 °C. Furthermore, the weight loss of approximately 150 °C suggested that the heating desorption of HFIP (boiling point 58 °C for a single chemical entity) and the desorption of HFIP from SF materials at a temperature greater than the boiling point had been considered as the formation of specific hydrogen bonds between SF and HFIP [29,30]. The 10 wt% loss temperature, except for the mass of volatile components (water and HFIP), was 258 °C, and the decomposition temperature was observed in the range of 260–320 °C. The temperatures corresponding to the maximum degradation rate are 285.43 °C (random coil), 281.38 °C (silk-I structure), and 313.73 °C (silk fiber) according to the literature [31]. The current findings backed up these claims.

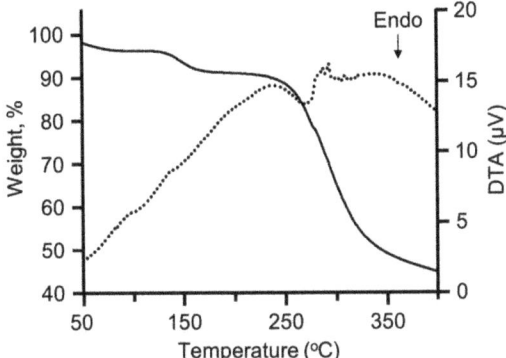

Figure 6. Thermogravimetric–differential thermal analysis curve of as-spun (nondrawing) regenerated silk fibroin fiber.

4. Conclusions

In this study, the dry spinning of SF was shown as a proof of concept for a process that uses ILs as dissolution aids and plasticizers in the development of natural polymeric materials. The mechanical properties of the RSF fibers reached 159.9 MPa and 31.5% at maximum stress and elongation, respectively. ILs were uniformly distributed in the fiber, indicating that they can effectively function as a plasticizing additive. Further studies are needed to clarify the function–property relationship between natural polymeric materials and ILs. Currently, there is no way to produce such materials more cheaply than petroleum-derived materials. However, in the future, protein materials may be produced more cheaply than petroleum-derived materials. In such a development situation, the concept of this research, which pioneered a new method of producing raw materials obtained from natural sources, will become even more important.

5. Patents

The research reported in this article is based on the following patent: Sato, T.; Morinaga, T.; Satoh, R. Method for producing polymer substance molding. Japan patent JP2021028434A. Publication date, 25 February 2021.

Author Contributions: Conceptualization, T.S. and T.M.; methodology, R.S. and T.M.; formal analysis, R.S. and T.M.; investigation, R.S. and T.M.; resources, R.S and T.M.; data curation, R.S.; writing—original draft preparation, R.S.; writing—review and editing, R.S, T.M. and T.S.; visualization, R.S.; supervision, T.S.; project administration, T.S.; funding acquisition, R.S. All authors have read and agreed to the published version of the manuscript.

Funding: This research was funded by Tsuruoka-Kosen Technology Promotion Association (2018, No. A-8).

Institutional Review Board Statement: Not applicable.

Informed Consent Statement: Not applicable.

Data Availability Statement: The data that support the findings of this study are available from the corresponding author, T.M., upon reasonable request.

Acknowledgments: The authors are deeply grateful to Tsuruoka Silk (Yamagata, Japan) for the donation of silkworm silk cocoons.

Conflicts of Interest: The authors declare no conflict of interest.

References

1. IPCC. Climate Change 2022: Impacts, Adaptation, and Vulnerability. Contribution of Working Group II to the Sixth Assessment Report of the Intergovernmental Panel on Climate Change. Summary for Policymakers. 2022. Available online: https://www.ipcc.ch/report/ar6/wg2/downloads/report/IPCC_AR6_WGII_SummaryForPolicymakers.pdf (accessed on 21 May 2022).
2. Van Wormer, E.; Mazet, J.A.K.; Hall, A.; Gill, V.A.; Boveng, P.L.; London, J.M.; Gelatt, T.; Fadely, B.S.; Lander, M.E.; Sterling, J.; et al. Viral emergence in marine mammals in the North Pacific may be linked to Arctic sea ice reduction. *Sci. Rep.* **2019**, *9*, 15569. [CrossRef] [PubMed]
3. Northrup, D.L.; Basso, B.; Wang, M.Q.; Morgan, C.L.S.; Benfey, P.N. Novel technologies for emission reduction complement conservation agriculture to achieve negative emissions from row-crop production. *Proc. Natl. Acad. Sci. USA* **2021**, *118*, e2022666118. [CrossRef] [PubMed]
4. Chen, J.M. Carbon neutrality: Toward a sustainable future. *Innovation* **2021**, *2*, 100127. [CrossRef]
5. Chen, W.; Meng, J.; Han, X.; Lan, Y.; Zhang, W. Past, present, and future of biochar. *Biochar* **2019**, *1*, 75–87. [CrossRef]
6. Temesgen, S.; Rennert, M.; Tesfaye, T.; Nase, M. Review on Spinning of Biopolymer Fibers from Starch. *Polymers* **2021**, *13*, 1121. [CrossRef] [PubMed]
7. King, F.W. Process for Wet Spinning Aromatic Polyamides. U.S. Patent US3079219A, 6 December 1960.
8. Sekiyama, K.; Sekiyama, K.; Ishikawa, M.; Sato, R.; Murata, S. Artificial Polypeptide Fiber and Method for Producing Same. World Intellectual Property. Organization Patent WO2012165476A1, 1 June 2011.
9. Rizzo, G.; Presti, M.L.; Giannini, C.; Sibillano, T.; Milella, A.; Guidetti, G.; Musio, R.; Omenetto, F.G.; Farinola, G.M. Bombyx mori Silk Fibroin Regeneration in Solution of Lanthanide Ions: A Systematic Investigation. *Front. Bioeng. Biotechnol.* **2021**, *9*, 653033. [CrossRef] [PubMed]
10. Koroskenyi, B.; McCarthy, S.P. Synthesis of Acetylated Konjac Glucomannan and Effect of Degree of Acetylation on Water Absorbency. *Biomacromolecules* **2001**, *2*, 824–826. [CrossRef] [PubMed]
11. Kim, S.-J.; Jang, J. Effect of degree of polymerization on the mechanical properties of regenerated cellulose fibers using synthesized 1-allyl-3-methylimidazolium chloride. *Fibers Polym.* **2013**, *14*, 909–914. [CrossRef]
12. Abdelghafour, M.M.; Orbán, Á.; Deák, Á.; Lamch, Ł.; Frank, É.; Nagy, R.; Ádám, A.; Sipos, P.; Farkas, E.; Bari, F.; et al. The Effect of Molecular Weight on the Solubility Properties of Biocompatible Poly(ethylene succinate) Polyester. *Polymers* **2021**, *13*, 2725. [CrossRef]
13. Heinze, T.; Schwikal, K.; Barthel, S. Ionic Liquids as Reaction Medium in Cellulose Functionalization. *Macromol. Biosci.* **2005**, *5*, 520–525. [CrossRef]
14. Phillips, D.M.; Drummy, L.F.; Conrady, D.G.; Fox, D.M.; Naik, R.R.; Stone, M.O.; Trulove, P.C.; De Long, H.C.; Mantz, R.A. Dissolution and Regeneration of *Bombyx mori* Silk Fibroin Using Ionic Liquids. *J. Am. Chem. Soc.* **2004**, *126*, 14350–14351. [CrossRef]
15. Hauru, L.K.J.; Hummel, M.; Nieminen, K.; Michud, A.; Sixta, H. Cellulose regeneration and spinnability from ionic liquids. *Soft Matter* **2016**, *12*, 1487–1495. [CrossRef]
16. Shamshina, J.L.; Zavgorodnya, O.; Berton, P.; Chhotaray, P.; Choudhary, H.; Rogers, R.D. Ionic Liquid Platform for Spinning Composite Chitin–Poly(lactic acid) Fibers. *ACS Sustain. Chem. Eng.* **2018**, *6*, 10241–10251. [CrossRef]
17. Togo, A.; Suzuki, S.; Kimura, S.; Iwata, T. Wet Spinning of α-1,3-glucan using an Ionic Liquid. *J. Fiber Sci. Technol.* **2021**, *77*, 213–222. [CrossRef]
18. Vocht, M.P.; Beyer, R.; Thomasic, P.; Müller, A.; Ota, A.; Hermanutz, F.; Buchmeiser, M.R. High-performance cellulosic filament fibers prepared via dry-jet wet spinning from ionic liquids. *Cellulose* **2021**, *28*, 3055–3067. [CrossRef]
19. Al Aiti, M.; Das, A.; Kanerva, M.; Järventausta, M.; Johansson, P.T.; Scheffler, C.; Göbel, M.; Jehnichen, D.; Brünig, H.; Wulff, L.; et al. Dry-Jet Wet Spinning of Thermally Stable Lignin-Textile Grade Polyacrylonitrile Fibers Regenerated from Chloride-Based Ionic Liquids Compounds. *Materials* **2020**, *13*, 3687. [CrossRef]
20. Sato, T.; Morinaga, T.; Satoh, R. Method for Producing Polymer Substance Molding. Japan Patent JP2021028434A, 25 February 2021.
21. Facas, G.G.; Maliekkal, V.; Neurock, M.; Dauenhauer, P.J. Activation of Cellulose with Alkaline Earth Metals. *ACS Sustain. Chem. Eng.* **2022**, *10*, 1943–1950. [CrossRef]
22. Hawes, C.S. Coordination sphere hydrogen bonding as a structural element in metal–organic Frameworks. *Dalton Trans.* **2021**, *50*, 6034–6049. [CrossRef]
23. Magaz, A.; Roberts, A.; Faraji, S.; Nascimento, T.R.; Medeiros, E.S.; Zhang, W.; Greenhalgh, R.; Mautner, A.; Li, X.; Blaker, J.J. Porous, Aligned, and Biomimetic Fibers of Regenerated Silk Fibroin Produced by Solution Blow Spinning. *Biomacromolecules* **2018**, *19*, 4542–4553. [CrossRef]
24. Zhang, J.; Tominaga, K.; Yamagishi, N.; Gotoh, Y. Comparison of Regenerated Cellulose Fibers Spun from Ionic Liquid Solutions with Lyocell Fiber. *J. Fiber Sci. Technol.* **2020**, *76*, 257–266. [CrossRef]

25. Zhang, B.; Hoagland, D.A.; Su, Z. Ionic Liquids as Plasticizers for Polyelectrolyte Complexes. *J. Phys. Chem. B* **2015**, *119*, 3603–3607. [CrossRef] [PubMed]
26. Koyanagi, R.; Zhu, Z.; Asakura, T. Regenerated Bombyx mori silk fiber with enhanced biodegradability. *J. Insect Biotechnol. Sericol.* **2010**, *79*, 27–30. [CrossRef]
27. Ajisawa, A. Dissolution of silk fibroin with calciumchloride/ethanol aqueous solution. *J. Sericultural Sci. Jpn.* **1998**, *67*, 91–94. [CrossRef]
28. Kaewpirom, S.; Boonsang, S. Influence of alcohol treatments on properties of silk-fibroin-based films for highly optically transparent coating applications. *RSC Adv.* **2020**, *10*, 15913–15923. [CrossRef]
29. Drummy, L.F.; Phillips, D.M.; Stone, M.O.; Farmer, B.L.; Naik, R.R. Thermally Induced α-Helix to β-Sheet Transition in Regenerated Silk Fibers and Films. *Biomacromolecules* **2005**, *6*, 3328–3333. [CrossRef]
30. Yoshioka, T.; Tashiro, K.; Ohta, N. Molecular Orientation Enhancement of Silk by the Hot-Stretching-Induced Transition from α-Helix-HFIP Complex to β-Sheet. *Biomacromolecules* **2016**, *17*, 1437–1448. [CrossRef]
31. Zhao, M.; Qi, Z.; Tao, X.; Newkirk, C.; Hu, X.; Lu, S. Chemical, Thermal, Time, and Enzymatic Stability of Silk Materials with Silk I Structure. *Int. J. Mol. Sci.* **2021**, *22*, 4136. [CrossRef]

Article

Improved On-Site Characterization of Arsenic in Gypsum from Waste Plasterboards Using Smart Devices

Masamoto Tafu [1,*], Juna Nakamura [2], Momoka Tanii [2], Saori Takamatsu [1] and Atsushi Manaka [1]

[1] Department of Applied Chemistry and Chemical Engineering, National Institute of Technology, Toyama College, Toyama 939-8630, Japan; stakamatsu@nc-toyama.ac.jp (S.T.); manaka@nc-toyama.ac.jp (A.M.)
[2] ECOdesign Engineering Program, Advanced Course, National Institute of Technology, Toyama College, Toyama 939-8630, Japan; h1611334@mailg.nc-toyama.ac.jp (J.N.); momoka19990828@icloud.com (M.T.)
* Correspondence: tafu@nc-toyama.ac.jp; Tel.: +81-76-493-5402

Abstract: The impurities in waste plasterboards, a product of ethical demolition, are a serious problem for their recycling. Plasterboards, the wall materials used in old buildings, are often recycled into gypsum powder for various applications, including ground stabilization. However, this powder contains various chemical impurities from the original production process of the gypsum itself, and such impurities pose a risk of polluting the surrounding soil. Here, we present a simple method for verifying the presence of arsenic, a harmful element in recycled gypsum that is suitable for use at demolition sites. First, we developed a simple pretreatment method using a cation-exchange resin to dissolve insoluble gypsum suspended in water by exploiting a chemical equilibrium shift, and we estimated the quantity suitable for releasing the arsenic from arsenic-containing gypsum. This pretreated solution could then be tested with a conventional arsenic test kit by observing the color changes in the test paper using the image sensor of a smart device. This simple method could determine a wide range of arsenic quantities in the gypsum, which would be helpful for monitoring arsenic in recycled gypsum powder, thereby supporting the development of a safe circular economy for waste plasterboards.

Keywords: plasterboard; arsenic; recycling; on-site determination

Citation: Tafu, M.; Nakamura, J.; Tanii, M.; Takamatsu, S.; Manaka, A. Improved On-Site Characterization of Arsenic in Gypsum from Waste Plasterboards Using Smart Devices. *Materials* **2022**, *15*, 2446. https://doi.org/10.3390/ma15072446

Academic Editors: Ana Mladenovic and Claudio Ferone

Received: 26 January 2022
Accepted: 22 March 2022
Published: 26 March 2022

Publisher's Note: MDPI stays neutral with regard to jurisdictional claims in published maps and institutional affiliations.

Copyright: © 2022 by the authors. Licensee MDPI, Basel, Switzerland. This article is an open access article distributed under the terms and conditions of the Creative Commons Attribution (CC BY) license (https://creativecommons.org/licenses/by/4.0/).

1. Introduction

Plasterboards consisting of solidified gypsum (calcium sulfate dihydrate, $CaSO_4 \cdot 2H_2O$) between paper sheets are widely used as wall materials in houses constructed using the 2 × 4 (two-by-four) method. In Japan, the lifetime of houses is approximately 40 years, and when houses are demolished, the plasterboards are collected for recycling. Specifically, the reclaimed gypsum is pulverized, treated, and used in new plasterboards. However, this recycling process is limited, and most gypsum in plasterboards is derived from mining (natural gypsum) as well as the byproducts of various chemical plant processes (chemical gypsum) and flue gas desulfurization (FGD), as shown in Figure 1. Specifically, chemical gypsum originates from phosphate and fluoride production and smelting, whereas FGD gypsum is a byproduct of thermal power plants using coal and heavy oil. Moreover, it potentially contains chemical impurities, including fluoride, arsenic, and cadmium, derived from the raw materials used in these chemical processes. As shown in this figure, gypsum is also is widely used as a component of Portland cement, as well as a ground stabilizer to improve ground hardness, which has been exhaustively studied. However, because gypsum in plasterboards is supplied from various sources, the recycled gypsum from waste plasterboards poses a risk of soil pollution by potentially releasing fluoride, arsenic, and cadmium into the surrounding soil.

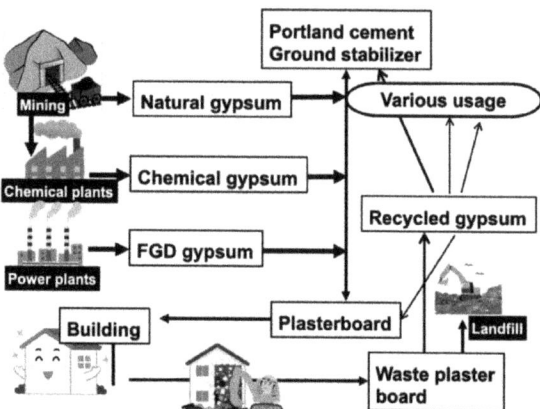

Figure 1. Material flow of gypsum in Japan. FGD: flue gas desulfurization.

In particular, the arsenic in waste gypsum has severe environmental effects. As a hazardous waste product of the metallurgical industry, arsenic-bearing gypsum (ABG) is derived from the lime neutralization of waste acid liquor [1,2]. Some amount of ABG was used in plasterboards in Japan from 1973 to 1997, and in 2017, approximately 1.5×10^4 tons of ABG were separated from abandoned buildings in Japan [3]. Waste plasterboards containing ABG must be carefully collected during demolition, and ABG-containing waste plasterboards must be identified to safely recycle gypsum for ground stabilization to avoid polluting the soil. Ideally, this identification is carried out directly at construction sites, which could lead to the development of a safer circular economy for recycled gypsum from waste plasterboards.

One on-site determination method for analyzing the arsenic in gypsum is X-ray fluorescence (XRF) [4,5], but this method requires skilled handling and/or a license to operate the radiation apparatus. To overcome this problem, we focused on adapting facile commercial test kits for determining the arsenic contents in an aqueous solution, a method that can be employed in the field. Because of the low solubility of gypsum, determining its arsenic levels requires a pretreatment to dissolve the gypsum into a homogeneous solution through a pyrolysis process involving harmful chemicals (such as hydrochloric acid [6] or perchloric acid [4]). After pretreatment, arsenic is released into the solution in a form suitable for detection, and conventional analytical methods can then be applied. We previously demonstrated that gypsum was easily dissolved in water containing cation- and anion-exchange resins and that the fluoride content in gypsum could be successfully determined in the resulting solution using a simple colorimetric method [7]. Based on these earlier findings, we hypothesized that the arsenic in waste gypsum could also be released by a pretreatment technique using only a cation-exchange resin because arsenic forms arsenate anions.

Therefore, in this study, we aimed to develop a simple pretreatment method for determining the arsenic levels in the gypsum from waste plasterboards to facilitate its use as a ground stabilizer. We adopted the following approach. First, we determined the suitable quantity of cation-exchange resin required to dissolve gypsum and release the arsenic it contains into water. The volume of arsenic in the resultant solution could then be determined using a conventional arsenic determination test kit based on Gutzeit's method. We also endeavored to interpret the color change in the test paper from the arsenic test kit based on the concentration of arsenic using data from the image sensors in smart devices, such as smartphones and/or tablets. The results suggest that the proposed method can rapidly determine the amount of arsenic in gypsum. We expect this innovative technique to facilitate the monitoring of harmful pollutants in recycled gypsum powder obtained from waste plasterboards for environmental safety.

2. Materials and Methods

2.1. Materials and Samples

First, we prepared arsenic-containing gypsum samples for characterization instead of using existing ABG from waste plasterboards. Calcium sulfate dihydrate (FUJI FILM Wako Pure Chemical, Bellwood Rd, VA, USA) was used to prepare this arsenic-containing gypsum by mixing 0.3 g of reagent gypsum and 10 cm^3 of aqueous solutions containing various amounts of sodium arsenite. Each mixture was then ultrasonicated for 5 min and dried in a convection oven at 80 °C for 24 h. The water used in all experiments was prepared via ion exchange and ultrapurification using a Milli-Q water purification system (Milli-Q A10, Merck-Millipore, Burlington, MA, USA).

2.2. Dissolution of Gypsum by Cation-Exchange Resin

A total of 300 mg of the arsenic-containing gypsum samples was mixed with 20 cm^3 of water, and various amounts of a cation-exchange resin were added (Amberlite IR120 H, DuPont Water Solutions, Wilmington, DE, USA). The mixture was shaken at 200 strokes per minute for 5 min using a reciprocal shaker. The temperature was adjusted to 298 K. The liquid phase was separated via pressure filtration through a cartridge membrane filter (pore size: 0.45 µm). The amount of gypsum dissolved in the water was analyzed by determining the calcium and sulfur content using inductively coupled plasma atomic emission spectrometry (ICP-AES, 720ES, Agilent Technologies, Inc., Santa Clara, CA, USA) with argon plasma.

2.3. Determination of Arsenic Content in the Gypsum

In order to measure the amount of arsenic present in an aqueous solution, two types of arsenic test kits based on Gutzeit's colorimetric method for lower and higher contents (MQuant Arsenic tests, model 1.01747 and 1.17927, respectively, Merck KgaA, Darmstadt, Germany) were selected. The arsenic-containing gypsum samples were dissolved by using the method described above. After the pretreatment, the obtained water samples were tested using the arsenic test kits, which indicate the arsenic contents through changes in the color of the test paper. This color change was determined using the image sensor of a tablet device (ZenPad 8.0, ASUSTeK Computer, Taipei, Taiwan).

In order to confirm these results, the volume of arsenic released from the arsenic-containing gypsum samples was also characterized by ICP-AES, as follows: each sample was mixed with water, and the cation-exchange resin using the method above, and the arsenic content in the obtained solution was analyzed. The determination limit of arsenic by ICP-AES was approximately 0.05 mg/L.

3. Results and Discussion

3.1. Suitable Volume of Cation-Exchange Resin for Gypsum Dissolution

First, the required amount of cation-exchange resin that adequately dissolves gypsum was determined. Figure 2 shows the sulfur and calcium concentrations in water, as measured by ICP-AES after treating the pristine gypsum reagent with the cation-exchange resin. Because of the low solubility of gypsum in water, the calcium and sulfur concentrations after mixing them in water without the cation-exchange resin differed from the values obtained by dissolving all the gypsum in water using the resin (blue and red lines in the figure, respectively). Specifically, adding the cation-exchange resin increased the sulfur concentration and decreased the calcium concentration. This phenomenon indicates that shifting the chemical equilibrium (Equation (1)) to the right successfully dissolved the gypsum in water, which was attributed to a decrease in the calcium concentration as a result of using the ion-exchange resin (Equation (2)).

$$CaSO_4 \cdot 2H_2O \rightleftharpoons Ca^{2+} + SO_4^{2-} + 2H_2O \quad (1)$$

$$R\text{-}H^+ + Ca^{2+} \rightarrow [R\text{-}Ca^{2+}]^+ \quad (2)$$

In Equation (2), R represents the cation-exchange resin. The experimental results demonstrated that 3.0 g or more of the cation-exchange resin in 20 cm^3 of water was required to dissolve 0.3 g of gypsum reagent.

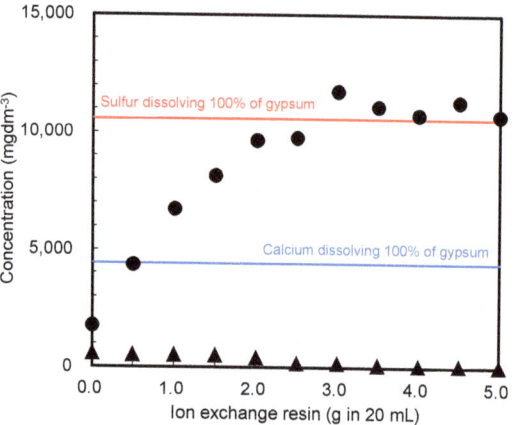

Figure 2. Sulfur and calcium concentrations in the water sample, as measured by ICP-AES after treating the gypsum reagent with a cation-exchange resin. Closed circles (●): sulfur concentration, triangles (▲): calcium concentration. Colored lines: sulfur (red) or calcium (blue) after dissolving 100% of gypsum in water using the resin.

The release of arsenic from the gypsum sample was then examined under the pretreatment conditions. In this study, we prepared arsenic-containing gypsum samples with predetermined amounts of arsenic instead of using existing ABG; thus, the arsenic contents were known and did not require further determination. Figure 3 shows the change in the arsenic concentration in water, as determined by ICP-AES after dissolving the gypsum samples containing various amounts of arsenic.

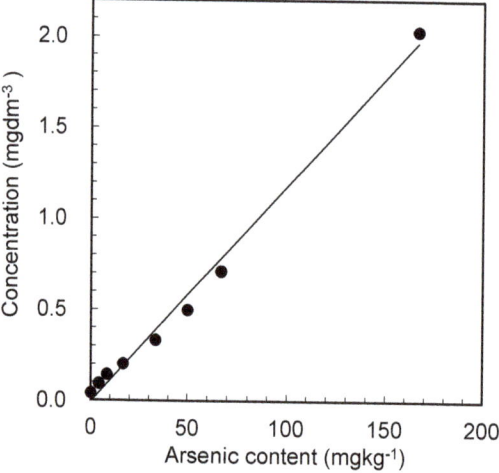

Figure 3. Arsenic concentrations in the water sample, as measured by ICP-AES, after treating the suspensions of the arsenic-containing gypsum samples with a cation-exchange resin as a function of the known arsenic content in the prepared samples. Experimental condition: gypsum: 0.3 g, cation-exchange resin: 4 g in 20 mL of water.

Figure 3 shows a strong linear relationship between the arsenic contents in the gypsum sample and the arsenic concentration in the treated water. Evidently, the arsenic in the gypsum is successfully released into the water after the cation-exchange resin pretreatment, thus eliminating the need for conventional treatment techniques that employ harmful chemicals. However, the amounts of arsenic released into the treated water were slightly lower than the values estimated from the arsenic used to prepare the gypsum samples. This result suggests that some arsenic ions were adsorbed on the ion-exchange resin. The pretreatment conditions, including the selection of ion-exchange resin, must therefore be optimized in future research.

3.2. Improved Determination of Arsenic Concentration Using Conventional Tests and Image Processing

Next, the amount of arsenic was determined using the arsenic test kit for higher arsenic content by analyzing the arsenic content in the solution obtained after the resin pretreatment, as described in the previous section. Figure 4 shows a photograph of the test paper from the kit after detecting arsenic in different water samples. As shown in the figure, the color change is not easily recognizable without skilled observation. In order to better quantify these results, changes in the output signal from a tablet image sensor after capturing images of the test paper color were plotted as a function of the arsenic concentration, as shown in Figure 5. Evidently, the blue output value from the image sensor strongly correlates with the arsenic content in the solution over a concentration range from 0.05 to 0.15 mgdm^{-3}. A similar relationship was obtained using the test kit for lower arsenic contents (data not shown).

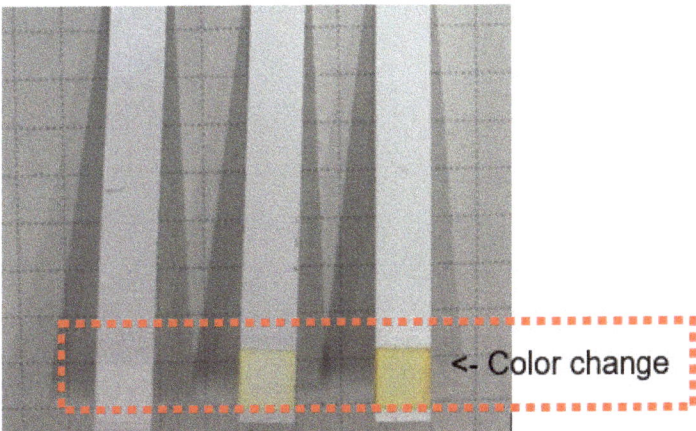

Figure 4. Photograph of the test paper for higher arsenic contents in the presence of different levels of arsenic in the water samples.

Based on these findings, we further investigated the arsenic levels in the simulated arsenic-containing gypsum. First, each gypsum sample was dissolved in water containing the cation-exchange resin, and the obtained solutions were tested using the arsenic test kit and the tablet for imaging, as described above. The results shown in Figure 6 indicate that the imaging with the tablet enables accurate determination of the arsenic concentration in the solution over the range of 10–100 mg kg^{-1} (Figure 6a) and 4–80 mg kg^{-1} (Figure 6b) using the arsenic test kits for higher and lower arsenic contents, respectively. Based on the results of the kit for lower contents, the arsenic content was thereafter determined using the arsenic test kit for higher arsenic contents, as the results obtained from the arsenic test kit for lower arsenic contents were quantitatively limited. However, these results indicated a higher sensitivity for detecting low arsenic contents in the gypsum.

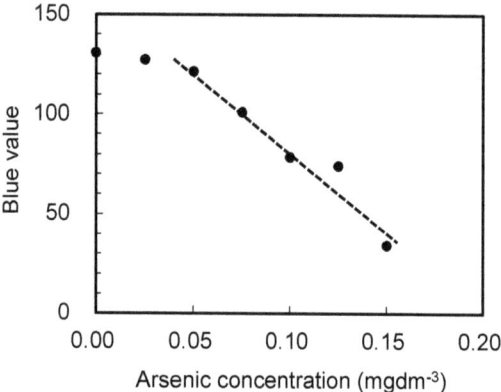

Figure 5. Change in the blue value of the image sensor as a function of the arsenic concentration in the solution.

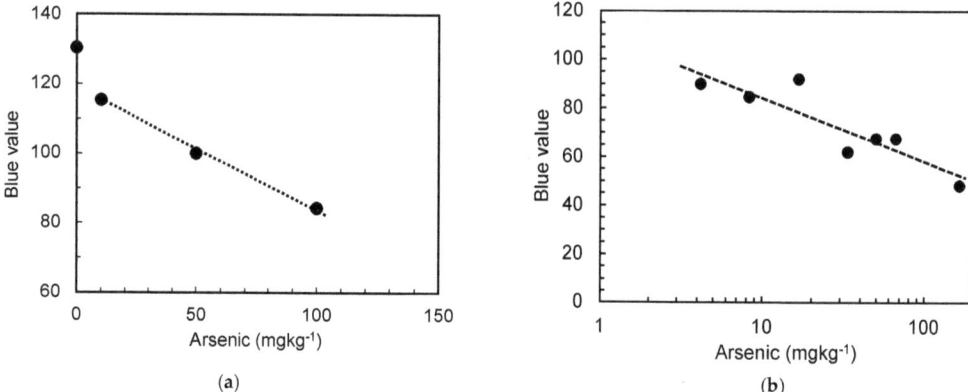

Figure 6. Arsenic levels in arsenic-containing gypsum, as determined using the proposed method, using the test kits for (**a**) high; (**b**) low arsenic contents.

3.3. Benefits of the Study Results in Gypsum Recycling

The results of this study were used to evaluate the benefits of recycling gypsum. Certain properties of gypsum make it suitable for use as a fertilizer [8]; therefore, recycled gypsum from waste plasterboards could be used in agricultural applications if the impurity concentrations are controlled. Because arsenic is harmful to agricultural activities [9], arsenic-containing gypsum should not be recycled for this purpose to avoid soil pollution. However, conventional methods for monitoring arsenic require specific analytical methods, skills, and equipment and involve lengthy processes. Consequently, it can be difficult to determine the volume of arsenic-containing gypsum in waste plasterboards at intermediate waste treatment facilities. Indeed, for the safe recycling of gypsum, determining the presence or absence of arsenic is more significant than quantifying the exact amount of arsenic present. The results described in the previous section suggest that arsenic-bearing gypsum can be easily identified using simplified pretreatment and conventional arsenic test kits. Further, this novel, simple method can be used for the on-site determination of the arsenic content in waste gypsum. Our results could, therefore, be useful in identifying ABG before accepting it for further reprocessing based on its arsenic content.

Fluoride is also an important impurity in waste plasterboards. We previously reported a method for the on-site determination of the fluoride content in waste gypsum by pretreatment with cation- and anion-exchange resins [7]. We also developed a simplified stabilization method for fluoride in gypsum by adding dicalcium phosphate dihydrate (DCPD, $CaHPO_4 \cdot 2H_2O$) [10,11] using the transformation reaction of DCPD to stable fluorapatite (FAp, $Ca_{10}(PO_4)_6F_2$) [12].

The flowchart in Figure 7 shows the benefits of our combined findings in making gypsum safe for various recycling applications. If the arsenic content is sufficiently low, the gypsum can be saved from being landfilled, but the fluoride content must be checked. If the fluoride content is also sufficiently low, the gypsum can be used in agricultural applications. Alternatively, the fluoride can be stabilized using DCPD, making the gypsum suitable for ground stabilization. Thus, our approach leads to a reduction in the amount of waste gypsum disposed of in landfills, safer use of recycled gypsum in agricultural applications, and the efficient use of DCPD to stabilize fluoride in the recycled gypsum used for ground stabilization, which could prevent the release of fluoride into the surrounding soil.

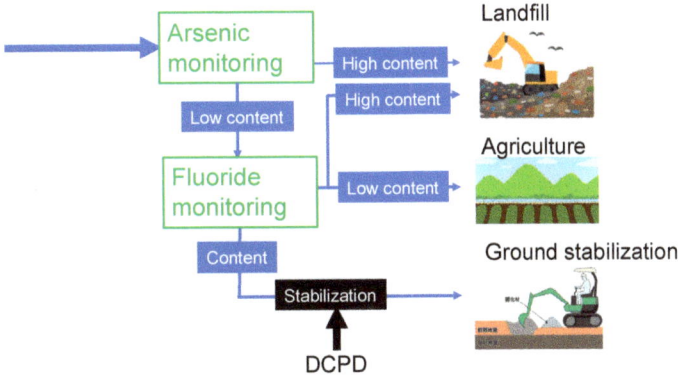

Figure 7. Potential advantages of our findings for the safe reprocessing of recycled gypsum obtained from waste plasterboards. DCPD: dicalcium phosphate dihydrate.

4. Conclusions

The results of this study suggest that the arsenic content in gypsum recycled from waste plasterboards could be determined via a pretreatment method employing an ion-exchange resin, which facilitates the release of arsenic into the solution. Then, the results of conventional colorimetric arsenic test kits can be monitored using a tablet to better quantify the arsenic concentration.

The key points can be summarized as follows:

(1) Although gypsum is a stable compound in water, its solubility was sufficiently enhanced by adding a cation-exchange resin.
(2) Using our proposed method, we accurately determined the arsenic concentration in the gypsum sample over a range of 5–100 mg kg^{-1} using different test kits for higher or lower arsenic contents.

The safety of recycled gypsum powder from the waste plasterboard is essential for various applications, particularly when the gypsum is used in soil. The results of this research are thus expected to be readily applied to the construction of a safe waste plasterboard recycling system that adheres to the concept of a circular economy.

Author Contributions: Conceptualization, M.T. (Masamoto Tafu), A.M. and J.N.; Methodology, M.T. (Masamoto Tafu), J.N. and M.T. (Momoka Tanii); Software for the tablet, A.M.; Writing—original draft preparation, M.T. (Masamoto Tafu) and J.N.; Writing—review and editing, M.T. (Masamoto Tafu), A.M. and S.T.; Project administration, S.T. and M.T. (Masamoto Tafu); Funding acquisition, M.T. (Masamoto Tafu). All authors have read and agreed to the published version of the manuscript.

Funding: This research was funded by the Gypsum Board Association of Japan.

Institutional Review Board Statement: Not applicable.

Informed Consent Statement: Not applicable.

Data Availability Statement: The data that support the findings of this study are available from the corresponding author, M.T. (Masamoto Tafu), upon reasonable request.

Acknowledgments: We are grateful to Haruka Tsunekawa for providing technical support with the arsenic monitoring and Nobuhito Motomura and Takayuki Motomura from the Recycle Factory Co. Ltd., Hokkaido, Japan for various useful suggestions on recycling waste plasterboards for agricultural use.

Conflicts of Interest: The authors declare no conflict of interest.

References

1. Li, X.; Zhu, X.; Qi, X.; Li, K.; Wei, Y.; Wang, H.; Hu, J.; Hui, X.; Zhang, X. Pyrolysis of arsenic-bearing gypsum sludge being substituted for calcium flux in smelting process. *J. Anal. Appl. Pyrolysis* **2018**, *130*, 19–28. [CrossRef]
2. Zhang, D.; Yuan, Z.; Wang, S.; Jia, Y.; Demopoulos, G.P. Incorporation of arsenic into gypsum: Relevant to arsenic removal and immobilization process in hydrometallurgical industry. *J. Hazard. Mater.* **2015**, *300*, 272–280. [CrossRef] [PubMed]
3. Yang, D.; Sasaki, A.; Endo, M. Reclamation of a waste arsenic-bearing gypsum as a soil conditioner via acid treatment and subsequent Fe(II) As stabilization. *J. Clean. Prod.* **2019**, *217*, 22–31. [CrossRef]
4. Potts, P.J.; Ramsey, M.H.; Carlisle, J. Portable X-ray fluorescence in the characterisation of arsenic contamination associated with industrial buildings at a heritage arsenic works site near Redruth, Cornwall, UK. *J. Environ. Monit.* **2002**, *4*, 1017–1024. [CrossRef] [PubMed]
5. Melamed, D. Monitoring arsenic in the environment: A review of science and technologies with the potential for field measurements. *Anal. Chim. Acta* **2005**, *532*, 1–13. [CrossRef]
6. Wang, S.; Zhang, D.; Li, X.; Zhang, G.; Wang, Y.; Wang, X.; Gomez, M.A.; Jia, Y. Arsenic associated with gypsum produced from Fe(III)-As(V) coprecipitation: Implications for the stability of industrial As-bearing waste. *J. Hazard. Mater.* **2018**, *360*, 311–318. [CrossRef] [PubMed]
7. Manaka, A.; Sawai, H.; Tafu, M.; Toshima, T.; Serikawa, Y.; Chohji, T. A simple preprocessing method using ion exchange resins for the analysis of fluoride content in gypsum. *J. Ecotechnol. Res.* **2011**, *16*, 47–50. [CrossRef]
8. Oster, J. Gypsum usage in irrigated agriculture: A review. *Nutr. Cycl. Agroecosystems* **1982**, *3*, 73–89. [CrossRef]
9. Jiao, W.; Chen, W.; Chang, A.C.; Page, A.L. Environmental risks of trace elements associated with long-term phosphate fertilizers applications: A review. *Environ. Pollut.* **2012**, *168*, 44–53. [CrossRef] [PubMed]
10. Tafu, M.; Chohji, T. Reaction between calcium phosphate and fluoride in phosphogypsum. *J. Eur. Ceram. Soc.* **2006**, *26*, 767–770. [CrossRef]
11. Tafu, M.; Chohji, T.; Morioka, I.; Hiwasa, M.; Nakano, H.; Fujita, T. Stabilization of Fluoride in Waste Gypsum by Using Surface-Modified Calcium Phosphate Particle. *Trans. Mater. Res. Soc. Jpn.* **2010**, *35*, 377–380. [CrossRef]
12. Tafu, M.; Chohji, T. Reaction of Calcium Hydrogenphosphate Dihydrate (DCPD) with a Solution Containing a Small Amount of Fluoride. *J. Ceram. Soc. Jpn.* **2005**, *113*, 363–367. [CrossRef]

Article

Tensile Examination and Strength Evaluation of Latewood in Japanese Cedar

Akihiro Takahashi [1,*], Naoyuki Yamamoto [1], Yu Ooka [2] and Toshinobu Toyohiro [1]

[1] Department of Mechanical Engineering, National Institute of Technology, Miyakonojo College, 473-1 Yoshio, Miyazaki 885-8567, Japan; naogen@cc.miyakonojo-nct.ac.jp (N.Y.); toyohiro@cc.miyakonojo-nct.ac.jp (T.T.)
[2] Architecture, National Institute of Technology, Miyakonojo College, 473-1 Yoshio, Miyazaki 885-8567, Japan; y-ooka@cc.miyakonojo-nct.ac.jp
* Correspondence: akihiro@cc.miyakonojo-nct.ac.jp; Tel.: +81-986-47-1172

Abstract: With the crisis awareness of global warming and natural disasters, utilization of local wood has drawn increasing attention in achieving the Sustainable Development Goals (SDGs). It is necessary to investigate the deformation and fracture of the structural tissue in wood in order to improve the safety and reliability of wood application. However, deformation and fracture mechanisms of the structural tissue in each annual ring are unknown. The mechanical characteristics of wood are reflected in the properties of earlywood and latewood. In the present study, microstructural observation and tensile tests were conducted to examine the relationship between the mechanical properties and fracture behavior of latewood in the growth direction in Japanese cedar. Brittle fracture behavior of the latewood specimen was confirmed based on the tensile stress–strain curve and features of the fracture surface. Moreover, two fracture modes, tensile fracture and shear fracture, were recognized. Weibull analysis of tensile strength in each fracture mode was performed to evaluate the reliability and utility of brittle latewood. Lastly, two fracture mechanisms were discussed based on the failure observation findings by a scanning electron microscope.

Keywords: Japanese cedar; latewood; mechanical property; fracture surface observation

1. Introduction

The increasing demand for materials made from renewable sources with a small environmental load has increased recently. This is due to several driving forces such as shortages of natural resources, changes in consumers and their concerns over social environmental issues, and the SDGs [1]. As a naturally grown material with carbon sequestration properties, wood has significant appeal as a sustainable material. The use of wood in industry and construction can reduce carbon in nature [2]. Therefore, wood is an environmentally friendly material that has been used for the construction of houses [3], marine environments [4], bridges [5–7], and wooden goods for many centuries. Regardless of species, engineered wood is a valuable construction material because of their highly desirable strength/density index. Significant progress in technology has been made for the last several decades to push the limit of wood construction with an advantage of higher strength/density index than other materials. As a result, there has been a noteworthy shift in public perception in terms of the acceptance of wood as a material for high-rise buildings by engineered wood such as glued laminated timber, GLT laminated veneer lumber, LVL, and cross-laminated timber, CLT [8]. There is already a growing list of high-rise wooden buildings that have been constructed in different countries [9], and the trend is expected to continue. In general, buildings up to 10 stories tend to use the CLT as the primary structure [10–15]. The longitudinal elastic modulus and tensile strength of the GLT beam and CLT wall panel are one of the important characteristic values that determine the suitability of the high-rise buildings.

Japanese cedar (*Cryptomeria japonica*) called Sugi, a kind of conifer, is the most produced wood in Japan, accounting for 57% in 2017 [15] and is expected as domestic lamina in all layers for the GLT and CLT frames [16] in Japan. The main supply of raw Sugi tree is shifting from medium-diameter logs (with a diameter between 140 and 220 mm) to large-diameter logs due to Japanese government's policies with respect to forestry management, and the production of large logs (over 300 mm in diameter) is increasing significantly within the timber manufacturing industry [17]. In an earthquake-prone country such as Japan, anti-seismic buildings with large-sized CLT are desired. For that reason, facilitating the collection of wide laminae from outside in a large-diameter log for large-sized laminated timbers is efficient, as shown in Figure 1.

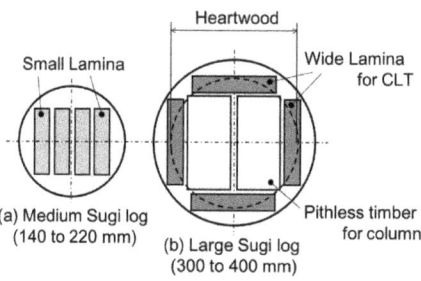

Figure 1. Cutout arrangement of lamina from different log diameters.

Although some information on Sugi timber quality has been obtained for medium Sugi logs, little is known about the quality in the large logs, the supply of which is expected to increase imminently. Studies have been carried out to evaluate mechanical properties of engineered woods such as GLT and CLT made of medium Sugi logs [18–22] or large Sugi logs [23–27] by Japanese researchers. According to research by Ido et al. [19] using lamina taken from large Sugi logs, the estimated tensile strength of CLT, as calculated using the Young's modulus of lamina of each layer, and the tensile strength of a lamina unit were found to be in good agreement with the measured tensile strength of CLT. On the other hand, the modulus of elasticity, MOE, and the modulus of rupture, MOR, in a Sugi sample by bending tests have widely been researched related to wood properties such as density, moisture content, microfibril angle (MFA), and so on [28–34], where these characteristics in medium-diameter Sugi logs were examined. However, tensile properties in the growth direction with respect to the microtome sample in the outer region where the lamina is cut have not been investigated yet. Wood is a hierarchically structured material with levels that might be termed as the tree structure, macroscopic (annual rings), microscopic (cellular), ultrastructural (cell walls), and biochemical levels (polymers such as cellulose, hemicellulose, and lignin). Thus, to utilize the functions of large-diameter Sugi logs, knowledge of their physical and mechanical properties is necessary as a whole. As above, for the engineered wood structure design such as large-sized GLT and CLT, knowledge of wood strength and rigidity is fundamental. However, previous studies have mostly focused on wood properties related to macro cross-sectional characteristics. Considering the current state of knowledge, it is a truism to claim that the mechanical strength of a wood material depends on its microstructure. Moreover, the microscopic mechanisms underlying the mechanical performance of wood need to be explained to meet the wood demand and improve technical skills and structural design technologies. To this end, in-depth studies on large-diameter Sugi logs for CLT have just begun.

An annual growth ring in Japanese cedar is shown in Figure 2. In general, latewood and earlywood are determined based on the plane cut of a tree log. The latewood region has a narrow wall thickness of 0.1 to 0.4 mm.

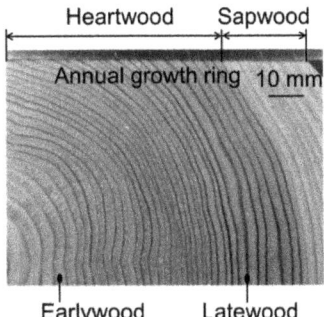

Figure 2. Annual growth ring in Japanese cedar used in present study.

At present, there are no material testing standards for tensile tests using miniature wood test pieces. However, a suitable tensile examination method for thin latewood in Japanese cedar was previously proposed in [35]. Moreover, it was clarified that the failure behavior of latewood follows two macroscopic fracture patterns [36]. However, their fracture mechanisms remain unclear. Fewer studies have shown the tensile properties and fracture behavior in latewood taken from the outer region in large-diameter Sugi logs and to accumulate fundamental data on wood properties. The aim of the present study is to investigate the tensile stress–strain behavior of a latewood as a simple substance specimen collected from the outer side in heartwood in a Sugi log. In addition, Weibull analysis is performed on the fracture strength of latewood, and different failure behaviors of the latewood are discussed based on experimental results in SEM observation. There is a lack of research into learning the from microtome sector to surely understand the tensile behavior in latewood. The present study will improve our knowledge and skills of holistic wood utilization and structural design techniques on manufacturing engineered woods such as large-sized GLT and CLT by Japanese cedar.

2. Materials and Methods

2.1. Experimental Material

Wood samples in this study were cedar from Nichinan city in southern Kyushu, Japan. The tree age was 40 years. The cut log had a body diameter of 300 mm (up to 4 m above the ground) and was sun-dried. The moisture content (MC) defined as the weight of water in the cut log was given as a percentage of the oven-dried weight [37]:

$$MC = \frac{moist\ weight - oven - dried}{oven - dried} \times 100(\%), \quad (1)$$

The MC in the cut log was 12%. The wood density calculated from the green volume and air-dry weight was 370 kg/m^3 [37].

Latewood samples were collected from rings 25 to 35 starting from the pith (i.e., outside heartwood). The manufacturing process and water immersion of the latewood tensile specimens in the growth direction taken from the cut log is shown in Figure 3. The specimen for tensile testing was cut in a straight grain orientation using a cutter tool and sandpaper. After cutting, the specimen was immersed in water for 24 h to remove residual strain on the specimen after cutting. After immersion, the specimen was dried in ambient air (relative humidity of 65% and room temperature of 298 K) for 72 h. Based on [38–40], the specimen thickness (radial) was 0.2 ± 0.05 mm, width (tangential) was 3.0 ± 0.5 mm, and length (longitudinal) was 130 mm.

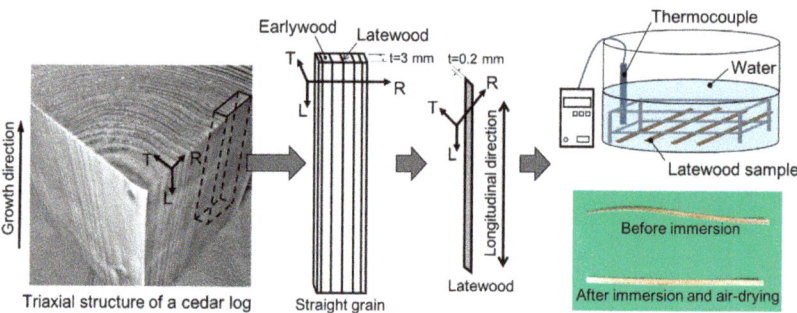

Figure 3. Manufacturing process and orientation of the latewood tensile specimen.

Microstructures of the latewood specimen in two orientations, T-R plane (a) and L-T plane (b), are shown in Figure 4. Tracheids with square or slightly rectangular cells of differing thickness were observed in the T-R plane. The thickness of the cell wall of tracheids ranged from 1 to 5 μm. The other region (three broken-line frames) running from the top to the bottom of the microstructure was the ray tissue. The tracheids and ray cells had a specific tilt angle, θ to the L direction in the L-T plane (b). The range was approximately 0 to 20°, measured by a protractor qualitatively. The density and *MC* in the latewood specimen measured by [37] were 897 kg/m^3 and 10%, respectively.

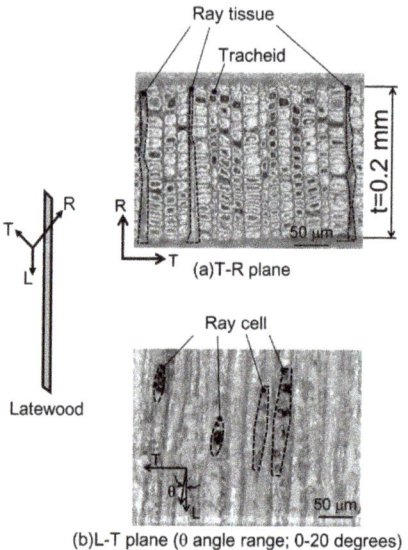

Figure 4. Microstructures in the two orientations in the latewood specimen: T-R plane (**a**) and L-T plane (**b**).

2.2. Tensile Examination Procedure

Tensile testing of the latewood specimen was performed in accordance with Japanese industrial standards, JIS Z2241 [41], at an initial loading speed of 1.0 mm/min. The universal testing system (EZ-SX, Shimadzu) was used to assess mechanical properties in the present study. Tensile load, *F*, was measured by a one-side loaded cell with a capacity of 500 N, and displacement and tensile strain of the tensile specimen in the growth direction were measured using a noncontact extensometer (DVE-101/201, Shimadzu, Kyoto, Japan) and one-side strain gauge (FLK-1-11, Tokyo Measuring Instruments Lab., Tokyo, Japan) [42]. For area measurement to calculate tensile stress, σ, and elastic modulus, E, the average value of three measured cross-sectional areas, A, of the specimen in the longitudinal axis before tensile examination was used. Therefore, engineering tensile stress, $\sigma = F/A$. E, was determined as a proportion of a regression line fitted to the stress–strain chart between 10 and 30 MPa. The 6 specimens attached to the strain gauge were turned out and 39 specimens without the gauge were prepared for tensile examination using a noncontact type extensometer, as shown in Figure 5. In advance, a tensile test was conducted using a thin metallic lead (Pb, thickness of 0.2 mm and elastic modulus of 16 GPa) attached to a strain gauge, and it was confirmed that the elastic modulus was measurable without the influence of the adhesive. A microphotograph of the specimen that failed along the frame of the attached strain gauge is shown in Figure 6. This failure was considered to be due to the hardness of the adhesive. As a result, the deformation in the elastic region was evaluated by a strain gauge and the noncontact type extensometer was used for fracture strain evaluation.

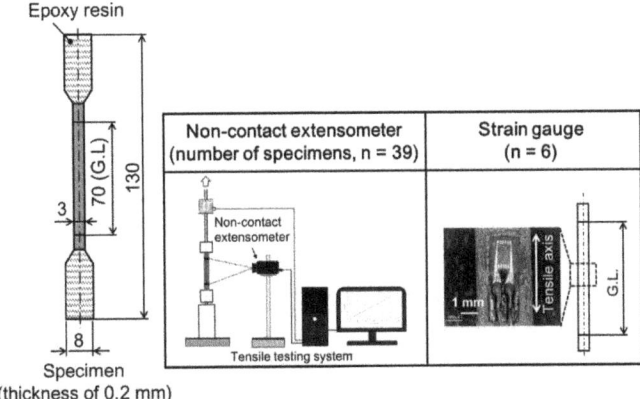

Figure 5. The experimental information e in the present study.

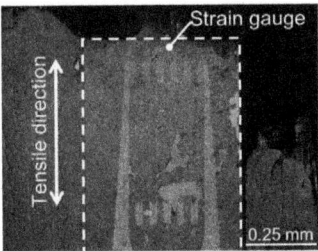

Figure 6. Failure along the edge of the strain gauge occurred during tensile testing.

An optical microscope (OM, VHX-2000, KEYENCE, Osaka, Japan) and electrical scanning microscopy (SEM) were used to observe the microstructural features. SEM (S-4800, Hitachi High-Technologies Corp., Ibaraki, Japan) was also used to examine the fractographic features. For interior views of the SEM observation, sputter coating with the Pt-Pd target was conducted.

2.3. Tensile Strength Evaluation by Weibull Statistics

The Weibull analysis was employed to assess a wide range of issues, including the mechanical properties of brittle materials and life-time testing [43]. The two-parameter continuous probability density function for tensile strength variables is expressed by:

$$P = \left(\frac{m}{\sigma_0}\right)\left(\frac{\sigma}{\sigma_0}\right)^{m-1} exp\left[-\left(\frac{\sigma}{\sigma_0}\right)^m\right], \qquad (2)$$

The mean density function is asymmetrical and will assume only positive values. The symbol m is the Weibull modulus and σ_0 is the scale parameter.

The cumulative distribution function that gives the probability of failure P at stress σ is expressed as:

$$P = 1 - exp\left[-\left(\frac{\sigma}{\sigma_0}\right)^m\right], \qquad (3)$$

where P is the probability of failure at stress, σ_B. In this study, the tensile strength of latewood was assessed by Weibull analysis.

3. Results and Discussion

The two dominant fracture modes observed in the present study, tensile fracture (a) and shear fracture (b), are shown in Figure 7. The macroscopic crack path of tensile fracture was perpendicular to the loading direction. On the other hand, the shear crack grew across to both ends of the specimen, causing a shear fracture, (b). In addition, the specimen with a θ of 0 to 11° exhibited a tensile fracture and another θ of 12 to 20° exhibited a shear fracture. This suggested a strong relationship between the θ and the fracture behavior of latewood.

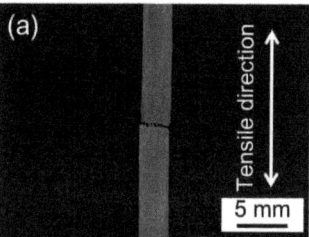 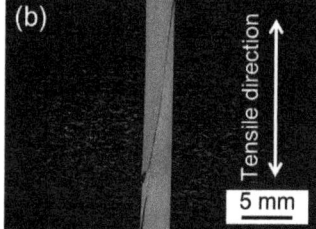

Figure 7. Classified failure modes were tensile fracture (**a**) and shear fracture (**b**).

Tensile stress–strain curves, which correspond to the two fracture modes, are shown in Figure 8. The curve of the tensile fracture was an approximately linear mechanical response, whereas that of the shear fracture followed a gentle curve from the elastic region. The applied stress was at a maximum (i.e., tensile strength, σ_B) and the tensile specimen quickly broke. This brittle fracture behavior also occurred at a low loading speed of 0.01 mm/min. Reiterer et al. reported that stress–strain curves of a latewood tensile specimen with MFA <5° and =20° in *Prica abies* [44]. Results on the fracture observation photograph for both were unknown in [44], but it is interesting that they were close to those in our experimental findings. The measurement of MFA in testing latewood was our next research task.

Several procedures have been suggested to determine Weibull parameters such as linear regression (the Weibull plot), weighted linear regression, and maximum likelihood [45–47]. The Weibull plot is the most common and simplest method to determine Weibull parameters. The tensile strength values are ranked from the minimum to the maximum and each value is assigned a probability of failure (P) based on its ranking, i, with i ranging from 1 to n, where n is the number of measurements of (tensile fracture, $n = 19$, and shear fracture, $n = 20$). The cumulative probability of failure (P) is calculated using the following equation:

$$P_i = \frac{i - 0.3}{n + 0.4},\quad (4)$$

where i is the rank and n is the total number of data. In this study, n was the total number of data for the measured tensile strength.

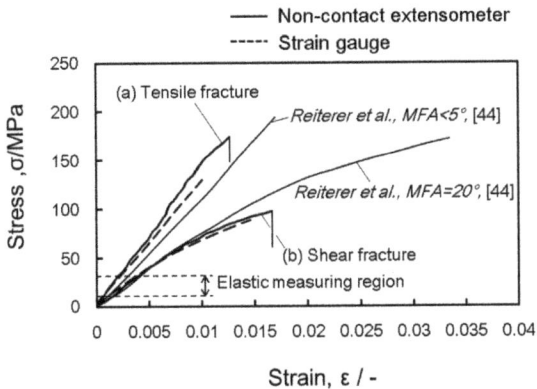

Figure 8. Stress–strain curves obtained from tensile examination. The curves correspond to the two respective failure modes.

The results of Weibull analysis by plotting $lnln(1/1-P)$ versus $ln\sigma_B$ for the tensile strength of the latewood specimen in the growth direction are shown in Figure 9. In the case, Equation (3) can be rearranged to give the following equation:

$$lnln\left(\frac{1}{1-P}\right) = mln\sigma - mln\sigma_0,\quad (5)$$

and the experimental tensile strength data plotted in Equation (5) give an approximate straight line from whose equation the parameters m and σ_0 can be estimated based on the linear regression. The Weibull parameter m of a tensile fracture was $m = 4.8$, with a correlation coefficient of $R = 0.98$, whereas that of a shear fracture was $m = 5.5$, with an $R = 0.95$. The mean tensile strength σ_{mean} can be expressed as [48]:

$$\sigma_{mean} = \sigma_0 \Gamma\left(1 + \frac{1}{m}\right),\quad (6)$$

where $\Gamma(\cdot)$ is a gamma function. The mean tensile strength σ_{mean} calculated by Equation (6) in the shear fractured specimen was approximately 29% lower than that in the tensile-fractured specimen. Weibull parameters and other mechanical properties are summarized in Table 1.

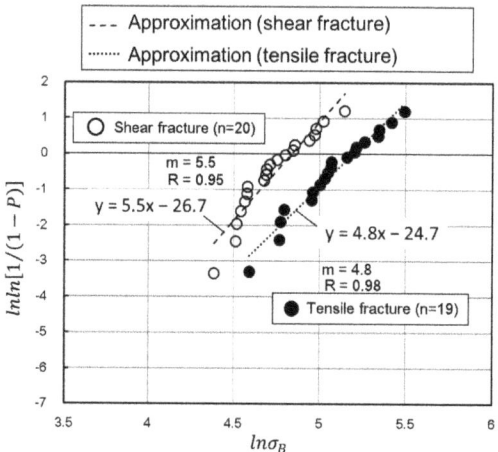

Figure 9. Weibull analysis plots corresponding to the two failure modes.

Table 1. Weibull parameters, elastic modulus, E, and fracture strain, ε_f, determined by tensile examination in the present study.

	Weibull Parameters				Elastic Modulus, E, GPa (Strain Gauge)	Fracture Strain, ε_f, (-)
	n	σ_0, MPa	m Value	Correlation Coefficient, R	Mean Tensile Strength, σ_{mean}, MPa		
Tensile fracture	19	180	4.8	0.98	165	13.3	0.015 ± 0.005
Shear fracture	20	126	5.5	0.95	116		0.014 ± 0.006

The simple composite mechanics rule that can be utilized to take into account the tracheids' orientation is the maximum stress criterion, assuming that the plane is modeled as a thin mat with fibers (analogous to the tracheids) oriented at an angle θ (analogous to the angle θ in Figure 4) to the fiber axis. Failure of the mat occurs either at a critical (local) stress value $\sigma_1 \geq \sigma_{1u}$ parallel to the fibers, $\sigma_2 \geq \sigma_{2u}$ perpendicular to the fibers, or at a shear stress $\tau_{12} \geq \tau_{12u}$ along the fibers. The (local) in-plane stresses working parallel and perpendicular to the fibers (σ_1, σ_2, τ_{12}) can then be expressed as the (global) stresses applied in the x- and y-directions of the fiber mat (σ_x, σ_y, τ_{xy}) according to [49]:

$$\left\{\begin{array}{c} \sigma_1 \\ \sigma_2 \\ \tau_{12} \end{array}\right\} = [T] \left\{\begin{array}{c} \sigma_x \\ \sigma_y \\ \tau_{xy} \end{array}\right\}, \tag{7}$$

where the transformation matrix is given by:

$$[T] = \begin{bmatrix} \cos^2\theta & \sin^2\theta & 2\cos\theta\sin\theta \\ \sin^2\theta & \cos^2\theta & -2\cos\theta\sin\theta \\ -\cos\theta\sin\theta & \cos\theta\sin\theta & \cos^2\theta - \sin^2\theta \end{bmatrix}, \tag{8}$$

If uniaxial tension ($\sigma_2 = \tau_{12} = 0$) is assumed, the stress σ_{xu} (i.e., applied tensile stress) to cause failure in the material can be expressed for each of the three failure modes as:

$$\sigma_{xu} = \frac{\sigma_{1u}}{\cos^2\theta}, \tag{9}$$

$$\sigma_{xu} = \frac{\sigma_{2u}}{\sin^2\theta}, \tag{10}$$

$$\sigma_{xu} = \frac{\tau_{12u}}{\cos\theta\sin\theta}, \tag{11}$$

where σ_{xu} in Equations (9)–(11) indicate axial stress, transverse stress, and shear stress, respectively. Under the assumption of independent modes of failure with no interaction between each other and using experimental data for σ_{1u}, σ_{2u}, and τ_{12u}, Equations (9)–(11) can be used to calculate the maximum tensile strength of latewood material with tracheids oriented at a given angle, θ. Conversely, data of the maximum (global) tensile strength can be predicted together with information on θ to calculate the local properties σ_{1u}, σ_{2u}, and τ_{12u}, for a material. As shown in Figure 10, the predicted critical tensile strength for tensile fracture (solid line, AB) is given by

$$\sigma_{ux} = \frac{165}{\cos^2\theta}; \ (0° \leq \theta < 12°), \tag{12}$$

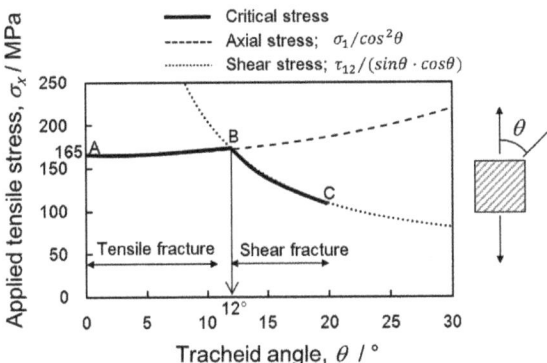

Figure 10. Predicted critical tensile strength depending on angle, θ, of the applied stresses for the onset of two failure modes.

σ_{mean} is 165 MPa when indicating tensile fracture, and the angle, $\theta = 12°$, is a specific tilt angle when transiting from tensile fracture to shear fracture. The two dashed lines of Equations (9) and (10) cross at point B. The stress value at point B is obtained at 173 MPa from Equation (12). Next, the shear fracture value, τ_{12u} at point B can be calculated by substituting $\sigma_{ux} = 173$ MPa in Equation (11), and a certain value, $\tau_{12u} = 35$ MPa, is determined. Therefore, the predicted critical tensile strength for shear fracture (solid line, BC) for angles from 12 to 20° can be drawn, as shown in Figure 10. Further, it can be seen that Figure 10 qualitatively corresponds to tensile strength depending on the angle of the tracheid and therewith the prediction of the fracture behavior.

A photograph of the nearby crack path in the tensile fracture observed from the L-T plane is shown in Figure 11. Microscopic crack deflection was frequent during fracture propagation of the tensile fracture. The fracture surface observed from above is shown in Figure 12. Fracture modes of tracheids were observed, and the enlarged view of frame A shows fibrils in the tip of the fractured tracheid (Fs indicated by an arrow), the separation of cells at the middle lamella (IC indicated by an arrow), a crack path cut through the tracheids (TW indicated by an arrow), and an hierarchical crack propagation within the cell wall (IW indicated by two arrows). According to Côté and Hanna [50], three kinds of cell fractures in many species are recognized: intercell failure (IC), transwall failure (TW), and intrawall failure (IW). Intercell failure takes place at the middle tracheid lamella and is simply the interfacial debonding between tracheids at these junctions. Transwall failure is the complete rupture when the fracture path cuts across the wall. Intrawall failure occurs within the secondary wall and, in most instances, it is at the S_1/S_2 interface or close to it. These fracture characteristics were confirmed in the fracture surface in the tensile fracture. These fracture modes tended to produce a highly rough fracture surface, as shown in Figure 11. Figure 12 also shows fracture of the ray cell observed on the tensile fracture surface. It is well known that the structure and distribution of the ray cell have a strong relationship with the compressive mechanical property and its fracture behavior in Sugi timber [51]. In present study, the influence of the ray cell on the tensile fracture behavior is unknown. However, it is suggested that the tensile fracture is partly related because the fracture of the ray cell was included in a part of the crack path.

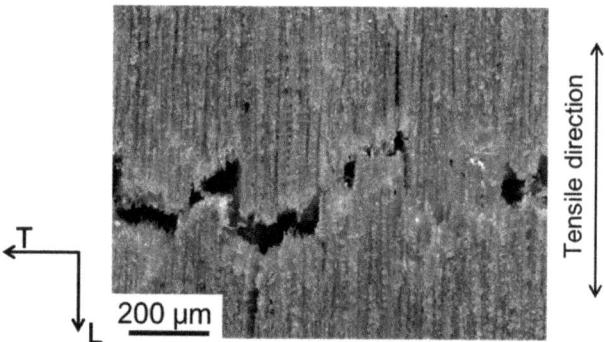

Figure 11. Serrated crack path in the tensile fracture observed in the L-T plane.

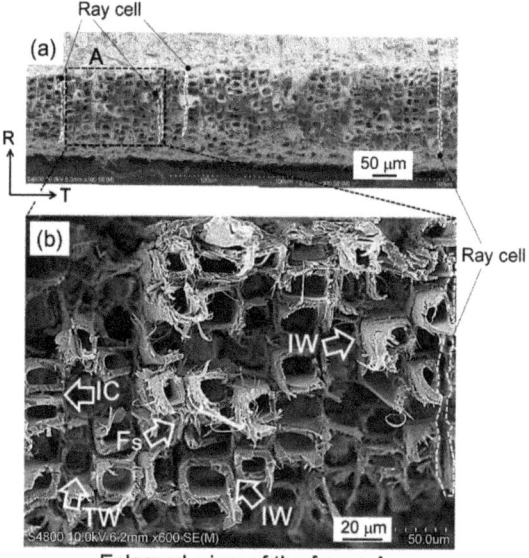

Figure 12. SEM image of the T-R fracture surface indicating the behavior of the tensile fracture. The enlarged view (**b**) is section A surrounded by the dashed lines in (**a**).

Ifju et al. [52] and Kifetew et al. [53] also observed fractures in the latewood region where the fibrils appeared at the tips of tracheids, as shown in Figure 12b. Approximately 97% of the Japanese cedar's cells are tracheids. Tracheids in the latewood have thick-walled cells, which consist of the S_2 layer of 86% in the cedar [54]. The microfibrils of the S_2 layer are almost fully aligned along the direction of the tracheid [53] and, therefore, its internal structure plays a key role in determining the mechanical properties, especially under applied load parallel to the grain. The microfibril angle (MFA) is the angle between helical windings of microfibrils in the S_2 layer of the tracheid and the longitudinal cell axis; on research in Sugi, MFA was found to be a crucial factor in obtaining mechanical properties such as stiffness [31–33,54–57] and bending load–deflection behavior [34,58]. A large MFA shows low stiffness, on the other hand, and a small MFA in wood shows high stiffness. In general, each cell was considerably stiffer and stronger parallel to its axis than perpendicular. Therefore, the elastic modulus and tensile strength for the specimen with the tensile fracture were higher than those of the specimen with the shear fracture.

A photograph of a single shear crack in the shear fracture observed from the L-T plane is shown in Figure 13. Macroscopic shear fractures occurred due to brittleness and at an angle of 12 to 20° in the tensile direction. This corresponds with the growth direction of tracheids being tilted at θ of 12 to 20° in the L direction. A microphotograph taken perpendicular to the shear fracture surface is shown in Figure 14. The shear crack propagated through the tracheid interface and intercellular layer into the ray tissue. We confirmed that the crack propagated in a stepwise manner, as shown in the upper part of Figure 14. In addition, the fracture of tracheids was mainly due to interfacial debonding along the lamellar structure of the tracheid. It is known that the dry wood cell interface between tracheids is filled with deposits such as lignin, gum, resin, and tylose [59]. The shear strength was significantly lower than the fibril strength in the tracheid. For this reason, interfacial debonding at the tracheid interface readily occurred. As shown in Figure 4b, the ray tissue configuration in the L-T plane had a high aspect ratio and its tips were sharp. Although the ray tissue and cell structures were not clarified, the shear resistance (i.e., shear modulus and shear strength) of the intercellular layer of ray tissue was weak based on fracture surface observation in this study. Miyoshi et al. measured the breaking length of ray tissue after the lateral tensile test [60] and reported that the mechanical properties of wood in the lateral direction are significantly affected by the structural features such as deformation of cell shapes and arrangement of ray tissue or tracheids [61]. Figure 15 shows another microphotograph of the shear fracture surface. Intercell failure predominated in the entire fracture surface in the tracheid region. Several twisting and tearing fibrils (indicated by arrows) were observed in the region of failed tracheids above and below the microphotograph in Figure 15a. Those of the fibrils were spread out in response to the shear direction, while an open plane without fibrils was viewed in the region of the ray cell (see Figure 15b). The strength characteristics in the region of the ray cell were lower than those of tracheids with fragments spiraling out. The observational finding is obviously evidence to decide the existence of different fracture strength levels on the interface tracheids and ray cell. The crack origin site of the shear fracture in the present study is unknown, but considering the earlier occurrence of shear cracks in latewood, ray cells may be involved. This result suggests that nonlinearity in the stress–strain curve due to shear loading as illustrated in Figure 8 was caused by accumulation fracture at ray cells occurring during testing.

Based on this study, the mechanical properties of Sugi latewood were closely related to the tilt of tracheid grains. Many previous studies demonstrated the influence of the slope of wood grain on the MOE and MOR as follows: Hankinson's formula is well known as a prediction equation for the strength as a function of the grain angle [62]. Xavier et al. [63] and Bilko et al. [64] reported that grain deviation in the testing force axis causes a degradation in shear strength using the shear arcan test, in which the angle of the grain ranges from 0 to 90°. Gupta et al. revealed the effects of the grain angle on the shear strength by the shear block test [65]. Mania et al. demonstrated that the grain deviation angle has the greatest influence on mechanical parameters, such as elastic energy and work until maximum load, using the bending test with different wood species [66]. Although these studies were conducted using samples containing earlywood and latewood, their conclusions are consistent with the present study in that the tilt of tracheids, θ, is important for mechanical properties.

Figure 13. A single shear crack path in the shear fracture observed in the L-T plane.

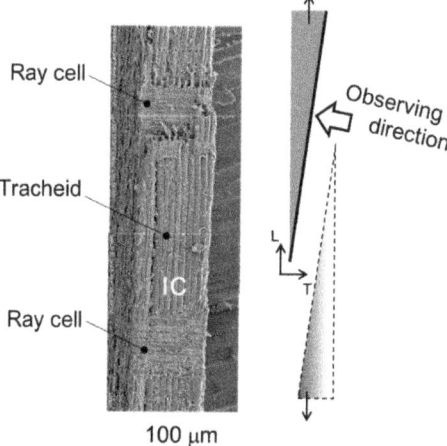

Figure 14. Microfractography of the shear fracture behavior along the tracheid interface and intercellular layer into the ray tissue.

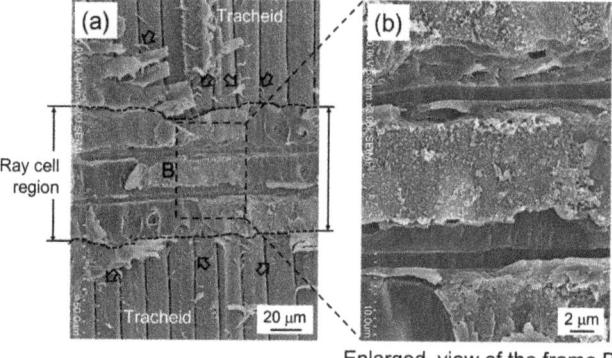

Figure 15. SEM image showing the behavior of the shear fracture. Arrows indicate tearing fibril fragments in the intercell-failed tracheids. The enlarged view (**b**) is section B surrounded by the dashed lines in (**a**).

In the present study, although the fracture mechanisms were not clarified in detail, our results suggest that the low mechanical property in ray cells have a distinct effect on fracture morphologies in Sugi latewood. These results might also influence the other properties such as cutting and processing of timber and durability in wood products in large-diameter Sugi logs because GLT and CLT beams are subject to shearing. On the other hand, large-diameter Sugi trees are equivalent to aged Sugi. For effective utilization of Sugi wood resources in the future, it is important to understand the properties of wood derived from aged-Sugi trees.

4. Conclusions

In this paper, the tensile stress–strain behavior of latewood as a simple substance specimen collected from the outer side in heartwood in a large-diameter Sugi log was investigated. Based on this study, including tensile examination, Weibull statistics analysis, and fracture surface observation by SEM, we made the following conclusions:

1. Two fracture modes in the L-T plane of latewood, tensile fracture and shear fracture, were revealed by tensile examination. This study suggested that the two fracture modes depend on the tilt of tracheids observed in the L-T plane.
2. The average tensile strength by Weibull statistics analysis of the shear-fractured specimen was approximately 29% lower than that in the tensile-fractured specimen.
3. Fibrils from within the tracheid were closely fully related to tensile fractures in latewood. On the other hand, the shear crack occurred at an angle of 12 to 20° in the tensile direction. There were two features of shear fractures: interfacial debonding of tracheids and crack propagation of the intercellular layer in ray cells. Moreover, it was found as evidence to decide the existence of different fracture strength levels on the interface tracheids and ray cells under shearing.

Author Contributions: Conceptualization, A.T. and Y.O.; methodology, A.T. and N.Y.; formal analysis, A.T. and Y.O.; performing and experiments, A.T., Y.O. and N.Y.; data curation, A.T. and T.T.; writing—original draft preparation, A.T.; writing—review and editing, A.T. and T.T. All authors have read and agreed to the published version of the manuscript.

Funding: The present work was partially supported by the ICHIJU Industrial Science and Technology Promotion Foundation, Japan of FY2021.

Institutional Review Board Statement: Not applicable.

Informed Consent Statement: Not applicable.

Data Availability Statement: Data will be made available upon reasonable request.

Acknowledgments: The authors would like to thank S. Yuki and H. Yamamoto for technical assistance with the experiments.

Conflicts of Interest: The authors declare no conflict of interest.

References

1. Food and Agriculture Organization of the United Nations. *Transforming Food and Agriculture to Achieve the SDGs*; Technical Reference Document; FAO: Rome, Italy, 2018.
2. Gold, S.; Rubik, F. Consumer Attitudes towards Timber as a Construction Material and towards Timber Frame Houses—Selected Findings of a Representative Survey among the German Population. *J. Clean. Prod.* **2009**, *17*, 303–309. [CrossRef]
3. Chuki, S.; Sarkar, R.; Kurar, R. A Review on Traditional Architecture Houses in Buddhist Culture. *Am. J. Civil Eng. Arch.* **2017**, *5*, 113–123. [CrossRef]
4. Treu, A.; Zimmer, K.; Brischke, C.; Larnøy, E.; Gobakken, L.R.; Aloui, F.; Cragg, S.M.; Flæte, P.; Humar, M.; Westin, M.; et al. Durability and Protection of Timber Structures in Marine Environments in Europe: An Overview. *BioResources* **2019**, *14*, 10161–10184. [CrossRef]
5. Smith, R. Comparison of Perception versus Reality in Timber Bridge Performance. *J. Mater. Civil Eng.* **1998**, *10*, 238–243. [CrossRef]
6. Yazdani, N. Parametric Study on Behavior of Stress-Laminated Southern Pin Bridge Decks. *J. Transp. Res. Rec.* **2000**, *1740*, 85–95. [CrossRef]

7. Liuzzi, M.A.; Fiore, A.; Gieco, R. Some Structural Design Issues on a Timber Bridge for Pedestrians. *Procedia Manuf.* **2020**, *44*, 583–590. [CrossRef]
8. Iqbal, A. Developments in Tall Wood and Hybrid Buildings and Environmental Impacts. *Sustainability* **2021**, *13*, 11881. [CrossRef]
9. Salvadori, V. An Overview of the Tallest Timber Buildings in the World. In *8° Forum dell'Edilizia in Legno, Multistory Timber Build*; 2019; pp. 1–10. Available online: https://www.researchgate.net/publication/338913741 (accessed on 15 January 2022).
10. Van de Kuilen, J.W.G.; Ceccotti, A.; Xia, Z.; He, M. Very Tall Wooden Buildings with Cross Laminated Timber. *Procedia Eng.* **2011**, *14*, 1621–1628. [CrossRef]
11. Fleming, P.; Smith, S.; Rmage, M. *Measuring-Up in Timber: A Critical Perspective on Mid- and High-Rise Timber Buildings Design*; Cambridge University Press: Cambridge, UK, 2014; Volume 18, pp. 20–30.
12. Foster, R.M.; Reynolds, T.P.S.; Ramage, M.H. Proposal for Defining a Tall, Timber Building. *J. Struc. Eng.* **2016**, *142*, 02516001. [CrossRef]
13. Foster, R.M.; Reynolds, T.P.S.; Ramage, M.H. Rethinking Height Criteria in the Context of Tall Timber. *CTBUH J.* **2017**, *4*, 28–33.
14. Michaela, E.; Olivier, A.; John, W.C.D.; Joanna, H.; Lennart, S. Experimental Micromechanical Characterization of Wood Cell Walls. *Wood Sci. Technol.* **2013**, *17*, 163–182.
15. Shaheda, T.A.; Michael, S.; Erik, S.; Thomas, K.B. A Numerical Study of the Stiffness and Strength of Cross-Laminated Timber Wall-to-Floor Connections under Compression Perpendicular to the Grain. *Buildings* **2021**, *11*, 442.
16. Ministry of Agriculture, Forestry and Fisheries in Japan. Available online: https://www.maff.go.jp/e/index.html (accessed on 15 January 2022).
17. Iwanaga, S.; Hayafune, M.; Tanaka, W.; Ikami, Y. Domestic Large-Diameter Log Use in the Japanese Lumber Manufacturing Industry: Focusing on Regional Differences. *J. For. Res.* **2021**, *27*, 8–14. [CrossRef]
18. Ukyo, S.; Ido, H.; Nagao, H.; Kato, H. Simultaneous Determination of Shear Strength and Shear Modulus in Glued-Laminated Timber using a Full-Scale Shear Block Specimen. *J. Wood. Sci.* **2010**, *56*, 262–266. [CrossRef]
19. Ido, H.; Nagao, H.; Kato, H.; Ogiso, J.; Miyatake, A. Effects of the Width and Lay-up of Sugi Cross-Laminated Timber (CLT) on its Dynamic and Static Elastic Moduli, and Tensile Strength. *J. Wood Sci.* **2016**, *62*, 101–108. [CrossRef]
20. Totsuka, M.; Aoki, K.; Inayama, M.; Morita, K. Partial Compressive Properties of Cross Laminated Timber I. Experimental Study of Effects on Mechanical Properties of CLT of Japanese Cedar. *Mokuzai Gakkaishi* **2020**, *66*, 8–15. (In Japanese) [CrossRef]
21. Yamamoto, K.; Fujita, K.; Watanabe, Y.; Miyatake, A.; Shibusawa, T.; Tanaka, S.; Kanayama, K. Flexural Property in Out-of–Plane Bending Test of Cross Laminated Timber (CLT) Composed of Different Grade Lamina of Sugi. *J. Soc. Mater. Sci. Jpn.* **2021**, *70*, 561–566. (In Japanese) [CrossRef]
22. Ukyo, S.; Shindo, K.; Miyatake, A. Evaluation of Rolling Shear Modulus and Strength of Japanese Cedar Cross-Laminated Timber (CLT) Laminae. *J. Wood. Sci.* **2019**, *65*, 31. [CrossRef]
23. Matsumura, Y.; Murata, K.; Ikami, U.; Matsumura, J. Influence of Sawing Patterns on Lumber Quality and Yield in Large Sugi (*Cryptomeria japonica*) Logs. *For. Prod. J.* **2012**, *62*, 25–31. [CrossRef]
24. Murata, K.; Ikami, U.; Matsumura, Y.; Todoroki, C. Sawing Patterns for Large-Diameter Sugi (*Cryptomeria japonica D. Don*) Sawlogs: Current Status and Future Outlook. *Wood Mater. Sci. Eng.* **2013**, *8*, 26–36. [CrossRef]
25. Matsumura, Y.; Ikami, Y.; Murata, K.; Matsumura, J. Quality of Squared Lumber without Pith Sawn from Large-Diameter Sugi (*Cryptomeria japonica*) Logs. *Mokuzai Gakkaishi* **2013**, *59*, 138–145. (In Japanese) [CrossRef]
26. Matsumura, Y.; Murata, K.; Ikami, Y.; Ohmori, M. Application of the Wood Properties of Large-Diameter Sugi (*Cryptomeria japonica*) Logs to Sorting Logs and Sawing Patterns. *J. Wood. Sci.* **2013**, *59*, 271–281. [CrossRef]
27. Ido, H.; Kato, H.; Nagao, H.; Harada, M.; Ikami, Y.; Matsumura, Y.; Matsuda, Y.; Saito, S. Grades and Mechanical Properties of Dimension Lumber for Wood Frame Construction Obtained from Large-Diameter Sugi (*Cryptomeria japonica*) Logs. *Mokuzai Gakkaishi* **2017**, *63*, 282–290. (In Japanese) [CrossRef]
28. Wang, S.-Y.; Ko, C.-Y. Dynamic Modulus of Elasticity and Bending Properties of Large Beams of Taiwan-Grown Japanese Cedar from Different Plantation Spacing Sites. *J. Wood. Sci.* **1998**, *44*, 62–68. [CrossRef]
29. Yamashita, K.; Hirakawa, Y.; Fujisawa, Y.; Nakada, R. Effects of Microfibril Angle and Density on Variation of Modulus of Elasticity of Sugi (*Cryptomeria japonica*) Logs among Eighteen Cultivars. *Mokuzai Gakkaishi* **2000**, *46*, 510–522. (In Japanese)
30. Chuang, S.-T.; Wang, S.-Y. Evaluation of Standing Tree Quality of Japanese Cedar Grown with Different Spacing using Stress-Wave and Ultrasonic-Wave Methods. *J. Wood. Sci.* **2001**, *47*, 245–253. [CrossRef]
31. Zhu, J.; Tadooka, N.; Tanaka, K. Growth and Wood Quality of Sugi (*Cryptomeria japonica*) Planted in Akita Prefecture (II). Juvenile/Mature Wood determination of Aged Trees. *J. Wood. Sci.* **2005**, *51*, 95–101. [CrossRef]
32. Ishidoh, M.; Ishiguri, F.; Iizuka, K.; Yokota, S.; Ohno, H.; Yoshizawa, N. The Evaluation of Modulus of Elasticity at an Early Stage of Growth in Sugi (*Cryptomeria japonica*) Wood using S_2 Microfibril Angle of Latewood Tracheids as a Wood Quality Indicator. *Mokuzai Gakkaishi* **2009**, *55*, 10–17. (In Japanese) [CrossRef]
33. Fukui, T.; Yanase, Y.; Sawada, Y.; Fujii, Y. Estimations of the Moisture Content above the Fiber Saturation Point in Sugi Wood using the Correlation Between the Specific Dynamic Young's Modulus and Tangent Loss. *J. Wood Sci.* **2020**, *65*, 35. [CrossRef]
34. Takahashi, Y.; Ishiguri, F.; Aiso, H.; Takashima, Y.; Hiraoka, Y.; Iki, T.; Ohshima, J.; Iizuka, K.; Yokota, S. Inheritance of Static Bending Properties and Classification of Load-Deflection Curves in *Cryptomeria japonica*. *Holzforschung* **2021**, *75*, 105–113. [CrossRef]

35. Yuki, S.; Takahashi, A.; Yamamoto, N.; Toyohiro, T. Study on Miniature Specimen for Wood Tensile Test. In Proceedings of the International Workshop on Fundamental Research for Science and Technology 2018, Bangkok, Thailand, 28–29 August 2018; pp. 86–87.
36. Yuki, S.; Takahashi, A.; Yamamoto, N.; Toyohiro, T. Fracture Mode and Tensile Strength on Latewood in Japanese Cedar. In Proceedings of the International Conference on Applied Electrical and Mechanical Engineering 2019, Nakhon Phanom, Thailand, 4–6 September 2019; pp. 39–42.
37. Japanese Standards Association. *JIS Z2101-2009 Methods of Test for Woods*; Japanese Standards Association: Tokyo, Japan, 2009.
38. Jeong, G.Y.; Zink-Sharp, A.; Hindman, D.P. Tensile Properties of Earlywood and Latewood from Loblolly Pine (*Pinus taeda*) using Digital Image Correlation. *Wood Fiber Sci.* **2009**, *41*, 51–63.
39. Roszyk, E.; Moliński, W.; Kamiński, M. Tensile Properties along the Grains of Earlywood and Latewood of Scots Pine (*Pinus sylvestri* L.) in Dry and Wet State. *BioResources* **2016**, *11*, 3027–3037. [CrossRef]
40. Yang, N.; Zhang, L. Investigation of Elastic Constants and Ultimate Strengths of Korean Pine from Compression and Tension Tests. *J. Wood Sci.* **2018**, *64*, 85–96. [CrossRef]
41. *JIS Z2241-2009*; Metallic Materials-Tensile Testing-Method of Test at Room Temperature. Japanese Standards Association: Tokyo, Japan, 2011.
42. Yuki, S.; Takahashi, A.; Yamamoto, N.; Toyohiro, T. A Study on Evaluation for Tensile Properties of Latewood Cedar. In Proceedings of the 8th International Joint Symposium on Engineering Education, Busan, Korea, 26–27 December 2018; pp. 111–116.
43. Weibull, W. A Statistical Distribution Function of Wide Applicability. *J. Appl. Mech.* **1951**, *18*, 293–305. [CrossRef]
44. Reiterer, A.; Lichteneggerz, H.; Tscheggy, S.; Fratzl, P. Experimental Evidence for a Mechanical Function of the Cellulose Microfibril Angle in Wood Cell Walls. *Phill. Mag. A* **1999**, *79*, 2173–2184. [CrossRef]
45. Babero, E.; Fernández-Sáez, J.; Navarro, C. Statistical Distribution of the Estimator of Weibull Modulus. *J. Mater. Sci. Lett.* **2001**, *20*, 847–849. [CrossRef]
46. Deng, B.; Jiang, D. Determination of the Weibull Parameters from the Mean Value and the Coefficient of Variation of the Measured Strength for Brittle Ceramics. *J. Adv. Ceram.* **2017**, *6*, 149–156. [CrossRef]
47. Lu, C.; Danzer, R.; Fischer, D. Fracture Statistics of the Brittle Materials: Weibull or Normal Distribution. *Phys. Rev.* **2002**, *E65*, 067102. [CrossRef]
48. Olkin, I.; Gleser, L.J.; Derman, C. *Probability Models and Applications*, 2nd ed.; World Scientific: Singapore, 2020.
49. Hull, D.; Vlyne, T.W. *An Introduction to Composite Materials*; Cambridge University Press: New York, NY, USA, 1996.
50. Côté, W.A.; Hanna, R.B. Ultrastructure Characteristics of Wood Fracture Surfaces. *Wood Fiber. Sci.* **1983**, *15*, 135–163.
51. Kitahara, K. *Mokuzai Butsuri*; Morikita Publishing Co., Ltd.: Tokyo, Japan, 1977. (In Japanese)
52. Ifju, G.; Kennedy, R.W. Some Variables Affecting Microtensile Strength of Douglas-Fir. *For. Prod. J.* **1962**, *12*, 213–217.
53. Kifetew, G.; Thuvander, F.; Berglund, L.; Jindberg, H. The Effect of Drying on Wood Fracture Surfaces from Specimens Loads in Wet Condition. *Wood Sci. Technol.* **1998**, *32*, 83–94. [CrossRef]
54. Ohgama, T.; Yamada, T. Young's Moduli of Earlywood and Latewood in Transverse Direction of Softwoods. *J. Soc. Mater. Sci. Jpn.* **1981**, *30*, 707–711. (In Japanese) [CrossRef]
55. Kujidani, Y.; Kitahara, R. Wood Properties of *Cryptomeria japonica* in Southern Kyusyu II–Characteristics of Obi-sigi Cultivars. *J. Soc. Mater. Sci. Jpn.* **2003**, *52*, 336–340. (In Japanese) [CrossRef]
56. Tsushima, S.; Koga, S.; Oda, K.; Shiraishi, S. Growth and Wood Properties of Sugi (*Cryptomeria japonica*) Cultivars Plamted in the Kyusyu Region. *Mokuzai Gakkaishi* **2005**, *51*, 394–401. (In Japanese) [CrossRef]
57. Kujidani, Y.; Kitahara, R. Variation of Wood Properties with Height Position in the Stems of Obi-Sigi Cultivars. *Mokuzai Gakkaishi* **2009**, *55*, 198–206. (In Japanese) [CrossRef]
58. Hirakawa, Y.; Yamashita, K.; Nakada, R.; Fujikawa, Y. The Effects of S_2 Microfibril Angles of Latewood Tracheids and Densities on Mudulus of Elasticity Variations of Sugi Tree (*Cryptomeria japonica*) Logs. *Mokuzai Gakkaishi* **1997**, *43*, 717–724.
59. U.S. Department Agriculture, Forest Service, Forest Product Laboratory. *Wood Handbook, Wood as an Engineering Material*; General Technical Report FPL-GTR-190; U.S. Department Agriculture, Forest Service, Forest Product Laboratory: Madison, WI, USA, 2010.
60. Miyoshi, Y.; Furuta, Y. Rheological Consideration in Fracture of Wood in Lateral Tension. *J. Wood Sci.* **2016**, *62*, 138–145. [CrossRef]
61. Miyoshi, Y.; Kojiro, K.; Furuta, Y. Effects of Density and Anatomical Feature on Mechanical Properties of Various Wood Species in Lateral Tension. *J. Wood Sci.* **2018**, *64*, 509–514. [CrossRef]
62. Hankinson, R.L. Investigation of Crushing Strength of Spruce at Varying Angles of Grain. *Air Serv. Inf. Circ.* **1921**, *3*, 130.
63. Xavier, J.C.; Garrido, N.M.; Oliveira, M.; Morais, J.L.; Camanho, P.P.; Pierron, F. A comparison between the Iosipescu and Off-Axis Shear Test Methods for the Characterization of *Pinus Pinaster Ait*. *Compos. Part A Appl. Sci. Manuf.* **2004**, *35*, 827–840. [CrossRef]
64. Bilko, P.; Skoratko, A.; Rurkiewicz, A.; Malyszko, L. Determination of the Shear Modulus of Pine Wood with the Arcan Test and Digital Image Correlation. *Materials* **2021**, *14*, 468. [CrossRef] [PubMed]
65. Gupta, R.; Sinha, A. Effect of Grain Angle on Shear Strength of Douglas-Fir Wood. *Holzforschung* **2012**, *66*, 655–658. [CrossRef]
66. Mania, P.; Siuda, F.; Roszyk, E. Effects of Slope Grain on Mechanical Properties of Different Wood Species. *Materials* **2020**, *13*, 1503. [CrossRef] [PubMed]

Article

Athermal ω Phase and Lattice Modulation in Binary Zr-Nb Alloys

Mitsuharu Todai [1], Keisuke Fukunaga [2] and Takayoshi Nakano [2,*]

[1] Department of Environmental Materials Engineering, National Institute of Technology, Niihama College, 7-1 Yagumo-cho, Niihama 792-8580, Ehime, Japan; m.todai@niihama-nct.ac.jp
[2] Division of Materials and Manufacturing Science, Graduate School of Engineering, Osaka University, 2-1 Yamada-oka, Suita 565-0871, Osaka, Japan; keisuke.fukunaga@mat.eng.osaka-u.ac.jp
* Correspondence: nakano@mat.eng.osaka-u.ac.jp; Tel.: +81-6-6879-7505

Abstract: To further explore the potential of Zr-based alloys as a biomaterial that will not interfere with magnetic resonance imaging (MRI), the microstructural characteristics of Zr-xat.% Nb alloys ($10 \leq x \leq 18$), particularly the athermal ω phase and lattice modulation, were investigated by conducting electrical resistivity and magnetic susceptibility measurements and transmission electron microscopy observations. The 10 Nb alloy and 12 Nb alloys had a positive temperature coefficient of electrical resistivity. The athermal ω phase existed in 10 Nb and 12 Nb alloys at room temperature. Alternatively, the 14 Nb and 18 Nb alloys had an anomalous negative temperature coefficient of the resistivity. The selected area diffraction pattern of the 14 Nb alloy revealed the co-occurrence of ω phase diffraction and diffuse satellites. These diffuse satellites were represented by $g_\beta + q$ when the zone axis was [001] or [113], but not [110]. These results imply that these diffuse satellites appeared because the transverse waves consistent with the propagation and displacement vectors were $q = <\zeta \bar{\zeta} 0>^*$ for the $\zeta \sim 1/2$ and <110> directions. It is possible that the resistivity anomaly was caused by the formation of the athermal ω phase and transverse wave. Moreover, control of the athermal ω-phase transformation and occurrence of lattice modulation led to reduced magnetic susceptibility, superior deformation properties, and a low Young's modulus in the Zr-Nb alloys. Thus, Zr-Nb alloys are promising MRI-compatible metallic biomaterials.

Keywords: magnetic resonance imaging (MRI); transmission electron microscopy (TEM); phase transition; electrical resistivity; metallic biomaterials

1. Introduction

Magnetic resonance imaging (MRI) is a very useful tool in the fields of orthopedic and brain surgery, because it can yield various cross-sectional images of the human body [1]. However, the magnetic properties of metallic implants must be taken into consideration during their design, because MRI entails the application of a high magnetic field to obtain the images. Unsharp parts (artifacts) occur in the MRI data around stainless-steel and Co-Cr alloy implants, because these alloys have high magnetic susceptibility. The artifact is caused by the heat generation from and displacement of the metallic implants [2] in a high magnetic field. Thus, to reduce artifacts and obtain clear images, it is necessary to develop a metallic implant with low magnetic susceptibility [3].

Zr-based alloys have emerged as promising candidates for application as biomaterials that would result in fewer MRI artifacts [4]. This is because of their good biocompatibility [5,6], superior corrosion resistance [7], and lower magnetic susceptibility than stainless steel and Co-Cr alloys. Nomura et al. reported on the relationship between the phase constitution and magnetic susceptibility of as-cast Zr-Nb alloys at room temperature; they suggested the magnetic susceptibility of Zr-Nb alloys can be reduced by controlling the volume fraction of the athermal ω phase [8,9].

The ω phase is formed when Ti, Zr, and Hf are subjected to a high hydrostatic pressure [10,11]. This phase was also found to be metastable when β-stabilizing elements such as Nb and Ta, etc., were added to the Ti and Zr [12]. The crystal structure of this phase belongs to the space group P-3m1 [13,14], and it is known to occur as the athermal ω phase [15–19], isothermal ω phase [19,20], or stress-induced ω phase [12,21]. The athermal ω phase is associated with rapid cooling from the β phase region to room temperature. The isothermal ω phase is associated with isothermal aging within the approximate temperature range of 303–577 K. The precipitation of the ω phase is known to be related to the magnetic susceptibility and mechanical properties of Zr-Nb alloys. Particularly, the precipitation of the ω phase in Zr-based alloys leads to extreme brittleness and must therefore be controlled to realize high resistance to fracture and good fatigue performance [12,20]. In addition, the Young's modulus can be decreased by suppressing the athermal ω phase. In Ti single crystals with a low e/a (i.e., low average number of valence electron per atom in the free-atom configuration), the Young's modulus in the <001> direction in the unstable β phase has been found to be similar to that of human bone [22–25]. The lattice softening of c' in the unstable β phase in Ti alloys has been reported to reduce the Young's modulus in the <001> direction. In previous papers, we also reported on the occurrence of lattice modulation in Ti-Nb alloys under the combined effects of an unstable β phase and a low e/a [25,26]. Considering the similarity between the athermal ω phase transformation in Ti-Nb alloys and Zr-Nb alloys, we expected that, in the case of Zr-Nb alloys, an unstable β phase with lattice modulation would be associated with the lattice softening of c' and a low Young's modulus. The value of e/a decreases with decreasing Nb content in Zr-Nb alloys. Furthermore, the volume fraction of athermal ω phase increases with decreasing Nb content [8,9]. Thus, it is essential to find a Zr-Nb alloy with a low e/a, lattice modulation and no ω phase. Although the realization of Zr-Nb alloys with acceptable magnetic properties and good mechanical properties requires an understanding of the significance of the volume fraction of athermal ω phase and the details of the lattice modulation, the athermal ω phase transformation and the occurrence of lattice modulation in Zr-xNb alloys have yet to be clarified. Nomura et al. reported that the α + β + ω phase exists in Zr-6 at.% Nb alloys, and β + ω phase exists in the upper 9 at.% Nb alloys [8]. They also reported that the ω phase was not detected in the XRD patterns of the upper 12 at.% Nb alloys.

Thus, the purpose of this study was to clarify the relationship between the metastable transition and magnetic susceptibility of Zr-x at.% Nb alloys ($10 \leq x \leq 18$) by conducting electrical resistivity and magnetic susceptibility measurements and transmission electron microscopy (TEM) observations to examine their physical properties. Considering that the precipitation of the athermal ω phase affects the magnetic susceptibility, which is closely related to the electronic structure in Zr-xNb alloys, electrical resistivity measurements is appropriate for investigating athermal ω phase transformation. The Vickers hardness test was also applied, because it can sensitively detect the precipitation of the athermal ω phase.

2. Materials and Methods

Master ingots of the Zr-x at.% Nb (x = 10, 12, 14 and 18) alloys were prepared by an arc melting method. The ingots were melted on a water-cooled coper hearth in a high-purity Ar gas atmosphere. The chemical compositions of the ingots are listed in Table 1.

Table 1. Chemical composition of Zr-x at.% Nb (x = 10, 12, 14, and 18) alloy master ingots.

Nomial Composition	Chemical Composition (at.%)	
	Zr	Nb
Zr-10at.%Nb	Bal.	9.88
Zr-12at.%Nb	Bal.	12.12
Zr-14at.%Nb	Bal.	13.54
Zr-18at.%Nb	Bal.	17.88

The compositions of the master ingots were analyzed by inductively coupled plasma–optical spectroscopy. The actual compositions of all examined alloys were confirmed to be nearly the same as the corresponding nominal compositions. These master ingots were remelted at least five times to prevent segregation. The ingots were subjected to a homogenization heat treatment at 1273 K for 24 h in quartz tubes, followed by quenching in ice water. The specimens to which the electrical resistivity measurements, magnetic susceptibility measurements, Vickers hardness tests, and TEM observations were to be applied were cut from the master ingots in quartz tubes prior to being subjected to a solution heat treated at 1273 K for 1 h. The specimens also were quenched in ice-cold water after solution heat treatment. According to the phase diagram for the Zr-Nb alloy system, only the β phase exists in these alloys at 1273 K. Note that, in this paper, each alloy is referred to based on its Zr content. For example, the Zr-10 at.% Nb alloy is referred to as the 10 Nb alloy.

The specimens of all examined measurements were polished with emery paper up to #2000 and then electropolished with a solution of 92 vol.% methanol and 8 vol.% perchloric acid at approximately 250 K. The electrical resistivity was measured by applying a standard four-probe method under the conditions of cooling from 300 to 4.2 K at a rate of approximately 1 K/min. The dimensions of the electrical resistivity measurement specimens were 2 mm × 10 mm × 0.2 mm. The magnetic susceptibility (χ) was measured by operating a superconducting quantum inference device magnetometer (Quantum Design; SQUID) with an external magnetic field of $\mu_0 H = 2$ T and a constant cooling rate of approximately 1 K/min. The dimensions of magnetic susceptibility measurement specimens were 10 mm × 10 mm × 10 mm. The micro-Vickers hardness tests were performed at room temperature. The applied load was 2.94 N, and the loading time was 15 s. Specimens for TEM observations cut into a circle with a diameter of 3 mm from the ingots and then were polished to a thickness below 200 μm by emery paper. Finally, a thin foil was prepared for the TEM observations by using the twin-jet technique. TEM was performed by operating a JEM 3010 (JEOL) at 300 kV.

3. Results

3.1. Electrical Resistivity Measurements

Figure 1 shows the temperature dependence of the electrical resistivity of various Zr-x at.% Nb alloys (x = 10, 12, 14, and 18). The resistivity of the 10 Nb and 12 Nb alloys monotonically decreased with decreasing temperature. The analysis of the resistivity curve for the 14 Nb alloy revealed negative temperature coefficients (NTCs) at temperatures below room temperature. This NTC has previously been confirmed for several Ti alloys [27–30]. In previous studies, an anomalous NTC in the resistivity curve was interpreted as the growth of the athermal ω phase and occurrence of lattice modulation during the cooling process [29,31]. This result indicated that lattice modulation would occur in the β phase of the 14 Nb alloy. The resistivity of the 14 Nb alloy began decreasing again at T_{max} = 100 K, as indicated by the double arrow in Figure 1c. The resistivity curve for the 18 Nb alloy was found to have a local minimum at T_{min} = 240 K, as indicated by the arrow, and an NTC below T_{min}. The resistivity of this alloy began decreasing again at T_{max} = 70 K. These results revealed that, T_{min} and T_{max} decreased with increasing Nb content. Furthermore, the composition dependence and transformation behavior observed in this study were found to be similar to those previous reports for some β phase in Ti alloys [27,29,31].

Additionally, a sharp decrease in the electrical resistivity was observed for all alloys at temperatures below 15 K. These drops in resistivity were induced by the superconductive transition. As shown in Figure 2, the superconductive transition temperature, T_c, of these alloys increased with increasing Nb temperature. The error bar is almost the same as in the plots. Generally, the T_c of Zr-Nb alloys is higher than that of Ti-Nb alloys [26,31]; here, its value changed as the Nb content increased. These results indicate that the phase constitution gradually changed as the Nb content increased.

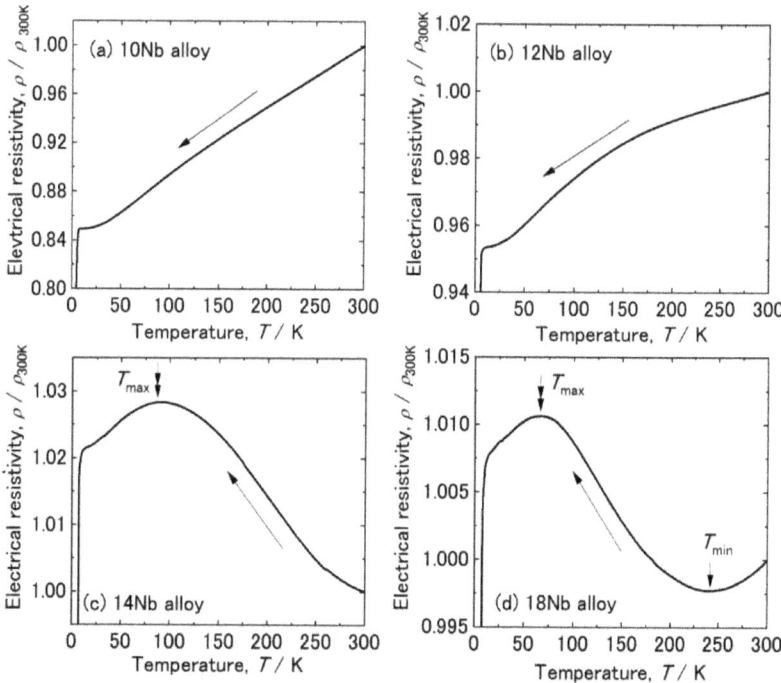

Figure 1. Electrical resistivity of various Zr-x at.% Nb alloys (x = 10, 12, 14, and 18) during the cooling process: (**a–d**) x = 10, 12, 14, and 18, respectively.

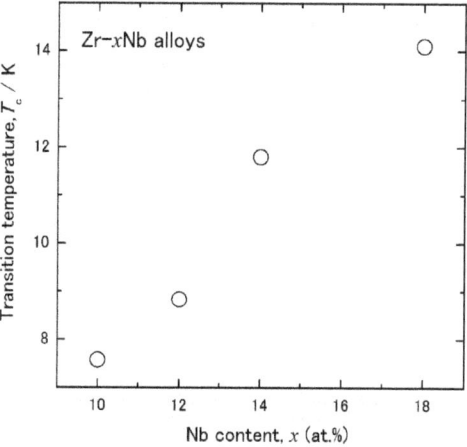

Figure 2. Nb content dependence of the superconductive transition temperature (T_c) of Zr-x at.% Nb ($10 \leq x \leq 18$) alloys.

3.2. *TEM Observations*

Figure 3a–d show the selected area diffraction patterns (SADPs) of the Zr-x at.% Nb ($10 \leq x \leq 18$) alloys, as obtained under the conditions of a beam direction of [113] at room temperature.

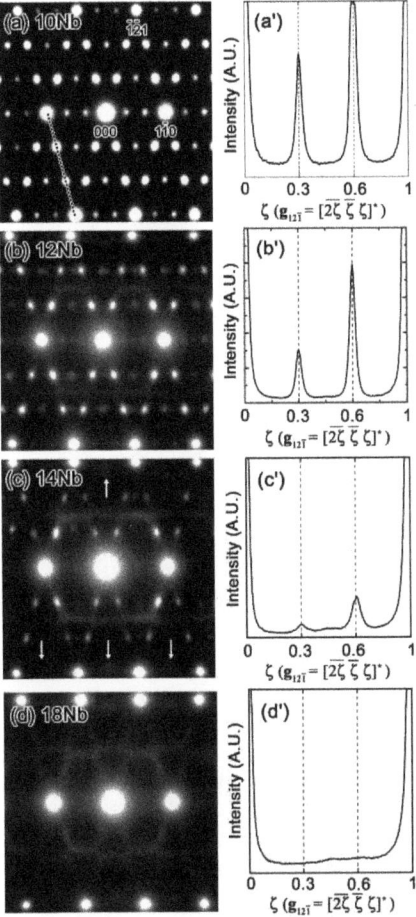

Figure 3. Results for Zr-x at.% Nb alloys ($10 \leq x \leq 18$), as obtained by applying the incident beam in the [113] direction at room temperature: (**a–d**) diffraction patterns for x = 10, 12, 14, and 18, respectively; (**a′–d′**) corresponding intensity profiles.

The diffraction pattern of the 10 Nb alloy indicated the existence of the β phase and additional spots at $g_\beta + 1/3<2\zeta\ \zeta\ \zeta>^*$ and $g_\beta + 2/3<2\zeta\ \zeta\ \zeta>^*$. Here, g_β is the β-phase reciprocal lattice vector, and the asterisk (*) indicates the orientation in the reciprocal space. Additional spots have been reported to be associated with the diffraction pattern of the athermal ω phase [12]; here, they appeared in the SADPs of the 10 Nb, 12 Nb and 14 Nb alloys. It should be noted that the intensity of the ω-phase spot decreased with increasing Nb content. Additionally, these diffraction patterns did not show clearly defined spots, and only diffuse streaks were observed in the case of the 18 Nb alloy. This behavior can be better understood by comparing the one dimensional intensity profiles along $[\bar{2}1\bar{1}]^*$, as shown in Figure 3a′–d′. These intensity profiles show that additional spots appeared at the commensurate $g_\beta + 1/3<2\zeta\ \zeta\ \zeta>^*$ and $g_\beta + 2/3<2\zeta\ \zeta\ \zeta>^*$ positions, and that their intensity decreased with increasing Nb content. In addition, Figure 4 shows that HV decreases with increasing Nb content. These results indicate that the volume fraction of the athermal ω phase gradually decreased with increasing Nb content; moreover, these results were in good agreement with the results of the resistivity measurements.

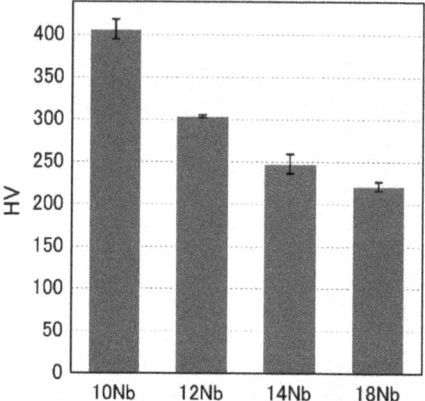

Figure 4. HV variation of the Zr-x at.% Nb alloys.

To understand the morphology of the athermal ω phase, dark-filed images of the ω phase spot in the 14 Nb alloy were obtained; the results are shown in Figure 5.

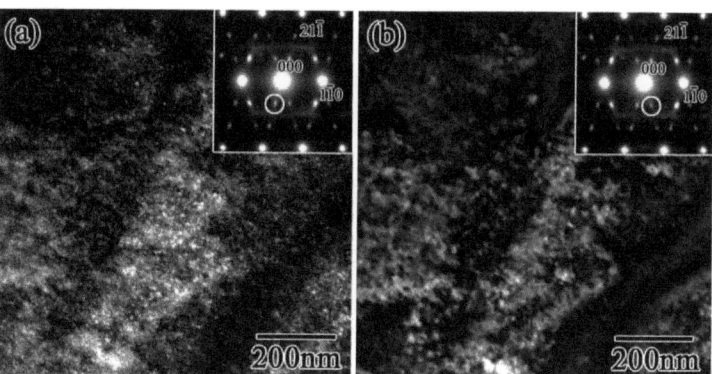

Figure 5. Dark-field images of Zr-14 at.% Nb alloy were observed by a circle in diffraction patterns in (**a**,**b**), respectively.

The athermal ω phase appeared as an approximately 10 nm–diameter sphere, which is in agreement with a previous reported results [32]. The density of the precipitated athermal ω phase decreased with increasing Nb concentration, although the size showed little Nb dependence. The shape of precipitated athermal ω phase in Zr-x at.% Nb alloys was found to be the same as that in Ti alloys [12] and other Zr based alloys [16]. It should also be noted that, in the 14 Nb alloy, the rod-like streaks along $[1\bar{1}0]^*$, as indicated by the white arrows in Figure 3c, and the weak diffuse satellite near $g_{12\bar{1}} + 1/2[\zeta\bar{\zeta}0]^*$ are indicated by the appearance of lattice modulation.

4. Discussion

To investigate the occurrence of lattice modulation in the β phase, TEM was applied to the 14 Nb alloy and observed by TEM, according to the results of Ti alloys [31,33,34].

To begin, the transverse movement of the atoms of β phase in the 14 Nb alloy was confirmed. Figure 6a,b shows the SADPs for the 14 Nb alloy.

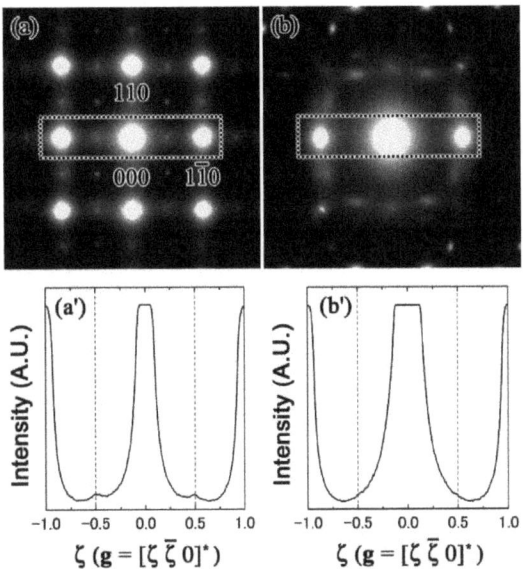

Figure 6. SADPs of the Zr-14 at.% Nb alloy: (**a**) SADP showing the zone axis of [001]; (**b**) SADP showing the systematic condition; (**a′,b′**) corresponding intensity profiles for the respective areas enclosed in the dotted boxes.

Note that the pattern shown in Figure 6b is an approximately 7° tilt of that shown in Figure 6a. Interestingly, rod-like streaks and diffuse satellites at $\pm 1/2[1\bar{1}0]^*$ can be observed in Figure 6a, but not in Figure 6b. The disappearance of the diffuse satellites in Figure 6b implies that double diffraction led to the appearance of these satellites. The disappearance of the diffuse satellites at $\pm 1/2[1\bar{1}0]^*$ became more apparent after comparing the line intensity profiles of the dotted boxed regions, which are shown in Figure 6a′,b′. These findings indicate that the movement of the atoms of the β phase in the 14 Nb alloy was a transverse wave.

It should be noted that these satellites and rod-like streaks were not observed in the (110) SADP, as can be seen in Figure 7a. The disappearance of these satellites is evident in the intensity profile results shown in Figure 7b.

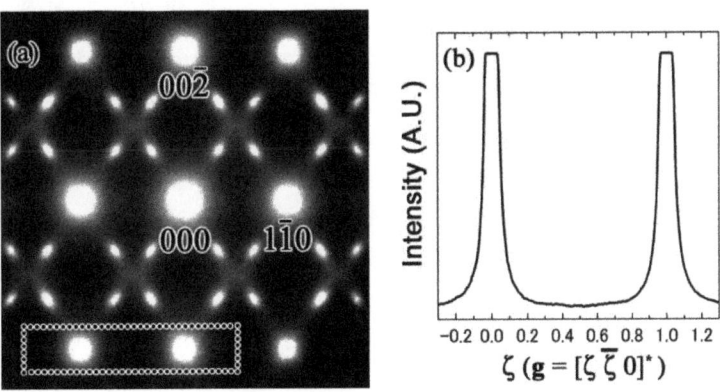

Figure 7. (**a**) SADP of the Zr-14 at.% Nb alloy (zone axis of [110]); (**b**) intensity profile along [1$\bar{1}$0].

From these results, the orientation of the atom movement was determined. Particularly, considering that the movement of atoms was responsible for the occurrences of the diffuse satellites and rod-like streaks, we expected that the diffuse satellites and rod-like streaks would be missing in the special case. In this case, the inner product of the scattering vector of diffuse satellites, g, and the displacement of atoms, R, is zero. This means that $g \cdot R = 0$. We can rewrite the scattering vector of diffuse satellite as $g = g_\beta + q$, where g_β is the scattering vector of the β phase and q is the propagation vector for the displacement of atoms R. In the case of the 14 Nb alloy, R is given by a transverse wave, as has been previously described. The inner product of $q \cdot R$ is always zero because this condition defined a transverse wave. Thus, the relation $g \cdot R = 0$ is only satisfied when $g_\beta \cdot R = 0$. If R is the [001] direction, no diffuse reflections appear in the SADP when zone axis is [001], as shown in Table 2.

Table 2. Appearance or disappearance of diffuse satellites for combinations of examined zone axis and [001], [110], and [111] displacement direction R.

Direction of R	Zone Axis // [113]	Zone Axis // [001]	Zone Axis // [110]
[001]	Present	Absent	Present
[110]	Present	Present	Absent
[111]	Present	Present	Present

However, diffuse satellites and rod-like streaks appeared in the SADP when the zone axis is [001], as can be seen in Figure 6a. This implies that R was not [001] direction. When the displacement of atoms occurred in the [110] direction, the satellite reflections and streaks were not observed in the SADPs showing the [110] zone axis; however, they were present in the SADPs when the zone axis are [113] or [001], as can be seen in Figure 7. This is more apparent in the line intensity profiles corresponding to the boxed region in Figure 7a (i.e., in Figure 7b). Thus, the satellites at $g_\beta + <\zeta\bar{\zeta}0>^*$ were present because of the transverse wave consistent with the displacement of atoms in the <110> direction, which was perpendicular to the propagation vector $q = <\zeta\bar{\zeta}0>^*$, for $\zeta \sim 1/2$.

Figure 8 shows the SADPs of the 18 Nb alloy at room temperature; in this case, a positive temperature dependence of the resistivity can be observed.

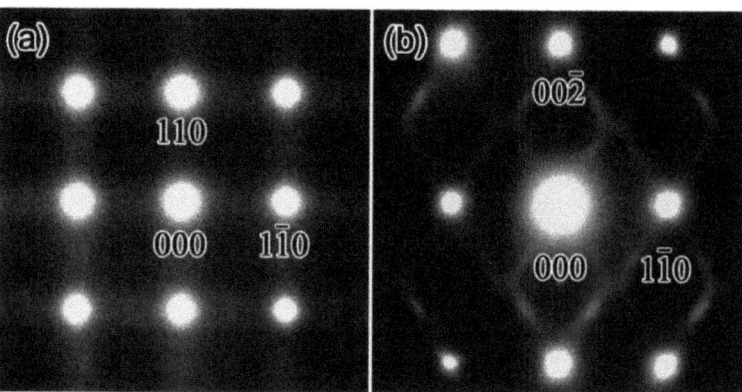

Figure 8. Diffraction patterns of Zr-18 at.%Nb alloy. The zone axis is [001] in (**a**) and [110] in (**b**).

The results shown in Figure 8 confirm weak lattice modulation in the 18 Nb alloy, as rod-like streaks in the $<1\bar{1}0>^*$ direction can be observed in the (001) SADP, but not in the (110) SADP. However, no diffuse satellites were observed at $g_\beta + <\zeta\bar{\zeta}0>^*$ in the (001) SADP. It is possible that, in the case of Zr-Nb alloys, the diffuse satellites at $g_\beta + <\zeta\bar{\zeta}0>^*$ are also related to the NTC, which can be determined from the resistivity curve, and have a temperature dependence similar to that observed for Ti-Nb alloys [31]. As previously mentioned, the lattice modulation that occurs in Zr-Nb alloys is similar to the β phase in Ti alloys [26,31]. In fact, when the diffraction patterns in the 18 Nb alloy were investigated below T_{min} (= 240 K), the intensities of the ω phase reflections and diffuse satellites were stronger, as can be seen in Figure 9.

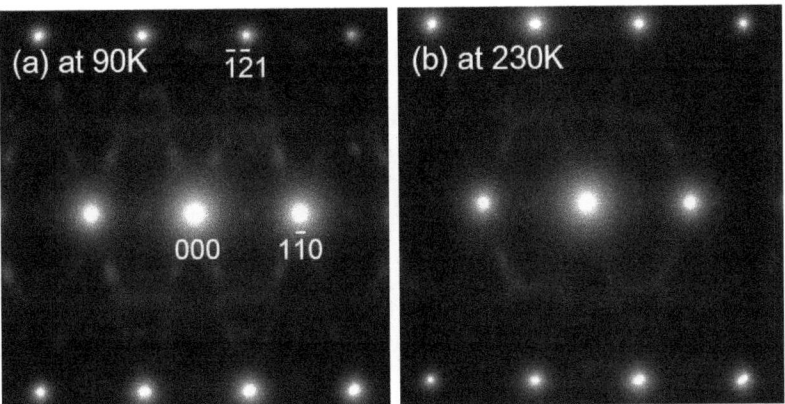

Figure 9. Temperature dependence of diffraction patterns of Zr-18 at.%Nb alloy: (**a**) taken at 90 K and (**b**) taken at 230 K.

In Ti-Nb alloys, the β phase with lattice modulation presents as nanoscale-domain-like structures (i.e., nanodomains) [35]. It is possible that increasing the interface, such as the athermal ω phase and/or nanodomains, and the matrix β phase as the temperature is decreased coincidentally increases the electrical resistivity during the cooling process. Thus, the NTC is considered to have been attributable to the presence of an athermal ω phase and the lattice modulation in the β phase. It has previously been reported that the lattice modulation that occurs in the β phase in Ti alloys is related to the β phase stability, and is thus indicative of the softening of the elastic stiffness, c', which contributes to reduce the Young's modulus in the <001> direction [22–26]. In this study, transverse lattice modulation was found to occur in the β phase of the Zr-xNb alloys. The occurrence of this transverse wave may imply the softening of the elastic stiffness c'. It is necessary to minimize the Young's modulus of metallic biomaterials to prevent bone degradation and absorption; this undesirable outcome is also known as stress shielding, and it occurs when there is a difference between the Young's modulus of metallic implants and that of human bone [36–38].

By controlling the phase stability and matching the loading axis orientation with the <001> orientation, Zr-Nb-based metallic biomaterials with low Young's moduli can be developed. Recently, additive manufacturing has enabled the simultaneous control of the shapes and textures of metallic materials [39–44]. Applying such processes in combination with the phase stability control of Zr-Nb alloys has led to the development of custom-made implants with low Young's moduli.

Figure 10 shows the room-temperature magnetic susceptibilities of 10 Nb, 14 Nb and 18 Nb alloys. The magnetic susceptibilities of several other metallic biomaterials are also shown [8]. As can be seen, the magnetic susceptibility values for the Zr-Nb alloys were lower than those for the Co-Cr-Mo and Ti alloys.

The magnetic susceptibility values for the 10 Nb, 14 Nb, and 18 Nb alloys were found to be in good agreement with those reported by Nomura et al. [8]. Nomura et al. also reported that the magnetic susceptibility decreased when the thermal ω phase precipitated in the Zr-16 mass% Nb alloy. In this study, the lowest magnetic susceptibility was found in the 10 Nb alloys with the most ω phase precipitation. Thus, the low magnetic susceptibility of Zr alloys has been attributed to the high volume fraction of the ω phase. However, because the precipitation of the ω phase significantly embrittled the Zr-Nb alloy, its magnetic susceptibility and mechanical properties must be well balanced. Figure 11 shows the temperature dependence of electrical resistivity and magnetic susceptibility of the 18 Nb alloy.

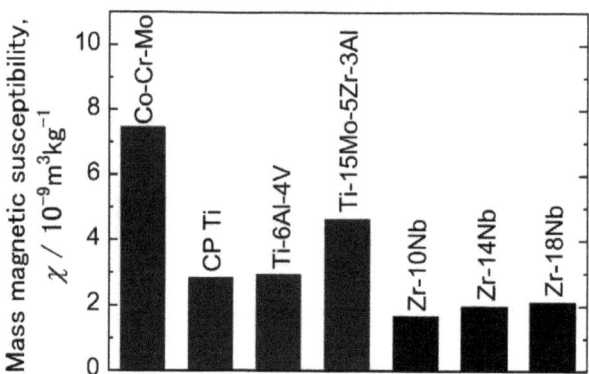

Figure 10. Magnetic susceptibilities of several alloys at room temperature [8].

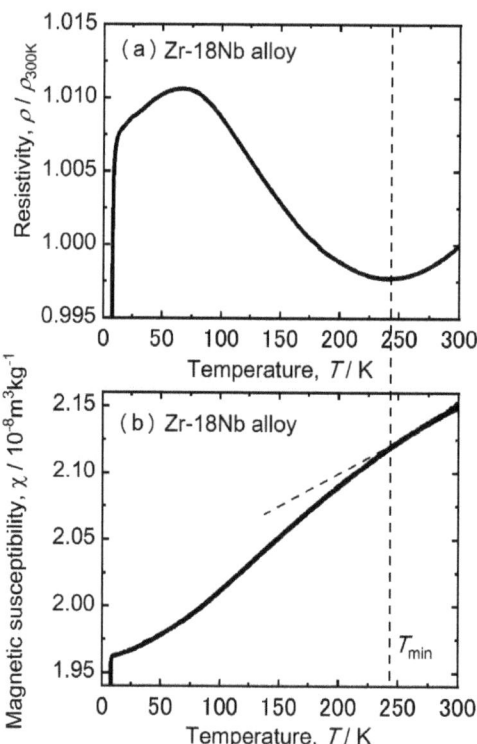

Figure 11. Electrical resistivity (**a**) and magnetic susceptibility (**b**) of the Zr-18 at.% Nb alloy.

It should be noted that the magnetic susceptibility decreased below T_{min}. The temperature dependence of the athermal ω phase and lattice modulation likely contributed to the decrease in the magnetic susceptibility. Thus, Zr-Nb alloys with low magnetic susceptibility, superior mechanical properties, and low Young's modulus can be realized by controlling the ω phase transformation and the lattice modulation. It was previously reported that as-cast alloys with $14 \leq x \leq 20$ have superior ductility and a low Young's modulus [9]. In this study, the occurrences of athermal ω phase transformation and lattice modulation in Zr-Nb alloys were found to be dependent on temperature and Nb content. This finding suggests that the volume fraction of the athermal ω phase and lattice modulation in Zr-Nb alloys can be controlled with relative ease. Thus, Zr-Nb alloys can be considered to be good MRI-compatible materials.

5. Conclusions

The relationship between the metastable phase and magnetic susceptibility of Zr-Nb alloys was investigated by conducting electrical resistivity measurements, magnetic susceptibility measurements, and TEM observations, focusing on an anomalous NTC in the resistivity curve, and the occurrence of the athermal ω phase and lattice modulation. The resistivity curves for the 10 Nb and 12 Nb alloy revealed a positive temperature coefficient at temperatures below 300 K. The resistivity curves for the 14 Nb and 18 Nb alloys revealed an anomalous NTC. Diffuse satellites were observed at $g_\beta + q$ in the SADPs of the 14 Nb alloy when zone axis was [113] and [001]; however, this was not the case when the zone axis was [110]. This result implies that the movement of atoms can be described as a transverse wave with a propagation vector of $q = <\zeta\bar{\zeta}0>^*$, for $\zeta \sim 1/2$ and displacement in the <110> direction. Furthermore, these diffuse satellites did not appear when the temperature coefficient was positive. Overall, the results indicate that low magnetic susceptibility and a relatively low Young's modulus can be achieved by controlling the volume fraction of the athermal ω phase and the occurrence of lattice modulation. This means that Zr-Nb alloys can be manufactured to have low magnetic susceptibility, good deformability, and a low Young's modulus by controlling the microstructure. Thus, Zr-Nb alloys are expected to be a next-generation MRI-compatible metallic biomaterial for medical devices.

Author Contributions: Conceptualization, M.T., K.F. and T. N.; investigation, M.T. and K.F.; project administration, M.T. and T.N.; supervision, T.N.; writing—original draft preparation, M.T.; writing—review and editing, M.T. and T.N. All authors have read and agreed to the published version of the manuscript.

Funding: This research was funded by JSPS KAKENHI, grant numbers 18H05254 and 20K05141; the Council for Science, Technology and Innovation (CSTI); the Cross Ministerial Strategic Innovation Promotion Program (SIP); the Innovative Design/Manufacturing Technologies Program (Establishment and Validation of the Base for 3D Design and Additive Manufacturing Standing on the Concepts of "Anisotropy" and "Customization") of the New Energy and Industrial Technology Development Organization (NEDO); and the Light Metal Educational Foundation, Inc., Japan.

Institutional Review Board Statement: Not applicable.

Informed Consent Statement: Not applicable.

Data Availability Statement: Not applicable.

Conflicts of Interest: The authors declare no conflict of interest.

References

1. Olsen, O.E. Practical body MRI—A paediatric perspective. *Eur. J. Radiol.* **2008**, *68*, 299–308. [CrossRef] [PubMed]
2. Olsrud, J.; Lätt, J.; Brockstedt, S.; Romner, B.; Björkman-Burtscher, I.M. Magnetic resonance imaging artifacts caused by aneurysm clips and shunt valves: Dependence on field strength (1.5 and 3 T) and imaging parameters. *J Magn. Reson. Imag.* **2005**, *22*, 433–437. [CrossRef]

3. New, P.F.; Rosen, B.R.; Brady, T.J.; Buonanno, F.S.; Kistler, J.P.; Burt, C.T.; Hinshaw, W.S.; Newhouse, J.H.; Pohost, G.M.; Taveras, J.M. Potential hazards and artifacts of ferromagnetic and nonferromagnetic surgical and dental materials and devices in nuclear magnetic resonance imaging. *Radiology* **1983**, *147*, 139–148. [CrossRef] [PubMed]
4. Thomsen, P.; Larsson, C.; Ericson, L.E.; Sennerby, L.; Lausmaa, J.; Kasemo, B. Structure of the interface between rabbit cortical bone and implants of gold, zirconium and titanium. *J. Mater. Sci. Mater. Med.* **1997**, *8*, 653–665. [CrossRef] [PubMed]
5. Eisenbarth, E.; Velten, D.; Müller, M.; Thull, R.; Breme, J. Biocompatibility of beta-stabilizing elements of titanium alloys. *Biomaterials* **2004**, *25*, 5705–5713. [CrossRef] [PubMed]
6. Yamamoto, A.; Honma, R.; Sumita, M. Cytotoxicity evaluation of 43 metal salts using murine fibroblasts and osteoblastic cells. *J. Biomed. Mater. Res.* **1998**, *39*, 331–340. [CrossRef]
7. Zhou, F.Y.; Wang, B.L.; Qiu, K.J.; Lin, W.J.; Li, L.; Wang, Y.B.; Nie, F.L.; Zheng, Y.F. Microstructure, corrosion behavior and cytotoxicity of Zr-Nb alloys for biomedical application. *Mater. Sci. Eng. C* **2012**, *32*, 851–857. [CrossRef]
8. Nomura, N.; Tanaka, Y.; Kondo, R.; Doi, H.; Tsutsumi, Y.; Hanawa, T. Effect of phase constitution of Zr-Nb alloys on their magnetic susceptibilities. *Mater. Trans.* **2009**, *50*, 2466–2472. [CrossRef]
9. Kondo, R.; Nomura, N.; Tsutsumi, Y.; Doi, H.; Hanawa, T. Microstructure and mechanical properties of as-cast Zr-Nb alloys. *Acta Biomater.* **2011**, *7*, 4278–4284. [CrossRef]
10. Xia, H.; Parthasarathy, G.; Luo, H.; Vohra, Y.K.; Ruoff, A.L. Crystal structures of group IVa metals at ultrahigh pressures. *Phys. Rev. B Condens. Matter* **1990**, *42*, 6736–6738. [CrossRef]
11. Xia, H.; Duclos, S.J.; Ruoff, A.L.; Vohra, Y.K. New high-pressure phase transition in zirconium metal. *Phys. Rev. Lett.* **1990**, *64*, 204–207. [CrossRef]
12. Sikka, S.K.; Vohra, Y.K.; Chidambaram, R. Omega phase in materials. *Prog. Mater. Sci.* **1982**, *27*, 245–310. [CrossRef]
13. Silcock, J.M. An X-ray examination of the ω phase in TiV, TiMo and TiCr alloys. *Acta Metall.* **1958**, *6*, 481–493. [CrossRef]
14. Hatt, B.A.; Roberts, J.A. The ω-phase in zirconium base alloys. *Acta Metall.* **1960**, *8*, 575–584. [CrossRef]
15. De Fontaine, D.; Paton, N.E.; Williams, J.C. The omega phase transformation in titanium alloys as an example of displacement controlled reactions. *Acta Metall.* **1971**, *19*, 1153–1162. [CrossRef]
16. Sass, S.L. The ω phase in a Zr-25 at.% Ti alloy. *Acta Mater.* **1969**, *17*, 813–820. [CrossRef]
17. Benites, G.M.; Guillermet, F.; Cuello, G.J.; Campo, J. Structural properties of metastable phases in Zr-Nb alloys: I. Neutron diffraction study and analysis of lattice parameter. *J. Alloys Comp.* **2000**, *299*, 183–188. [CrossRef]
18. Nomura, N.; Oya, K.; Tanaka, Y.; Kondo, R.; Doi, H.; Tsutsumi, Y.; Hanawa, T. Microstructure and magnetic susceptibility of as-cast Zr-Mo alloys. *Acta Biomater.* **2010**, *6*, 1033–1038.
19. Okunishi, E.; Kawai, T.; Mitsuhara, M.; Farjami, S.; Itakura, M.; Hara, T.; Nishida, M. HAADF-STEM studies of athermal and isothermal ω-phase in β-Zr alloy. *J. Alloy. Comp.* **2013**, *577*, S713–S716. [CrossRef]
20. Kondo, R.; Tsutsumi, Y.; Doi, H.; Nomura, N.; Hanawa, T. Effect of phase constitution on magnetic susceptibility and mechanical properties of Zr-rich Zr-Mo alloys. *Acta Biomater.* **2011**, *7*, 4259–4266.
21. Dey, G.K.; Tewari, R.; Jyoti, G.; Gupta, S.C.; Joshi, K.D.; Sikka, S.K. Formation of a shock deformation induced ω phase in Zr 20 Nb alloy. *Acta Mater.* **2004**, *53*, 5243–5254. [CrossRef]
22. Lee, S.H.; Todai, M.; Tane, M.; Hagihara, K.; Nakajima, H.; Nakano, T. Biocompatible low Young's modulus achieved by strong crystallographic elastic anisotropy in Ti-15Mo-5Zr-3Al alloy single crystal. *J. Mech. Behav. Biomed. Mater.* **2012**, *14*, 48–54. [CrossRef] [PubMed]
23. Tane, M.; Akita, S.; Nakano, T.; Hagihara, K.; Umakoshi, Y.; Niinomi, M.; Nakajima, H. Peculiar elastic behavior of Ti-Nb-Ta-Zr single crystals. *Acta Mater.* **2008**, *56*, 2856–2863. [CrossRef]
24. Tane, M.; Akita, S.; Nakano, T.; Hagihara, K.; Umakoshi, Y.; Niinomi, M.; Mori, H.; Nakajima, H. Low Young's modulus of Ti-Nb-Ta-Zr alloys caused by softening in shear modului c' and c44 near lower limit of body-centered cubic phase stability. *Acta Mater.* **2010**, *58*, 6790–6798. [CrossRef]
25. Wang, P.; Todai, M.; Nakano, T. Beta titanium single crystal with bone-like elastic modulus and large crystallographic elastic anisotropy. *J. Alloy. Comp.* **2019**, *782*, 667–671. [CrossRef]
26. Wang, P.; Todai, M.; Nakano, T. ω-phase transformation and lattice modulation in biomedical β-phase Ti-Nb-Al alloys. *J. Alloy. Comp.* **2018**, *766*, 511–516. [CrossRef]
27. Wang, P.; Todai, M.; Nakano, T. β phase instability in Binary Ti-xNb Biomaterial single crystal. *Mater. Trans.* **2013**, *54*, 156–160.
28. Ames, S.L.; McQuillan, A.D. The resistivity-temperature-concentration relationships in the system niobium-titanium. *Acta Metall.* **1954**, *2*, 831–836. [CrossRef]
29. Ikeda, M.; Komatsu, S.; Sugimoto, T.; Kamei, K. Temperature range of formation of athermal ω phase in quenched β Ti-Nb alloys. *J. Jap. Ins. Metals.* **1988**, *52*, 1206–1211. [CrossRef]
30. Ho, J.C.; Collings, E.W. Anomalous electrical resistivity in titanium-molybdenum alloys. *Phys. Rev. B* **1972**, *6*, 3727–3738. [CrossRef]
31. Todai, M.; Fukuda, T.; Kakeshita, T. Relation between negative temperature coefficient in electrical resistivity and athermal ω phase Ti-xNb ($26 \leq x \leq 29$ at.%) alloys. *J. Alloys Compd.* **2013**, *577*, S431–S434. [CrossRef]
32. Kuan, T.S.; Sass, S.L. The direct imaging of a linear defect using diffuse scattering in Zr-Nb b.c.c. solid solutions. *Philos. Mag.* **1977**, *36*, 1473–1498. [CrossRef]

33. Todai, M.; Fukuda, T.; Kakeshita, T. Temperature dependence of diffuse satellites in Ti–(50−x)Pd–xFe (14 ≤ x ≤ 20 (at.%)) alloys. *J. Alloys Compd.* **2014**, *615*, 1047–1051. [CrossRef]
34. Todai, M.; Fukuda, T.; Kakeshita, T. Direction of atom displacement in incommensurate state of Ti–32Pd–18Fe shape memory alloy. *Mater. Lett.* **2013**, *108*, 293–296. [CrossRef]
35. Tahara, M.; Kim, H.Y.; Inamura, T.; Hosoda, H.; Miyazaki, S. Lattice modulation and superelasticity in oxygen-added β-Ti alloys. *Acta Mater.* **2011**, *59*, 6208–6218. [CrossRef]
36. Noyama, Y.; Miura, T.; Ishimoto, T.; Itaya, T.; Niinomi, M.; Nakano, T. Bone Loss and Reduced Bone Quality of the Human Femur after Total Hip Arthroplasty under Stress-Shielding Effects by Titanium-Based Implant. *Mater. Trans.* **2012**, *53*, 565–570. [CrossRef]
37. Nakano, T.; Kaibara, K.; Ishimoto, T.; Tabata, Y.; Umakoshi, Y. Biological apatite (BAp) crystallographic orientation and texture as a new index for assessing the microstructure and function of bone regenerated by tissue engineering. *Bone* **2012**, *51*, 741–747. [CrossRef]
38. Matsugaki, A.; Aramoto, G.; Nakano, T. The alignment of MC3T3-E1 osteoblasts on steps of slip traces introduced by dislocation motion. *Biomaterials* **2012**, *33*, 7327–7335. [CrossRef]
39. Ishimoto, T.; Ozasa, R.; Nakano, K.; Weinmann, M.; Schnitter, C.; Stenzel, M.; Matsugaki, A.; Nagase, T.; Matsuzaka, T.; Todai, M.; et al. Development of TiNbTaZrMo bio-high entropy alloy (BioHEA) super-solid solution by selective laser melting, and its improved mechanical property and biocompatibility. *Scr. Mater.* **2021**, *194*, 113658. [CrossRef]
40. Ishimoto, T.; Hagihara, K.; Hisamoto, K.; Sun, S.H.; Naknao, T. Crystallographic texture control of beta-type Ti–15Mo–5Zr–3Al alloy by selective laser melting for the development of novel implants with a biocompatible low Young's modulus. *Scr. Mater.* **2021**, *132*, 34–38. [CrossRef]
41. Gokcekaya, O.; Hayashi, N.; Ishimoto, T.; Ueda, K.; Narushima, T.; Nakano, T. Crystallographic orientation control of pure chromium via laser powder bed fusion and improved high temperature oxidation resistance. *Addit. Manuf.* **2020**, *36*, 101624. [CrossRef]
42. Nagase, T.; Hori, T.; Todai, M.; Sun, S.H.; Nakano, T. Additive manufacturing of dense components in beta-titanium alloys with crystallographic texture from a mixture of pure metallic element powders. *Mater. Des.* **2019**, *173*, 107771. [CrossRef]
43. Todai, M.; Nakano, T.; Liu, T.; Yasuda, H.Y.; Hagihara, K.; Cho, K.; Ueda, M.; Takeyama, M. Effect of building direction on the microstructure and tensile properties of Ti-48Al-2Cr-2Nb alloy additively manufactured by electron beam melting. *Addit. Manuf.* **2017**, *13*, 61–70. [CrossRef]
44. Nomura, N.; Kawasaki, A. Development of low magnetic zirconium-based alloys and the additive manufactured builds for biomedical applications. *J. Jpn. Soc. Pow. Pow. Metall.* **2021**, *68*, 431–435. [CrossRef]

Article

Application of the Sinter-HIP Method to Manufacture Cr–Mo–W–V–Co High-Speed Steel via Powder Metallurgy

Kazuyuki Furuya [1,*], Shiro Jitsukawa [2] and Takayuki Saito [1]

[1] Department of Industrial System Engineering, National Institute of Technology, Hachinohe College, 16-1 Uwanotai, Tamonoki, Hachinohe, Aomori 039-1192, Japan; saito-c@hachinohe-ct.ac.jp
[2] Former-National Institute of Technology, Fukushima College, 30 Nagao, Hirauearagawa, Iwaki, Fukushima 970-8034, Japan; jitsukawa.shiro@gmail.com
* Correspondence: kazuyuki-m@hachinohe-ct.ac.jp; Tel.: +81-178-27-7263

Abstract: 1.2C–4Cr–4Mo–10W–3.5V–10Co–Fe high-speed steel (JIS SKH57; ISO HS10-4-3-10) is often manufactured via casting and forging. By applying powder metallurgy, the properties of the abovementioned material can be improved. In this study, the effects of sintering conditions on the formation of precipitates and pores are evaluated. Additionally, strength with and without hydrostatic pressure during sintering is evaluated via static bending and impact tests. Sintering via hot isostatic pressing (HIP) at 1463 K can effectively eliminate pores and prevent the coarsening of precipitates. Toughness and strength improved by 50% by applying HIP.

Keywords: powder metallurgy; high-speed steel; SKH57; hot isostatic pressing; sintering

1. Introduction

High-speed steels (hereinafter referred to as HSSs) play an important role in industries because they are widely used, especially in cutting tools [1–4]. 1.2C–4Cr–4Mo–10W–3.5V–10Co–Fe HSS (JIS SKH57; ISO HS10-4-3-10) is used for cutting tools, as well as for the wear-resistant parts of dies. Carbides containing the alloying elements play an important role to keep hardness at elevated temperatures during cutting [5].

The casting and forging process often results in the coarsening of the carbides to reduce the toughness of SKH57 [6,7]. To improve toughness of the alloy, powder metallurgy (PM) technique is applied. PM is effective for the fine and uniform distribution of the carbides [8–14].

Sintering with and without isostatic pressing is used. Sintering during isostatic pressure (HIP; hot isostatic pressing) is effective to remove pores in the compacts [15–19]. The metal capsule method is often applied for sintering by HIP [10,20,21]. In this method, metal powder is degassed and encapsulated, and then subjected to HIP.

Often water-atomized powders are used as the raw powders for sintered HSS. The finer the powders, the easier it is to sinter because the specific surface area increases. However, when fine powders are formed in a metal mold, problems with moldability may occur, such as galling due to the fine powder, and poor powder fluidity. Therefore, it was thought that the sintering and moldability could be improved by granulating fine water-atomized powder [22,23].

Moreover, in HIP using the metal capsule method, cost increase is inevitable because of the processes required for capsule fabrication, degassing, encapsuling, capsule removal, and so forth. Therefore, HIP treatment without using capsules was attempted. Normally, in this method, the pores in the compact are closed by other methods before HIP. However, this time, by devising the heat treatment conditions, we decided to close the pores with HIP equipment and continue the HIP treatment in the same equipment. In this way, we aimed to reduce costs by simplifying the process and to accommodate various shapes of products. The quenching and tempering conditions after the HIP process were studied based on the report by Ando et al. [24–26]. If the coarsening of precipitates, which leads to a decrease in fracture toughness, can be restricted by the above methods, chipping can be reduced, and it is expected to be applicable to large dies and metal press molds.

In this report, the possibility of improving the properties of SKH57 HSS by applying the PM method is discussed by measuring density, observing microstructure, and conducting hardness, bending, and impact tests on sintered SKH57.

2. Materials and Methods

To achieve better performance in HSS manufactured via PM, the size of the raw powder used must be reduced. In this study, raw powder was subjected to water atomization (PF-5F; Epson Atmix Co., Ltd., Hachinohe, Japan), as the powder yielded from this method is finer than that produced via gas atomization [27]. The size of the raw powder was #4000–#3000, with an average particle size of approximately 5 µm. The chemical composition of the raw powder is shown in Table 1.

Table 1. Chemical compositions of SKH57 raw powders (mass%).

	C	Si	Mn	Ni	Cr	Mo	W	V	Co	O	Fe
Measured	1.27	0.34	0.31	0.10	4.28	3.52	9.86	3.45	9.37	0.24	bal.
JIS	1.20–1.35	max0.40	max0.40	max0.25	3.80–4.50	3.00–4.00	9.00–11.00	3.00–3.70	9.00–11.00	–	bal.

The surface of the raw powder was oxidized during water atomization; however, oxygen was released during sintering and was not detected in the sintered material. An SEM image of the raw powders is shown in Figure 1a. As the raw powder particles were extremely fine, only a 10% polyvinyl alcohol (binder) aqueous solution was added to the raw powder to increase the size via spray drying (to prevent the galling of the metal mold). The SEM image of the granulated powder is shown in Figure 1b. The granulated powders were pressed at 600 MPa in the metal mold to form the specimens (compacts). An OM picture in a cross-section of the compact is shown in Figure 1c. The elliptically deformed granulated powders can be clearly seen. The compacts prepared for the bending test measured 36.3 mm long, 10.1 mm wide, and 5.55 mm thick, whereas those prepared for the impact test measured 55.5 mm long, 10.1 mm wide, and 11.1 mm thick.

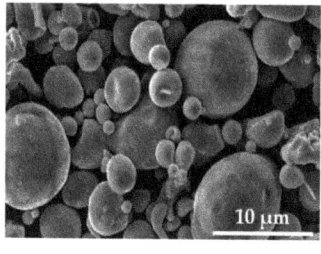
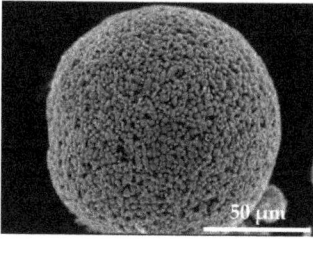

(a)　　　　　　　　　　　(b)　　　　　　　　　　　(c)

Figure 1. SEM images of water-atomized powders and granulated powder, and OM picture in cross-section of compact. (**a**) Raw powders; (**b**) granulated powder; (**c**) granulated powders after pressing.

Figure 2 shows the relationship between time and temperature for (a) sintering without S-HIP and (b) sintering with S-HIP. Sintering without and with S-HIP were performed using a firing furnace and HIP furnace, respectively. Schematic illustrations of the HIP furnace and the sintering process involving S-HIP are illustrated in Figure 3. In the S-HIP process, the compacts were directly hot isostatically pressed without canning. Typically, the powder is placed in a metal capsule, the inside of the capsule is degassed (canning), and the capsule is hot isostatically pressed with the capsule sealed. In this case, however, by proceeding with sintering in a vacuum prior to pressurization, independent closed pores were formed inside the sample and were not connected to the surface; subsequently, the pores were extinguished via pressurization.

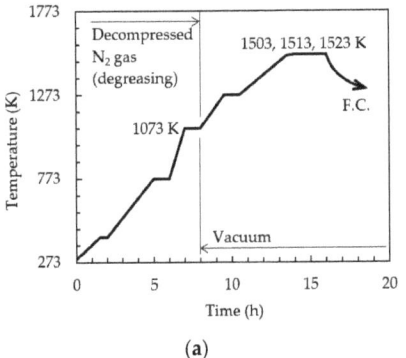 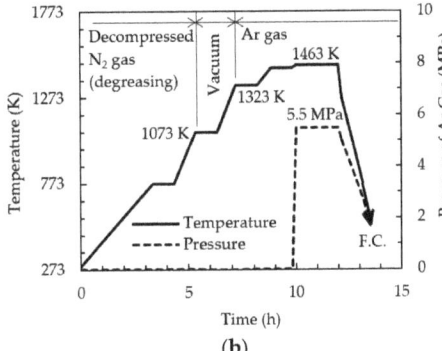

Figure 2. Relationship between time and temperature for sintering (**a**) without S-HIP and (**b**) with S-HIP.

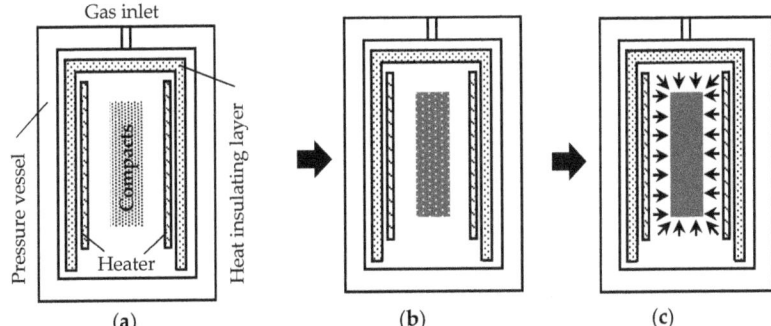

Figure 3. Schematic of HIP furnace and progress of heat treatment. (**a**) Degasing in decompressed N_2 gas; (**b**) sintering in vacuum; (**c**) S-HIPping in Ar gas.

The compacts were heated to 1073 K in a 0.01 Pa N_2 atmosphere for degassing and removing the binder. Sintering without S-HIP was conducted at temperatures of 1503, 1513, and 1523 K for 2 h in vacuum. Because the sintering temperature often affects the microstructure and mechanical properties, several sintering temperatures were applied. For sintering via S-HIP, heating to 1073 and 1323 K was conducted in a vacuum for degassing the open pores. Subsequently, argon gas was injected into the HIP furnace, and S-HIP was performed at 1463 K for 2 h under a pressure of 5.5 MPa. The sintering temperature with S-HIP was lower than that without S-HIP, which prevented the growth of precipitates and afforded higher strength. After sintering was performed, the compacts were cooled in a furnace.

The sintered bodies were normalized in vacuum at 1133 and 1023 K for 3 and 1 h, respectively, followed by oil quenching after heating in a salt bath at 1513 K for 110–120 s. After quenching was performed, tempering at 833 K for 2 h was performed a few times on the sintered bodies.

Several types of specimens, as shown in Table 2, were prepared. Specimen A was a cast-forged material prepared using the conventional method. Specimens B–D and E were prepared via PM without and with S-HIP, respectively. The immersion density of the specimens was measured, and the results are summarized in the table.

Table 2. Summary of prepared specimens.

Specimen ID	Method	Sintering Temperature	S-HIP	Immersion Density (g/cm³)
A	Casting and Forging	N/A	N/A	8.27
B	PM	1503 K	N/A	8.00
C	PH	1513 K	N/A	8.16
D	PM	1523 K	N/A	8.25
E	PM	1463 K	Applicable	8.27

The Rockwell hardness (HRC), bending strength, and Charpy impact value of the specimens were obtained. The hardness was measured with a load of 1471 N and a holding time of 10 s. Meanwhile, the bending strength and impact value were measured using the JIS Z2511 and Z2242 methods for powder metallurgy products, respectively.

The configurations of the bend and impact specimens are illustrated in Figure 4. Panel (a) of the figure shows a schematic illustration of the jig for the bending test, and the dimensions of the bend and impact specimens are shown in panel (b). The shape of the bending specimen was rectangular without notches, and the impact specimen had a notch in the center of the specimen, with a depth of 2 mm and a radius of curvature of 10 mm.

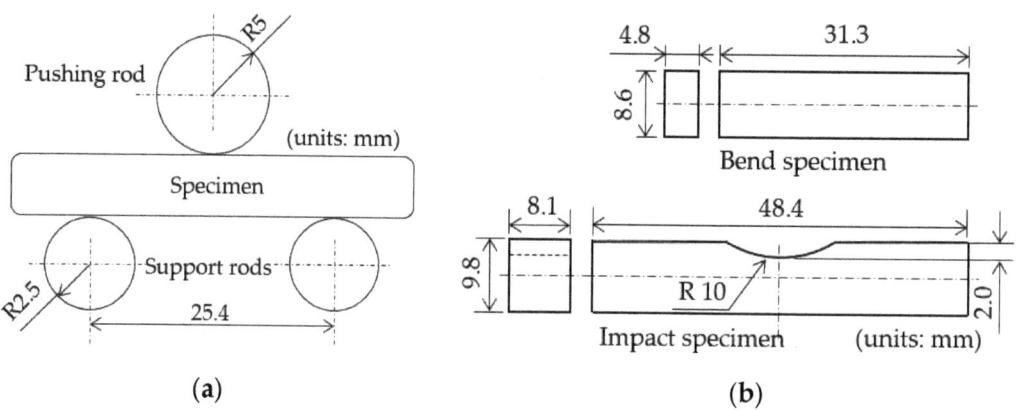

Figure 4. Configuration of jig for bending test and dimensions of bend and Charpy impact specimens. (a) Schematic of jig; (b) dimensions of specimens.

3. Results and Discussions

3.1. Immersion Density and Microstructure

The relationship between immersion density and temperature is shown in Figure 5. Pores and precipitates in the sintered specimens prepared without S-HIP (sintered bodies B, C, and D) were observed under an optical microscope, as shown in Figure 6. Meanwhile, the pores and precipitates in the specimens that underwent S-HIP (sintered body E) and those in the cast-forged material (A) are shown in Figure 7. The precipitates in sintered bodies B (without S-HIP) and E (with S-HIP), and in the cast-forged material (A) are shown in Figure 8.

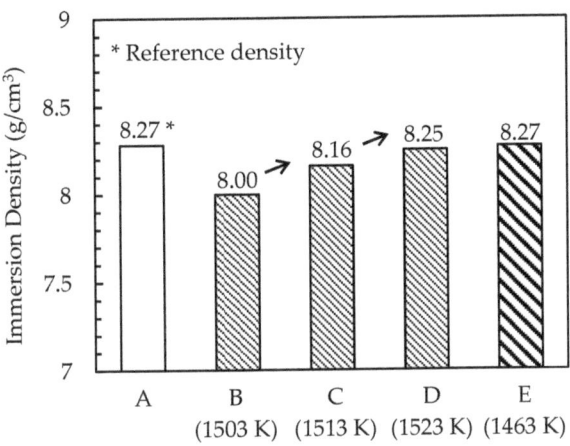

Figure 5. Immersion density change in specimens. Labels A–E represent IDs of each specimen.

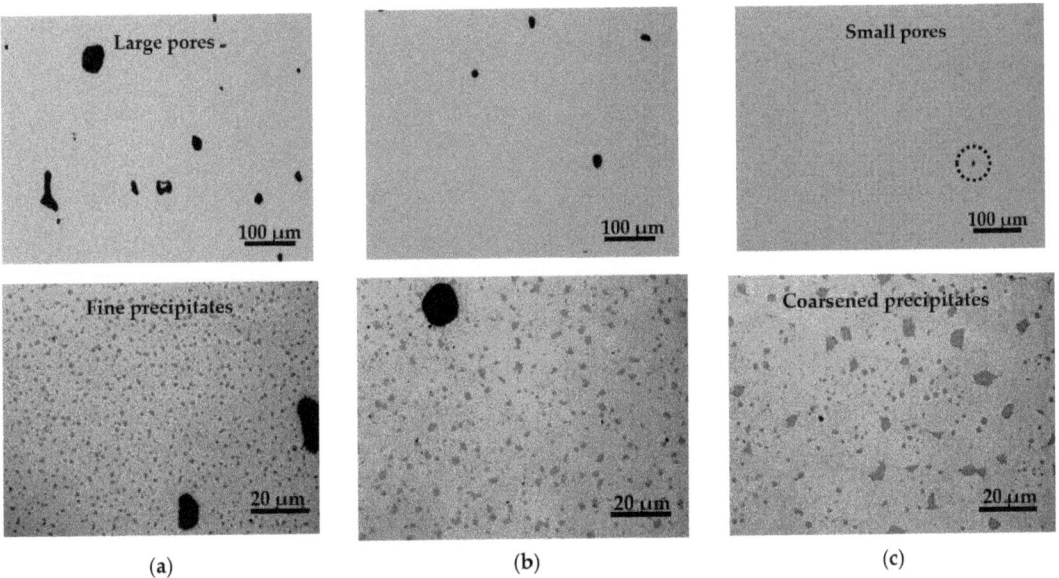

Figure 6. Changes in appearance of pores and precipitates with respect to sintering temperature. Panels (a) to (c) show that higher sintering temperatures correspond to fewer and smaller pores, but coarser precipitates. (a) Specimen B (1503 K); (b) specimen C (1513 K); (c) specimen D (1523 K).

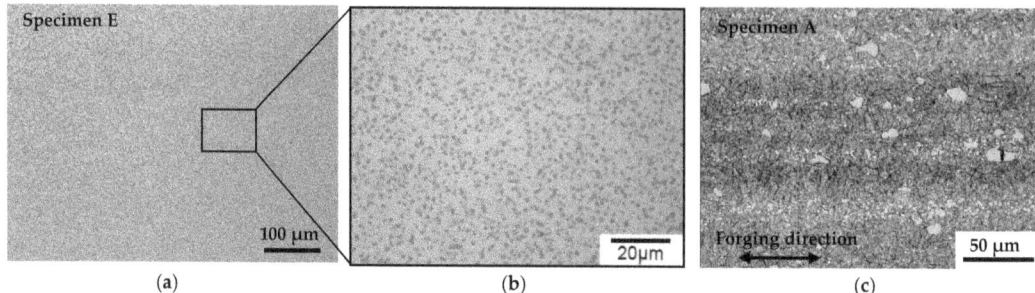

Figure 7. Microstructures of specimens E and A. No pores and fine precipitates were observed in specimen E. In specimen A, large precipitates segregated along forging direction. (**a**) 1463 K sintering with S–HIPping; (**b**) partial magnified view of panel (**a**); (**c**) cast–forged material.

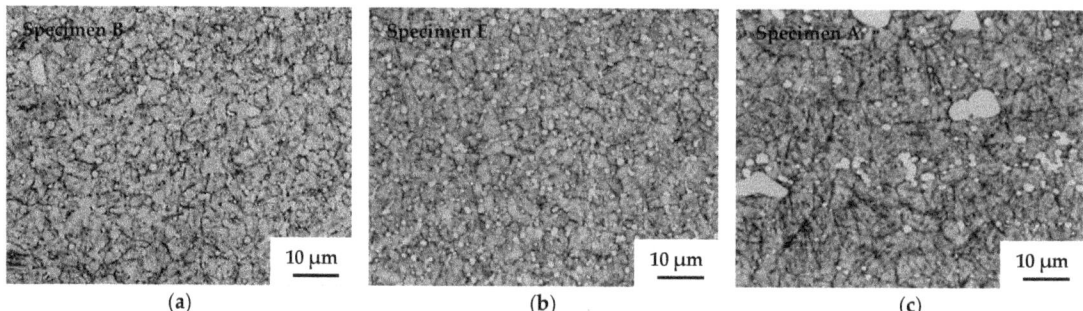

Figure 8. Microstructure of specimens B, E, and A. Size of grains and precipitates between specimens B and E did not differ significantly. In contrast, grains and precipitates of specimen A were larger than those of specimen E. (**a**) 1503 K sintering; (**b**) 1463 K sintering with S–HIPping; (**c**) cast–forged material.

As shown in Figure 5, the densities of sintered bodies B, C, and D approached that of cast-forged material (A) as the sintering temperature increased. The density of the cast-forged material (A) and sintered–hot isostatically pressed body (E) was 8.27 g/cm^3. The trend mentioned above is consistent with the volume fraction of pores shown in Figure 6. In addition, almost no pores were detected in the cast-forged material (A) and sintered–hot isostatically pressed body (E), indicating that the S-HIP treatment effectively reduced the number of pores. Figure 6 shows that the precipitate size increased with the sintering temperature. As shown in Figures 6 and 7, the precipitates in the cast-forged material (A) were the largest and occupied the largest volume fraction. Meanwhile, Figure 8 shows that the grain size and precipitate size between sintered body B and the hot isostatically pressed body (E) did not differ significantly.

3.2. Mechanical Properties

The cast-forged material (A), sintered bodies B, C, and D, and the sintered–hot isostatically pressed body (E) exhibited similar hardness values of 67–68 HRC.

Results from the three-point bend tests and impact tests are summarized in Figure 9. The bending strength, deflection, and impact values of the sintered–hot isostatically pressed body (E) increased by 44%, 40%, and 58%, respectively, compared with those of the cast-forged material (A). The sintered–hot isostatically pressed body (E) indicated the highest values for all properties, whereas the properties of sintered body B with large pores were inferior to those of the cast-forged material (A); the pores in sintered body B were larger than 50 μm. These findings suggest that the pores in the sintered body reduced the bending strength, deflection, and impact values. In fact, the stress intensity factor K for a 50-μm-long crack was approximately 15 MPa m$^{1/2}$, which corresponds to the typical value of fracture toughness for cast-forged high-speed steels [28].

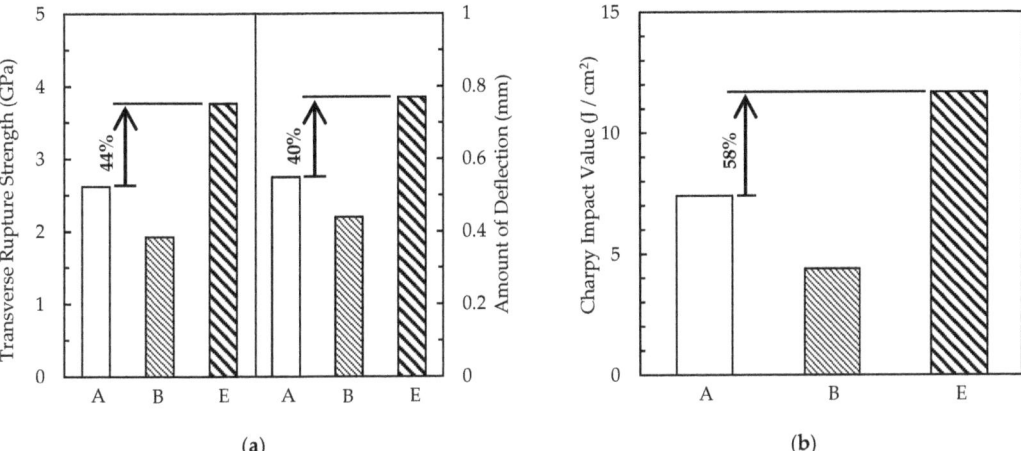

Figure 9. Results of bend and impact tests on specimen A (cast-forged material), B (sintered body), and E (sintered–hot isostatically pressed body). S–HIP significantly improved strength and toughness. (**a**) Comparison of strength and deflection among specimens A, B, and E; (**b**) comparison of toughness among specimens A, B, and E.

The stress intensity factor K (=$\sigma(\pi a)^{1/2}$) was calculated based on a tensile stress of approximately 2000 MPa (measured bending strength of sintered body B) and a crack length of 50 μm (average size of large pores in sintered body B). As failure is dictated by the largest defect in the specimen, and the maximum size of pores in the sintered body is considered to exceed 50 μm, the K value due to pores should be smaller than that for a crack of the same size; however, the calculated K value should correspond to the lower limit.

The bending strength, deflection, and impact values of the cast-forged material (A) were lower than those of the sintered–hot isostatically pressed body (E). Larger precipitates tend to result in a more significant decrease in the strength and toughness [29–31], as evidenced by the cast-forged material [29–31].

4. Summary

The effects of sintering temperature on density, pores, and microstructure were investigated to improve the performance of high-speed steel SKH57 manufactured via powder metallurgy. Meanwhile, the S–HIP method was applied, and the strength and toughness afforded were analyzed. The main conclusions obtained were as follows:

(1) Pores and precipitates in SKH57-sintered HSS depended significantly on the sintering temperature;
(2) S–HIP can sufficiently remove pores even at the lowest sintering temperature (1463 K) in the range investigated in the current experiment;
(3) Significant improvements in terms of strength and toughness were achievable via S–HIP.

Author Contributions: K.F. conceptualized and designed the experiments; K.F. and T.S. performed the experiments; K.F. and S.J. analyzed the data; K.F. and S.J. wrote the manuscript. All authors have read and agreed to the published version of the manuscript.

Funding: This research was funded by the Hachinohe Regional Advance Technology Promotion Center Foundation, Japan.

Institutional Review Board Statement: Not applicable.

Informed Consent Statement: Not applicable.

Data Availability Statement: The data presented herein are available upon request from the corresponding author.

Acknowledgments: The author would like to express his sincere appreciation to Y. Kato, K. Asaka, K. Abe, and H. Nakamura for devising a plan for this study; T. Iwaoka for assisting with the test; and Y. Fukushima, K. Doi, R. Sakata, and F. Toyama for assisting with various heat treatments, including S–HIP. We also would like to acknowledge the Hachinohe Regional Advance Technology Promotion Center Foundation for its continuous funding through this work.

Conflicts of Interest: The authors declare no conflict of interest.

References

1. Mesquita, R.A.; Barbosa, C.A. High-Speed Steels Produced by Conventional Casting, Spray Forming and Powder Metallurgy. *Mater. Sci. Forum* **2005**, *498–499*, 244–250. [CrossRef]
2. Wei, S.; Zhu, J.; Xu, L. Research on wear resistance of high speed steel with high vanadium content. *Mater. Sci. Eng. A* **2005**, *404*, 138–145. [CrossRef]
3. Kim, C.K.; Kim, Y.C.; Park, J.I.; Lee, S.; Kim, N.J.; Yang, J.S. Effects of alloying elements on microstructure, hardness, and fracture toughness of centrifugally cast high-speed steel rolls. *Met. Mater. Trans. A* **2005**, *36*, 87–97. [CrossRef]
4. Leskovšek, V.; Ule, B. Improved vacuum heat-treatment for fine-blanking tools from high-speed steel M2. *J. Mater. Process. Technol.* **1998**, *82*, 89–94. [CrossRef]
5. Kasak, A.; Dulis, E.J. Powder-metallurgy tool steels. *Powder Met.* **1978**, *21*, 114. [CrossRef]
6. Uchida, N.; Nakamura, H. Influence of Carbide Fraction and Carbon Content on Properties of Sintered High-speed-steels. *Tetsu Hagane* **1998**, *84*, 315–320. (In Japanese) [CrossRef]
7. Rapatz, F. *Die Edelstähle*; Springer: Berlin/Heidelberg, Germany, 1962; p. 803.
8. Kawai, N.; Takada, H.; Yukawa, T. Present Status of Hot Isostatic Pressing. *Tetsu Hagane* **1981**, *67*, 1431–1438. (In Japanese) [CrossRef]
9. Fujiwara, Y.; Ueda, H.; Masatomi, H.; Suzuki, H. Fatigue of Hot Isostatic Pressing WC-12% Co Alloys. *J. Jpn. Soc. Powder Powder Metall.* **1980**, *27*, 181–184. (In Japanese) [CrossRef]
10. Kawai, N. On the Application Technology of Hot Isostatic Pressing. *J. Jpn. Soc. Powder Powder Metall.* **1994**, *41*, 209–214. (In Japanese) [CrossRef]
11. Kawai, N.; Hirano, M.; Honma, K.; Tatsuno, T. Effect of Nitrogen and Carbon Equivalent on Properties of Power Metallurgical W-Mo Series High-speen Steel. *Tetsu Hagane* **1986**, *72*, 1921–1928. (In Japanese) [CrossRef]
12. Várez, A.; Levenfeld, B.; Torralba, J.; Matula, G.; Dobrzanski, L. Sintering in different atmospheres of T15 and M2 high speed steels produced by a modified metal injection moulding process. *Mater. Sci. Eng. A* **2004**, *366*, 318–324. [CrossRef]
13. Trabadelo, V.; Giménez, S.; Iturriza, I. Microstructural characterisation of vacuum sintered T42 powder metallurgy high-speed steel after heat treatments. *Mater. Sci. Eng. A* **2009**, *499*, 360–367. [CrossRef]
14. Wießner, M.; Leisch, M.; Emminger, H.; Kulmburg, A. Phase transformation study of a high speed steel powder by high temperature X-ray diffraction. *Mater. Charact.* **2008**, *59*, 937–943. [CrossRef]
15. Alessandro, S.; Raja, H.U.K.; Konstantinos, G.; Martina, M.; Advenit, M.; Moataz, M.A. Powder HIP of pure Nb and C-103 alloy: The influence of powder characteristics on mechanical properties. *Int. J. Refract. Met. Hard Mater.* **2022**, *104*, 105803.
16. Westerlund, J. HIPing advances to benefit PM. *Met. Powder Rep.* **1999**, *2*, 26.
17. Kawai, N.; Hirano, M.; Tatsuno, T.; Homma, K. Effects of VC or VN on the Properties of High Speed Steels Produced by Powder Metallurgy. *J. Jpn. Soc. Powder Metall.* **1986**, *33*, 402–407. (In Japanese) [CrossRef]
18. Fuyuan, Y.; Central Iron and Steel Research Institution. Study of HIP applied to T15 HSS. *Met. Powder Rep.* **1992**, *47*, 51.
19. Saller, H.A.; Paprocki, S.J.; Dayton, R.W.; Hodge, E.S. A Method of Bonding. Canadian Patent 680160, 18 February 1964.
20. Nanes, H.D.; Seifert, D.A.; Watts, C.R. *Hot Isostatic Processing*; MCIC-Report MCIC-77-34; MCIC Center: Colombus, OH, USA, 1977.
21. Nohara, A.; Soh, T.; Nakagawa, T. Numerical Simuration of the Hot Isostatic Pressing Process. In Proceedings of the 3rd International Conference on Isostatic Pressing, 2, London, UK, 10–12 November 1986.
22. Newitt, D.M.; Conway-Jones, J.M. A contribution to the theory and practice of granulation. *Trans. Inst. Chem. Eng.* **1958**, *36*, 422–442.
23. Rumpf, H. Grundlagen und Methoden des Granulierens. *Chem. Ing. Tech.* **1958**, *30*, 144–158. [CrossRef]
24. Ando, H.; Okayama, A.; Soeno, K. Some Properties of Sintered and Hot-Forged High Speed Steels. *Tetsu Hagane* **1975**, *61*, 130–138. (In Japanese) [CrossRef]
25. Ando, H.; Okayama, A.; Soeno, K. Microstructures and Mechanical Properties of Sintered and Hot-Forged High-Carbon High-Vanadium Tool Steels. *Tetsu Hagane* **1975**, *61*, 2629–2638. (In Japanese) [CrossRef]
26. Ando, H.; Okayama, A.; Soeno, K. Microstructures and Mechanical Properties of Sintered, Hot-Forged High-Carbon High-Molybdenum and High-Carbon High-Chromium Tool Steels. *Tetsu Hagane* **1977**, *63*, 1000–1007. (In Japanese) [CrossRef]

27. Takigawa, H.; Manto, H.; Kawai, N.; Homma, K. Properties of high-speed steels produced by powder metallurgy. *Powder Metall.* **1981**, *24*, 196–201. [CrossRef]
28. Jovičević-Klug, P.; Puš, G.; Jovičević-Klug, M.; Žužek, B.; Podgornik, B. Influence of heat treatment parameters on effectiveness of deep cryogenic treatment on properties of high-speed steels. *Mater. Sci. Eng. A* **2022**, *829*, 142157. [CrossRef]
29. Curry, D.A.; Knott, J.F. Effects of microstructure on cleavage fracture stress in steel. *Met. Sci.* **1978**, *12*, 511–514. [CrossRef]
30. Curry, D.A.; Knott, J.F. Effect of microstructure on cleavage fracture toughness of quenched and tempered steels. *Met. Sci.* **1979**, *13*, 341–345. [CrossRef]
31. Lin, H.; Deis, M.; Woodard, T.; Combs, D.; Binoniemi, R.R. Cyclic Deformation, Fatigue and Fracture Toughness of a Nano-Composite High Strength Steel. *SAE Trans.* **2005**, *114*, 285. [CrossRef]

Article

Simple Methods for Evaluating Acid Permeation and Biofilm Formation Behaviors on Polysiloxane Films

Nobumitsu Hirai [1,*], Masaya Horii [2], Takeshi Kogo [3], Akiko Ogawa [1], Daisuke Kuroda [3], Hideyuki Kanematsu [4], Junji Nakata [5] and Shigeru Katsuyama [5]

1. Department of Chemistry and Biochemistry, National Institute of Technology (KOSEN), Suzuka College, Shiroko-Cho, Suzuka 510-0294, Japan; ogawa@chem.suzuka-ct.ac.jp
2. Advanced Engineering Course of Science and Technology for Innovation, National Institute of Technology (KOSEN), Suzuka College, Shiroko-Cho, Suzuka 510-0294, Japan; r03a22@ed.cc.suzuka-ct.ac.jp
3. Department of Materials Science and Engineering, National Institute of Technology (KOSEN), Suzuka College, Shiroko-Cho, Suzuka 510-0294, Japan; kougo@mse.suzuka-ct.ac.jp (T.K.); daisuke@mse.suzuka-ct.ac.jp (D.K.)
4. Joint Research Center between Academia and Industries, National Institute of Technology (KOSEN), Suzuka College, Shiroko-Cho, Suzuka 510-0294, Japan; kanemats@mse.suzuka-ct.ac.jp
5. Division of Materials and Manufacturing Science, Osaka University, Yamadaoka, Suita 565-0871, Japan; nakata@mat.eng.osaka-u.ac.jp (J.N.); katsuyama@mat.eng.osaka-u.ac.jp (S.K.)
* Correspondence: hirai@chem.suzuka-ct.ac.jp

Abstract: The sulfuric acid permeation and biofilm formation behaviors of polysiloxane films have been investigated, and simple methods for evaluating the sulfuric acid permeation and biofilm formation behaviors have been proposed in this paper. The polysiloxane films used in these experiments were practically impermeable to the aqueous sulfuric acid solution, and the amount of biofilm formation varied depending on the composition of the films. Further, the amount of sulfuric acid permeation can be estimated by measuring the polarization curves of polysiloxane films with different thicknesses formed on iron electrodes. By measuring the adhesion work of pure water and simulated biofilm droplets on polysiloxane films of different compositions, we can estimate the resistance of biofilm formation on the polysiloxane films.

Keywords: silane coating; silane compound; biofilm formation; acid permeation; sewerage systems; sewage; contact angle; adhesion work; wettability

1. Introduction

Since the 1970s, Japan's sewerage systems have grown remarkably, and the total length of sewerage culverts nationwide as of the end of the fiscal year 2019 was approximately 480,000 km. In 2019, sewerage culverts that have exceeded the standard service life of 50 years accounted for approximately 22,000 km (5% of the total length), 76,000 km (16% of the total length) 10 years later, and 170,000 km (35% of the total length) 20 years later [1]. The incidence of road cave-ins increases 30 years after the installation of sewage pipes [2], and the number of road cave-ins caused by pipeline facilities in Japan was approximately 2900 per year at the end of the fiscal year 2019. Road cave-ins caused by sewage pipes not only interfere with the sewerage system, such as sewage collection and drainage, but also have a major impact on human life and road traffic if large-scale culvert damage occurs. As a result, evaluating the condition of buried sewage pipes through various methods, such as in-drain surveys, and reconstructing or repairing them to prevent damage to old sewage pipes is important.

The formation of a coating on the inner walls of sewers using a sealant is one method for preventive sewer repair, preventing road cave-ins [3,4]. Preventing sulfuric acid penetration is a requirement for maintaining the performance of the coating films. Concrete sewage walls react with sulfuric acid to form gypsum dihydrate and ettringite, which drastically reduce the strength of the concrete walls. Therefore, the coating film must be impervious to sulfuric acid to prevent corrosion by sulfuric acid solution.

In addition to corrosion prevention, biofilm (sludge, dirt) on the inner walls is desirable [5,6]. When sulfate-reducing bacteria are present in the biofilm, they convert sulfuric acid into hydrogen sulfide, which induces hydrogen sulfide-induced metal corrosion (e.g., corrosion of manholes) in the absence of oxygen, in addition to the toxicity of hydrogen sulfide. Because of their ability to absorb a wide range of inorganic ions, biofilms can also cause scale. The scale is the precipitation of inorganic salts, such as calcium carbonate, calcium sulfate, and silica precipitated in water on the inner wall. Therefore, biofilm prevention performance is sought in the coating film for sewer interior walls.

As previously stated, the coating film for the inner wall must be capable of preventing sulfuric acid penetration and biofilm formation. However, because of the need to handle bacteria, the evaluation of sulfuric acid penetration prevention performance requires long-term experiments in a controlled environment, whereas the evaluation of biofilm prevention performance requires an appropriate experimental environment and experimental equipment. Therefore, simple test methods that may be used for screening are required.

Based on the foregoing, this study proposes simple methods for evaluating sulfuric acid penetration prevention performance and biofilm prevention performance that can be applied to coating films for sewer interior walls. Using the proposed methods, the sulfuric acid penetration prevention performance and the biofilm prevention performance can be evaluated in a short time and without the need for bacteria, respectively.

2. Materials and Methods

2.1. Polysiloxane Film

Polysiloxane films containing various components were applied to mortar, carbon steel (SS400), and glass substrates. The films used were of the alkoxysilane compounds, called "Permeate" [7] (D&D, Yokkaichi, Japan), and owing to their low viscosity (61.5 mPa·s), they easily permeated the pores of coatings. After interpenetrating pores and coating substrates, the films react with the moisture in the ambient atmosphere and harden, owing to the formation of inorganic polymers.

2.2. Immersion Experiment of Polysiloxane Film in Sulfuric Acid Solution

A sample of 200 μm-thick polysiloxane film formed on the mortar was immersed in a 5% sulfuric acid aqueous solution for 120 days. Following that, the sample was broken to expose the film's cross-section, the broken sample was embedded in resin, and the sample's surface was dry-polished with emery paper. Furthermore, the sample's surface was coated with osmium tetroxide to make it conductive before being observed with an electron probe microanalyzer (EPMA). The EPMA used was JXA-8800 (JEOL, Tokyo, Japan).

2.3. Polarization Experiment of Polysiloxane Film Formed on Carbon Steel

On carbon steel (SS400) plates, samples of polysiloxane film with thicknesses 0, 15, and 30 μm were formed. The sample was then placed into an electrochemical cell. A platinum wire served as the counter electrode and a mercury/mercury sulfate electrode as the reference electrode. The electric potential shown in this paper was based on the mercury/mercury sulfate electrode. The electrolyte was a 5% solution of sulfuric acid. The potential was held at -1.1 V for 10 min after the electrolyte was introduced into the electrochemical cell. Then, the cell was held at rest potential for 10 min without energizing. After that, the potential was scanned from the resting potential to 0.8 V at a rate of 20 mV/min, and the corrosion current of the sample was measured. The corrosion current in the active state ranged from -0.6 to -0.4 V, allowing for a simple evaluation of sulfuric acid permeation through the films. The polarization curve was performed using the method described in the Japanese Industrial Standards [8] for measuring the anodic polarization curve of stainless steel.

2.4. Biofilm Formation Experiment on Polysiloxane Film

The amount of real biofilm produced by *Aliivibrio fischeri* [9] (JCM18803, RIKEN BioResource Center (BRC), Tsukuba, Japan) was investigated as follows: *Aliivibrio fischeri* was cultivated in Marine Broth at 22 °C for two days. Following that, the cultivated media were diluted fourfold with phosphate-buffered saline, and glass plates coated with polysiloxane films of eight compositions were placed on them at 22 °C for two days to form the biofilm. The polysiloxane films evaluated in this study were prepared using oligomer B and oligomer A, the latter of which contains more phenyl groups. Experiments were conducted on eight compositions of oligomer B/oligomer A ratios ranging from 100/0 to 30/70 in 10 increments. Finally, the plates were dipped in 0.1% crystal violet aqueous solution for biofilm staining, the stained biofilm was dissolved in pure water, and the absorbance was measured three times on each substrate using an ultraviolet (UV)–visible (vis) spectrophotometer at a wavelength of 580.0 nm (UV-1800, Shimadzu, Kyoto, Japan). The Student's t-test was used to assess considerable differences between substrates of different compositions. Statistical significance was evaluated at $p < 0.05$.

2.5. Wettability Experiment of Pure Water and Simulated Biofilm Droplet on Polysiloxane Film

The following experiments were conducted with a DMo-501 contact angle meter (Kyowa Surface Science, Niiza, Japan). Glass plates were coated with polysiloxane films of eight compositions, which were the same as for the biofilm formation experiment. On each substrate, the contact angle was measured four times using 2 μL of pure water droplets or 1% alginate aqueous solution droplets. Alginate is the main component of *Pseudomonas aeruginosa* biofilm [10] and is one of the biofilm model substances [11]. The different adhesion properties of pure water droplets or 1% alginate aqueous solution droplets on each substrate were used to estimate the simple evaluation of biofilm formation on the films. The Student's t-test was used to assess considerable differences between substrates of different compositions. Statistical significance was evaluated at $p < 0.05$, $p < 0.01$, and $p < 0.001$.

3. Results and Discussion

3.1. Immersion Experiment of Polysiloxane Film in Sulfuric Acid Solution

Figure 1 shows a microscopic image of the sample with three lines describing how the EPMA line analysis was performed. The resin for encapsulation is shown in the upper part of the picture, the mortar of the base material is shown in the lower part of the picture, and the coating film is shown in the middle. The coating film was uniformly thick with a thickness of 180 μm.

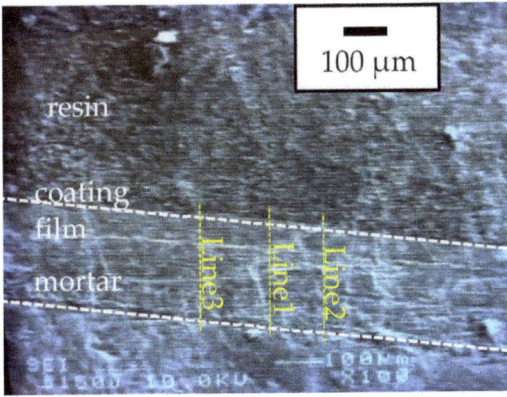

Figure 1. Microscopic image of the sample with three lines describing where EPMA line analysis was performed.

Figure 2 shows the results of the EPMA line analysis, which revealed that no sulfur was detected in the coating film. As a result, sulfate ions barely diffused through the coating film. Notably, cement was present in the base material's mortar, and a trace amount of sulfur was detected on the base material side owing to the sulfate ions in the cement.

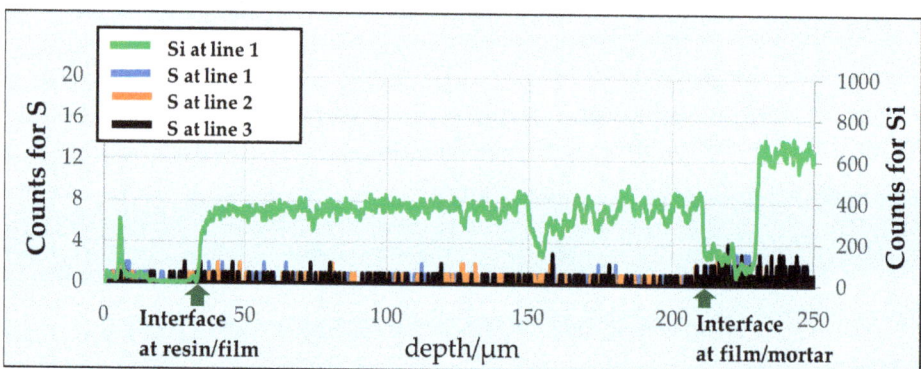

Figure 2. Line analysis of silicon (Si) and sulfur (S) by EPMA. The measured lines are shown in Figure 1.

3.2. Polarization Experiment of Polysiloxane Film Formed on Carbon Steel

Figure 3 shows the polarization curves of polysiloxane films with thicknesses of 0, 15, or 30 μm formed on a carbon steel plate. The active current of corrosion of carbon steel was observed between approximately −0.3 and −0.9 V, as indicated in the curve for carbon steel plate without polysiloxane coating, and the passive retention current was observed above approximately −0.3 V. The flame potential was approximately −0.3 V. Table 1 shows a straightforward evaluation of sulfuric acid permeation through the films based on corrosion current at an active state ranging from −0.6 to −0.4 V. Table 1 shows that as the layer thickness increased by 15 μm, the corrosion current decreased by a factor of approximately 2500–3000. As a result, the polarization curves can be used to measure the coating effect quantitatively.

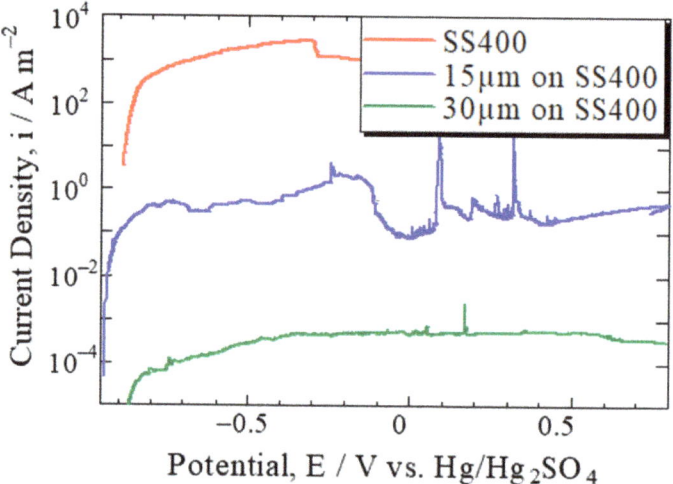

Figure 3. Polarization curves of polysiloxane film with 0, 15, or 30 μm thickness formed on the carbon steel plate.

Table 1. Simple evaluation of sulfuric acid permeation through the films estimated from the corrosion current at active state from −0.6 to −0.4 V.

	At −0.6V	At −0.5V	At −0.4V
SS400	1.4×10^3	1.9×10^3	2.4×10^3
15 μm on SS400	4.2×10^{-1}	5.5×10^{-1}	5.4×10^{-1}
30 μm on SS400	1.6×10^{-4}	2.9×10^{-4}	4.0×10^{-4}

3.3. Biofilm Formation Experiment on Polysiloxane Film

Figure 4 shows the absorbance of pure water where the stained biofilms, which are formed on polysiloxane films with oligomer B/oligomer A compositions ranging from 100/0 to 30/70, were dissolved. The standard error is shown as an error bar. The figure shows that the amount of biofilm formation varies depending on the oligomer B/oligomer A composition ratio. Although the results obtained were not sufficiently significant, the amount of biofilm formation tended to be lower when the oligomer B/oligomer A composition ratio was 100/0 or 60/40. As ethanol reacts with the film when used as the solvent, pure water was used as a solvent for extracting crystal violet in this study. The use of pure water instead of ethanol as the solvent for extracting crystal violet can be one of the reasons for the greater variability than in biofilm quantification by crystal violet staining in previous studies [12,13].

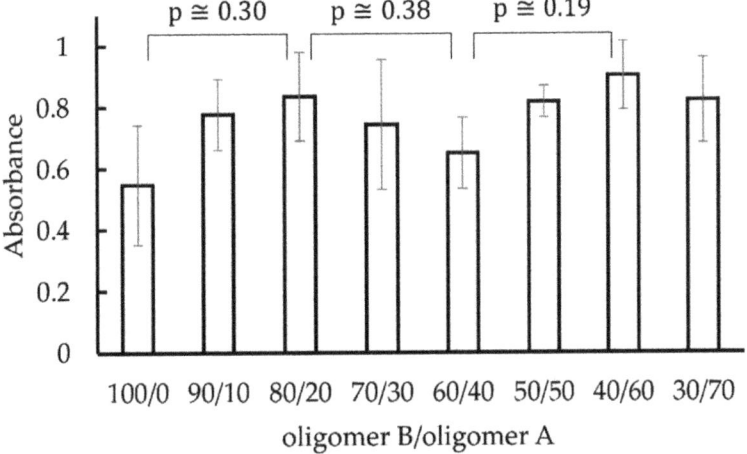

Figure 4. Absorbance of pure water where the stained biofilm, which is formed on polysiloxane films with various compositions of oligomer B/oligomer A ranging from 100/0 to 30/70, was dissolved. The standard error is shown as an error bar.

3.4. Wettability Experiment of Pure Water and Simulated Biofilm Droplet on Polysiloxane Film

Figure 5 shows the contact angles of 2 μL of pure water droplets on polysiloxane films with various compositions of oligomer B/oligomer A ranging from 100/0 to 30/70. The standard error is shown as an error bar. The figure shows that there is a difference in contact angles depending on the composition ratio of oligomer B/oligomer A. In particular, when the composition ratio of oligomer B/oligomer A was 100/0 or 60/40, the contact angle was smaller than for the other composition ratios.

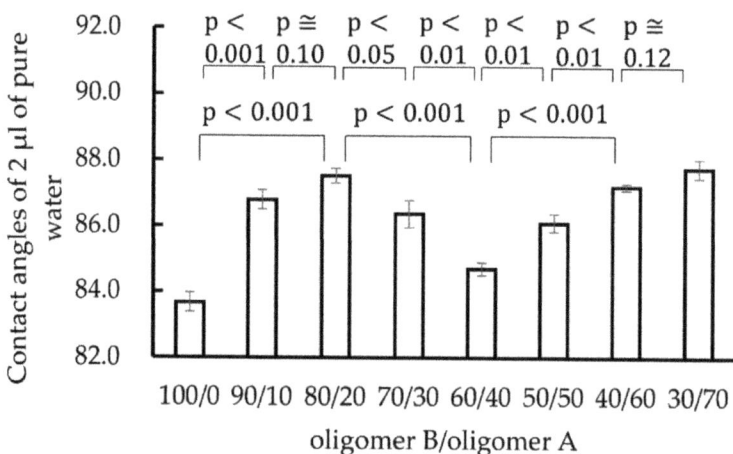

Figure 5. Contact angles of 2 μL of pure water droplets on polysiloxane films with various compositions of oligomer B/oligomer A ranging from 100/0 to 30/70. The standard error is shown as an error bar.

Figure 6 shows the contact angles of 2 μL of 1% alginate aqueous solution droplets on polysiloxane films with various oligomer B/oligomer A compositions ranging from 100/0 to 30/70. The standard error is shown as an error bar. The figure shows that contact angles differ depending on the oligomer B/oligomer A composition ratio. The contact angle was higher when the oligomer B/oligomer A composition ratio was 100/0 or 70/30 (60/40).

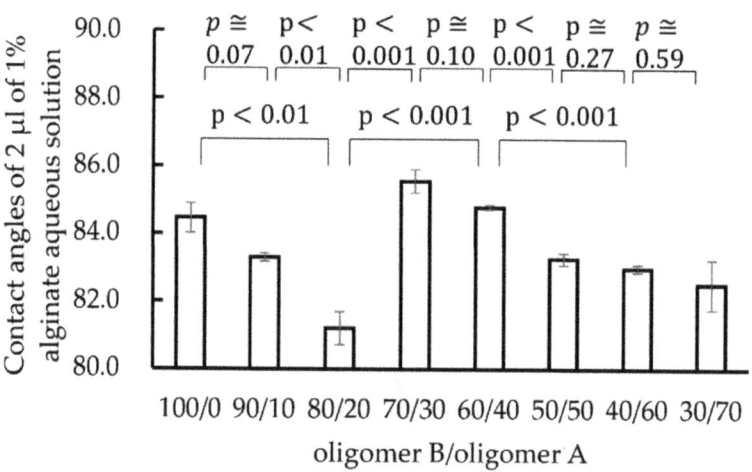

Figure 6. Contact angles of 2 μL of 1% alginate aqueous solution droplets on polysiloxane films with various compositions of oligomer B/oligomer A ranging from 100/0 to 30/70. The standard error is shown as an error bar.

Figures 4–6 are comparable. The substrate having a smaller contact angle with pure water and a larger contact angle with a biofilm-simulating alginate aqueous solution is less likely to form a biofilm. The causes of the above tendency were discussed from the perspective of adhesion work. Antimicrobial qualities can influence resistance to biofilm formation in water, but because all the films in this study were polysiloxane films, no substantial difference in antimicrobial properties would be observed. For example, a study [14] showed that when *Pseudomonas aeruginosa* biofilms were generated on a silver substrate with antimicrobial properties versus a tin substrate without antimicrobial properties, the former produced more biofilms than the latter. The authors focused on biofilm adhesion work as one of the influences on the resistance to biofilm development other than antibacterial capabilities. Biofilms are known to have some barrier effect against the outside world [15], and the authors reasoned that whether a biofilm spreads easily on the substrate affects the biofilm's development behavior. From the experimental results, we calculated the adhesion work of pure water and that of 1% alginate solution as a simulated biofilm to each substrate and evaluated whether the value of the former minus the latter was related to the difficulty of biofilm adhesion. Figure 7 shows the effect of substrate species on the adhesion work of an alginate solution minus the adhesion work of pure water. In Figure 4, the substrate with less biofilm development tended to have a positive value in Figure 7, supporting the hypothesis that the difference in adhesion work between pure water and simulated biofilm is related to the difficulty in biofilm formation. The greater variation in Figure 4 and the smaller variation in Figures 5 and 6 suggest that the prediction of biofilm adhesion behavior from adhesion work of an alginate solution and pure water are adequate. Other effects, such as the structure and unevenness of the substrate, as well as the interfacial energy between water and biofilm, are known to affect biofilm adhesion difficulty in general. In this experiment, all substrates were formed by specifying a mixture of oligomer A and oligomer B. Since the difference in other effects owing to the difference in substrate type was small, we believe that these effects were not noticeable.

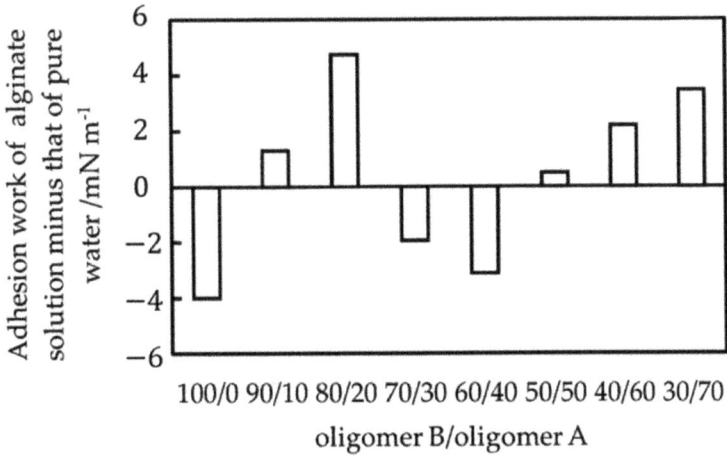

Figure 7. Effect of substrate species on the adhesion work of an alginate solution minus the adhesion work of pure water.

4. Conclusions

In this study, the sulfuric acid permeation behavior and biofilm formation behavior of polysiloxane films have been investigated, and simple evaluation methods for the sulfuric acid permeation and biofilm formation behaviors have been proposed. The results obtained are as follows:

(1) The polysiloxane films used in these experiments are almost impermeable to the sulfuric acid aqueous solution, and the amount of biofilm formation varies with the film composition.

(2) The amount of sulfuric acid permeation can be estimated by measuring the polarization curves of polysiloxane films with different thicknesses formed on iron electrodes.

(3) We can estimate the resistance of biofilm formation on the polysiloxane films by measuring the adhesion work of pure water droplets and simulated biofilm droplets on polysiloxane films of different compositions.

Author Contributions: Conceptualization, N.H.; EPMA analysis, N.H., J.N. and S.K.; polarization curve analysis, N.H.; biofilm formation, M.H.; wettability analysis, M.H.; consideration for EPMA and polarization curve, N.H., T.K. and D.K.; consideration for biofilm formation, N.H., A.O. and H.K.; consideration for wettability, N.H. and M.H. All authors have read and agreed to the published version of the manuscript.

Funding: This research was partially supported by JSPS KAKENHI Grant Numbers JP16K06819 and JP 20K05185.

Institutional Review Board Statement: Not applicable.

Informed Consent Statement: Not applicable.

Acknowledgments: The authors would like to thank Y. Sawada, K. Sano, and A. Suzuki of D&D for providing the materials.

Conflicts of Interest: The authors declare no conflict of interest.

References

1. Sewerage Maintenance and Management. Available online: https://www.mlit.go.jp/mizukokudo/sewerage/crd_sewerage_tk_000135.html (accessed on 28 January 2022).
2. Yokota, T.; Fukatani, W.; Miyamoto, T. The Present Situation of the Road Cave in Sinkholes Caused by Sewer Systems (FY2006-2009). *Tech. Note Natl. Inst. Land Infrastruct. Manag.* **2012**, *668*, 88–89.
3. Roghanian, N.; Banthia, N. Development of a Sustainable Coating and Repair Material to Prevent Bio-Corrosion in Concrete Sewer and Waste-Water Pipes. *Cem. Concr. Compos.* **2019**, *100*, 99–107. [CrossRef]
4. Merachtsaki, D.; Tsardaka, E.C.; Anastasiou, E.; Zouboulis, A. Evaluation of the Protection Ability of a Magnesium Hydroxide Coating against the Bio-Corrosion of Concrete Sewer Pipes, by Using Short and Long Duration Accelerated Acid Spraying Tests. *Materials* **2021**, *14*, 4897. [CrossRef] [PubMed]
5. Santo Domingo, J.W.; Revetta, R.P.; Iker, B.; Gomez-Alvarez, V.; Garcia, J.; Sullivan, J.; Weast, J. Molecular Survey of Concrete Sewer Biofilm Microbial Communities. *Biofouling* **2011**, *27*, 993–1001. [CrossRef] [PubMed]
6. Auguet, O.; Pijuan, M.; Batista, J.; Borrego, C.M.; Gutierrez, O. Changes in Microbial Biofilm Communities during Colonization of Sewer Systems. *Appl. Environ. Microbiol.* **2015**, *81*, 7271–7280. [CrossRef] [PubMed]
7. Ogawa, A.; Kanematsu, H.; Sano, K.; Sakai, Y.; Ishida, K.; Beech, I.B.; Suzuki, O.; Tanaka, T. Effect of Silver or Copper Nanoparticles-Dispersed Silane Coatings on Biofilm Formation in Cooling Water Systems. *Materials* **2016**, *9*, 632. [CrossRef] [PubMed]
8. G0579; Method of Anodic Polarization Curves Measurement for Stainless Steels. Jpn. Ind. Stand: Tokyo, Japan, 2007.
9. Visick, K.L. An Intricate Network of Regulators Controls Biofilm Formation and Colonization by Vibrio fischeri. *Mol. Microbiol.* **2009**, *74*, 782–789. [CrossRef] [PubMed]
10. Davies, D.G.; Geesey, G.G. Regulation of the Alginate Biosynthesis Gene AlgC in *Pseudomonas aeruginosa* during Biofilm Development in Continuous Culture. *Appl. Environ. Microbiol.* **1995**, *61*, 860–867. [CrossRef] [PubMed]
11. Cao, B.; Christophersen, L.; Kolpen, M.; Jensen, P.Ø.; Sneppen, K.; Høiby, N.; Moser, C.; Sams, T. Diffusion Retardation by Binding of Tobramycin in an Alginate Biofilm Model. *PLoS ONE* **2016**, *11*, e0153616. [CrossRef] [PubMed]
12. Kanematsu, H.; Shindo, K.; Barry, D.M.; Hirai, N.; Ogawa, A.; Kuroda, D.; Kogo, T.; Sano, K.; Ikegai, H.; Mizunoe, Y. Electrochemical Responses of Graphene with Biofilm Formation on Various Metallic Substrates by Using Laboratory Biofilm Reactors. *ECS Trans.* **2018**, *85*, 491–498. [CrossRef]
13. Tanaka, N.; Kogo, T.; Hirai, N.; Ogawa, A.; Kanematsu, H.; Takahara, J.; Awazu, A.; Fujita, N.; Haruzono, Y.; Ichida, S.; et al. In-situ Detection Based on the Biofilm Hydrophilicity for Environmental Biofilm Formation. *Sci. Rep.* **2019**, *9*, 8070. [CrossRef] [PubMed]
14. Kanematsu, H.; Ikigai, H.; Yoshitake, M. Evaluation of Various Metallic Coatings on Steel to Mitigate Biofilm Formation. *Int. J. Mol. Sci.* **2009**, *10*, 559–571. [CrossRef] [PubMed]
15. Hoyle, B.D.; Alcantara, J.; Costerton, J.W. *Pseudomonas aeruginosa* Biofilm as a Diffusion Barrier to Piperacillin. *Antimicrob. Agents Chemother.* **1992**, *36*, 2054–2056. [CrossRef] [PubMed]

Article

Characteristics of Vibrating Fluidization and Transportation for Al₂O₃ Powder

Koichiro Ogata [1,*], Tsutomu Harada [1], Hideo Kawahara [2], Kazuki Tokumaru [1], Riho Abe [1], Eiji Mitani [3] and Koji Mitani [3]

1. Department of Mechanical Engineering, National Institute of Technology, Oita College, 1666 Maki, Oita 870-0152, Japan; tsutomu.3_eleven@icloud.com (T.H.); k-tokumaru@oita-ct.ac.jp (K.T.); r-abe@oita-ct.ac.jp (R.A.)
2. Shipping Department, National Institute of Technology, Oshima College, 1091-1 Komatsu, Suo-Oshima, Oshima 742-2193, Japan; kawahara@oshima-k.ac.jp
3. SSC Corporation Limited, 26-216 Ushiroji, Tomomachi, Hiroshima 720-0202, Japan; e-mitani@ssc-hvaf.co.jp (E.M.); k-mitani@ssc-hvaf.co.jp (K.M.)
* Correspondence: k-ogata@oita-ct.ac.jp; Tel.: +81-975-552-6927

Abstract: This study focused on the vibrating fluidized-bed-type powder feeder used in HVAF thermal spraying equipment. This feeder has been used in thermal spraying equipment and industrial applications. However, particulate materials' flow mechanism and stable transport characteristics have not been fully understood. This study experimentally investigated the fluidization characteristics, powder dispersion state, and powder transportation characteristics of Al_2O_3 particles during vertical vibration fluidization. The material used was Al_2O_3 particles of 2.9 μm and 3808 kg/m³, classified as the group C particles in the Geldart diagram. As experimental conditions, the fluidized air velocity to the bottom of the powder bed and the vibration intensity in the vertical direction changed. The critical fluidization air velocity was defined to evaluate the generating powder flow by vertical vibrating fluidization. As a result, good fluidization of the powder bed of Al_2O_3 was obtained by the vertical vibration, as well as an airflow that was higher than the critical fluidization air velocity. Regarding powder transportation characteristics, it was clarified that the fluidized air velocity at the bottom of the powder dispersion vessel and the pressure difference from the powder dispersion vessel to the transportation part significantly affect the mass flow rate.

Keywords: cohesive powder; vertical vibration; fluidization; dispersion; transportation

Citation: Ogata, K.; Harada, T.; Kawahara, H.; Tokumaru, K.; Abe, R.; Mitani, E.; Mitani, K. Characteristics of Vibrating Fluidization and Transportation for Al₂O₃ Powder. *Materials* 2022, *15*, 2191. https://doi.org/10.3390/ma15062191

Academic Editors: Hideyuki Kanematsu, Yoshikazu Todaka and Takaya Sato

Received: 29 January 2022
Accepted: 13 March 2022
Published: 16 March 2022

Publisher's Note: MDPI stays neutral with regard to jurisdictional claims in published maps and institutional affiliations.

Copyright: © 2022 by the authors. Licensee MDPI, Basel, Switzerland. This article is an open access article distributed under the terms and conditions of the Creative Commons Attribution (CC BY) license (https://creativecommons.org/licenses/by/4.0/).

1. Introduction

In recent years, ceramic materials with particle sizes from micron- to nano-size have been developed and applied to various products such as semiconductors, automobiles, and medical equipment. The reason is that ceramic materials have excellent mechanical, thermal and surface properties such as wear resistance, heat resistance, and corrosion resistance. Thermal spraying is an effective method to utilize the mechanical properties of ceramics. Thermal spraying is a surface treatment method in which metal particles are melted or softened on the surface of a product and then sprayed at high speed and high temperature to form a film and improve the material's surface modification [1]. This research focuses on the thermal spraying method in which the powder material is sprayed onto the substrate with compressed gas.

The High-Velocity Oxygen Flame (HVOF) method, High-Velocity Air Flame (HVAF) method, cold spray (CS) method, and aerosol deposition (AD) method have been developed as devices for spraying with compressed gas [1–14]. Much research and development have been conducted on HVOF and HVAF spraying equipment [1–5]. HVOF spraying equipment has the powder material sprayed onto the substrate at high speed and high temperature with high-pressure oxygen [1]. Regarding HVAF thermal spraying equipment, the results show that the quality of the film on the substrate is higher than that of HVOF thermal spraying. The reasons for this are that the HVAF sprayed coating had higher mechanical properties such as elastic modulus, higher fracture toughness, and equal or higher abrasion compared to its HVOF-sprayed counterparts [3]. Furthermore, the compressed gas used in HVAF is air, which is cheaper than the oxygen used in HVOF. Therefore, it is expected to reduce operating costs.

Next, in the CS method [6–11], the working gas temperature is lower than the material's melting point or softening temperature. A supersonic flow accelerates the feeding particle material through the Laval nozzle. At that time, the material collides with the substrate at high speed in the solid state to form a film [6]. The cold spray also has some disadvantages compared to other thermal spray techniques. The deposition of coatings using CS is based on the plastic deformation capacity of the particles and the substrate. Consequently, CS requires the substrates to have a minimum ductility to produce well-bonded coatings [10]. In addition, the consumption of a large amount of He and N_2 gas used in the system and expensive equipment are also problems [6]. The AD method is a technology that enables film formation even under conditions close to room temperature [12]. Recent research has achieved high-quality alumina coatings [13,14]. On the other hand, the AD method also has restrictions such as low vacuum conditions for forming a film on the substrate and the high cost of the equipment.

For this reason, we focused on the HVAF thermal spraying equipment and the research and development conducted in this area [15,16]. The HVAF thermal spraying under development uses kerosene as fuel, which is considered to have a cost advantage over the AD and CS. The equipment includes a swirl-type combustion chamber using sprayed fuel and a fluidized-bed-type powder feeder. However, the flow mechanism and performance of the combustion part and powder feed part have not been clarified. Therefore, in the previous study [17], the relationship between the ignition characteristics, the stability and transition characteristics of the flame, and the equivalence ratio with the combustion temperature and the length of the combustion flame was clarified using a curved impinging spray combustor. On the other hand, the flow mechanism and transport characteristics of the vibrating fluidized-bed-type powder feeder, which is supposed to be used in the development equipment, have not been clarified.

In this research, we focused on the vibrating fluidized-bed-type powder feeder, which is used for feeding a powder material with compressed gas and can use the functions of the material as it is. For example, acoustic sound fluidization mixes two kinds of powder in the CS method [18]. In the AD method, after the powder filled in the vessel is mixed with gas to form an aerosol by vibration, the aerosol particles are transported by the gas flow generated via the pressure difference between the feeding chamber and the deposition chamber [13,14]. However, micron-sized particles' flow mechanisms and stable transport characteristics have not been clarified.

There is much research on vibrating fluidized powder beds [19–24]. These studies have investigated the flow pattern [20,24] and flow characteristics [19–24] to the particulate materials of micron- and nano-size when the vibrating amplitude, frequency, and fluidized air velocity are changed. Regarding the particle dispersion in the vibrated powder bed, there is also an experimental study of the release of dust from cohesive powder by vibrating fluidization [25,26]. On the other hand, there are no research reports on the transportation characteristics of powder using a vibrating fluidized bed.

Based on the above background, in this study, we attempted to investigate the operating conditions and elucidate the flow characteristics of a micron-sized ceramic powder dispersed and transported by vibration fluidization. An alumina powder that is classified as being in Group C in the Geldart diagram [27] was used. As for the experimental conditions, the secondary air velocity to the powder feeder was made constant, and the fluidized air velocity to the bottom of the powder bed and the vibration intensity in the vertical direction were changed. This paper reports the results obtained for the fluidization characteristics, powder dispersion state, and powder transportation characteristics of the powder bed during vertical vibration fluidization.

2. Experiment

2.1. Materials

This study used Al_2O_3 (Sumitomo Chemical Co., Ltd., Tokyo, Japan, AM-27) as the cohesive ceramic powder. The particle size distribution of alumina powder was measured by a laser-diffraction particle size analyzer (HORIBA, LA-950). As for the measurement conditions, the refractive index of the alumina was 1.76, and the refractive index of the distilled water was 1.333. The particles were led to the measurement cell with distilled water. At that time, the particles and distilled water were circulated in the measurement cell while stirring. However, ultrasonic dispersion treatment was not performed to maintain the aggregated particles. The particle size was measured three times at each experimental condition and averaged. Figure 1 shows the result of the particle size distribution of the Al_2O_3 used. It can confirm that the primary particle size was less than 1 µm, and the agglomerated particle appeared at around 10 µm. The median particle diameter D_{50} was 2.92 µm, and the particle density was 3808 kg/m^3. Table 1 indicates the measurement result of the flowability of Al_2O_3 particles. These particle properties were measured by the Powder Tester (HOSOKAWA MICRON, PT-X). The compressibility and the cohesiveness denoted a high value, and these flowabilities were evaluated as significantly worse. As a result, the flowability index was 26.5, as shown in Table 1. The flowability was assessed as Very Poor based on Carr's flowability index [28–30].

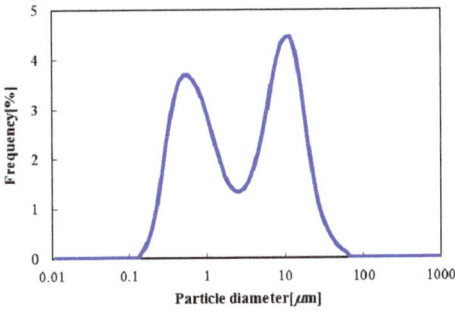

Figure 1. Particle size distribution of Al_2O_3 powder used.

Table 1. Measurement results of Carr's flowability index.

Powder Characteristics	Al_2O_3	
	Measured Value	Points
Angle of Repose (deg)	46	14.5
Compressibility (%)	51.8	0
Angle of Spatula (deg)	66.9	12
Cohesiveness (%)	84.1	0
Flowability Index (-)	26.5	
Flowability	Very Poor	

2.2. Test Equipment

Figure 2 shows a schematic diagram of the test equipment for the powder transportation system that used vibrating fluidization. The test equipment consisted of a powder dispersion vessel, a venturi-type powder feeder, and a powder recovery unit. A powder dispersion vessel was made using an acrylic cylindrical pipe with an inner diameter of

50 mm, a height of 300 mm, and a thickness of 8 mm. A polyurethane tube with an inner diameter of 4.23 mm was used for the air supply and the powder feeder.

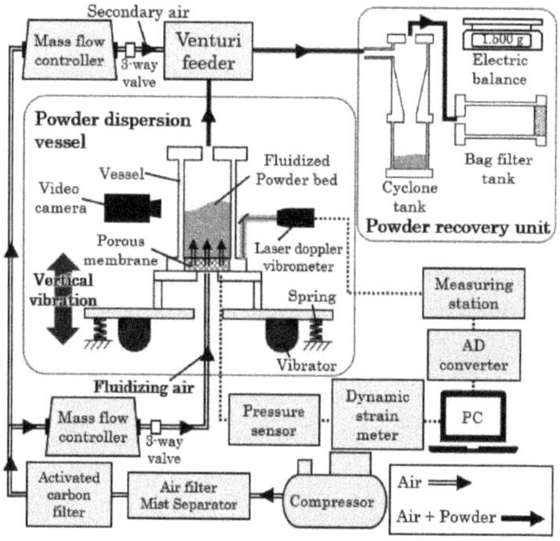

Figure 2. Test equipment for vibrating fluidization and particle transportation.

A porous membrane of 6 mm thickness was installed at the air supply part of the bottom of the powder dispersion vessel to provide uniform air for fluidization. Furthermore, two vibration motors were connected at the bottom of the dispersion vessel with an iron plate and the vibration spring to give vertical vibration to the powder dispersion vessel. The vibration motor (URAS TECHNO, SEE-0.5) had a weight inside and could be adjusted from 0% to 100%. In this experiment, the motor weights were 54%, 71%, 86%, and 100%.

A Venturi-type feeder was installed above the powder dispersion vessel to assist in transporting powder from the powder dispersion vessel to the receiving tank. The fluidized air from the bottom of the powder bed into the powder dispersion vessel and the secondary air from the upstream side of the Venturi feeder were exhausted through the filter at the end of the powder-receiving tank. Here, the internal shape of the Venturi-type powder feeder was 7 mm in diameter on the upstream side, e.g., the secondary air introduction section, the flow path was tapered toward the downstream direction, and the throttle section was 4 mm in diameter. Further, the flow path was restored by taper from the throttle part, and the diameter on the downstream side was 7 mm, which was the same as the upstream side. The powder-receiving unit consisted of a cyclone and a receiving tank with a bug filter.

2.3. Experimental Conditions

In this study, the vibrating fluidization experiments and the powder transportation experiments were carried out. The alumina powder used in the fluidization and the powder transportation experiments had a median particle diameter of 2.92 μm, a particle density of 3808 kg/m^3, and an initial filling amount of 100 g in the dispersion vessel. Table 2 shows the experimental conditions of the vibrating fluidization for Al_2O_3 particles, where f is the frequency, u_b is the fluidizing air velocity at the bottom of the powder dispersion vessel, u_t is the secondary air velocity on the upstream side of the Venturi-type feeder, and Λ is the vibration strength.

Table 2. Experimental conditions of vibrating fluidization.

Frequency f (Hz)	Fluidizing Air Velocity u_b (m/s)	Secondary Air Velocity u_t (m/s)	Vibration Strength Λ (−)
60	0.0017~0.055	3.03	0~9.23

Table 3 shows the conditions of the powder transportation experiment using vibrating fluidization. Although the parameters were the same as those in the fluidization experiment, this experiment was conducted with fluidized air velocity and vibration intensity, which enabled powder transportation.

Table 3. Experimental conditions for powder transportation.

Frequency f (Hz)	Fluidizing Air Velocity u_b (m/s)	Secondary Air Velocity u_t (m/s)	Vibration Strength Λ (–)
60	0.068~0.510	3.03	6.45~9.23

The vibration strength Λ was calculated from Equation (1) after measuring the vibration amplitude A. Here, g is the gravitational acceleration.

$$\Lambda = \frac{A(2\pi f)^2}{g} \quad (1)$$

Regarding the amplitude, the maximum value of the vibration amplitude, measured at intervals of 10 s using a laser Doppler vibrometer, was used. The amplitude A of vertical vibration at 60 Hz was obtained using a laser Doppler vibrometer, as shown in Figure 3. The measurement point was chosen for the lower flange of the dispersion vessel to obtain accurate data of vertical vibration to the particles inside the dispersion vessel. The recorded result displayed that the maximum amplitude reached 0.637 mm in the case of the motor weight of 100%, and the vibration strength was 9.23.

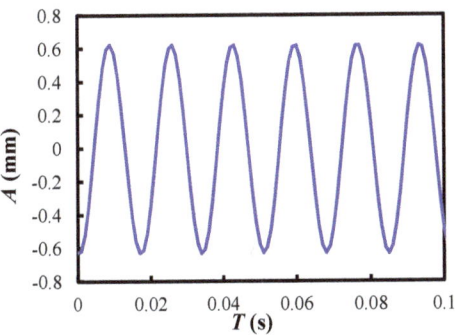

Figure 3. Relationship between vibration amplitude and elapsed time up to 0.1 s, where Λ = 9.23.

Table 4 indicates the vibration strength when changing the weight of the vibration motor in the present experiment. The vibration strength could be set from 5.0 to 9.23.

Table 4. Calculation results of vibration strength.

Motor Weight (%)	54	71	86	100
Amplitude, A (mm)	0.345	0.445	0.565	0.637
Vibration strength, Λ (–)	5.0	6.45	8.19	9.23

2.4. Experimental Method

2.4.1. Fluidization Experiment

The fluidization characteristics of Al_2O_3 particles were investigated via a de-fluidization experiment, which is less susceptible to inter-particle forces [20]. In the experiment, after filling the dispersion vessel with Al_2O_3 particles, vertical vibration was applied to simultaneously supply fluidized air and secondary air at the bottom of the powder bed to fluidize the powder bed for 5 min. Next, the fluidizing air velocity gradually reduced from the set velocity, and the pressure at the bottom of the dispersion vessel was also measured at each air velocity. The pressure drop ΔP inside the powder bed was calculated using the following equation.

$$\Delta P = p_d - p_a \tag{2}$$

Here, p_d is the pressure when the powder is filled inside the dispersion vessel, and p_a is the air pressure without the powder in the dispersion vessel.

In this study, the fluidization characteristics of Al_2O_3 particles were evaluated from the pressure drop and the fluidization state of the powder bed when the fluidization air velocity of the dispersion vessel was reduced at regular intervals. A digital video camera was used to visualize the fluidized state.

2.4.2. Powder Transportation Experiment

Here, the procedure for powder transportation experiments using the vibration fluidization operation is described. Al_2O_3 particles were poured naturally into a cylindrical dispersion vessel. The dispersion vessel was vibrated vertically by turning on the vibration motors attached to an iron plate at the bottom of the dispersion vessel. At the same time, the three-way valve opened to provide the secondary air, and then the valve of the fluidized air was also introduced to supply air toward the porous membrane at the bottom of the dispersion vessel. The Al_2O_3 particles entering the dispersion vessel could be dispersed and supplied based on this operation.

The mass of the particles was collected in the cyclone and the tank with a bag filter. The transportation mass of the powder was measured an electronic balance (Shimadzu, UX-2200H) every minute after starting the vibrating fluidization. The experiment was repeated five times, and the mass flow rate G_s was calculated by the relationship between the transportation mass of the powder M_p and the elapsed time T.

Furthermore, in this experiment, the particle size inside a dispersion vessel after the powder transportation experiment was measured to confirm the particle agglomeration before and after vibrating fluidization. Samples were taken from the dispersion vessel, and the particle size was measured with a laser-diffraction-type particle size analyzer as in Section 2.1.

3. Results

3.1. Fluidization Characteristics

Generally, the group C particles in the Geldart diagram have poor flowability. Therefore, it was necessary to confirm the fluidization characteristics of the alumina powder used in this study. For this reason, this section describes the fluidization characteristics of alumina powder when there was no vibration and when vibration was applied. Here, the de-fluidization experiment was conducted as described in Section 2.4.1. The pressure drop was measured, and the fluidized state of the powder bed inside a dispersion vessel was also visualized during the fluidization experiment.

Figure 4a shows the relationship between the powder bed ΔP and the fluidizing air velocity at the bottom of the dispersion vessel u_b in the case of no vibration. It was found that the pressure drop gradually decreased with the decreasing of the fluidized air velocity. In general, the group A and B particles in the Geldart classification have easy fluidization characteristics. The pressure drop curve showed that the pressure drop took a constant value under the high-velocity condition, and the pressure decreased linearly with the decreasing of the fluidization velocity. At that time, the minimum fluidization velocity could be defined from the pressure drop curve. However, the pressure drop curve of the alumina powder, which belongs to the group C particle, showed different tendencies.

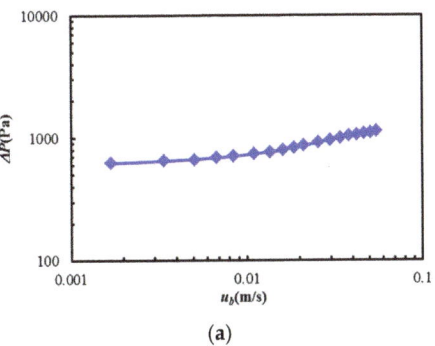 

(a) (b)

Figure 4. Flow characteristics of the powder bed of Al$_2$O$_3$ due to the fluidization without vibration. (a) Relation of the pressure drop and the fluidizing velocity. (b) Visualization of the powder bed (u_b = 0.055 m/s).

To confirm this, we visualized the fluidized state of alumina powder without vibration. Figure 4b shows a snapshot of the flow state of the powder bed under the condition of fluidizing air velocity u_b = 0.055 m/s. It can be observed that fluidization of the powder bed did not occur, and that fixed thin channels were formed inside the bed. This was because the adhesive force on the particles had a strong effect, and the entire powder bed could not be fluidized only with fluidizing air. As a result, as shown in Figure 4b, the fluidized state was confirmed over the applied fluidizing air velocity range. The result indicated that the fluidization of the alumina powder used in the present study was difficult in non-vibration conditions. Therefore, it was concluded that a vibration fluidization operation is required to easily fluidize and disperse the alumina powder for the group C particles in the Geldart diagram.

Next, the results of vertical vibrating fluidization are described. Figure 5 shows the relationship between the pressure drop of the powder bed ΔP and the fluidizing air velocity at the bottom of the powder dispersion vessel u_b when the vertical vibration was applied. In this experiment, the powder bed was vibrated vertically and fluidized for 5 min at a fluidized air velocity of u_b = 0.055 m/s. Afterwards, four kinds of measurement were conducted with the air velocity us set from 0.030 m/s to 0.055 m/s, as shown in Figure 5a–d, because of the examination of the effect of the starting air velocity on the vertical vibrating fluidization in this study. The pressure drop of the powder bed of Al$_2$O$_3$ was measured at each fluidization velocity in the case of the same vibration strength. Here, the vibration strength Λ was from 5 to 9.23, as shown in Table 4.

As shown in Figure 5a, the pressure drop of Λ = 5.0 gradually decreased up to the air velocity of u_b = 0.01 m/s, and the pressure increased rapidly with the decreasing of the velocity, and then the pressure drop took a maximum value close to the velocity of u_b = 0.007 m/s. Regarding the other vibration strength conditions of Figure 5a, the pressure drop increased with the reducing of the fluidization velocity and reached a maximum value at specific fluidizing air velocities for each vibration strength. This was because the powder bed of Al$_2$O$_3$ in the vessel transferred from the fluidization state to the expansion state as the fluidizing air velocity decreased. Additionally, the pressure drop after the maximum pressure decreased with the decreasing of the fluidizing air velocity at all the vibration strength conditions, and then the state of the powder bed became close to the fixed bed. The pressure drop curve seemed to have some variability when the starting velocity was changed, as shown in Figure 5b–d. However, the pressure drop curve was almost similar to Figure 5a. Therefore, the same powder behavior appeared in this study's vertical vibration experiments.

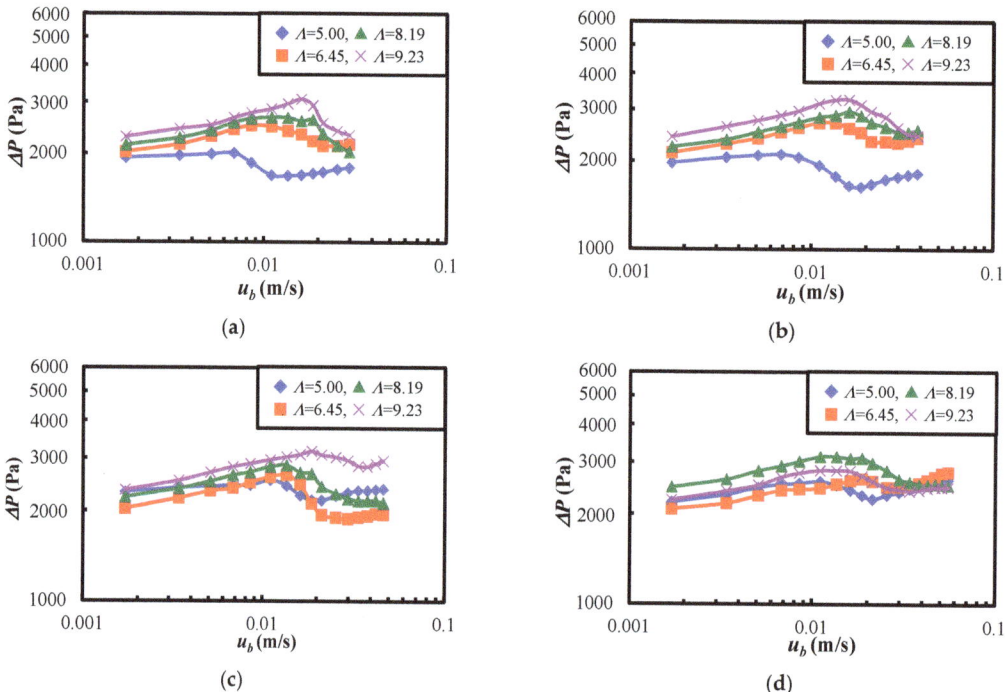

Figure 5. Relationship between the pressure drop of the powder bed of Al_2O_3 and the fluidizing air velocity at the bottom of vessel when the fluidization velocity at the start of measurement and the vibration strength were changed. (**a**) u_s = 0.030 m/s. (**b**) u_s = 0.038 m/s. (**c**) u_s = 0.047 m/s. (**d**) u_s = 0.055 m/s.

In addition, the critical fluidization velocity u_c was defined to discuss the critical conditions for vibrating fluidization of Al_2O_3 powder. Here, the critical velocity was estimated as the fluidizing air velocity at the maximum pressure in Figure 5a–d. Figure 6 shows the relationship between the critical fluidization air velocity u_c and the vibration strength Λ. Here, the fluidization velocity was changed at the start of measurement u_s. The figure shows that the critical fluidization air velocity increased as the vibration strength increased, although there was some variability. The reason was presumed to be that the powder bed was strongly compressed by vertical vibration when the vibration intensity increased. On the other hand, there was no significant effect of the fluidized velocity at the start of measurement in the experimental conditions of this study.

Figure 7 shows the relationship between the normalized pressure drop $\Delta P/\Delta P_i$ and the normalized fluidizing air velocity u_b/u_c when the air velocity at the start of measurement u_s was changed. Here, u_c is the above-mentioned critical fluidization air velocity. The pressure ΔP_i, defined as the gravity at the initial packing of the powder bed, is divided by the cross-sectional area, as expressed by Equation (3). In the equation, M_{pi} is the initial filling mass of the powder, A is the cross-sectional area of the dispersion vessel, and D is the inner diameter of the particle dispersion vessel.

$$\Delta P_i = \frac{M_{pi}g}{A} = \frac{4M_{pi}g}{\pi D^2} \qquad (3)$$

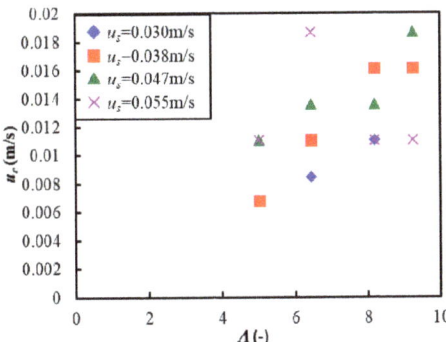

Figure 6. Critical fluidization air velocity during the vibrating fluidization against the vibration strength where the fluidization velocity at the start of measurement u_s = 0.03~0.055 m/s.

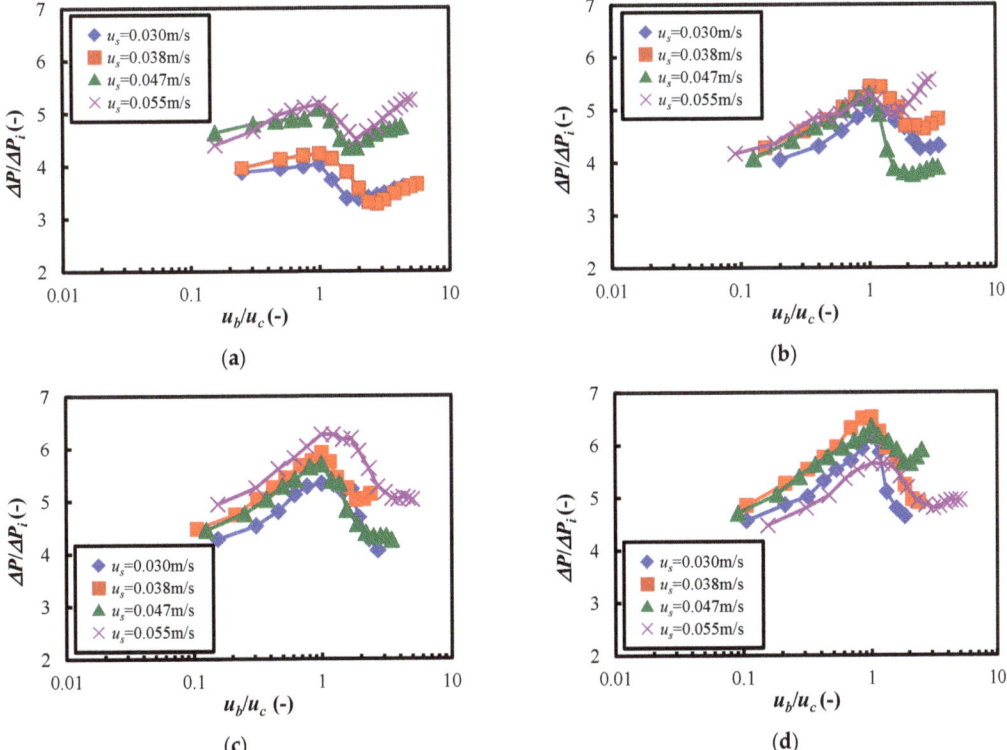

Figure 7. Relationship between the normalized pressure drop and the normalized fluidizing velocity at each vibration strength when the fluidization velocity at the start of measurement and the vibration strength were changed. (**a**) Λ = 5.0. (**b**) Λ = 6.45. (**c**) Λ = 8.19. (**d**) Λ = 9.23.

The pressure drop acting on the powder bed ΔP is the adhesive force of the Al_2O_3 powder used in this study and the compressive force due to vibration. The maximum value of ΔP is the required pressure to transfer the powder flow from the expanded state to the fluidization state. Therefore, $\Delta P \Delta Pi$ can be interpreted as an index showing the conditions for generating powder flow due to the vibrating fluidization used in this study.

In Figure 7a–d, it can be seen that $\Delta P \Delta P_i$ decreased sharply as u_b/u_c exceeded 1.0, regardless of the fluidized air velocity at the start of measurement. This trend showed that the inter-particle and the particle-wall surface forces were reduced by the vertical vibration and the supplied air velocity to the powder bed. At this time, it was observed that the flow pattern of the powder bed was transferred from the expansion bed to the fluidization bed. Furthermore, $\Delta P \Delta P_i$ tended to reach a constant state or increase when u_b/u_c increased. Here, the finding that $\Delta P \Delta P_i$ was close to constant shows the expanded and fluidized state of the powder bed as seen in the group A and B particles in the Geldart diagram. On the other hand, as for the increase in $\Delta P \Delta P_i$, it could be speculated that the reaggregation of particles may have occurred in the powder bed due to vibrating fluidization. From the above results, it is clear that to obtain the satisfactory fluidization state of the Al_2O_3 powder used in this study, it is necessary to apply vertical vibration and, at the same time, supply an airflow higher than the critical fluidization air velocity.

Next, the flow state of the powder bed during vertical vibration fluidization obtained from the experiment is summarized. Figure 8a shows the typical flow pattern of the transition state of the powder bed caused by the decrease in the fluidized air velocity u_b at the bottom of the dispersion vessel as a schematic diagram for each vibration strength. The evaluation criteria for the flow state in the figure were as follows.

(I) Fixed bed: The powder bed was close to a stationary state, and agitation or expansion due to the vibration and fluidization of air was not observed.
(II) Expansion bed: Expansion of the powder bed was confirmed by vibrating fluidization. There were no bubbles in the powder bed.
(III) Fluidized bed: The powder bed was fluidized, and air bubbles were also observed, as shown in Figure 8b.
(IV) Dispersion bed: The powder bed was intensely fluidized, and the powder was dispersed toward the upside of the powder dispersion vessel, as shown in Figure 8c.

Complete fluidization could not obtain all the investigated experimental conditions in this study. Figure 8a confirmed that the powder bed expanded and fluidized under the condition of the minimum vibration intensity ($\Lambda = 5.0$) in this experiment, but a fixed bed was generated again when a high fluidized air velocity was applied. This behavior occurred when the vibration strength given to the powder bed was low. In this condition, the powder flow became unstable. As a result, the stable fluidization and dispersion operation of the Al_2O_3 powder was complex in the condition. On the other hand, it was clarified that relatively good fluidization and dispersion could be obtained when the vibration intensity $\Lambda \geq 6.45$ and the fluidized air velocity was high. Based on this result, the powder transportation experiment was conducted by means of the vibrating fluidization of Al_2O_3 powder under the condition of a vibration intensity $\Lambda \geq 6.45$.

3.2. Powder Transportation Characteristics

Figure 9 shows the relationship between the transported mass of the powder M_p and the elapsed time T when the fluidizing air velocity u_b was changed. Figure 9a–c show the transported mass of the powder in the case of the different vibration strengths. Here, the fluidized air velocity was varied within the range of Table 3. From these figures, it can be seen that the transportation amount increased with the increasing of the fluidizing air velocity at the bottom of the powder bed. Furthermore, except for the conditions of $\Lambda = 6.45$ and $u_b = 0.510$ m/s in Figure 9a, the amount of powder transportation increased almost linearly with the elapsed time. It was confirmed that the stable transportation of Al_2O_3 powder was realized using the vertical vibration and fluidization operation, as shown in Figure 9a–c. On the other hand, in the case of $u_b = 0.510$ m/s in Figure 9a, the powder transportation amount increased significantly after the start of the experiment.

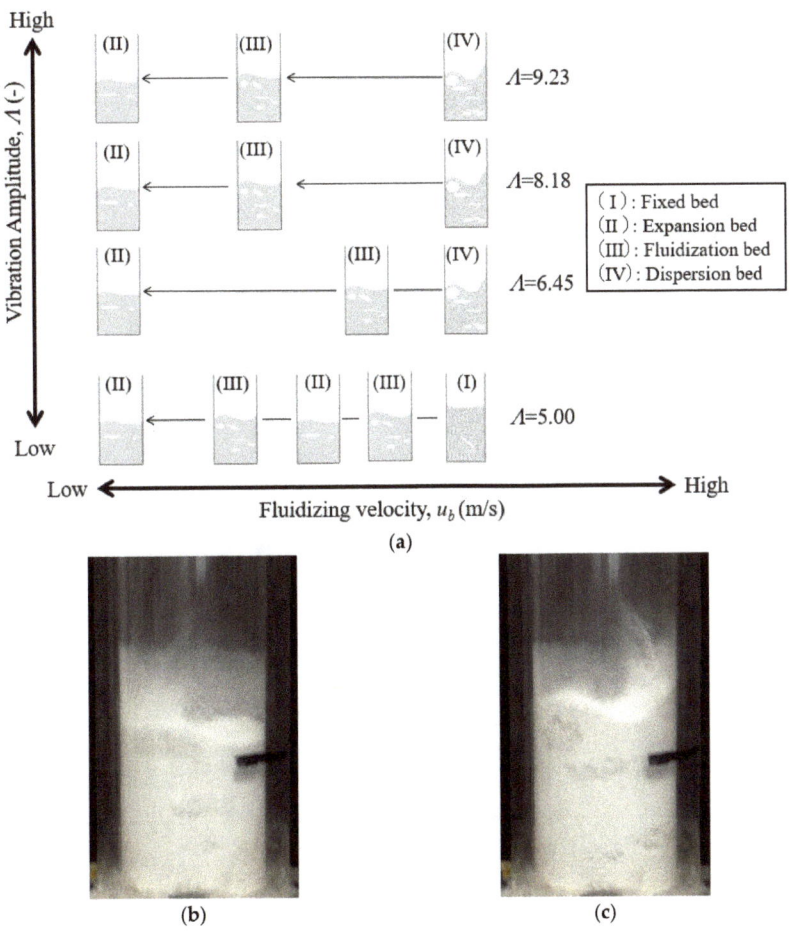

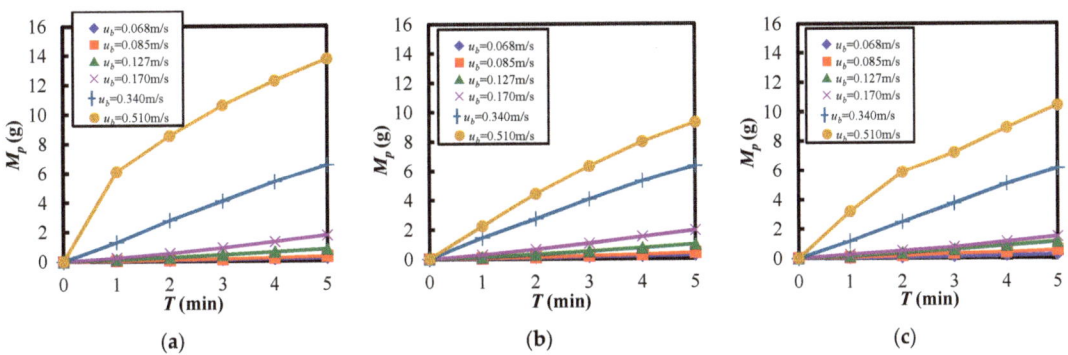

Figure 8. Classification of flow pattern and snapshots of Al_2O_3 by vertical vibrating fluidization. (**a**) Classification of flow pattern. (**b**) Fluidization bed (Region III). (**c**) Dispersion bed (Region IV).

Figure 9. Transported mass of the powder against the elapsed time in the case of different fluidizing air velocities at each vibration strength. (**a**) $\Lambda = 6.45$. (**b**) $\Lambda = 8.19$. (**c**) $\Lambda = 9.23$.

Figure 10 shows the snapshot of the powder flow in the dispersion vessel at this time. From the figure, it can be confirmed that the entire powder bed rose to the upside part of the dispersion vessel after the fluidized air was supplied. Then, the powder bed collided with the top of the powder dispersion vessel, some powder was supplied to the collector, and the remaining powder fell to the bottom of the dispersion vessel to be dispersed and supplied. From this, it was found that the quantitative supply for this study could not be realized under these conditions. Furthermore, it was suggested that the powder bed might become unstable if a high fluidized air velocity is applied under low vibration intensity. Therefore, it was confirmed that care must be taken when setting the conditions for stable transportation of the cohesive powder, such as the group C powder in the Geldart diagram.

Figure 10. Ascending phenomenon of the powder bed during the vibrating fluidization in the case of $\Lambda = 6.45$, $u_b = 0.510$ m/s.

Figure 11 shows the results of the mass flow rate of the Al_2O_3 powder G_s when the fluidized air velocity u_b and vibration strength Λ were changed. Here, the mass flow rate defined the increment of the powder transport amount per minute and was calculated as the average value for 5 min. However, when $\Lambda = 6.45$ and $u_b = 0.510$ m/s, the powder transport amount increased suddenly, as described above in Figure 9a, so the mass flow rate was calculated as an average of the data over 4 min, excluding the data for the first 1 min. The result indicated that the mass flow rate of the powder increased as the fluidizing air velocity increased when the vibration strength was constant. Furthermore, the mass flow rate of the powder took the almost same value when the fluidizing air velocity was constant, and the vibration strength was varied. This means that the vibration strength did not significantly affect the change of the mass flow rate in the experimental condition of this study. On the other hand, it was found that the fluidizing air velocity significantly affected the stable powder transportation of the cohesive Al_2O_3 powder.

Figure 12 indicates the relationship between the fluidizing air velocity u_b and the pressure difference ΔP_a inside the powder transportation device. The pressure is the differential pressure between the exit of the venturi feeder and the inlet of the cyclone in the case of no filling powder in Figure 2. The pressure difference tended to increase with the increasing of the fluidizing air velocity. In addition, the relationship between the mass flow rate G_s and the pressure difference ΔP_a was examined.

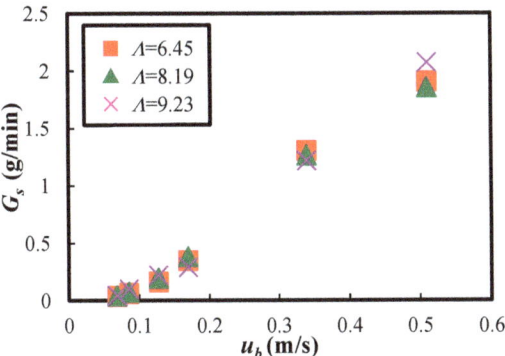

Figure 11. Relationship between the mass flow rate of the powder and the fluidizing air velocity at the bottom of the dispersion vessel when the vibration strength was changed.

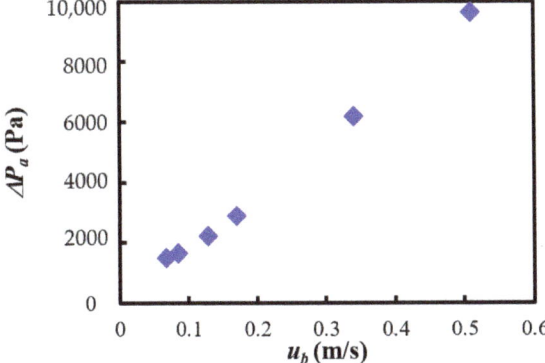

Figure 12. Pressure difference between the exit of venturi feeder and the inlet of cyclone against fluidizing air velocity.

Figure 13 shows the result of the mass flow rate and the pressure difference when the vibration strength Λ was changed. It was revealed that the mass flow rate was increased linearly with the increasing of the pressure difference between the feeder and the cyclone. Therefore, it was necessary to increase the pressure difference between the feeder and the cyclone to increase the mass flow rate of the Al_2O_3 powder. Furthermore, this device anticipated the operating conditions of the mass flow rate of 1 g/min or more. Regarding this requirement, it was necessary to maintain a pressure difference over 5000 Pa (Figure 13). Figures 11–13 show that the powder's mass flow rate was strongly related to the fluidizing velocity and the pressure. In future work, we will try to introduce a mathematical model for the powder transportation of fluidized-type powder feeders.

3.3. Dispersion Characteristics

Figure 14a–c indicate the particle size distribution in the powder dispersion vessel after the vertical vibration fluidization when the vibration strength and the fluidization velocity were changed. From all the results, it was found that the distribution had two identical peaks that were the same as the original particle diameter of Figure 1. The trend was almost similar to that observed before vibrating fluidization in that the primary particle size was also less than 1 μm, and the agglomerated particle appeared at around 10 μm. However, the distinction of the particle diameter in the case of changing experimental conditions, such as vibration strength and the fluidizing velocity, was not clear. Therefore, the median particle diameter was plotted against the fluidization velocity at each vibration strength.

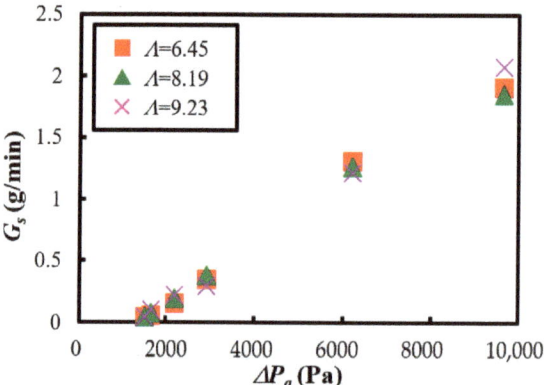

Figure 13. Relationship between the mass flow rate and the pressure difference in the case of the vibration strength Λ = 6.45~9.23.

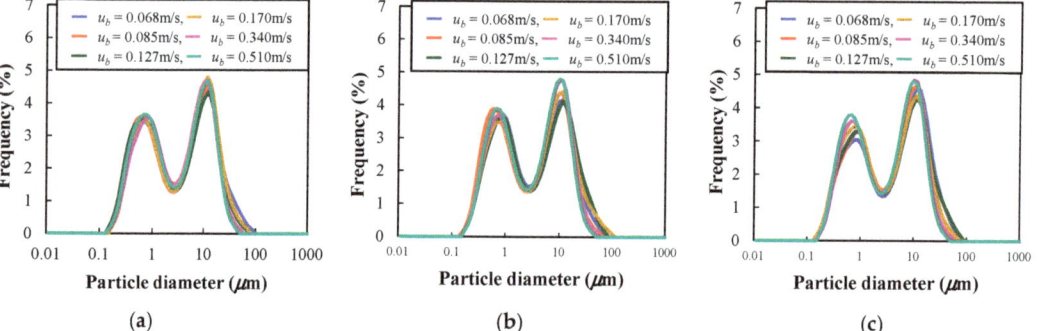

Figure 14. Particle size distribution after vertical vibrating fluidization at each vibration strength and fluidization velocity. (**a**) Λ = 6.4. (**b**) Λ = 8.19. (**c**) Λ = 9.23.

After the vibrating fluidization experiment, the particles were collected from the powder dispersion vessel and the particle diameter was measured at each experimental condition. Figure 15 shows the relationship between the median particle diameter D_{50} of Al_2O_3 and the fluidized air velocity u_b. Although there are some variations in the figure, the median particle diameter was about 3 to 5 μm. As described in Section 2.1, the median particle diameter of Al_2O_3 powder used in this study was 2.92 μm. This means that the particle diameters before and after vibrating fluidization did not change significantly. Therefore, it is suspected that the Al_2O_3 powder was well dispersed and transported to the receiving tank. However, the variations of the particle diameter became prominent in the case of the lower fluidizing air velocity. It is possible that the fluidization of the powder bed was insufficient when the fluidization air velocity was low. In addition, there was no correlation between median particle diameter and vibration strength in the range of experimental conditions in this study.

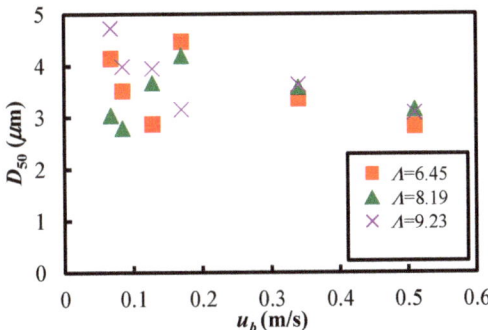

Figure 15. Relationship between the median particle diameter after the vibrating fluidization in the powder dispersion vessel and the fluidizing air velocity (Λ = 6.45~9.23).

4. Conclusions

This study evaluated the fluidization characteristics, dispersion characteristics, and transportation characteristics of Al_2O_3 powder using vertical vibration fluidization. The results obtained are shown below.

(1) It was confirmed that fluidization was difficult for the Al_2O_3 used in this study based on the measurement of the pressure drop and the visualization of the powder bed. Therefore, vibration was necessary for easy fluidization and dispersion in the case of the material used in this system.

(2) In order to evaluate the powder flow of Al_2O_3 generated by vertical vibration fluidization, the critical fluidization air velocity was defined. As a result, to start the fluidization of the powder bed of Al_2O_3, it was necessary to supply an airflow higher than the critical fluidization air velocity together with the vertical vibration. Therefore, it was important to understand the critical fluidization velocity of the materials used. In addition, it was clarified that favourable fluidization and dispersion in the powder bed could be obtained under the conditions of vibration intensity $\Lambda \geq 6.45$ and high fluidization air velocity within the range of this experiment.

(3) The powder transport experiment confirmed that the stable transportation of Al_2O_3 was realized by using vertical vibration and fluidization operations. Furthermore, it was clarified that the fluidized air velocity at the bottom of the powder dispersion vessel and the pressure difference from the powder dispersion vessel to the transportation part significantly affected the mass flow rate. On the other hand, it was found that when a high fluidized air velocity was applied under the condition of low vibration strength, the powder bed could become unstable, and thus, care must be taken when setting the operating conditions.

(4) From the particle size distribution analysis results in the particle dispersion vessel, it was found that the median particle diameter underwent no significant change before and after vibrating fluidization in the experimental conditions of this study. The results show that favourable powder flow could be obtained in the present experimental system.

Author Contributions: Conceptualization, K.O., H.K., E.M. and K.M.; methodology, K.O. and H.K.; formal analysis, investigation, data curation, K.O. and T.H.; writing—original draft preparation, K.O. and T.H.; writing—review and editing, K.T. and R.A.; visualization, K.O. and K.T.; supervision and project administration, K.O., H.K., E.M. and K.M. All authors have read and agreed to the published version of the manuscript.

Funding: This research received no external funding.

Institutional Review Board Statement: Not applicable.

Informed Consent Statement: Not applicable.

Data Availability Statement: Not applicable.

Acknowledgments: A part of this work was supported by the GEAR5.0 project for education and advanced resources at the National Institute of Technology, KOSEN.

Conflicts of Interest: The authors declare no conflict of interest.

References

1. Kuroda, S.; Watanabe, M.; Kim, K.; Katanoda, H. Current status and future prospects of warm spray technology. *J. Therm. Spray Technol.* **2011**, *20*, 653–676. [CrossRef]
2. Matikainen, V.; Koivuluoto, H.; Vuoristo, P. A study of Cr3C2-based HVOF- and HVAF-sprayed coatings: Abrasion, dry particle erosion and cavitation erosion resistance. *Wear* **2020**, *446–447*, 203188. [CrossRef]
3. Bolelli, G.; Bursi, M.; Lusvarghi, L.; Manfredini, T.; Matikainen, V.; Rigon, R.; Sassatelli, P.; Vuoristo, P. Tribology of FeVCrC coatings deposited by HVOF and HVAF thermal spray process. *Wear* **2018**, *394–395*, 113–133. [CrossRef]
4. Baiamonte, L.; Bjorklund, S.; Mulone, A.; Klement, U.; Joshi, S. Carbide-laden coatings deposited using a hand-held high-velocity air-fuel (HVAF) spray gun. *Surf. Coat. Technol.* **2021**, *406*, 126725. [CrossRef]
5. Govande, A.R.; Chandak, A.; Sunil, B.R.; Dumpala, R. Carbide-based thermal spray coatings: A review on performance characteristics and post-treatment. *Int. J. Refract. Met. Hard Mater.* **2022**, *103*, 105772. [CrossRef]
6. Sakaki, K. The progress state and problem of in new thermal spray technology "Cold Spray". *Surface Technol.* **2012**, *63*, 541–547. (In Japanese)
7. Stoltenhoff, T.; Borchers, C.; Gartner, F.; Kreye, H. Microstructures and key properties of cold-sprayed and thermally sprayed copper coatings. *Surf. Coat. Technol.* **2006**, *200*, 4947–4960. [CrossRef]
8. Gartner, F.; Stoltenhoff, T.; Voyer, J.; Kreye, H.; Riekehr, S.; Kocak, M. Mechanical properties of cold-spray and thermal sprayed cooper coatings. *Surf. Coat. Technol.* **2006**, *200*, 6770–6782. [CrossRef]
9. Sun, J.; Yamanaka, K.; Zhou, S.; Saito, H.; Ichikawa, Y.; Ogawa, K.; Chiba, A. Adhesion mechanism of cold-sprayed Sn coatings on carbon fiber reinforced plastics. *Appl. Surf. Sci.* **2022**, *579*, 151873. [CrossRef]
10. Poza, P.; Garrido-Maneiro, M.A. Cold-sprayed coatings: Microstructure, mechanical properties, and wear behaviour. *Prog. Mater. Sci.* **2022**, *123*, 100839. [CrossRef]
11. Gwalani, B.; Song, M.; Silverstein, J.; Escobar, J.; Wang, T.; Pole, M.; Johnson, K.; Jasthi, B.K.; Devaraj, A.; Ross, K. Thermal stability and mechanical properties of cold sprayed Nickel-Yttria coating. *Scr. Mater.* **2022**, *207*, 114281. [CrossRef]
12. Akedo, J. Aerosol deposition of ceramic thick films at room temperature: Densification mechanism of ceramic layers. *J. Am. Ceram. Soc.* **2006**, *89*, 1834–1839. [CrossRef]
13. Kim, I.-S.; Ko, P.-J.; Cho, M.-Y.; Lee, Y.-S.; Sohn, H.; Park, C.; Shin, W.H.; Koo, S.-M.; Lee, D.-W.; Oh, J.-M. Fabrication of high-quality alumina coating through novel, dual-particle aerosol deposition. *Ceram. Int.* **2020**, *46*, 23686–23694. [CrossRef]
14. Yu, H.K.; Oh, S.; Choi, K.H.; Jeon, J.; Dong, X.; Lee, S.H.; Choi, J.-Y.; Akedo, J.; Park, J.H. Al$_2$O$_3$ coated glass by aerosol deposition with excellent mechanical properties for mobile electronic displays. *Ceram. Int.* **2021**, *47*, 30531–30535. [CrossRef]
15. Mitani, K.; Mitani, E. HVAF thermal spraying equipment for low-temperature spraying. JP 2017, 2017-8394, 1–10.
16. Mitani, K.; Mitani, E.; Kawahara, H. Combustion mechanism that enables ignition and flame stability in HVAF spraying equipment. JP 2020, 2020-29583, 1–10.
17. Kawahara, H.; Furukawa, K.; Ogata, H.; Mitani, E.; Mitani, K. Experimental study on the stabilization mechanism of diffusion flames in a curved impinging spray combustion field in a narrow region. *Energies* **2021**, *14*, 7171. [CrossRef]
18. Viscusi, A.; Ammendola, P.; Astarita, A.; Raganati, F.; Scherillo, F.; Squillance, A.; Chirone, R.; Carrino, L. Aluminum foam made via a new method based on cold gas dynamic sprayed powders mixed through sound assisted fluidization technique. *J. Mater. Processing Technol.* **2016**, *231*, 265–276. [CrossRef]
19. Mawatari, Y.; Tatemoto, Y.; Noda, K. Prediction of minimum fluidization velocity for vibrated fluidized bed. *Powder Technol.* **2003**, *131*, 66–70. [CrossRef]
20. Mawatari, Y.; Tsunekawa, M.; Tatemoto, Y.; Noda, K. Favorable vibrated fluidization conditions for cohesive fine particles. *Powder Technol.* **2005**, *154*, 54–60. [CrossRef]
21. Hakim, L.F.; Portman, J.L.; Casper, M.D.; Weimer, A.W. Aggregation behavior of nanoparticles in fluidized beds. *Powder Technol.* **2005**, *160*, 149–160. [CrossRef]
22. Barletta, D.; Donsi, G.; Ferrari, G.; Poletto, M.; Russo, P. The effect of mechanical vibration on gas fluidization of a fine aeratable powder. *Chem. Eng. Res. Des.* **2008**, *86*, 359–369. [CrossRef]
23. Barletta, D.; Poletto, M. Aggregation phenomena in fluidization of cohesive powders assisted by mechanical vibrations. *Powder Technol.* **2012**, *225*, 93–100. [CrossRef]
24. Cano-Pleite, E.; Shimizu, Y.; Acosta-Iborra, A.; Mawatari, Y. Effect of vertical vibration and particle size on the solids hold-up and mean bubble behavior in a pseudo-2D fluidized bed. *Chem. Eng. J.* **2016**, *304*, 384–398. [CrossRef]
25. Kahrizsangi, H.S.; Sofia, D.; Barletta, D.; Poletto, M. Dust generation in vibrated cohesive powders. *Chem. Eng. Trans.* **2015**, *43*, 769–774.
26. Salehi, H.; Lotrecchiano, N.; Barletta, D.; Poletto, M. Dust release from aggregative cohesive powders subjected to vibration. *Ind. Eng. Chem. Res.* **2017**, *56*, 12326–12336. [CrossRef]

27. Geldart, D. Type of gas fluidization. *Powder Technol.* **1973**, *7*, 285–292. [CrossRef]
28. Carr, R.L. Evaluating flow properties of solids. *Chem. Eng.* **1965**, *72*, 163–168.
29. Carr, R.L. Classifying flow properties of solids. *Chem. Eng.* **1965**, *72*, 69–72.
30. Carr, R.L. Properties of solids. *Chem. Eng.* **1969**, *13*, 7–16.

Article

Mechanical and Lubrication Properties of Double Network Ion Gels Obtained by a One-Step Process

Hiroyuki Arafune *[ ], Yuma Watarai, Toshio Kamijo, Saika Honma and Takaya Sato *

Department of Creative Engineering, National Institute of Technology, Tsuruoka College, 104 Sawada Ino-oka, Tsuruoka 997-8511, Yamagata, Japan; s180018@edu.tsuruoka-nct.ac.jp (Y.W.); kamijo@tsuruoka-nct.ac.jp (T.K.); saika@tsuruoka-nct.ac.jp (S.H.)
* Correspondence: harafune@tsuruoka-nct.ac.jp (H.A.); tsato@kosen-k.go.jp (T.S.); Tel.: +81-235-25-9054 (H.A.)

Abstract: Human joints support us to reduce the impact on our body and move them smoothly. As they are composed of gel-like structures, gel materials with soft and resilient properties are expected, as lubricants, to provide high efficiency and a long lifetime for mechanical parts. While double network gels including ionic liquids as swelling agents possess high mechanical strength and stable low friction under high temperature or vacuum, their fabrication process is complex and time-consuming. In this study, we applied one-pot synthesis to a double network ion gel (DNIG) to obtain a thin gel film by a simple coating method and examined its thermal, mechanical and tribological properties. The DNIG was obtained by one-pot synthesis (DNIG-1) combining polycondensation of tetraethoxysilane and radical polymerization of methyl methacrylate to form silica and poly(methyl methacrylate) as a 1st and 2nd network, respectively. Such obtained DNIG-1 was characterized and compared with DNIG obtained by a conventional two-step process (DNIG-2). Thermogravimetric analysis and the compressive stress–strain test showed high thermal stability and mechanical strength of DNIG-1. As friction at the glass/DNIG-1 interface showed high friction compared with that at glass/DNIG-2, various counterface materials were applied to examine their effect on the friction of DNIG-1. As SUS304/DNIG-1 showed much lower friction compared with glass/DNIG-1, the difference in the friction was presumably due to the different adsorption forces and compatibility between the materials.

Keywords: ionic liquid; double network gel; one-pot synthesis; tribology

1. Introduction

Friction is a ubiquitous phenomenon that is often seen in everyday life. After getting up in the morning, we rub the sleep from our eyes, eat breakfast and brush our teeth, where friction is included in all of these motions. In these motions, human cartilages support us to reduce the impact on human joints and move smoothly [1]. They are composed of gel-like structures with collagen fiber backbones and macromolecules, proteins and phospholipids, and can maintain low friction coefficients in the range 0.001–0.01 under tens of MPa for several tens of years without replacement [2]. Gel materials have fascinated the researchers of tribology who study the principle and application of friction, lubrication and wear to invent novel lubricants with soft and resilient properties [3–5]. Such development of soft and resilient tribomaterials (SRT materials [6]) is distinguished from the conventional process to obtain hard materials such as diamond-like carbon (DLC). Hard coatings can reduce friction based on the idea of reducing real contact area; furthermore, such a hard surface has the potential risk to hurt contact materials and induce severe wear and increase friction immediately [7]. Thus, SRT materials are expected, as a novel lubricant strategy, to provide high efficiency and a long lifetime for mechanical parts where low friction is maintained under high pressure without abrasive wear of counterface materials.

Double network hydrogels (DN hydrogels) based on a brittle 1st gel network and resilient 2nd gel network possess both high solvent content (mostly >80%), high mechanical strength and low friction, whose physicochemical properties can be tuned by selecting an appropriate polymer backbone with various properties [8]. Based on their high mechanical strength and toughness [9], high water content and lubricious surface [10], they are expected to be used as artificial biomaterials [11]. However, as they are easily dried, losing their surface characteristics, under harsh conditions such as long time exposure, high temperature or vacuum, their industrial application is still limited without substituting water for a hardly volatile solvent [12].

One effective approach to overcome such disadvantages of hydrogels is to substitute water with ionic liquids (ILs) to obtain ion gels [13]. ILs are molten salts wholly composed of cations and anions whose melting temperatures are lower than 100 °C [14]. Thanks to the high ionic conductivity, thermal stability and negligible volatility of ILs, ion gels have attracted much attention for various research fields such as electrochemical devices, lubricants, CO_2 separation membrane, and so on [15–17]. Our research group has developed double network ion gels (DNIGs) where the polymer backbone and swelling agent are composed of ILs. We applied an IL, *N,N*-diethyl-*N*-(2-methoxyethyl)-*N*-methylammonium bis (trifluoromethylsulfonyl)amide (DEME-TFSA) as a swelling agent and lubricant, IL-type polymer composed of the derivative of DEME-TFSA with a polymerizable moiety, *N,N*-diethyl-*N*-(2-methacryloylethyl)-*N*-methylammonium bis (trifluoromethylsulfonyl)amide (DEMM-TFSA) as a 1st network and methyl methacrylate (MMA) as a 2nd network. DNIG showed not only high compressive fracture stress (30 MPa) with high solvent content (85 wt%), but also could maintain low friction under high temperature (80 °C) or vacuum (2.4×10^{-4} Pa) [18]. These results showed that DNIGs can be applied as robust gel lubricants under harsh conditions where DN hydrogels are difficult. For example, as a conventional rubber seal scarifies lubrication property instead of its sealing properties, DNIGs are expected to be utilized as a novel sealing gel lubricant possessing both sealing and lubrication properties under high temperature or vacuum, which is applicable in outer space.

Despite these advantages, the fabrication process of DNIG is complex and time-consuming because the conventional stepwise photoradical polymerization process was adapted. In this process, the precursor solution of the 1st network is polymerized by UV irradiation or heat to obtain the 1st gel. Second, the obtained 1st gel is immersed into a precursor solution of the 2nd network until it reaches swelling equilibrium, followed by UV irradiation or heat to obtain a DN gel. The swelling speed of the 1st gel is determined by the cooperative diffusion and its time length is proportional to the square of its thickness. Thus, tensile stress is induced by immediate swelling in the case of thin gel film, leading to breaking itself. Songmio et al. fabricated ultrathin DN hydrogels of poly (2-acrylamide-2-methylpropane sulfonic acid)/polyacrylamide (PAMPS/PAAm) of ~30 μm by controlling swelling behavior using the salt effect and the pre-reinforced technique [19,20]. They immersed the thin PAMPS gel into a precursor solution of AAm containing 0.08 M NaCl to avoid fracture of the thin PAMPS gel film by reducing osmotic pressure. Such obtained PAMPS gel was pre-reinforced with PAAm and can be fully stretched in AAm solution without salt to obtain a thin film of PAMPS/PAAm hydrogel. However, the complexity of the fabrication process remains in such a multi-step approach.

For the industrial application of DNIG, the development of a one-pot synthesis process is effective due to its simplicity and low cost. Such obtained gel film lubricant can easily be attached to mechanical parts by a simple fixation method. This process opens up the way to fabricate thin gel films by a simple coating method, which possesses both lubricity and sealing performance. However, radical polymerization of both 1st and 2nd networks at the same time results in the formation of random polymers, losing characteristics of the DNIG. Recently, Kamio et al. reported the one-pot synthesis of DNIG by combining polycondensation of tetraethoxysilane (TEOS) and radical polymerization of dimethyl acrylamide. Such independent reaction prevents the reaction between the 1st and 2nd monomer, which led to the one-pot synthesis of DNIG [21]. In this case, a brittle inorganic network of silica particles contributed to the 1st network to dissipate loaded energy and gave high fracture energy for DNIG with an inorganic/organic network of TEOS and poly (dimethyl acrylamide). Such a simple one-pot process opened up the way to form free-shapeable robust ion gels, which are expected to be utilized as novel sealing gel lubricants with both sealing and lubrication properties. However, as they studied the mechanical properties and showed the robustness of DNIGs obtained by one-pot synthesis, their tribological properties are still unclear.

In this study, one-pot synthesis of DNIG combining polycondensation of TEOS and radical polymerization of MMA was examined to obtain DNIG with silica as the 1st network and poly (methyl methacrylate) (PMMA) as the 2nd network. As DNIG in a previous study composed of poly (DEMM-TFSA) and PMMA showed high mechanical strength and low friction, we also utilized PMMA to obtain such properties in this study. The thermal, mechanical and tribological properties of DNIG obtained by one-pot synthesis were obtained and compared with conventional DNIG. In the case of tribological characterization, the effect of counterface materials was also studied to examine the performance of DNIG as a gel lubricant.

2. Materials and Methods

2.1. Fabrication of DNIGs

One-pot synthesis of DNIG was performed by combining polycondensation and thermal radical polymerization. Acetonitrile was used as received from Kanto Chemical Co. Ltd. (Tokyo, Japan) N-(2-methoxyethyl)-N-methylpyrrolidiniium trifluoromethylsulfonylamide (MEMP-TFSA) was used as received from Nisshinbo HD. 2,2'-azobis(isobutyronitrile) (AIBN), formic acid, tetraethoxysilane (TEOS) and triethylene glycol dimethacrylate (TEGDMA) were used as received from Fujifilm Wako pure chemical corporation. Methyl methacrylate was purchased from Nakalai Tesque and purified to remove the polymerization inhibitor. One-pot synthesis of double network ion gels was performed by combining polycondensation and thermal radical polymerization. TEOS (8.0 wt%) as a 1st monomer, formic acid (6.5 wt%) for catalysis of the 1st network formation, MMA (22 wt%) as a 2nd monomer, AIBN (0.37 wt%) as an initiator, TEGDMA (0.87 wt%) as a crosslinker, MEMP-TFSA (57 wt%) and acetonitrile (5.5 wt%) were mixed and charged into a Schlenk tube, followed by argon bubbling for 5 min to deoxygenate the solution. The solution was poured into a reaction cell made from a pair of glass substrates separated by 0.2-mm-thick silicone rubber by using an injection syringe. The sample solution was heated at 50 °C under an Ar atmosphere for 24 h and then heated at 50 °C under vacuum for 24 h. Such thermal process was effective to promote polycondensation of TEOS and thermal radical polymerization of MMA to form the 1st network (silica) and the 2nd network (polymethyl methacrylate, PMMA) to form DNIG by a one-step process (DNIG-1).

As a control experiment, a single network gel of TEOS or MMA was also synthesized. A precursor solution of TEOS gel was obtained by mixing TEOS (19 wt%), MEMP-TFSA (58 wt%), formic acid (16 wt%) and acetonitrile (7 wt%). A precursor solution of MMA gel was obtained by mixing MMA (43 wt%), MEMP-TFSA (49 wt%), AIBN (0.73 wt%), TEGDMA (1.6 wt%) and acetonitrile (5.5 wt%). Both solutions were heated in the same manner as above. Typical stepwise radical polymerization of DNIG was also performed to obtain a DNIG formed by a two-step process (DNIG-2) shown in our previous paper [19].

2.2. Characterization of DNIG

Surface observation and elemental analysis of DNIG were performed by using a scanning electron microscope (SEM; JSM-7100F, JEOL, Tokyo, Japan) with energy-dispersive X-ray spectroscopy (EDX; JED-2300, JEOL). The thermal stability of DNIG was examined by thermogravimetric analysis (TG8120, Rigaku Co., Tokyo, Japan) under a nitrogen atmosphere to measure the weight loss from room temperature to 500 °C with a heating rate of 10 °C/min. We also examined the stability of DNIG against moisture. DNIG was placed in a Petri dish and then floated on the water bath in a closed environment. The water bath was heated at 50 °C for 48 h to provide moisture and water uptake was measured by thermogravimetric analysis. The mechanical strength of DNIG was examined through a compressive stress–strain test by using a universal testing system (Instron 3342, Instron). We set a cylindrical gel sample of 2 mm thickness and 4 mm diameter on the lower anvil to compress the upper anvil at 10%strain/min. Tribological properties of DNIG were examined by using a ball-on-plate type reciprocating tribometer (Tribogear type-38, Shinto Scientific Co. Ltd., Tokyo, Japan). A 10 mm ϕ ball sample was set in the upper holder connected to a load cell, and the sample gel substrate was fixed on a lower sliding stage. MEMP-TFSA was dropped onto the sample gel as a liquid lubricant and the friction force between a tribopair of ball and gel sample described as ball/sample was measured under 0.98 N at 0.5~50 mm/s to evaluate the lubrication properties. Glass, SUS304 and poly (tetrafluoroethylene) (PTFE) were chosen as counterface materials.

3. Results and Discussion

3.1. Physicochemical Properties of DNIG

3.1.1. Structural and Chemical Analysis of DNIG

Figure 1 shows the chemical composition of DNIG-1 and its photograph. DNIG-1 could be easily fabricated as a thin film (thickness: 200 μm), which showed that the one-pot synthesis process can be applied as a simple coating process with a thin film of lubricant gel on the various surface. It should be noted that such a thin film cannot easily be obtained by the conventional stepwise radical polymerization process, where immediate swelling of the thin 1st gel film in the precursor solution of the 2nd network leads to the collapse of the gel itself.

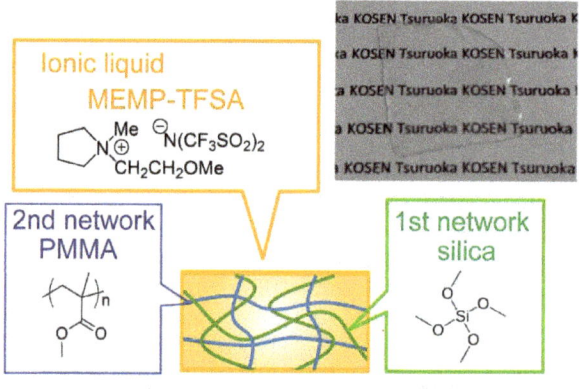

Figure 1. Schematic illustration of DNIG obtained by one-pot synthesis. A photograph of DNIG thin film (thickness: 200 μm) was also inserted in the upper right.

We next applied SEM and EDX to characterize the chemical composition of DNIG-1 (Figure 2). EDX mapping data show elements based on the 1st network silica (Si, O), 2nd network PMMA (C, O) and ionic liquid MEMP-TFSI (C, N, O, F, S). Although quantitative analysis of these components is difficult due to the overlap of elements, these mapping data support that each component was distributed evenly in DNIG.

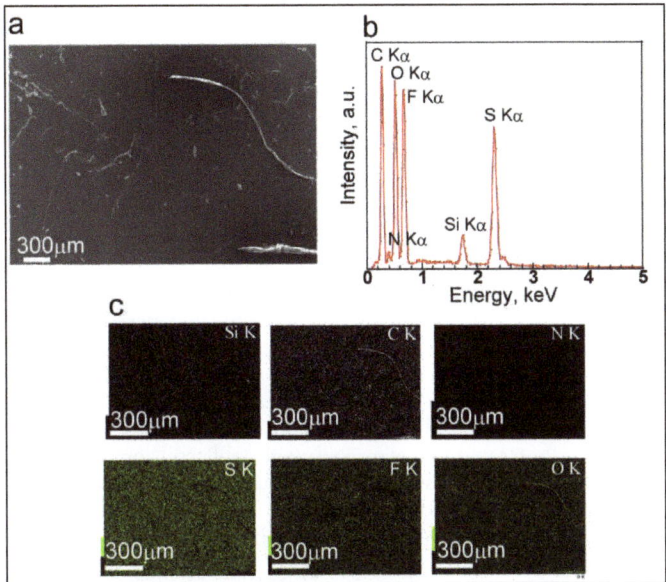

Figure 2. Surface SEM image (**a**), EDX spectra (**b**) and EDX mapping analysis data (**c**).

3.1.2. Thermal Stability of DNIG

We examined the thermal stability of DNIG-1 from TGA curves and compared it with conventional lubricant, poly α olefine (PAO). Figure 3 shows the TGA curves of DNIG-1 and PAO. DNIG-1 did not show thermal degradation until 300 °C. The temperatures of 5% weight loss (T_5) of DNIG-1 and PAO were 330 °C and 229 °C, respectively. Since DNIG-1 showed higher T_5, DNIG-1 is expected to be an efficient gel lubricant at high temperatures. The obtained T_5 of DNIG-1 was relatively higher than that of DNIG-2 (320 °C) [18], probably due to the higher thermal stability of the silica network compared with the 1st network of DNIG-2 (ionic liquid-type polymer). In addition, DNIG-1 also showed high resistance to moisture where the TGA curve did not show any change after exposure to 100% humidity for 48 h due to the hydrophobic property of MEMP-TFSA. These data show that DNIG-1 is expected to be utilized as a gel lubricant stable under high temperature or long time exposure where conventional oil lubricants cannot be applied.

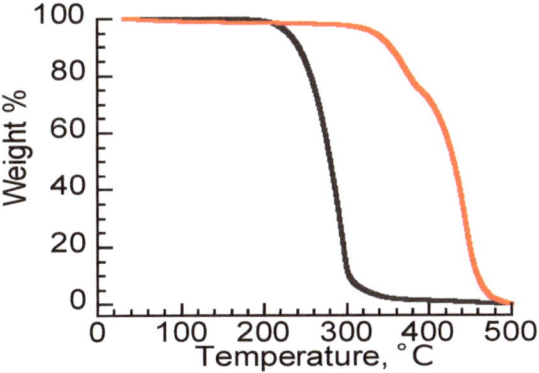

Figure 3. TGA curves of PAO (black line) and DNIG-1 (red line).

3.1.3. Mechanical Property of DNIG-1

Compressive stress–strain curves were obtained to evaluate the mechanical properties of DNIG-1 (Figure 4). The fracture stress of DNIG-1(41 MPa) was much higher than that of single network gels obtained from TEOS (0.04 MPa) or MMA (10 MPa). It should be noted that the obtained value was higher than that of conventional DNIG-2 (30 MPa) obtained by the 2-step method. Kamio et al. showed that the mechanical strength of DNIG-1 is dominated by the microstructure of the 1st and 2nd networks [21]. When a silica nanoparticle network is synthesized before the 2nd polymer network, a spatially continuous silica nanoparticle network is formed to obtain DN structure. On the other hand, the fast formation of the 2nd polymer network works as a large diffusion barrier for the 1st network where silica nanoparticles form spatially dispersed clusters, resulting in μ-DN structure. They examined controlling polymerization rates of silica and PMMA networks by changing reaction temperature or radical initiator. As they examined the formation time of the 1st and 2nd networks by dynamic viscoelastic properties, the 1st network formed faster than the 2nd network when they applied a lower temperature with the AIBN initiator, which resulted in the formation of DN structure. Though it was difficult to compare directly the microstructure of DNIG with a different component, the obtained DNIG-1 in this study was supposed to possess DN structure since it showed higher mechanical strength than that of conventional DNIG-2.

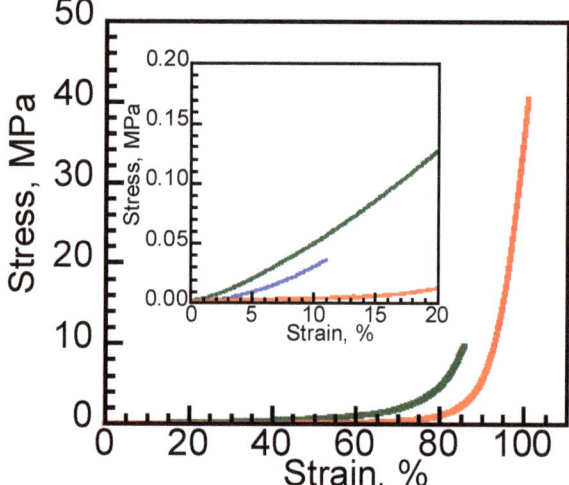

Figure 4. Compressive stress–strain curves of DNIG-1 (red line) and single network ion gel of silica (blue line) and PMMA (green line).

3.1.4. Lubrication Property of DNIG

The lubrication properties of DNIG-1 and DNIG-2 were compared by sliding speed dependence of the coefficient of friction (COF). Both DNIG-1 and DNIG-2 show a decrease in COF as the sliding speed decreased under the load of 0.98 N (Figure 5). Such behaviors represent the lubrication regime of elastic or hydrodynamic lubrication regime in soft materials [13]. In the case of hydrodynamic lubrication, counterface materials are in contact so then viscous resistance is the dominant factor, where COF is constant under the same bearing characteristic number. However, COF from 50 to 10 mm/s of DNIG-1 was higher compared with that of DNIG-2 with the same lubricants, indicating that the lubrication regime of these gels is in the elastic lubrication regime.

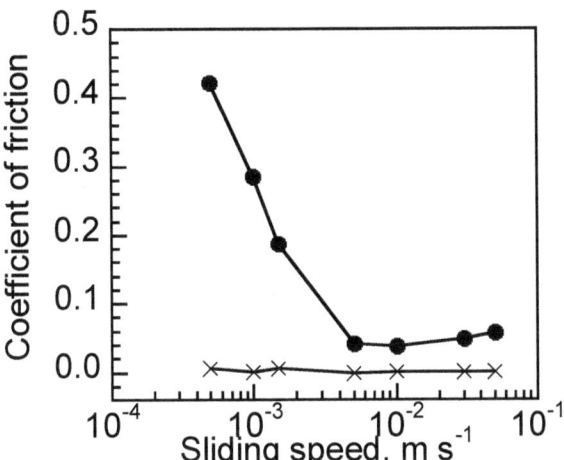

Figure 5. Sliding speed dependency of COF measured at glass/DNIG-1 (circle) and glass/DNIG-2 (triangle).

The COF increased immediately when the sliding speed became lower than 5.0 mm/s, indicating that they are in the mixed lubrication regime where glass ball and DNIG were in partial contact. Lubrication behavior in the mixed lubrication regime is dominated by the ratio of viscous resistance and solid contact. DNIG-1 showed higher friction especially at slower sliding speeds, indicating that boundary lubrication between glass/DNIG-1 was much higher than that of glass/DNIG-2. As the 1st network derived from TEOS and glass ball are both composed of a SiO_2 network, such high friction was supposed to be induced by high adsorption force due to the equal chemical potential of the same metal. We, therefore, changed the counterface material to examine the effect of the chemical composition of counterface materials on friction.

Figure 6 shows the sliding speed dependency of COF at DNIG-1 and various counterface materials. While the COF of glass/DNIG-1 and SUS304/DNIG-1 showed similar values at a sliding speed of 50~5.0 mm/s, a drastic reduction in COF of SUS304/DNIG-1 was observed at a lower sliding speed (1.5~0.5 mm/s) compared with that of glass/DNIG-1. Such reduction of the COF supports the theory that the same metal of the tribopair resulted in high friction due to the higher adsorption force. In contrast, PTFE/DNIG-1 showed less dependent behavior against sliding speed, where COF was similar in the measured sliding speed range. Such behavior is typical in the case of the boundary lubrication regime, where liquid film between the contact surfaces is thin and solid contact was dominant in friction. As PTFE showed high contact angle to MEMP-TFSA compared with other counterface materials, such behavior was supposed to be based on repellent PTFE surface against MEMP-TFSA. Such a repellent surface decreases the thickness of the liquid film of lubricant to induce the shift from a mixed to boundary lubrication regime, which resulted in high friction at a wide sliding speed range due to the increase of solid contact. Additionally, PTFE/DNIG-1 showed lower friction at 0.5 mm/s compared to glass/DNIG-1, indicating that boundary lubrication was lower in the case of PTFE/DNIG-1. The obtained result was supposed to be due to the self-lubrication effect of PTFE where worn particles of PTFE attached to the DNIG-1 surface to form the PTFE layer, inducing low friction of PTFE/PTFE [22].

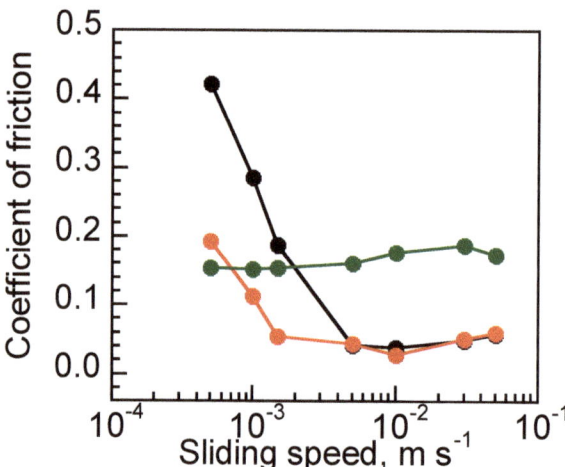

Figure 6. Sliding speed dependency of the COF measured at glass/DNIG-1 (black), SUS304/DNIG-1 (red) and PTFE/DNIG-1 (green).

4. Conclusions

One-pot synthesis of double network ion gel was successfully achieved by combining polycondensation of TEOS and radical polymerization of MMA. SEM and EDX mapping data of the obtained DNIG-1 supported that each component was distributed evenly in DNIG-1. TGA data show that DNIG-1 is expected to be utilized as a gel lubricant that is stable under high temperature or long time exposure where conventional oil lubricants cannot be applied. Compressive stress–strain curves showed that DNIG-1 obtained by one-pot synthesis possesses high compressive fracture stress (41 MPa) compared with that obtained by conventional two-step synthesis (30 MPa). The COF of the glass/DNIG-1 tribopair was relatively higher than that of glass/DNIG-2, probably due to the melting effect between the same metals. Change of counterface materials effectively reduced the COF at DNIG-1 which suggests the importance of the appropriate selection of tribopair. The obtained results open up the way for facile fabrication of lubricant gel with high mechanical strength and thermal stability for industrial application such as lubricant gel seal possessing both low friction and good sealing property.

Author Contributions: Conceptualization, H.A. and T.S.; Data curation, H.A. and Y.W.; Investigation, H.A., Y.W. and S.H.; Methodology, H.A., Y.W. and T.S.; Project administration, H.A. and T.S.; Resources, S.H.; Supervision, H.A., T.K. and T.S.; Writing—original draft, H.A. and T.S.; Writing—review and editing, Y.W., T.K., S.H. and T.S. All authors will be informed about each step of manuscript processing including submission, revision, revision reminder, etc. via emails from our system or assigned Assistant Editor. All authors have read and agreed to the published version of the manuscript.

Funding: This research was funded in part by the Grants-in-Aid for Scientific Research (No. 20K04249 and 20H02060).

Institutional Review Board Statement: Not applicable.

Informed Consent Statement: Not applicable.

Data Availability Statement: Data is contained within the article.

Conflicts of Interest: The authors declare no conflict of interest.

References

1. Klein, J. Hydration lubrication. *Friction* **2013**, *1*, 1–23. [CrossRef]
2. Klein, J. Molecular mechanisms of synovial joint lubrication. *Proc. Inst. Mech. Eng. Part J J. Eng. Tribol.* **2006**, *220*, 691–710. [CrossRef]
3. Gong, J.P.; Introduction, I. Friction and lubrication of hydrogels—Its richness and complexity. *Soft Matter* **2006**, *2*, 544–552. [CrossRef] [PubMed]
4. Katta, J.; Jin, Z.; Ingham, E.; Fisher, J. Biotribology of articular cartilage—A review of the recent advances. *Med. Eng. Phys.* **2008**, *30*, 1349–1363. [CrossRef] [PubMed]
5. Rudge, R.E.; Scholten, E.; Dijksman, J.A. Advances and challenges in soft tribology with applications to foods. *Curr. Opin. Food Sci.* **2019**, *27*, 90–97. [CrossRef]
6. Tsujii, Y.; Sakakibara, K.; Watanabe, H.; Kurihara, K.; Sato, T.; Nakano, K. Lubricant and SRT material. World Intellectual Property. Organization Patent WO2017171071A1, 5 October 2017.
7. Holmberg, K.; Ronkainen, H.; Matthews, A. Tribology of thin coatings. *Ceram. Int.* **2000**, *26*, 787–795. [CrossRef]
8. Haque, M.A.; Kurokawa, T.; Gong, J.P. Super tough double network hydrogels and their application as biomaterials. *Polymer* **2012**, *53*, 1805–1822. [CrossRef]
9. Ihsan, A.B.; Sun, T.L.; Kurokawa, T.; Karobi, S.N.; Nakajima, T.; Nonoyama, T.; Roy, C.K.; Luo, F.; Gong, J.P. Self-Healing Behaviors of Tough Polyampholyte Hydrogels. *Macromolecules* **2016**, *49*, 4245–4252. [CrossRef]
10. Mizukami, M.; Ren, H.Y.; Furukawa, H.; Kurihara, K. Deformation of contacting interface between polymer hydrogel and silica sphere studied by resonance shear measurement. *J. Chem. Phys.* **2018**, *149*, 163327. [CrossRef]
11. Fan, H.; Gong, J.P. Fabrication of Bioinspired Hydrogels: Challenges and Opportunities. *Macromolecules* **2020**, *53*, 2769–2782. [CrossRef]
12. Furukawa, H.; Kuwabara, R.; Tanaka, Y.; Kurokawa, T.; Na, Y.-H.; Osada, Y.; Gong, J.P. Tear Velocity Dependence of High-Strength Double Network Gels in Comparison with Fast and Slow Relaxation Modes Observed by Scanning Microscopic Light Scattering. *Macromolecules* **2008**, *41*, 7173–7178. [CrossRef]
13. Arafune, H.; Muto, F.; Kamijo, T.; Honma, S.; Morinaga, T.; Sato, T. Tribological properties of double-network gels substituted by ionic liquids. *Lubricants* **2018**, *6*, 89. [CrossRef]
14. Le Bideau, J.; Viau, L.; Vioux, A. Ionogels, ionic liquid based hybrid materials. *Chem. Soc. Rev.* **2011**, *40*, 907–925. [CrossRef] [PubMed]
15. Armand, M.; Endres, F.; MacFarlane, D.R.; Ohno, H.; Scrosati, B. Ionic-liquid materials for the electrochemical challenges of the future. *Nat. Mater.* **2009**, *8*, 621–629. [CrossRef] [PubMed]
16. Espejo, C.; Carrion, F.J.; Martinez, D.; Bermudez, M.D. Multi-walled carbon nanotube-imidazolium tosylate ionic liquid lubricant. *Tribol. Lett.* **2013**, *50*, 127–136. [CrossRef]
17. Zhang, J.; Kamio, E.; Matsuoka, A.; Nakagawa, K.; Yoshioka, T.; Matsuyama, H. Development of a Micro-Double-Network Ion Gel-Based CO_2 Separation Membrane from Nonvolatile Network Precursors. *Ind. Eng. Chem. Res.* **2021**, *60*, 12640–12649. [CrossRef]
18. Arafune, H.; Honma, S.; Morinaga, T.; Kamijo, T.; Miura, M.; Furukawa, H.; Sato, T. Highly Robust and Low Frictional Double-Network Ion Gel. *Adv. Mater. Interfaces* **2017**, *4*, 1700074. [CrossRef]
19. Liang, S.; Yu, Q.M.; Yin, H.; Wu, Z.L.; Kurokawa, T.; Gong, J.P. Ultrathin tough double network hydrogels showing adjustable muscle-like isometric force generation triggered by solvent. *Chem. Commun.* **2009**, 7518–7520. [CrossRef] [PubMed]
20. Liang, S.; Wu, Z.L.; Hu, J.; Kurokawa, T.; Yu, Q.M.; Gong, J.P. Direct Observation on the Surface Fracture of Ultrathin Film. *Macromolecules* **2011**, *44*, 3016–3020. [CrossRef]
21. Kamio, E.; Yasui, T.; Iida, Y.; Gong, J.P.; Matsuyama, H. Inorganic/Organic Double-Network Gels Containing Ionic Liquids. *Adv. Mater.* **2017**, *29*, 1704118. [CrossRef] [PubMed]
22. Puts, G.J.; Crouse, P.; Ameduri, B.M. Polytetrafluoroethylene: Synthesis and Characterization of the Original Extreme Polymer. *Chem. Rev.* **2019**, *119*, 1763–1805. [CrossRef] [PubMed]

Article

Solid-State-Activated Sintering of ZnAl$_2$O$_4$ Ceramics Containing Cu$_3$Nb$_2$O$_8$ with Superior Dielectric and Thermal Properties

Koichi Shigeno [1,*], Takuma Yano [1] and Hirotaka Fujimori [2]

[1] National Institute of Technology (KOSEN), Ube College, 2-14-1 Tokiwadai, Ube, Yamaguchi 755-8555, Japan; ntnm.894@gmail.com
[2] Graduate School of Sciences and Technology for Innovation, Yamaguchi University, 2-16-1 Tokiwadai, Ube, Yamaguchi 755-8611, Japan; hiro@hiro-fuji.net
* Correspondence: shigeno@ube-k.ac.jp; Tel./Fax: +81-836-35-5294

Abstract: Low-temperature co-fired ceramics (LTCCs) are dielectric materials that can be co-fired with Ag or Cu; however, conventional LTCC materials are mostly poorly thermally conductive, which is problematic and requires improvement. We focused on ZnAl$_2$O$_4$ (gahnite) as a base material. With its high thermal conductivity (~59 W·m^{-1}·K^{-1} reported for 0.83ZnAl$_2$O$_4$–0.17TiO$_2$), ZnAl$_2$O$_4$ is potentially more thermally conductive than Al$_2$O$_3$ (alumina); however, it sinters densely at a moderate temperature (~1500 °C). The addition of only 4 wt.% of Cu$_3$Nb$_2$O$_8$ significantly lowered the sintering temperature of ZnAl$_2$O$_4$ to 910 °C, which is lower than the melting point of silver (961 °C). The sample fired at 960 °C for 384 h exhibited a relative permittivity (ε_r) of 9.2, a quality factor by resonant frequency ($Q \times f$) value of 105,000 GHz, and a temperature coefficient of the resonant frequency (τ_f) of −56 ppm·K^{-1}. The sample exhibited a thermal conductivity of 10.1 W·m^{-1}·K^{-1}, which exceeds that of conventional LTCCs (~2–7 W·m^{-1}·K^{-1}); hence, it is a superior LTCC candidate. In addition, a mixed powder of the Cu$_3$Nb$_2$O$_8$ additive and ZnAl$_2$O$_4$ has a melting temperature that is not significantly different from that (~970 °C) of the pristine Cu$_3$Nb$_2$O$_8$ additive. The sample appears to densify in the solid state through a solid-state-activated sintering mechanism.

Keywords: ZnAl$_2$O$_4$ (gahnite); low-temperature co-fired ceramics (LTCCs); solid-state-activated sintering; microwave dielectric properties; thermal conductivity

1. Introduction

Low-temperature co-fired ceramics (LTCCs) are dielectric materials that can be co-fired with Ag or Cu, which are metals that exhibit low-resistance conduction at temperatures below their melting points (961 and 1084 °C, respectively) [1,2]. LTCCs have been widely used in small electronic devices, such as wiring substrates and integrated circuit packages for high-frequency communication. With the aim of improving energy efficiency, ultralow-temperature co-fired ceramics (ULTCCs) that can be co-fired with Al at 660 °C (below the melting point of Al) are currently being actively researched [3]. However, the poor heat-dissipation properties of LTCC materials are problematic because the temperature of semiconductors mounted on the LTCC materials rises, causing thermal runaway of these semiconductors. Aluminum-based oxide ceramics, such as Al$_2$O$_3$ (alumina), are relatively highly thermally conductive; however, a large amount (approximately 50% or more of the total) of a poorly thermally conductive low-softening-point glass needs to be added to achieve low-temperature sintering when used as the base material. Consequently, the majority of these conventional LTCC materials are poorly thermally conductive (approximately 2–7 W·m^{-1}·K^{-1}), which is a shortcoming [4,5]. The heat-generation densities of semiconductor-based electronic components, such as light-emitting diodes (LEDs), mounted on LTCC multilayer devices have recently been reported to be increasing [6,7];

hence, highly thermally conductive LTCC materials are in demand. We previously developed sintering additives for alumina using highly thermally conductive (~30 W·m^{-1}·K^{-1}) alumina as a base material that, when added in small amounts, enables alumina to be sintered at low temperatures. As a result, we developed a CuO–TiO$_2$–Nb$_2$O$_5$–Ag$_2$O additive that facilitates alumina sintering at 900 °C or less when added at 5 wt.%, thereby realizing highly thermally conductive (~18–20 W·m^{-1}·K^{-1}) low-temperature co-fired alumina (LTCA) [8,9].

Furthermore, we also focused on ZnAl$_2$O$_4$ (gahnite) as a base material; this aluminum-based oxide has good dielectric properties and is potentially more thermally conductive than alumina [10–16]. According to Surendran et al., a sample produced by firing a 0.83ZnAl$_2$O$_4$–0.17TiO$_2$ (molar ratio) composition at 1440 °C exhibited a thermal conductivity of 59 W·m^{-1}·K^{-1} [11]. Consequently, ZnAl$_2$O$_4$ is expected to be used as a novel highly thermally conductive LTCC substrate if it can be sintered at low temperature. Low-temperature sintering involving the addition of glass to ZnAl$_2$O$_4$ [14] has been studied. The use of crystalline additives other than glass has also been studied a few times [16]; however, densely sintered ZnAl$_2$O$_4$ has not been achieved below 1100 °C yet. Therefore, we aimed to develop a sintering additive that, when added in small amounts, enables ZnAl$_2$O$_4$ to be densely sintered below 1000 °C (a low temperature).

We previously found that the addition of only 5 wt.% CuO–Nb$_2$O$_5$ (in a 7:3 Cu:Nb molar ratio) led to a significantly lower sintering temperature (960 °C) [17]. The sample fired at 960 °C for 2 h exhibited a relative permittivity (ε_r) of 9.1, a quality factor by resonant frequency ($Q \times f$) value of 30,000 GHz (at a frequency of ~13 GHz), a temperature coefficient of the resonant frequency (τ_f) of −69 ppm·K^{-1}, and a thermal conductivity (κ) of 9.3 W·m^{-1}·K^{-1}, which are relatively satisfactory; however, these values can be further improved. Because the mixture of CuO and Nb$_2$O$_5$ is separated at the atomic level, we postulate that small amounts of unreacted components exacerbate the $Q \times f$ value of low-temperature sintered ZnAl$_2$O$_4$. In addition, the low-temperature sintering mechanism remains unknown. The composite oxide formed between a ZnO varistor and an Al$_2$O$_3$ substrate by low-temperature sintering has also been recently examined [7]; ZnO containing Bi$_2$O$_3$ is known to react with Al$_2$O$_3$ to produce ZnAl$_2$O$_4$ [18]. Therefore, understanding the sintering behavior of ZnAl$_2$O$_4$ from a practical perspective is also important.

This study has two objectives. The first involves further improving the dielectric properties and thermal conductivities of low-temperature sintered ZnAl$_2$O$_4$ ceramics. To achieve this goal, we examined adding small quantities of calcined Cu$_3$Nb$_2$O$_8$ [19], which has been reported to exhibit relatively good dielectric properties, instead of a mixture of CuO and Nb$_2$O$_5$ in the same molar ratio, prolonging the holding time at each firing temperature. The second involves gaining insight into the low-temperature sintering mechanism. Specifically, we examined whether densification is promoted only in the solid phase or in other phases, including the liquid phase. In addition, we examined the densification role of each additive component (Cu and Nb).

2. Materials and Methods

2.1. Fabricating Sintered Samples

Figure 1a shows the flow process used to prepare gahnite sintered bodies containing additives. First, commercially available Al$_2$O$_3$ and ZnO powders (99.99%, average particle size: 1 μm for each powder; Kojundo Chemical Laboratory Co., Ltd., Saitama, Japan) were mixed in a 1:1 molar ratio and ball-milled for 16 h, with water as the dispersion medium. The dried powder was then calcined at 1100 °C for 4 h in air to yield ZnAl$_2$O$_4$, which was then pulverized using a ball mill for 48 h, with water as the dispersion medium. The dried powder was used as the raw material, as shown in Figure 1b; its specific surface area was determined to be 8.65 m^2/g by the Brunauer–Emmett–Teller (BET) method (AUTOSORB-1, Spectris Co., Ltd., Kanagawa, Japan). According to the BET measurements and assuming particles were spherical [20], the average particle size was approximately 0.2 μm.

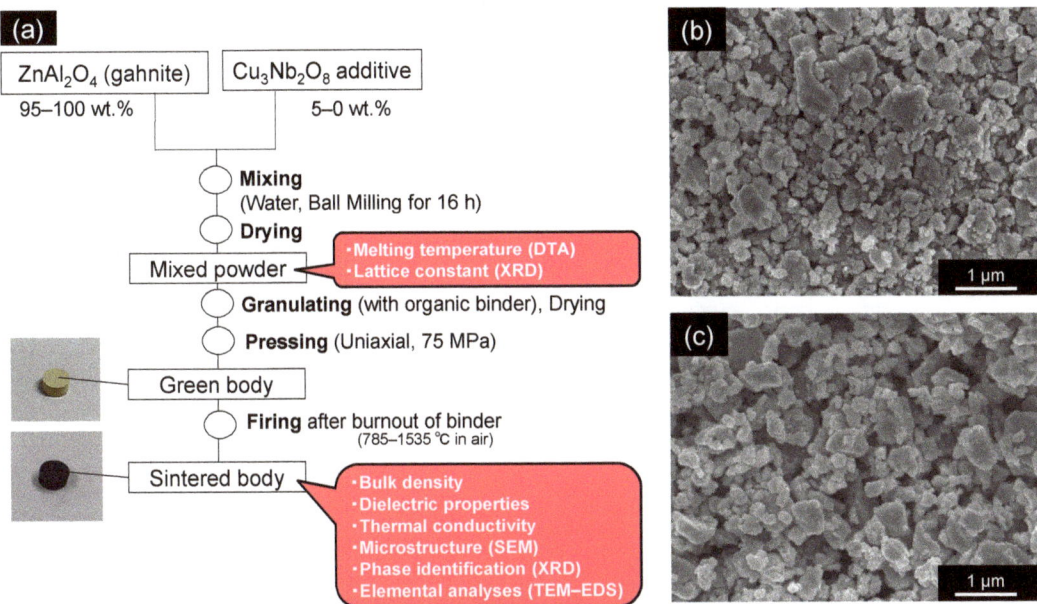

Figure 1. (a) Experimental process for preparing $ZnAl_2O_4$ ceramics containing the $Cu_3Nb_2O_8$ sintering aid. (b) Scanning electron microscopy (SEM) image of the $ZnAl_2O_4$ raw material powder. (c) SEM image of the $Cu_3Nb_2O_8$ raw material powder.

Table 1 lists the 14 main samples characterized in this study, as well as their synthesis conditions, and the purpose of each characteristic comparison. Commercially available CuO (99.3%, Nissin Chemco Ltd., Kyoto, Japan) and Nb_2O_5 (99.9%, Kojundo Chemical Laboratory Co., Ltd., Saitama, Japan) powders were mixed in a 3:2 Cu:Nb molar ratio and ball-milled for 16 h, with water as the dispersion medium. The dried powder was calcined at 835 °C for 2 h in air to yield $Cu_3Nb_2O_8$, which was pulverized using a ball mill for 48 h, with water as the dispersion medium. The dried powder was used as the additive, as shown in Figure 1c. $ZnAl_2O_4$ powder (100–95 wt.%) and the $Cu_3Nb_2O_8$ additive powder (0–5 wt.%) were mixed and ball-milled for 16 h, with water as the dispersion medium. We set the maximum additive concentration to 5 wt.% because we found that this quantity provided sufficiently well-sintered samples in our previous study [17]. For reference, ball-milling was conducted with yttria-stabilized zirconia (YSZ) balls in a polyethylene bottle at a rotation rate of 160 rpm.

The dried powder was granulated with polyvinyl alcohol (PVA, DKS Co., Ltd., Kyoto, Japan) binder and formed into a disc through uniaxial pressing at 75 MPa. The weight ratio of the dried powder–PVA binder was set to be 100:4. The green bodies were fired at 785–1535 °C in air for the required time (between 10 min and 384 h). The effects of sintering-aided calcination on sinterability and the dielectric and thermal properties were investigated. Specifically, $ZnAl_2O_4$ samples were fabricated by separately adding CuO and Nb_2O_5 (5 wt.% in total) in the same Cu:Nb molar ratio (3:2) as $Cu_3Nb_2O_8$ (referred to as "non-calcined" in Table 1). For reference, we also prepared a sintered body calcined at 885 °C for 2 h using only the $Cu_3Nb_2O_8$ additive powder. For reference, the heating and cooling rates used for calcination and firing were both set to 300 °C/h.

Table 1. ZnAl$_2$O$_4$ and additive quantities, additive conditions, firing temperatures, holding times, and comparison purposes of the main 14 samples examined in this study. Comparison purposes: *1) clarifying the effect of additive quantities for well-sintered samples; *2) clarifying the effect of firing temperature for samples of the same composition (ZnAl$_2$O$_4$ with 5 wt.% of calcined additive); *3) clarifying the effect of additive calcination; *4) clarifying the effect of prolonging holding times for samples fired at 910, 935, and 960 °C (retention times: 2 and 384 h).

Sample	ZnAl$_2$O$_4$ Quantity	Additive Quantity	Additive Conditions	Firing Temperature	Holding Time	Comparison Purpose
	wt.%			°C	h	
G01	100	0	Calcined	1485	2	*1)
G02	99	1	Calcined	1185	2	*1)
G03	98	2	Calcined	1085	2	*1)
G04	97	3	Calcined	1035	2	*1)
G05	96	4	Calcined	960	2	*1) *4)
G06	95	5	Calcined	785	2	*2)
G07	95	5	Calcined	960	2	*1) *2) *3)
G08	95	5	Non-calcined	960	2	*3)
G09	96	4	Calcined	910	2	*4)
G10	96	4	Calcined	910	384	*4)
G11	96	4	Calcined	935	2	*4)
G12	96	4	Calcined	935	384	*4)
G13	96	4	Calcined	960	384	*4)
G14	0	100	Calcined	885	2	*1)

2.2. Characterization

The properties of the prepared sintered bodies were evaluated on the basis of their ceramic densities, microwave dielectric properties, and thermal conductivities. The ceramic densities were measured using the geometrical method. The theoretical density (ρ) of each sample was calculated using the following equation:

$$\rho = \frac{W_1 + W_2}{\left(\frac{W_1}{\rho_1}\right) + \left(\frac{W_2}{\rho_2}\right)}, \quad (1)$$

where W_1 and W_2, are the weight percentages of ZnAl$_2$O$_4$ and Cu$_3$Nb$_2$O$_8$, respectively, and ρ_1 and ρ_2 are the densities of ZnAl$_2$O$_4$ (4.606 g/cm^3) and Cu$_3$Nb$_2$O$_8$ (5.655 g/cm^3), respectively [21]. Relative density was calculated by dividing the measured ceramic density by the theoretical density. Three major parameters that describe microwave dielectric properties, namely, the relative permittivity (ε_r), the quality factor by resonant frequency ($Q \times f$), and the temperature coefficient of the resonant frequency (τ_f), were measured using the Hakki–Coleman method [22] with a network analyzer (8720ES, Agilent Technologies, Santa Clara, CA, USA). τ_f values were calculated using the following equation:

$$\tau_f = \frac{1}{f(T_0)} \cdot \frac{f(T_1) - f(T_0)}{T_1 - T_0}, \quad (2)$$

where $f(T_0)$ and $f(T_1)$ are the resonant frequencies at 20 and 80 °C, respectively. Thermal conductivities were measured using the xenon flash method (LFA447, Netzsch, Selb, Germany). To discuss sinterability, as well as the microwave dielectric properties and thermal conductivities of the developed materials, we observed their microstructures by scanning electron microscopy (SEM; JSM-7600F, JEOL Ltd., Tokyo, Japan) at an accelerating voltage of 5 kV. To prepare samples for SEM, the samples were polished with 0.5 μm diamond abrasives, followed by thermal etching for 1 h at a temperature 50 °C lower than each firing temperature. The average grain size (D_g) was measured by the planimetric method [20,23] for at least 200 grains. In addition, we identified their crystalline phases via X-ray diffractometry (XRD; Ultima IV, Rigaku Co., Ltd., Tokyo, Japan) augmented with a Cu-Kα radiation source at 40 kV and 40 mA in the 2θ range of 20–60° with a step size of 0.02° and a rate of 4°·min^{-1}.

To further discuss the sintering mechanism, the melting temperatures of the mixed powders were determined by differential thermal analysis (DTA; Thermo Plus Evo II, Shimadzu Co., Ltd., Kyoto, Japan). The lattice constants of $ZnAl_2O_4$ and $Cu_3Nb_2O_8$ in the mixed powders were determined via a step scanning method of XRD augmented with a Cu-Kα radiation source at 40 kV and 30 mA in the 2θ range of 10–110° with a step size of 0.02° and a counting time of 7 s/step, with LaB_6 (SRM660b, NIST) as the internal standard. In addition, the fired samples were placed on Mo mesh and thinned by focused ion beam (FIB) milling (JEM-9320FIB, JEOL Ltd., Tokyo, Japan) with Ga ions to yield ~100 nm thick flakes; these flakes are half the average particle diameter of the raw $ZnAl_2O_4$ powder in size. For the samples thinned by FIB milling, transmission electron microscopy (TEM) at an accelerating voltage of 200 keV and energy-dispersive spectroscopy (EDS; JEM-2100, JEOL Ltd., Tokyo, Japan) were used to analyze the microstructures of the sintered samples and their elemental distributions.

3. Results and Discussion

3.1. Effect of the $Cu_3Nb_2O_8$ Additive and Its Calcination on the Properties of $ZnAl_2O_4$

We first evaluated the effect of the added $Cu_3Nb_2O_8$ and the properties of the $ZnAl_2O_4$ sample. Figure 2 shows the relationship between the firing temperature and ceramic densities of $ZnAl_2O_4$ alone, and $ZnAl_2O_4$ containing 1, 2, 3, 4, and 5 wt.% of the $Cu_3Nb_2O_8$ additive; samples were held for 2 h at each temperature. A firing temperature of 1485 °C was required for $ZnAl_2O_4$ to produce a dense sintered body in the absence of the sintering aid. On the other hand, significantly lower sintering temperatures were required for $ZnAl_2O_4$ containing the $Cu_3Nb_2O_8$ additive; that is, dense sintered bodies with relative densities of 95% or more were obtained for samples containing 4 wt.% or more of the sintering aid when fired at 960 °C for 2 h. This temperature is only 1 °C lower than the melting point of metallic silver (961 °C); therefore, these bodies do not fully meet LTCC material requirements. However, they can be improved by prolonging the holding time at a lower firing temperature (see Section 3.2).

Table 2 lists the additive quantities, additive conditions, firing temperatures, densities, average grain sizes (D_g), dielectric properties (ε_r, $Q \times f$, τ_f), and thermal conductivities (κ) of the main 14 samples (G01–14) examined in this study. For reference, typical data of a conventional LTCC (Al_2O_3 + glass) substrate are shown as sample R01. In addition, Figure 3 shows SEM images of polished surfaces that depict the microstructures of the sintered bodies. Figure 3a shows that the $ZnAl_2O_4$ sample fired at 1485 °C for 2 h (G01) contained grains approximately 1–2 μm in size (D_g: 2.10 μm), which are larger than the grains of the raw powder (Figure 1b). However, submicron-sized voids were observed at various locations. Figure 3b,c show $ZnAl_2O_4$ sintered bodies containing 5 wt.% of the $Cu_3Nb_2O_8$ additive. The sample fired at 785 °C for 2 h (G06) exhibited almost no necking between particles, with many voids present (Figure 3b). On the other hand, the sample fired at 960 °C for 2 h (G07) contained angular particles and few voids. The grains in Figure 3c (D_g: 0.55 μm) are larger than those in Figure 3b (D_g: 0.20 μm); however, they are still submicron in size. We conclude that the sample fired at 960 °C was densified. In addition, TEM–EDS (Figure 4) revealed that the Cu and Nb of the additive segregated at the grain boundaries; however, Cu was detected inside the $ZnAl_2O_4$ grains in some places.

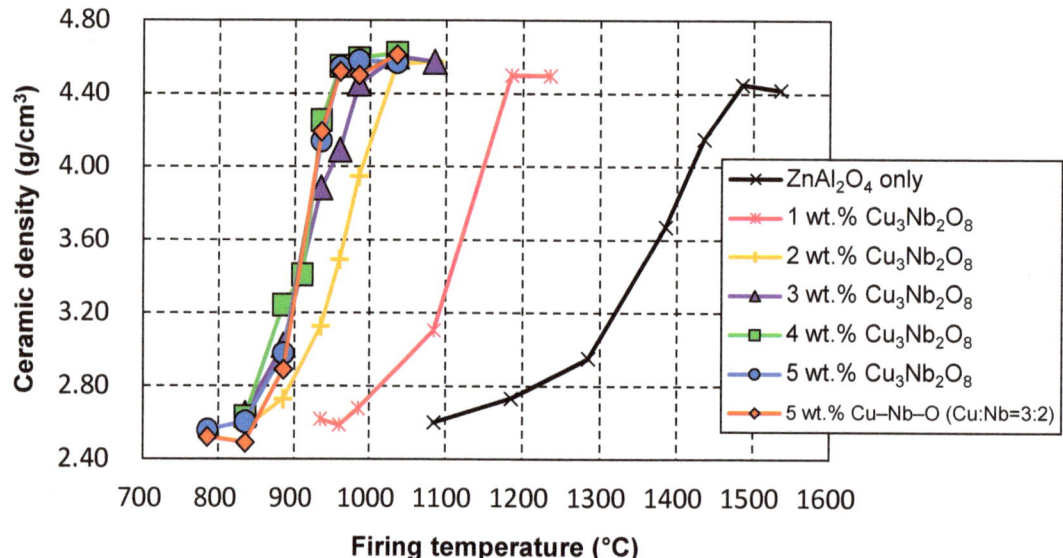

Figure 2. Ceramic densities of pristine $ZnAl_2O_4$, $ZnAl_2O_4$ containing 1, 2, 3, 4, and 5 wt.% of the $Cu_3Nb_2O_8$ additive, and $ZnAl_2O_4$ containing 5 wt.% of the $CuO–Nb_2O_5$ (Cu:Nb = 3:2 (molar ratio)) additive as functions of firing temperature. The holding time at each temperature: 2 h.

Table 2. Additive quantities and conditions, firing temperatures, densities, average grain sizes (D_g), dielectric properties (ε_r, $Q \times f$, τ_f), and thermal conductivities (κ) of the samples examined in this study; "×" indicates that the values of ε_r and $Q \times f$ of the sample (G06) were unable to be determined due to their significantly low values; "××" indicates that the values of τ_f and κ for the samples (G06, G09, and G11) with insufficient ceramic densities (relative densities less than 95%) were not measured. R01* is an example of a conventional LTCC substrate.

Sample	Additive Quantity	Additive Conditions	Firing Temp.	Holding Time	Ceramic Density	Relative Density	D_g	ε_r	$Q \times f$	τ_f	κ
	wt.%		°C	h	g/cm³	%	μm	—	GHz	ppm·K⁻¹	W·m⁻¹·K⁻¹
G01	0	Calcined	1485	2	4.47	97.0	2.10	9.0	16,100	−73	27.3
G02	1	Calcined	1185	2	4.50	97.5	3.10	9.0	98,200	−63	12.2
G03	2	Calcined	1085	2	4.57	98.9	1.36	9.1	81,700	−60	10.2
G04	3	Calcined	1035	2	4.60	99.4	0.89	9.3	77,000	−60	8.8
G05	4	Calcined	960	2	4.55	98.1	0.53	9.1	65,300	−54	8.9
G06	5	Calcined	785	2	2.56	55.0	0.20	×	×	××	××
G07	5	Calcined	960	2	4.55	97.8	0.55	9.2	60,800	−56	7.9
G08	5	Non-calcined	960	2	4.52	97.2	0.57	9.0	51,300	−63	8.5
G09	4	Calcined	910	2	3.41	73.5	0.30	6.0	2480	××	××
G10	4	Calcined	910	384	4.57	98.6	1.15	9.0	71,500	−54	11.3
G11	4	Calcined	935	2	4.01	86.4	0.41	8.0	7840	××	××
G12	4	Calcined	935	384	4.62	99.5	1.79	9.3	93,300	−61	10.2
G13	4	Calcined	960	384	4.57	98.5	2.60	9.2	105,000	−56	10.1
G14	100	Calcined	885	2	5.37	95.0	10.1	15.1	18,700	−70	6.1
R01*	—	—	900–1100 [4]	—	—	—	—	<15 [4,5]	>1000 [4]	<±100 [4]	2–7 [4,5]

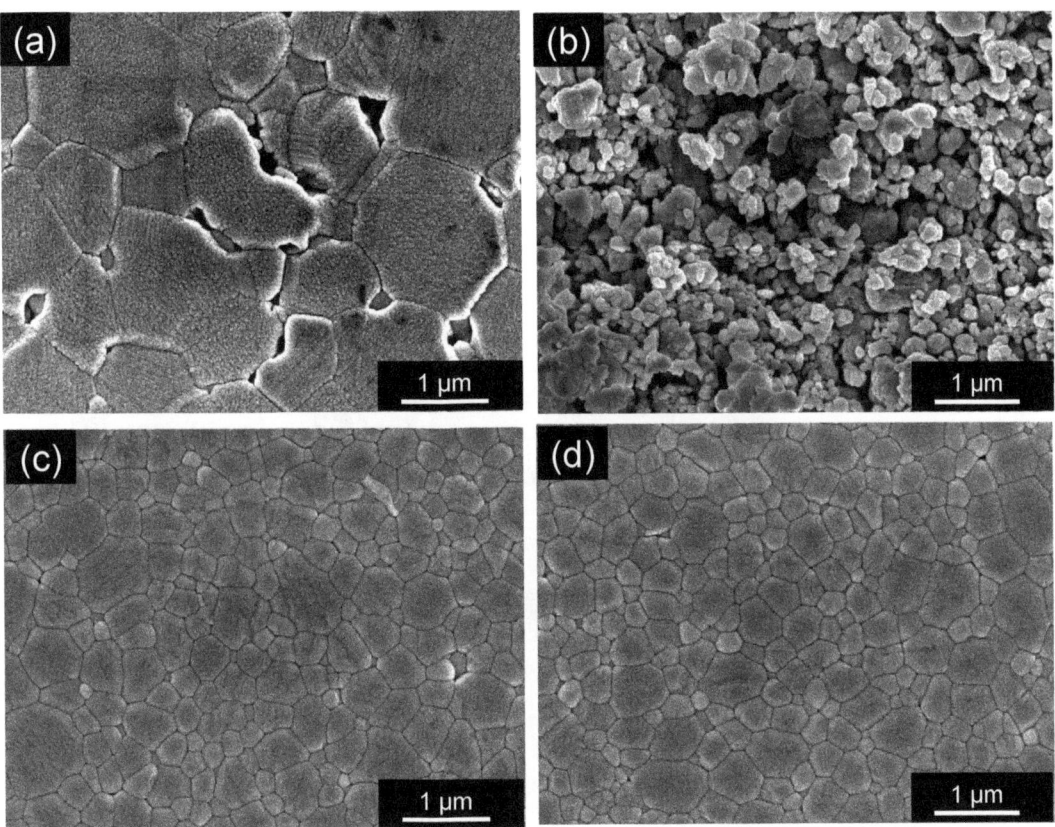

Figure 3. SEM images of polished surfaces of (**a**) pristine ZnAl$_2$O$_4$ fired at 1485 °C for 2 h (sample G01), (**b**) ZnAl$_2$O$_4$ containing 5 wt.% of the Cu$_3$Nb$_2$O$_8$ additive fired at 785 °C for 2 h (G06), (**c**) ZnAl$_2$O$_4$ containing 5 wt.% of the Cu$_3$Nb$_2$O$_8$ additive fired at 960 °C for 2 h (G07), and (**d**) ZnAl$_2$O$_4$ containing 5 wt.% of the CuO–Nb$_2$O$_5$ (Cu:Nb = 3:2 (molar ratio)) additive fired at 960 °C for 2 h (G08).

As summarized in Table 2, samples G01–G05 and G07 were all well sintered, with relative densities in excess of 95%. Little change was observed in ε_r with increasing amount of sintering aid. On the other hand, the $Q \times f$ of the ZnAl$_2$O$_4$ sample with 1 wt.% additive (G02, 98,200 GHz) was approximately sixfold higher than that of the additive-free sample (G01, 16,100 GHz), and decreased with further increases in additive quantity (G03–05 and G07). However, the $Q \times f$ value of sample G07 was observed to be 50,000 GHz or higher, even with 5 wt.% of the additive, which is a good value for low-temperature sintered ceramics. As has been extensively discussed, the main origins of the dielectric loss tangent (tan δ: reciprocal of the Q value) of an actual dielectric ceramic at microwave frequency is mainly determined by an intrinsic factor (the anharmonic terms of the crystal's potential energy) and extrinsic factors (lattice defects caused by impurities, disordered charge distribution in the crystal, grain boundaries, random grain orientation, microcracks, porosity, etc.) [24–26]. Considering the above viewpoint, the relative density of sample G01 was 97.0%; hence, it was 3.0% porous. In addition, Figure 3a shows grains and void diameters that are sufficiently large to be recognizable in the SEM image. In other words, we believe that this sample contained many extrinsic dielectric loss factors associated with its porosity. The addition of 1 wt.% of the sintering aid decreased the number of voids in the sintered ZnAl$_2$O$_4$ (observed in Figure 3a), which reduced lattice relaxation and increased its $Q \times f$ value. Penn et al. reported that dielectric loss depends strongly on pore volume, with only a small degree of porosity markedly affecting the loss of sintered alumina [26]. They also mentioned that loss may be related to the surface area of the pores, as the alumina at the surface of a pore is in a different environment to the alumina within the matrix. The presence of a free surface (such as a pore) leads to crystal lattice relaxation. The relaxed surface is effectively different from that of the bulk material and is, therefore, likely to have a different tan δ (one of the extrinsic factors of dielectric loss). This study also supports our views on $ZnAl_2O_4$ (above), an Al-based oxide. Further increases in additive quantity probably resulted in higher degrees of lattice defects in the $ZnAl_2O_4$ due to Cu-component incorporation (see Section 3.3.2); hence, lower $Q \times f$ values were observed. In addition, the $Q \times f$ values of the samples were also affected by the relatively low $Q \times f$ of the $Cu_3Nb_2O_8$ additive itself (G14, 18,700 GHz), which was presumably lower than that of $ZnAl_2O_4$ (106,000 GHz, reported by Zheng et al. [25]).

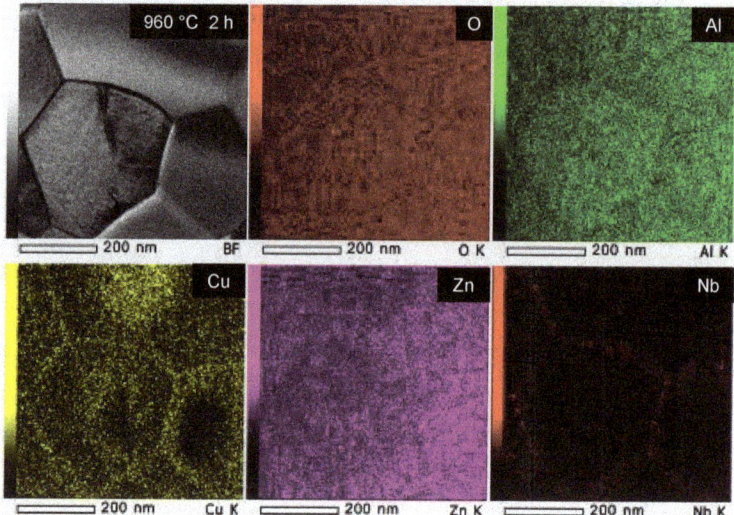

Figure 4. TEM image and EDS elemental maps of $ZnAl_2O_4$ containing 5 wt.% $Cu_3Nb_2O_8$ fired at 960 °C for 2 h (G07).

On the other hand, τ_f was observed to increase slightly with increasing additive quantity compared with the value for the pristine $ZnAl_2O_4$ sample; that is, the sample devoid of additive (G01) exhibited a τ_f value of −73 ppm·K^{-1}, whereas sample G07, with 5 wt.% of the additive, exhibited a value of −56 ppm·K^{-1}. The $Cu_3Nb_2O_8$ additive itself (G14) had a τ_f of −70 ppm·K^{-1}, which is almost the same as that of the additive-free sample (G01); the additive quantity was also small. Therefore, we conclude that the cause of the observed trend rested with the base material rather than the additive and presume that the solid solution of the Cu component in $ZnAl_2O_4$ was involved. κ was also observed to decrease monotonically with increasing additive quantity. In particular, the κ value of the sample containing 1 wt.% of additive (G02) was dramatically lower (by a factor of two) compared to that of the additive-free sample (G01). On the basis of our previous research, we believe that the solid solution of the Cu component in $ZnAl_2O_4$ was responsible for this observation [17].

We next examined the $ZnAl_2O_4$ sample containing 5 wt.% of the $CuO-Nb_2O_5$ (Cu:Nb = 3:2 (molar ratio)) sintering aid, and compared various characteristics of its calcined and uncalcined samples. As shown in Figure 2, we did not detect any difference in the sintering behavior of the uncalcined (G08) and calcined (G07) samples. The SEM image in Figure 3d reveals that the microstructure of the uncalcined sample (G08) was almost identical to that of calcined G07 shown in Figure 3c; both exhibited almost no voids and submicron-sized grains (for G08, D_g = 0.57 μm). The dielectric and thermal data summarized in Table 2 reveal no significant differences in the relative permittivities, τ_f values, and thermal conductivities of G07 and G08; however, calcined sample G07 showed a higher $Q \times f$ value (60,800 GHz) than that of uncalcined sample G08 (51,300 GHz). Furthermore, Figure 5 shows that the two exhibited almost no difference in relative permittivity, whereas the calcined sample exhibited higher $Q \times f$ values at firing temperatures of 935, 960, 985, and 1035 °C.

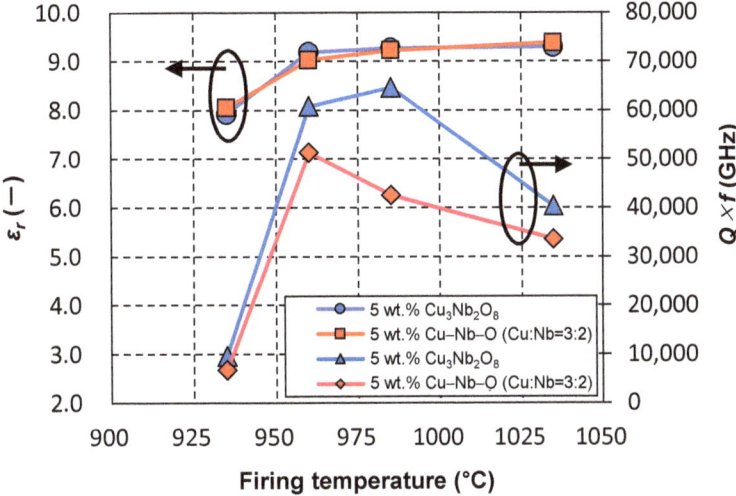

Figure 5. Effect of Cu–Nb–O additive (Cu:Nb = 3:2 (molar ratio)) calcination on the relationship linking firing temperature, relative permittivity, and the $Q \times f$ of $ZnAl_2O_4$ containing 5 wt.% of the additive. A 2 h holding time was applied at each temperature.

The $Q \times f$ value of $ZnAl_2O_4$ appeared to depend on whether or not the $CuO-Nb_2O_5$ additive was calcined; hence, the reason for this dependence was investigated by XRD. Figure 6a shows XRD patterns of the G01, G06, G07, G13, G08, and G14 sintered bodies, with Figure 6b showing enlarged patterns. The unknown peaks indicated by solid circles (●) near 2θ = 30° and around 2θ = 33° were also observed in the pattern of the pure $ZnAl_2O_4$ sample (G01). Therefore, we conclude that these peaks are not directly related to the observed characteristic change. Peaks corresponding to $Cu_3Nb_2O_8$ were observed for the calcined sample (G06) fired at 785 °C for 2 h, as well as for the sample (G07) fired at 960 °C for 2 h, although they were less intense in the case of the latter. In addition, unknown peaks, indicated by solid triangles (▲), were clearly observed. The XRD peaks observed for G06 and G07 were slightly shifted compared to those of the pure $Cu_3Nb_2O_8$ sintered body (G14). For example, the $1\bar{2}1$ reflection, which can be observed at 2θ = 32.6° in the XRD pattern of $Cu_3Nb_2O_8$, was observed at lower angles in the patterns of G06 and G07. According to Kim et al. [19], such a shift is consistent with a solid solution in which Zn occupies Cu sites in $Cu_3Nb_2O_8$. However, the same report revealed no significant change in the $Q \times f$ value due to the formation of the Zn solid solution. No peaks corresponding to $Cu_3Nb_2O_8$ were observed for the sample with the uncalcined additive (G08) fired at 960 °C for 2 h. In addition, peaks corresponding to $Zn_3Nb_2O_8$ were not observed. $Cu_3Nb_2O_8$ and $Zn_3Nb_2O_8$ reportedly exhibit relatively high $Q \times f$ values of approximately 50,000 and 80,000 GHz, respectively [19]. In other words, the formed $Cu_3Nb_2O_8$ and $Zn_3Nb_2O_8$ exhibit relatively high $Q \times f$ values when the sintered body is fabricated by the separate addition of sintering aids, such as CuO and Nb_2O_5. We, therefore, conclude that compounds related to CuO, Nb_2O_5, or ZnO, with lower $Q \times f$ values, were formed in the current work, which we presume to be the cause of the observed $Q \times f$ trend. Clearly, these results alone cannot identify the cause of the observed trend, and further detailed analysis is required.

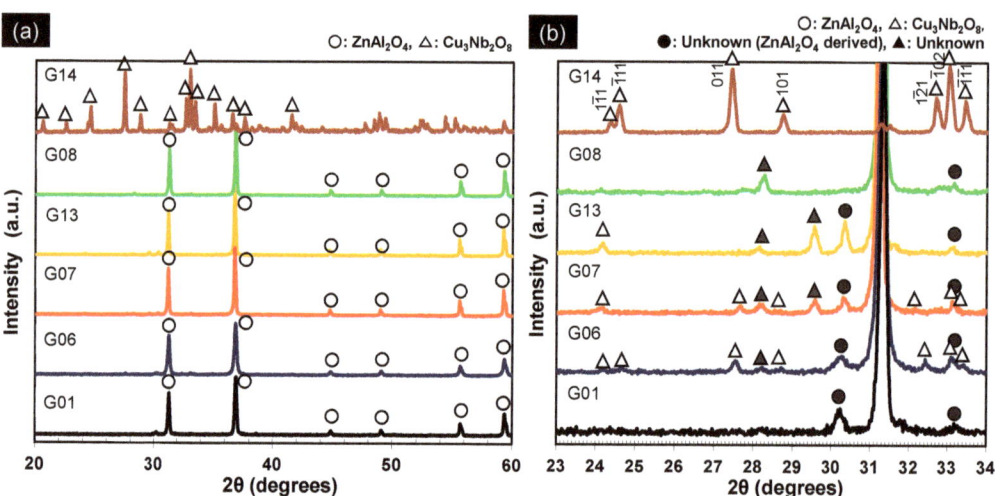

Figure 6. (**a**) XRD patterns of pristine $ZnAl_2O_4$ fired at 1485 °C for 2 h (G01), $ZnAl_2O_4$ containing 5 wt.% of the $Cu_3Nb_2O_8$ additive fired at 785 °C for 2 h (G06), $ZnAl_2O_4$ containing 5 wt.% of the $Cu_3Nb_2O_8$ additive fired at 960 °C for 2 h (G07), $ZnAl_2O_4$ containing 4 wt.% of the $Cu_3Nb_2O_8$ additive fired at 960 °C for 384 h (G13), $ZnAl_2O_4$ containing 5 wt.% of the CuO–Nb_2O_5 (Cu:Nb = 3:2 (molar ratio)) additive fired at 960 °C for 2 h (G08), and pristine $Cu_3Nb_2O_8$ fired at 885 °C for 2 h (G14). (**b**) Enlarged XRD patterns of G01, G06, G07, G13, G08, and G14 in panel (**a**). $Cu_3Nb_2O_8$ diffraction indices are indicated in the figure.

3.2. Effect of Holding Time on the Properties of $ZnAl_2O_4$

Figure 2 shows that $ZnAl_2O_4$ sinterability saturated when more than 4 wt.% $Cu_3Nb_2O_8$ was added. Therefore, the $Cu_3Nb_2O_8$ quantity was fixed at 4 wt.% and the effects of retention time at 910, 935, and 960 °C, which are below the melting point of Ag, were investigated. As discussed in detail in the next section (Section 3.3), these temperatures were also lower than the melting temperature of the additive. Figure 7a reveals that the ceramic densities of the samples in this system were significantly affected by the time held at the firing temperature; that is, dense sintered bodies were obtained by holding for an extended time (384 h) even at firing temperatures less than or equal to 960 °C (e.g., 910 °C). Figure 7b,c show relative permittivity and $Q \times f$ data, while Table 2 summarizes sample properties (G05, G09–G13). Relative permittivity (Figure 7b) exhibited almost identical behavior to sintered body density (Figure 7a), with relative permittivity and density observed to saturate concurrently. On the other hand, the $Q \times f$ value (Figure 7c) rose sharply to a value of 60,000 GHz or higher (e.g., for G05 and G10) when the ceramic density of the sample exceeded a threshold value of 4.4–4.5 g/cm^3. In addition, $Q \times f$ increased further with increasing holding time, even at the same density. The $Q \times f$ of the sample fired at 960 °C for 384 h (G13) was determined to be 105,000 GHz (at a frequency of ~13 GHz). This trend was also observed for samples fired for 384 h at lower temperatures (910 and 935 °C) (G10: 71,500, G12: 93,300 GHz). Moreover, the thermal conductivity of G13 exceeded 10 W·m^{-1}·K^{-1} (10.1 W·m^{-1}·K^{-1}), which is approximately 10% higher than that of the sample fired at 960 °C for 2 h (G05). This trend was also observed for the samples fired at lower temperatures (910, and 935 °C) for 384 h (G10: 11.3, G12: 10.2 W·m^{-1}·K^{-1}); hence, G10, G12, and G13 are excellent LTCC candidate materials.

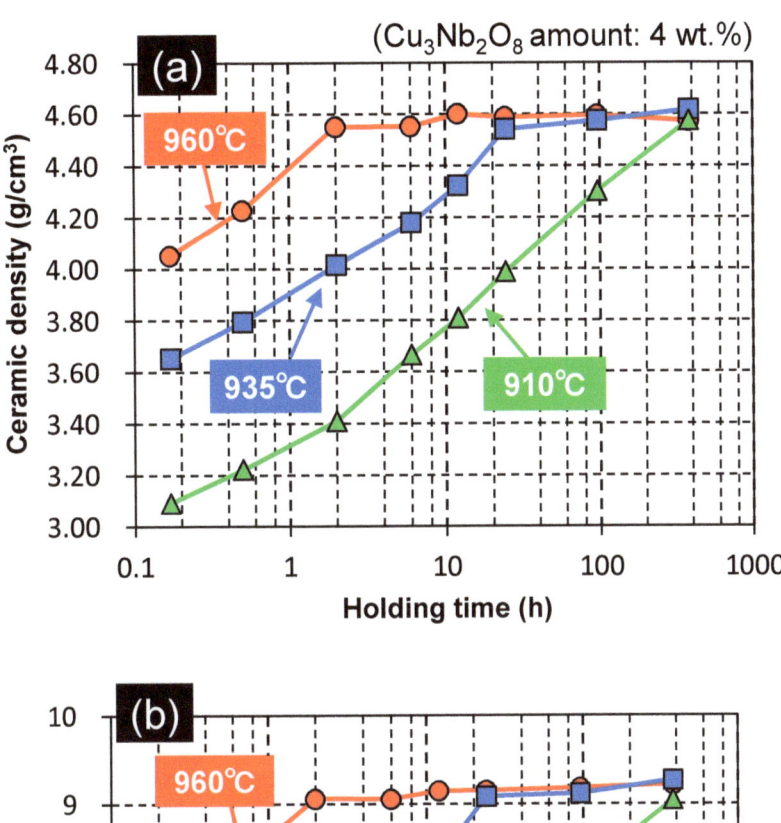

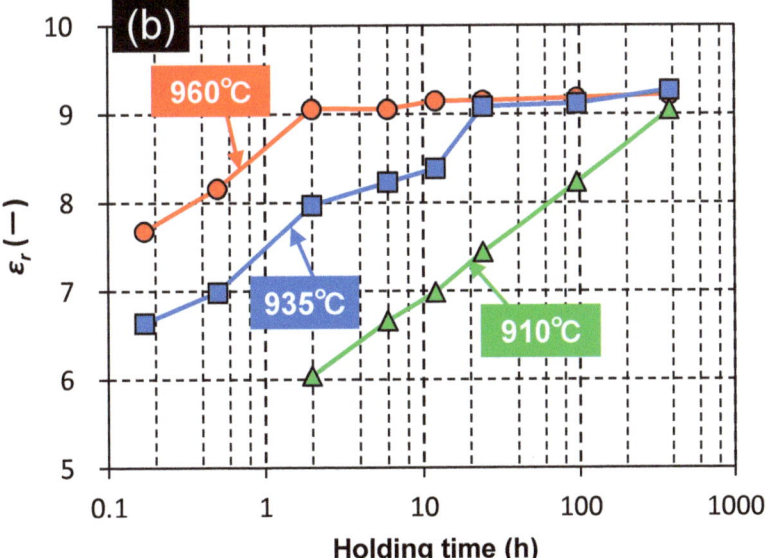

Figure 7. Cont.

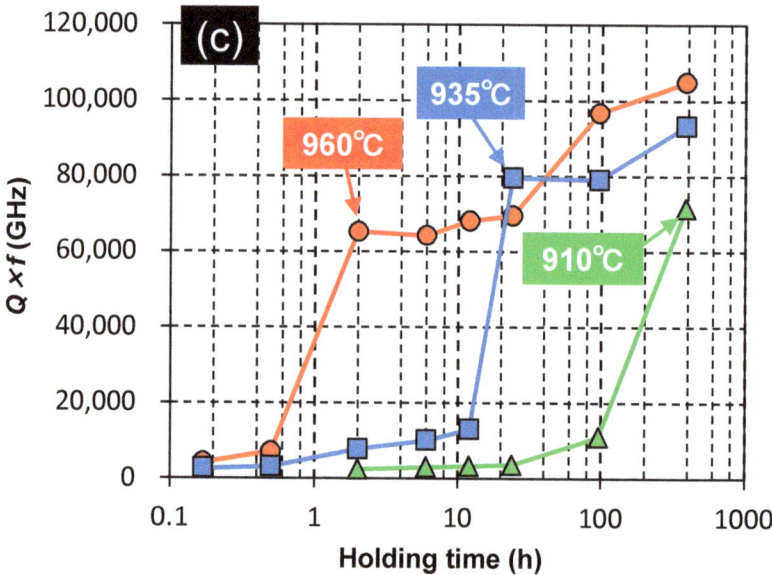

Figure 7. Ceramic densities and microwave dielectric and thermal properties of $ZnAl_2O_4$ containing 4 wt.% of $Cu_3Nb_2O_8$ as a function of holding time at firing temperatures of 910, 935, and 960 °C: (**a**) ceramic density; (**b**) relative permittivity ε_r; (**c**) $Q \times f$.

The reason for the observed increases in $Q \times f$ and thermal conductivity with prolonged holding time is discussed in terms of the SEM-observed sample microstructures. First, Figure 8a,c,e, reveal that all samples prepared with a retention time of 2 h (G09, G11, and G05) had submicron-sized grains, with higher firing temperatures leading to fewer voids. Almost no visible voids were observed in the sample fired at 960 °C for 2 h (G05). We presume that the porosity of the $ZnAl_2O_4$ sintered body suppresses crystal lattice relaxation (one of extrinsic factors of dielectric loss) [26]; hence, porosity is responsible for the higher $Q \times f$ of sample G05 compared with that of samples G09 and G11. Figure 8b,d,f reveal that the samples prepared with a retention time of 384 h (G10, G12, and G13) exhibited almost no voids. On the other hand, remarkable grain growth was observed, and many particles were larger than 1 μm; higher firing temperatures led to larger grains. Figure 8e,f compare the morphologies of the samples fired at 960 °C for 2 h (G05) and at 960 °C for 384 h (G13); hardly any voids and almost identical sintering densities were observed. On the other hand, sample G13 had the highest $Q \times f$ value in this study; it exhibited much higher grain growth (D_g: 2.60 μm) than G05 (D_g: 0.53 μm), with many particles larger than 2 μm. Therefore, larger grains may lead to fewer grain boundaries per unit volume (one of the extrinsic factors of dielectric loss) and higher $Q \times f$ values, which seem to have reduced the extrinsic dielectric loss factors and led to an equivalent $Q \times f$ value (106,000 GHz [25]) to that reported for pure $ZnAl_2O_4$. However, in the above study, the $Q \times f$ value was calculated to be 394,000 GHz when only the intrinsic factor was considered; hence, eliminating the influence of extrinsic factors remains an important issue. In addition, phonons have longer mean free paths that result in high heat conduction; we consider this to be related to the higher $Q \times f$ and thermal conductivity of the sample. For reference, the XRD pattern in Figure 6b shows that the phases generated in G13 (other than $ZnAl_2O_4$) at a retention time of 384 h were the same type as those observed for G08 (with a composition almost identical to that G05 with a retention time of 2 h), although their peak intensities were different. Identifying clear causes on the basis of these generated phases is a future objective.

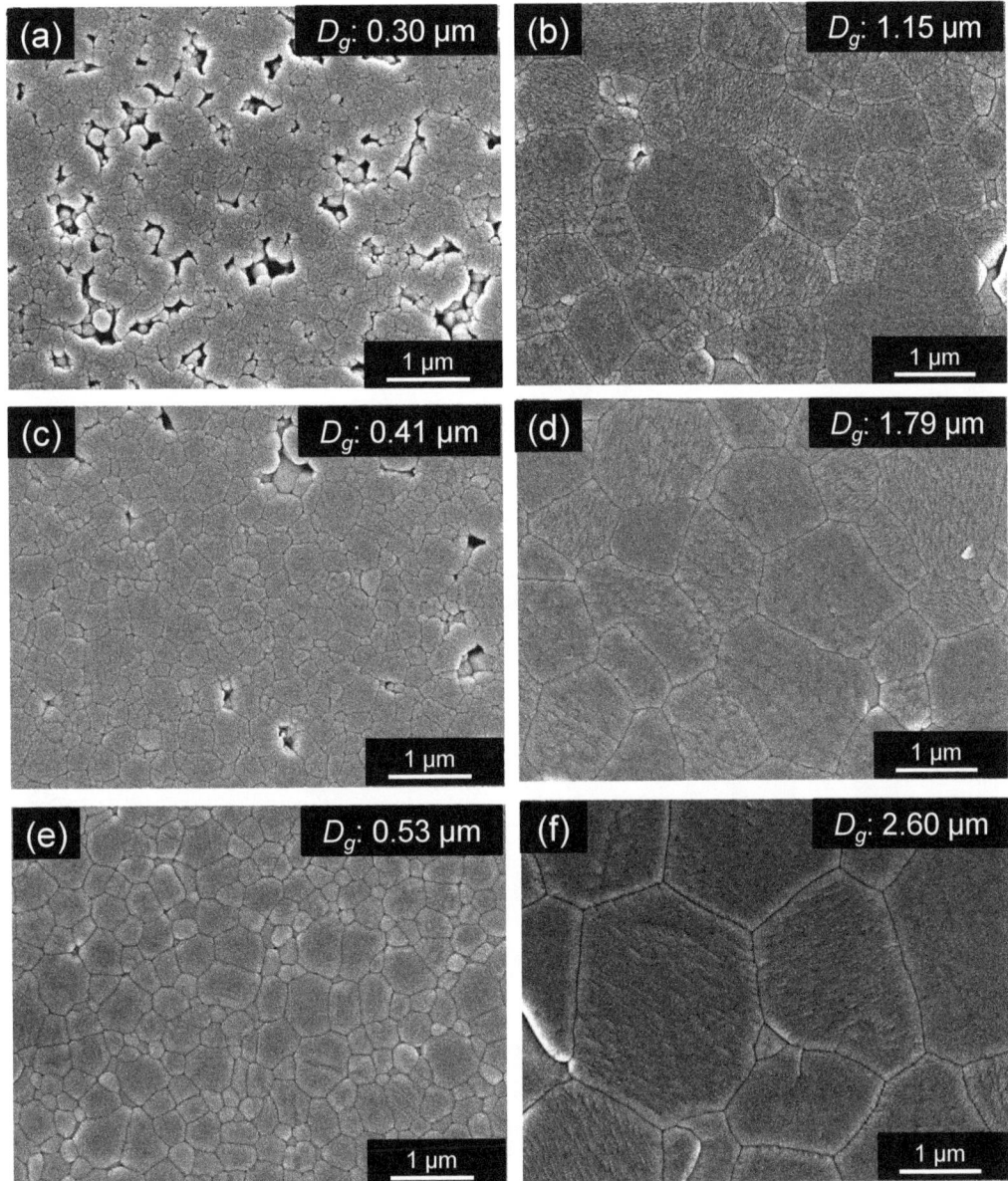

Figure 8. SEM images of polished surfaces of $ZnAl_2O_4$ containing 4 wt.% of the $Cu_3Nb_2O_8$ additive fired at (**a**) 910 °C for 2 h (G09), (**b**) 910 °C for 384 h (G10), (**c**) 935 °C for 2 h (G11), (**d**) 935 °C for 384 h (G12), (**e**) 960 °C for 2 h (G05), and (**f**) 960 °C for 384 h (G13). The average grain size (D_g) of each sample is also shown.

3.3. Proposed Sintering Mechanism

3.3.1. Melting Temperature

As discussed in Section 3.1, the $ZnAl_2O_4$ sample containing the $Cu_3Nb_2O_8$ additive began to densify at a low temperature (885 °C, Figure 2). LTCC materials normally contain large quantities of glass, and densification is assumed to occur through liquid-phase sintering. Therefore, densification normally commences at a temperature above the melting point (eutectic temperature); hence, the melting temperatures of $Cu_3Nb_2O_8$-additive/$ZnAl_2O_4$ mixtures need to be considered in addition to that of the $Cu_3Nb_2O_8$ additive. Figure 9a presents DTA curves of 100:0, 95:5, 90:10, 80:20, and 50:50 (w/w) $Cu_3Nb_2O_8$-additive/$ZnAl_2O_4$ mixtures when heated at 10 °C/min. The $Cu_3Nb_2O_8$ additive melts at 967 °C, which did not change dramatically for most $Cu_3Nb_2O_8$-additive/$ZnAl_2O_4$ powder mixtures. Moreover, this temperature is much higher than that at which densification begins (~885 °C, Figure 2), which suggests that densification occurs in the solid phase prior to liquid phase formation; this process was referred to as "solid-state-activated sintering" by German et al. [27,28]. The minimum melting temperature determined from the DTA curve of each sample was taken to be the liquid-phase formation temperature, T_m (K). Figure 9b depicts the relationship between T/T_m (where T (K) is the firing temperature) and the relative density of $ZnAl_2O_4$ containing 4 wt.% of $Cu_3Nb_2O_8$ held for 2, 24, and 384 h. A T/T_m ratio above 1.00 indicates the formation of a liquid phase, while a value below 1.00 reveals that densification occurs solely through a solid-state mechanism. Samples at all retention times had relative densities of 98% or higher when T/T_m = 0.99 (T = 1233 K). Even with a T/T_m of 0.94 (T = 1183 K), the sample held for 384 h achieved a relative density of 98% or higher. Hence, $ZnAl_2O_4$ clearly underwent solid-state densification before liquification of the sintering aid; this system, therefore, underwent solid-state-activated sintering.

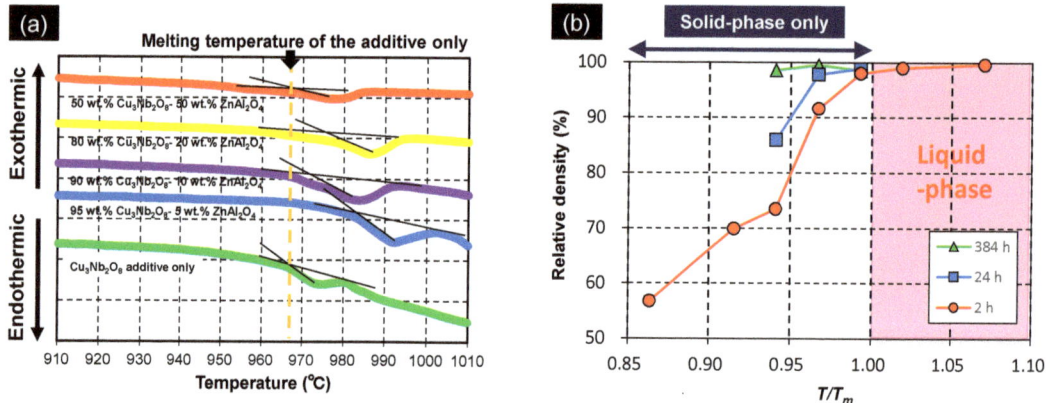

Figure 9. (a) Differential thermal analysis curves for $Cu_3Nb_2O_8/ZnAl_2O_4$ mixtures acquired at 10 °C/min. (b) Relative densities of $ZnAl_2O_4$ containing 4 wt.% of $Cu_3Nb_2O_8$ and held for 2, 24, and 384 h as a function of the firing-temperature/liquid-phase-formation temperature ratio (T/T_m).

3.3.2. Degree of Solid Solution

German et al. [27,28] proposed "an ideal binary phase diagram for promoting liquid-phase sintering (LPS)" (Figure 10). The composition and firing temperature of the actual liquid-phase sintered body correspond to the blue dot in Figure 10. An ideal combination of composition and temperature results in high solid solubility in the liquid (eutectic liquid in this diagram) and low solubility of the liquid in the solid. A significantly lower melting temperature compared to that of the single component affords a processing temperature advantage. Therefore, we conclude that the conditions for promoting liquid-phase sintering is satisfied in our case, with $ZnAl_2O_4$, as the base material, exhibiting higher solid solubility in the $Cu_3Nb_2O_8$ additive than that of the $Cu_3Nb_2O_8$ additive in $ZnAl_2O_4$.

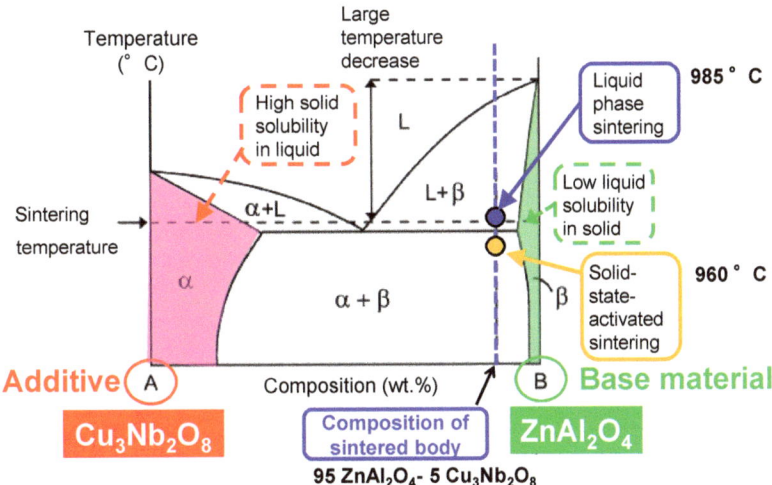

Figure 10. Ideal binary phase diagram for the liquid phase sintering of additive "A" and base material "B".

Although the proposed theory (above) is only intended to describe the behavior of samples when liquid phases form, we believe that the described behavior also occurred in the solid state in this study (solid-state-activated sintering; yellow dot in Figure 10). However, as the phase diagram for this system has not yet been reported, we assumed a pseudo-binary phase diagram for the $ZnAl_2O_4$–$Cu_3Nb_2O_8$ system. Accordingly, we determined the degrees of solid dissolution of the $Cu_3Nb_2O_8$ additive in $ZnAl_2O_4$, and $ZnAl_2O_4$ in the $Cu_3Nb_2O_8$ additive by monitoring lattice constant changes. Specifically, two types of mixed powder sample were heat-treated: 95 wt.% $Cu_3Nb_2O_8$ additive + 5 wt.% $ZnAl_2O_4$ (95C05Z), and 95 wt.% $Cu_3Nb_2O_8$ additive + 5 wt.% $ZnAl_2O_4$ (95Z05C). Changes in the XRD-measured lattice constants (unit cell volumes in this case) were determined as a function of heat-treatment temperature. The lattice constant of $Cu_3Nb_2O_8$ after calcination at 835 °C for 2 h was measured for the 95C05Z sample, while the lattice constant of $ZnAl_2O_4$ after calcination at 1100 °C for 4 h was measured for the 95Z05C sample, with values prior to heat treatment used as reference values for calculating change ratios.

Figure 11 displays the relationship between heat-treatment temperature and unit cell volume change ratios; the error bars show standard deviations (σ). Figure 11a reveals that the $Cu_3Nb_2O_8$ unit cell volume change ratio was approximately 0.25–0.55% higher than the reference value (at room temperature, 20 °C) when heat-treated at 585 °C or higher, although these results had relatively large error bars. This suggests that Zn^{2+}, which has a larger ionic radius (0.074 nm) than Cu^{2+} (0.073 nm) and Nb^{5+} (0.064 nm) when six-coordinated [29], was incorporated into $Cu_3Nb_2O_8$. It is not clear whether or not Al^{3+}, which has a smaller ionic radius (0.0535 nm when six-coordinated) than Cu^{2+} or Nb^{5+}, was incorporated in $Cu_3Nb_2O_8$. However, Al^{3+} was possibly an interstitial solid solution; in this case, a lower unit cell volume would not be expected for $Cu_3Nb_2O_8$.

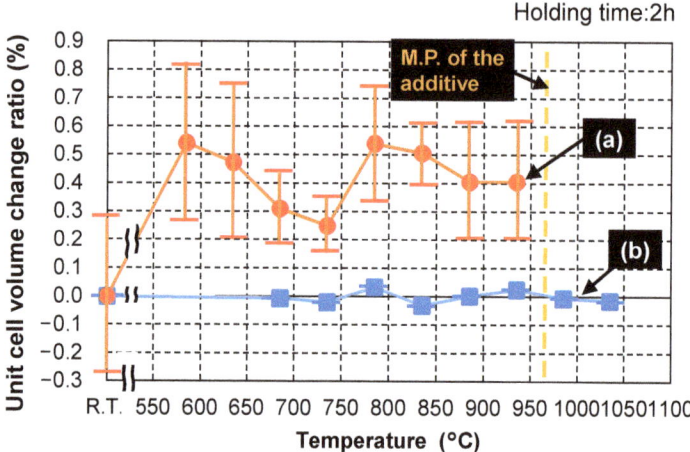

Figure 11. Unit cell volume change ratios of (**a**) $Cu_3Nb_2O_8$ in a $Cu_3Nb_2O_8$ sample containing 5 wt.% of $ZnAl_2O_4$ (95C05Z), and (**b**) $ZnAl_2O_4$ in a $ZnAl_2O_4$ sample containing 5 wt.% of the $Cu_3Nb_2O_8$ additive (95Z05C) as a function of heat-treatment temperature determined by XRD. Each particular temperature was held for 2 h.

On the other hand, Figure 11b shows almost no change in $ZnAl_2O_4$ unit cell volume ratio (less than 0.05%) when heat-treated at 685–1035 °C, which suggests that Cu^{2+}, with a smaller ionic radius than Zn^{2+}, but larger than Al^{3+} [29], was not significantly incorporated into the Zn and Al sites of $ZnAl_2O_4$. Alternatively, it is possible that Cu^{2+} was simultaneously incorporated into both the Zn sites of $ZnAl_2O_4$ and the Al sites of $ZnAl_2O_4$, which would offset any increase/decrease in the $ZnAl_2O_4$ unit cell volume. Shigeno et al. [9] reporteded a unit cell volume change ratio of only ~0.1% for Al_2O_3 when Cu^{2+} and Ti^{4+} were incorporated into Al_2O_3 in an alumina sample containing 5 wt.% of the Cu–Ti–Nb–O sintering aid. In addition, Phillips et al. [30] reported that the unit cell volume of Ti-doped sapphire was also only ~0.1% larger than that of pure sapphire after annealing at 1600 °C for 24 h, which is similar to the results obtained in this study. Therefore, even if Cu^{2+} was incorporated into the Al sites of $ZnAl_2O_4$, we estimate that the unit cell volume change ratio (increase ratio) was only about 0.1%, while the change ratio (decrease ratio) in the unit cell volume associated with incorporating Cu^{2+} into the Zn sites was estimated to be ~0.1%. In other words, the $ZnAl_2O_4$ unit cell volume change ratio (Figure 11b) was estimated to be smaller than that of $Cu_3Nb_2O_8$ (Figure 11a) (0.25–0.55%).

To gain further insight into the Cu and Nb distributions, the $ZnAl_2O_4$ particles were subjected to TEM–EDS before and after densification. Figure 12 shows TEM images and continuous EDS data for the intragranular additive elements (Cu and Nb) in sintered $ZnAl_2O_4$ containing 5 wt.% of $Cu_3Nb_2O_8$ fired at different temperatures (785 °C for 2 h, and 960 °C for 2 h). Each sample was subjected to six 30 min continuous EDS analyses at different intragranular locations; the values shown in parentheses denote standard deviations. Approximately 0.12 ± 0.04 at.% Cu was detected in the $ZnAl_2O_4$ sample (G06) fired at 785 °C for 2 h (Figure 12a). However, as shown in Figure 2a, this sample was not densified and is, therefore, considered to be a solid solution of Cu with the same valence as Zn (i.e., divalent). On the other hand, approximately 0.37 ± 0.23 at.% Cu was detected in the densified $ZnAl_2O_4$ sample (G07) fired at 960 °C for 2 h (Figure 12b). As evidenced by the standard deviations in Figure 12b and the EDS elemental map shown in Figure 4, Cu was not uniform and was sparsely distributed within the $ZnAl_2O_4$ grain. The reason for this sparse distribution remains unknown; however, we presume that one of the reasons is that the Cu diffuses slowly during solid-state sintering, unlike in liquid-phase sintering. Nonetheless, we can confirm that more Cu diffused into the densified

ZnAl$_2$O$_4$ grains than in grains before densification, suggesting that Cu was incorporated into ZnAl$_2$O$_4$ not only at Zn sites but also at some Al sites. In contrast, the amounts of Nb were determined to be much less than 0.1 at.%, which is regarded to be the detection limit of the EDS instrument used. No clear difference in this value was observed before and after densification; therefore, while we cannot provide a definitive conclusion because only local analyses were performed, we presume that Nb atoms were not present within the ZnAl$_2$O$_4$ particle grains.

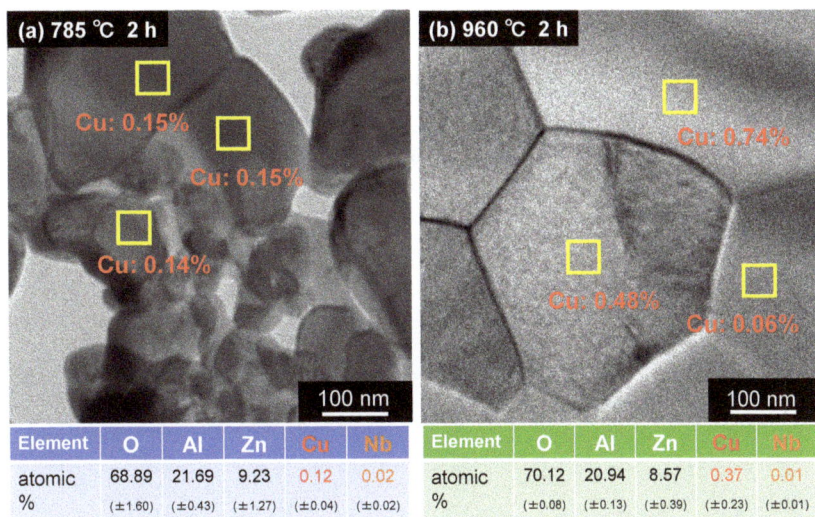

Figure 12. TEM images and EDS data (including standard deviation: σ) for intragranular atoms (O, Al, Zn, Cu, and Nb) in sintered ZnAl$_2$O$_4$ containing 5 wt.% of Cu$_3$Nb$_2$O$_8$ fired at (**a**) 785 °C for 2 h (G06), and (**b**) 960 °C for 2 h (G07).

The abovementioned results reveal that ZnAl$_2$O$_4$ mainly densified with the help of defects formed by the substitution of Cu^{2+} for Al^{3+} in the ZnAl$_2$O$_4$ lattice in the solid state at 835–960 °C, and they are believed to have been generated by the following reactions (Kröger–Vink notation):

$$2CuO\,Al_2O_3 \rightarrow 2Cu'_{Al} + V_{\ddot{O}} + 2O^{\times}_{O}, \tag{3}$$

and

$$6CuO + Al_2O_3 \rightarrow Al_2O_3 6Cu'_{Al} + 9O^{\times}_{O} + 2Al^{\cdots}_{i}. \tag{4}$$

Reaction (3) assumes defects due to vacancy formation, while Reaction (4) assumes defects due to interstitial ion formation. In the present study, oxygen vacancies ($V_{\ddot{O}}$ in Reaction (3)) or interstitial Al^{3+} ($Al^{\cdots}_{i}$ in Reaction (4)) were the main defects, as indicated by the significantly higher Cu concentration in the densified ZnAl$_2$O$_4$ grains. Nb atoms are assumed to assist in increasing the diffusion rate of Cu^{2+} by lowering the melting temperature of the sintering additive. The addition of Nb$_2$O$_5$ to CuO lowered the melting temperature from 1025 to 967 °C. Furthermore, although the ionic radius of Nb^{5+} is relatively close to that of Al^{3+}, they cannot substitute into Al^{3+} sites due to the +2 difference in valence.

The abovementioned results suggest that our system is qualitatively consistent with the "ideal binary phase diagram of LPS" advocated by German et al. [27,28]; therefore, they support the solid-state-activated sintering mechanism.

3.3.3. Solid-State-Activated Sintering Model in This Study

A schematic of the solid-state-activated sintering mechanism in this study is shown in Figure 13. Fine $ZnAl_2O_4$ particles are gradually incorporated into the additive at temperatures below 585 °C, without considerable densification (Figure 13a), and a solid solution forms at temperatures up to 835 °C, at which point saturation occurs and $ZnAl_2O_4$ reprecipitates. According to the Kelvin equation [31], reprecipitation occurs preferentially on the coarse particles, which are less soluble than the fine particles (Figure 13b) through a process known as "Ostwald ripening." Necks form between coarse particles as they grow. The neck region is less soluble than the coarse particles; hence, reprecipitation occurs preferentially at the necks. TEM–EDS (Figure 12) revealed that the deposited $ZnAl_2O_4$ was not pure, and those small quantities of additive components (Cu^{2+}) were incorporated into the Zn^{2+} sites, as well as the Al^{3+} sites of the $ZnAl_2O_4$ in neck regions. We presume that the incorporated atoms created defects (Al- or O-related defects) in the $ZnAl_2O_4$ lattice that promote solid-phase diffusion between $ZnAl_2O_4$ particles, thereby promoting sintering (Figure 13c). The solid solution particles of the additive probably formed an intergranular amorphous phase during densification [32], which reduced the grain-boundary energy of $ZnAl_2O_4$. As sintering occurs through solid-phase diffusion, long retention times are required; however, densification is guaranteed. Lastly, according to the TEM–EDS results (Figure 4), we conclude that most additive components segregated and remained in the intergranular amorphous phase and other crystal compounds at grain boundaries. Because the $ZnAl_2O_4$ grains were larger after sintering for 384 h than after sintering for 2 h (Figure 8b,d,f), we conclude that the grains increased in size during sintering.

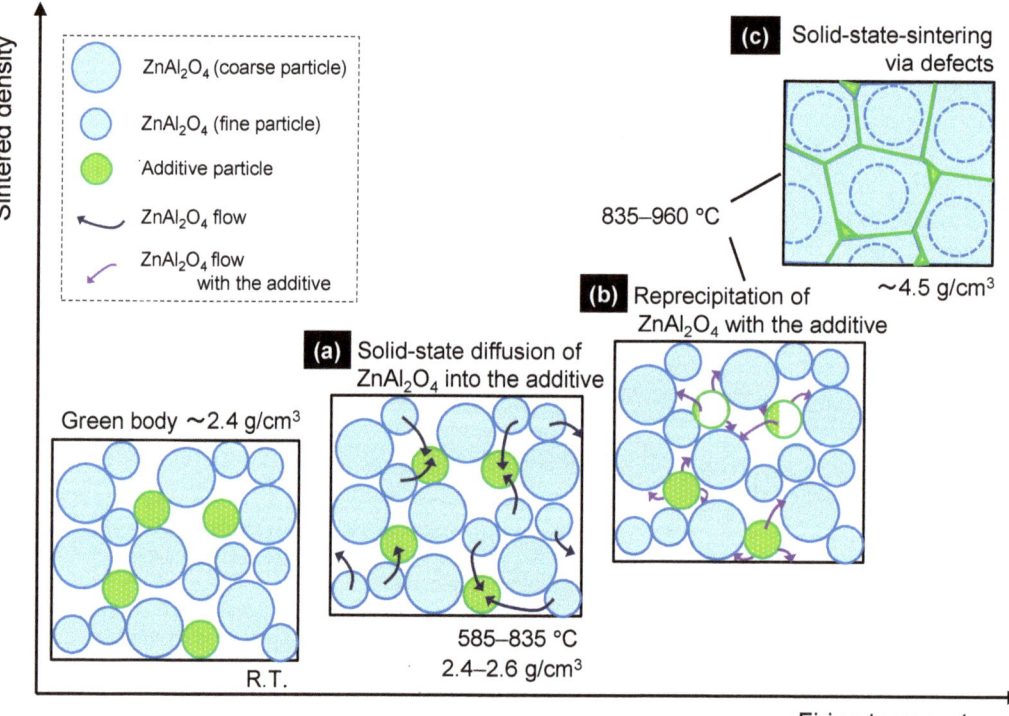

Figure 13. Solid-state-activated sintering model for $ZnAl_2O_4$ containing 5 wt.% of $Cu_3Nb_2O_8$. (a) Solid-state diffusion of $ZnAl_2O_4$ into the additive, (b) Reprecipitation of $ZnAl_2O_4$ with the additive, and (c) Solid-state-sintering via defects.

4. Conclusions

The addition of a small amount of the calcined $Cu_3Nb_2O_8$ sintering aid to $ZnAl_2O_4$, a highly thermally conductive dielectric material, successfully enabled $ZnAl_2O_4$ sintering at (low) temperatures below the melting point of silver (961 °C), resulting in better dielectric properties (in particular, a quality factor by resonant frequency product $Q \times f$) than those obtained by the addition of the non-calcined CuO–Nb_2O_5 (in a 3:2 Cu:Nb molar ratio) sintering aid. In addition, the sample fired at 960 °C exhibited more suitable dielectric characteristics, such as a relative permittivity ε_r of 9.2, a $Q \times f$ of 105,000 GHz, and a temperature coefficient of resonant frequency τ_f of −56 ppm·K^{-1} when the holding time was prolonged from 2 h to 384 h. The sample exhibited a thermal conductivity of 10.1 W·m^{-1}·K^{-1}, which exceeds that of conventional LTCCs (~2–7 W·m^{-1}·K^{-1}). Moreover, we suggest that this system underwent solid-state-activated sintering by specifically incorporating Cu^{2+} in the $ZnAl_2O_4$ lattice, in which densification was virtually complete even when the sintering aid remained in the solid state. The Nb component presumably indirectly affected the Cu^{2+} diffusion rate by lowering the melting temperature of the additive.

Future studies will further elucidate the mechanism associated with solid-state-activated sintering in this system and apply the acquired knowledge to the fabrication of various Al-based ceramic materials.

Author Contributions: K.S. conceptualized and designed the experiments; K.S. and T.Y. performed the experiments; K.S., T.Y. and H.F. analyzed the data. K.S. wrote the manuscript. All authors have read and agreed to the published version of the manuscript.

Funding: This research was funded by the Electric Technology Research Foundation of Chugoku.

Institutional Review Board Statement: Not applicable.

Informed Consent Statement: Not applicable.

Data Availability Statement: Data will be made available upon reasonable request.

Acknowledgments: We are indebted to Hirotaka Ogawa and Akinori Kan of Meijo University for their assistance with the dielectric property experiments. We also deeply thank Keisuke Ura of the Industrial Technology Institute, Miyagi Prefectural Government, for his assistance with calculating the unit cell volumes of $Cu_3Nb_2O_8$ from the XRD data.

Conflicts of Interest: The authors declare no conflict of interest.

References

1. Imanaka, Y. *Multilayered Low Temperature Cofired Ceramics (LTCC) Technology*; Springer: New York, NY, USA, 2005; pp. 1–18.
2. Sebastian, M.T. *Dielectric Materials for Wireless Communication*; Elsevier: New York, NY, USA, 2008; pp. 445–512.
3. Sebastian, M.T.; Wang, H.; Jantunen, H. Low temperature co-fired ceramics with ultra-low sintering temperature: A review. *Curr. Opin. Solid State Mater. Sci.* **2016**, *20*, 151–170. [CrossRef]
4. Sebastian, M.T.; Jantunen, H. Low loss dielectric materials for LTCC applications. *Int. Mater. Rev.* **2008**, *53*, 58–90. [CrossRef]
5. Induja, I.J.; Abhilash, P.; Arun, S.; Surendran, K.P.; Sebastian, M.T. LTCC tapes based on Al_2O_3–BBSZ glass with improved thermal conductivity. *Ceram. Int.* **2015**, *41*, 13572–13581. [CrossRef]
6. Yin, L.; Yang, L.; Yang, W.; Guo, Y.; Ma, K.; Li, S.; Zhang, J. Thermal design and analysis of multi-chip LED module with ceramic substrate. *Solid-State Electron.* **2010**, *54*, 1520–1524. [CrossRef]
7. Shimizu, M.; Tomioka, S.; Yoshida, N.; Kameyama, I.; Okuda, K. Varistor and Its Manufacturing Method. Japanese Patent Publication No. 2015-156406, 2015.
8. Shigeno, K.; Katsumura, H.; Kagata, H.; Asano, H.; Inoue, O. Preparation and characterization of low temperature sintered alumina by CuO-TiO_2-Nb_2O_5-Ag_2O additives. *Ferroelectrics* **2007**, *356*, 189–196. [CrossRef]
9. Shigeno, K.; Kuraoka, Y.; Asakawa, T.; Fujimori, H. Sintering mechanism of low-temperature co-fired alumina featuring superior thermal conductivity. *J. Am. Ceram. Soc.* **2021**, *104*, 2017–2029. [CrossRef]
10. Surendran, K.P.; Santha, N.; Mohanan, P.; Sebastian, M.T. Temperature stable low loss ceramic dielectrics in (1−x) $ZnAl_2O_4$–xTiO_2 system for microwave substrate applications. *Eur. Phys. J.* **2004**, *B41*, 301–306. [CrossRef]
11. Surendran, K.P.; Sebastian, M.T.; Manjusha, M.V.; Philip, J. A low loss, dielectric substrate in $ZnAl_2O_4$–TiO_2 system for microelectronic applications. *J. Appl. Phys.* **2005**, *98*, 044101. [CrossRef]
12. Roshni, S.; Sebastian, M.T.; Surendran, K.P. Can zinc aluminate-titania composite be an alternative for alumina as microelectronic substrate? *Sci. Rep.* **2017**, *7*, 40839. [CrossRef]

13. Fu, P.; Wang, Z.Y.; Lin, Z.D.; Liu, Y.Q.; Roy, V.A.L. The microwave dielectric properties of transparent $ZnAl_2O_4$ ceramics fabricated by spark plasma sintering. *J. Mater. Sci. Mater. Electron.* **2017**, *28*, 9589–9595. [CrossRef]
14. Thomas, S.; Sebastian, M.T. Effect of B_2O_3-Bi_2O_3-SiO_2-ZnO glass on the sintering and microwave dielectric properties of $0.83ZnAl_2O_4$-$0.17TiO_2$. *Mater. Res. Bull.* **2008**, *43*, 843–851. [CrossRef]
15. Takahashi, S.; Kan, A.; Ogawa, H. Microwave dielectric properties and cation distributions of $Zn_{1-3x}Al_{2+2x}O_4$ ceramics with defect structures. *J. Eur. Ceram. Soc.* **2017**, *37*, 3059–3064. [CrossRef]
16. Lan, X.-K.; Li, J.; Zou, Z.-Y.; Xie, M.-Q.; Fan, G.-F.; Lu, W.-Z.; Lei, W. Improved sinterability and microwave dielectric properties of $[Zn_{0.5}Ti_{0.5}]^{3+}$-doped $ZnAl_2O_4$ spinel solid solution. *J. Am. Ceram. Soc.* **2019**, *102*, 5952–5957. [CrossRef]
17. Shigeno, K.; Kaneko, S.; Nakashima, H.; Suenaga, K.; Fujimori, H. Low-temperature sintering of gahnite ceramic using Cu-Nb-O additive and evaluation of dielectric and thermal properties. *J. Electron. Mater.* **2020**, *49*, 6046–6054. [CrossRef]
18. Nunes, S.I.; Bradt, R.C. Grain growth of ZnO in ZnO-Bi_2O_3 ceramics with Al_2O_3 additions. *J. Am. Ceram. Soc.* **1995**, *78*, 2469–2475. [CrossRef]
19. Kim, D.-W.; Kim, I.-T.; Park, B.; Hong, K.-S.; Kim, J.-H. Microwave dielectric properties of $(1-x)Cu_3Nb_2O_8-xZn_3Nb_2O_8$ ceramics. *J. Mater. Res.* **2001**, *16*, 1465–1470. [CrossRef]
20. Matsui, K.; Matsumoto, A.; Uehara, M.; Enomoto, N.; Hojo, J. Sintering kinetics at isothermal shrinkage: Effect of specific surface area on the initial sintering stage of fine zirconia powder. *J. Am. Ceram. Soc.* **2007**, *90*, 44–49. [CrossRef]
21. Takada, T.; Wang, S.F.; Yoshikawa, S.; Jang, S.-J.; Newnham, R.E. Effect of glass addition on BaO-TiO_2-WO_3 microwave ceramics. *J. Am. Ceram. Soc.* **1994**, *77*, 1909–1916. [CrossRef]
22. Hakki, B.W.; Coleman, P.D. A dielectric resonator method of measuring inductive capacities in the millimeter range. *IRE Trans. Microw. Theory Tech.* **1960**, *8*, 402–410. [CrossRef]
23. Yamaguchi, T. Characterization techniques of ceramic: Properties of sintered bodies. *Ceram. Jpn.* **1984**, *19*, 520–529.
24. Tamura, H. Microwave dielectric losses caused by lattice defects. *J. Eur. Ceram. Soc.* **2006**, *26*, 1775–1780. [CrossRef]
25. Zheng, C.W.; Fan, X.C.; Chen, X.M. Analysis of infrared reflection spectra of $(Mg_{1-x}Zn_x)Al_2O_4$ microwave dielectric ceramics. *J. Am. Ceram. Soc.* **2008**, *91*, 490–493. [CrossRef]
26. Penn, S.J.; Alford, N.M.; Templeton, A.; Wang, X.; Xu, M.; Reece, M.; Schrapel, K. Effect of porosity and grain size on the microwave dielectric properties of sintered alumina. *J. Am. Ceram. Soc.* **1997**, *80*, 1885–1888. [CrossRef]
27. German, R.M.; Farooq, S.; Kipphut, C.M. Kinetics of liquid sintering. *Mater. Sci. Eng. A* **1988**, *105/106*, 215–224. [CrossRef]
28. German, R.M.; Suri, P.; Park, S.J. Liquid phase sintering. *J. Mater. Sci.* **2009**, *44*, 1–39. [CrossRef]
29. Shannon, R.D. Revised effective ionic radii and systematic studies of interatomic distances in halides and chalcogenides. *Acta Crystallogr. A.* **1976**, *32*, 751–767. [CrossRef]
30. Phillips, D.S.; Mitchell, T.E.; Heuer, A.H. Precipitation in star sapphire III. Chemical effects accompanying precipitation. *Philos. Mag.* **1980**, *A42*, 417–432. [CrossRef]
31. Kingery, W.D.; Bowen, H.K.; Uhlmann, D.R. *Introduction to Ceramics*, 2nd ed.; Wiley: New York, NY, USA, 1976; pp. 413–429.
32. Liu, H.-H.; Jean, J.-H. Processing and properties of a low-fire, high-thermal-conductivity alumina with $CuTiNb_2O_8$. *Int. J. Ceram. Eng. Sci.* **2020**, *2*, 38–45. [CrossRef]

Article

Properties of Magnetic Garnet Films for Flexible Magneto-Optical Indicators Fabricated by Spin-Coating Method

Ryosuke Hashimoto [1,*], Toshiya Itaya [1], Hironaga Uchida [2], Yuya Funaki [2] and Syunsuke Fukuchi [1]

[1] Department of Electronic and Information Engineering, National Institute of Technology, Suzuka College, Shiroko-cho, Suzuka 510-0294, Mie, Japan; itaya@info.suzuka-ct.ac.jp (T.I.); h29e37@ed.cc.suzuka-ct.ac.jp (S.F.)

[2] Department of Electrical and Electronic Information Engineering, Toyohashi University of Technology, Hibarigaoka, Tempaku-cho, Toyohashi 441-8580, Aichi, Japan; uchida@ee.tut.ac.jp (H.U.); funaki.yuya.xk@tut.jp (Y.F.)

* Correspondence: hashimoto@elec.suzuka-ct.ac.jp

Abstract: Non-destructive testing using a magneto-optical effect is a high-resolution non-destructive inspection technique for a metallic structure. It is able to provide high-spatial resolution images of defects. Previously, it has been difficult to fabricate flexible magneto-optical sensors because thermal treatment is necessary to crystallize the magnetic garnet. Therefore, it was not possible to apply magneto-optical imaging to complicated shapes in a test subject, such as a curved surface. In this study, we developed a new process for deposition of the magnetic garnet on the flexible substrate by applying the magnetic garnet powders that have already undergone crystallization. In this new process, as it does not require thermal treatment after deposition, flexible substrates with low heat resistance can be used. In this paper, we report our observations of the optical properties, magnetic hysteresis loop, crystallizability and density of the particles on the flexible substrate deposited by the spin-coating method.

Keywords: magneto-optical imaging; nondestructive testing; magnetic garnet; spin coating method; transmittance spectrum

1. Introduction

Currently, safety is maintained by detecting defects in a structure at an early stage with various non-destructive testing methods. Non-destructive testing using a magneto-optical (MO) effect is a high-resolution non-destructive inspection technique for a metallic structure [1–6]. The MO effect is a phenomenon where a polarization plane of light rotates when a linearly polarized light passes through the magnetized magnetic materials [7]. MO imaging is performed by placing a thin film, wherein a magnetic garnet is layered on a substrate on a test subject as an MO sensor [8–11].

Previously, the deposition process of the MO sensor required high-temperature thermal treatment [12] to crystallize the magnetic garnet because a sputtering method was used for the deposition of the magnetic garnet film [13,14]. Therefore, it was limited to a single crystalline substrate that was not able to deal with complicated shapes, such as a curved surface, which was contained in the test subject.

The purpose of our research is to enable the magnetic garnet film to form on flexible substrates and evaluate its physical properties in order to manufacture a flexible MO sensor capable of coping with curved surfaces. In our research group, we looked at a spin-coating method as the film formation method of the magnetic garnet. There are several methods to prepare the magnetic garnet films by using spin-coating methods. The metal-organic decomposition method [15–23] and the sol-gel method [24–30] are generally used, but heat treatment at the temperature from 600 to 900 °C. is necessary as with the sputtering method. Therefore, we developed a new process for deposition of the magnetic garnet materials on the flexible substrate by applying the magnetic garnet powders that have

already undergone crystallization. In this new process, as it does not require thermal treatment after deposition, flexible substrates with low heat resistance can be used.

In this paper, we report the observation results of the optical properties, magnetic hysteresis loop, crystallizability and density of the particles on the flexible substrate deposited by the spin-coating method.

2. Materials and Methods

The MO effect is a phenomenon related to the polarization state of the transmitted or the reflected light of a magnetized material. This effect contains some varieties, and this study describes MO imaging using the Faraday effect, which is the MO effect for light transmitted through a magnetic material. Figure 1 shows a schematic illustration of the Faraday effect. The Faraday effect is a magnetic phenomenon where a polarization plane rotates when linearly polarized light passes through a magnetic material. The angle of the polarization plane rotated due to the Faraday effect is referred to as the Faraday rotation angle (θ), and its magnitude is given in Equation (1).

$$\theta = F\,(M/MS)\,L \tag{1}$$

In Equation (1), MS is the saturation magnetization of a magnetic material, M is the magnitude of magnetization parallel to the traveling direction of light, and L is the distance where light passes through (i.e., the thickness of the magnetic material). It should be noted that F is a value specific to a material called a Faraday rotational coefficient. As seen in Equation (1), the Faraday rotational angle increases in proportion to the thickness of the magnetic material and the magnitude of magnetization.

The light intensity (I_1) of the MO image generated with an orthogonal analyzer method (Figure 1) is expressed as in the following Equation (2), where I_0 is the light intensity of the incident light, %T is the transmittance of the MO material, and θ is the Faraday rotational angle of the MO material.

$$I_1 = I_0 \times \%T \times \sin^2(\theta) \tag{2}$$

Therefore, it is important that the film has high transmittance and exhibits a large MO effect to obtain high light intensity in MO imaging.

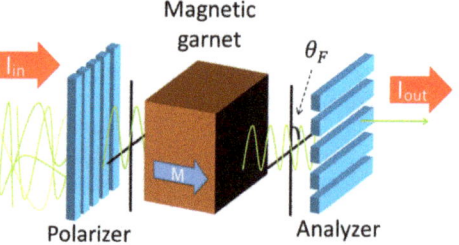

Figure 1. Schematic illustration of magneto-optical (MO) effect.

In this study, we developed a new spin-coating process for film formation of the magnetic garnet on a flexible substrate. Figure 2 shows preparing process of magnetic garnet sediment. We prepared the bismuth-substituted yttrium iron garnet (Bi:YIG, Kojundo Chemical Lab. Co., LTD., Kariya, Aichi, Japan). The Bi:YIG particles were prepared by the co-precipitation method [31–34]. The composition of Bi:YIG was $Bi_{0.5}Y_{2.5}Fe_{5.0}O_{12}$, and the purity of Bi:YIG was 99.9%. A polyvinyl alcohol (PVA, Kishida Chemical Co., LTD, Osaka, Osaka, Japan) aqueous solution was used as the organic binder. PVA concentrations in the PVA aqueous solution were 10 wt% and 15 wt%. Bi:YIG particles were mixed into the PVA aqueous solution with a stirrer at a weight ratio of 1:3 for 24 h. The mixture was deposited on the flexible substrate with the spin-coating method. After spin coating, the substrate was dried at room temperature to form the magnetic garnet sediment. Table 1 shows the film formation condition for a Bi:YIG-PVA sediment.

We investigated the transmittance, magnetic hysteresis loop and crystallizability of a Bi:YIG-PVA sediment on the flexible substrate deposited by this new spin-coating method. Transmittance of the Bi:YIG-PVA sediment obtained was measured with a visible spectrophotometer, the magnetic hysteresis loop was measured with a vibrating sample magnetometer (VSM, Tamakawa Co., LTD, Sendai, Miyagi, Japan) and crystallizability was evaluated with X-ray diffraction (XRD, Rigaku Co., LTD, Akishima, Tokyo, Japan).

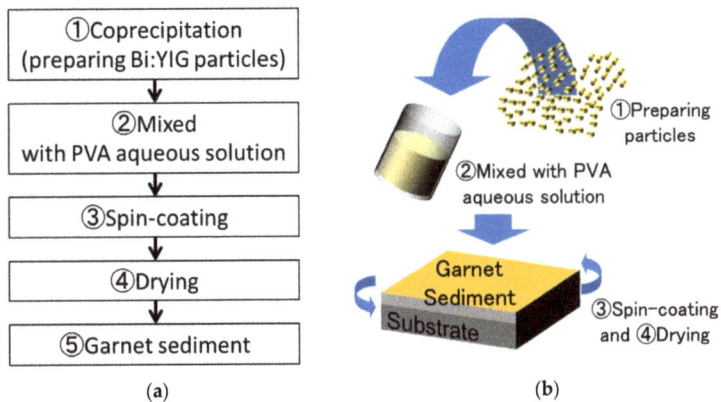

Figure 2. Process of preparing garnet sediment: (**a**) flow of process; (**b**) schematic illustration of spin-coating process.

Table 1. Film formation condition for bismuth-substituted yttrium iron garnet (Bi:YIG-PVA) sediments.

Sample Number	PVA Aqueous Solution Mass % Concentration (wt%)	Mass Ratio Bi:YIG-PVA	Spin Coating Condition (rpm/sec)	
			Step 1	Step 2
Sample 1	10	1:3	500/10	6000/60
Sample 2	15			

3. Results

3.1. Bismuth-Substituted Yttrium Iron Garnet Film

The bending image of a flexible Bi:YIG-PVA sediment is shown in Figure 3a. The magnetic garnet sediment could be bent freely, and the surface did not crack or break away even after it was bent, remaining in good shape. The thickness of the Bi:YIG-PVA sediment was approximately 30 μm which was estimated from cross-sectional observation by Field Emission Scanning Electron Microscope (SEM, Japan Electron Optics Laboratory Co., LTD, Akishima, Tokyo, Japan) as shown in Figure 3b. We confirmed the crystallization of the Bi:YIG-PVA sediment by XRD. A deposited Bi:YIG sediment showed typical peaks of magnetic iron garnet [35,36] as shown in Figure 3c. Figure 3d shows a hysteresis loop of a flexible Bi:YIG-PVA sediment by VSM. The Bi:YIG-PVA sediment showed a ferromagnetic property.

Figure 4 shows the transmittance spectrum of the magnetic garnet film measured with the visible spectrophotometer. Both Sample 1 and Sample 2 had almost 0% transmittance in the wavelength range from 400 nm to 1400 nm. As shown in Equation (2), MO imaging requires that light pass through it to some extent, and previous studies have shown that a transmittance of about 20% is necessary. There are several possible reasons for the 0% transmittance, such as thickness, multiple scattering and iron absorbed.

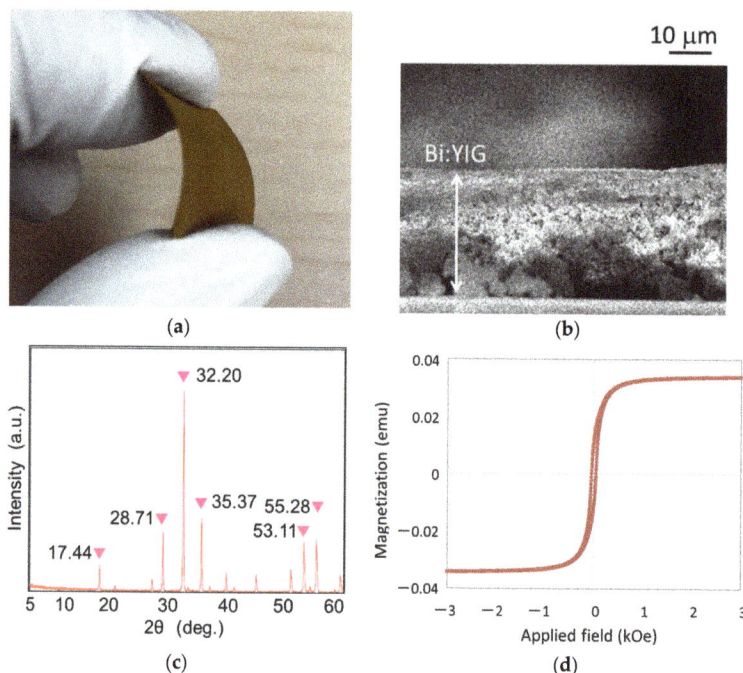

Figure 3. Result of surface examination: (**a**) bending image of a flexible Bi:YIG-PVA sediment; (**b**) cross-sectional observation by SEM; (**c**) X-ray diffraction (XRD) patterns; (**d**) hysteresis loop of a flexible Bi:YIG sediment by a vibrating sample magnetometer (VSM).

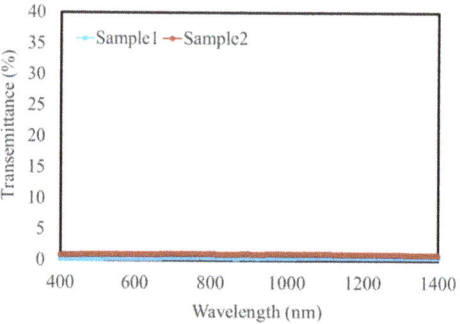

Figure 4. Transmittance spectrum of Bi:YIG-PVA film.

3.2. Yttrium Aluminum Garnet Film

Because Bi:YIG-PVA sediment had almost 0% transmittance in the visible range, we used yttrium aluminum garnet (YAG) particles, in which the iron sites of Bi:YIG were substituted with aluminum and have a small absorption coefficient, to form the film and investigate its optical properties. The composition of YAG was $Y_3Fe_5O_{12}$, and the purity of Bi:YIG was 99.9%. The film formation conditions are shown in Table 2. The transmittance spectrum of the deposited YAG-PVA sediments is shown in Figure 5. Based on the measurement results, we were able to confirm that the YAG-PVA sediment exhibited around 10% to 40% transmittance in all film formation conditions. A distortion in the wave form harmonic around the wavelength of 800 nm was due to noise caused by the detector of the spectrophotometer switching from the visible region to the near-infrared region.

Table 2. Film formation condition for YAG-PVA sediments.

Sample Number	PVA Aqueous Solution Mass % Concentration (wt%)	Mass Ratio YAG-PVA	Spin Coating Condition (rpm/sec)	
			Step 1	Step 2
Sample 1				1000/60
Sample 2	15	1:3	500/10	3000/60
Sample 3				6000/60

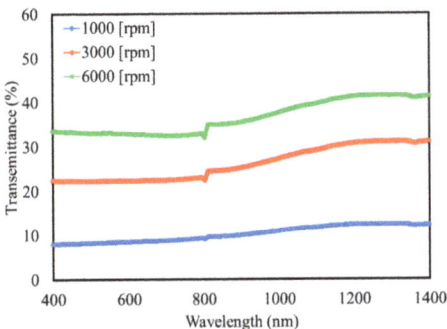

Figure 5. Transmittance spectrum of YAG-PVA film.

4. Discussion

The deposited layered with this process is a discontinuous sediment, wherein the particles are dispersed on the substrate. Therefore, it is expected that the light irradiated to the magnetic garnet particles will undulate by resonance or absorb the particles and behave as shown in Figure 6. In addition, the light intensity measured with the spectrophotometer may include the light passed between particles. Such light contributes to the transmittance but cannot be influenced by the Faraday effect because it does not pass through the magnetic material. Thus, we estimated the ratio of light that passed through the particle by analyzing the density of the magnetic garnet particle on the substrate.

The weight of the particles deposited on the substrate was measured to calculate the density of the magnetic garnet particles. The mass of the samples includes the mass of the substrate and PVA. Therefore, we measured the mass of the substrate before deposition and the mass of the sediment deposited with PVA alone. Then, the mass of the YAG-PVA sediment deposited on the substrate was measured by subtracting the mass of the substrate and PVA from the mass of the YAG-PVA sediment after deposition. The density of the YAG on the substrate was calculated by dividing the mass of the YAG-PVA sediment measured by the surface area of the substrate.

The calculated density of YAG per unit area on the substrate is shown by the solid line in Figure 7. As the rotational velocity of the spin coater increased, the density decreased. This is because the particles depositing on the substrate decreased as the rotational velocity increased to increase the centrifugal force. The density on the substrate became almost constant at a rotational velocity of 3000 rpm or more around 0.35 mg/cm^2. The calculation result in the case where the particles were densely deposited on the substrate in only one layer is shown by the broken line in Figure 7. The rotational velocity asymptotically approached the broken line with samples of 3000 rpm or more, suggesting that the particles were densely deposited on the substrate at around one layer. Therefore, the void space between the particles on the substrate could be assumed to be around 20%.

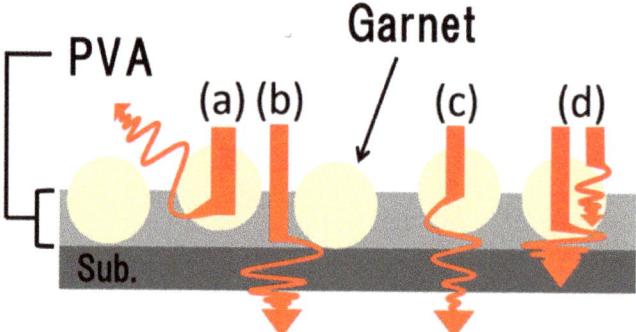

Figure 6. Behavior of the light irradiated to the magnetic garnet film: (**a**) reflected light; (**b**) the light passed between particles; (**c**) transmitted light; (**d**) absorption.

Figure 8a shows a bending image of flexible YAG-PVA sediment. The cross-sectional observations of the deposited YAG-PVA sediment by SEM were shown from Figure 8b to Figure 8c. In Figure 8b, it can be seen that the YAG-PVA particles appear to deposit on the substrate in multiple layers, while forming a cluster. Figure 8d shows the results of cross-sectional observation of sample 3 in Table 2. Meanwhile, it was confirmed that the thickness of the sediment decreased as the rotational velocity of the spin coating increased and was almost at the same level at both 3000 rpm and 6000 rpm. These observations closely correspond with the calculated results shown in Figure 7, and the validity of the calculation was confirmed.

Figure 9 shows the Faraday rotation angle of YAG-PVA sediment. The plot of the circle shows only substrate (without YAG-PVA sediment), and the square plot shows the Faraday rotation angle of the YAG-PVA sediment in Figure 9. This Faraday rotation result shows the paramagnetic material property. From the results in Figures 3 and 9, we confirmed that the garnet sediment was formed by a new spin-coating process.

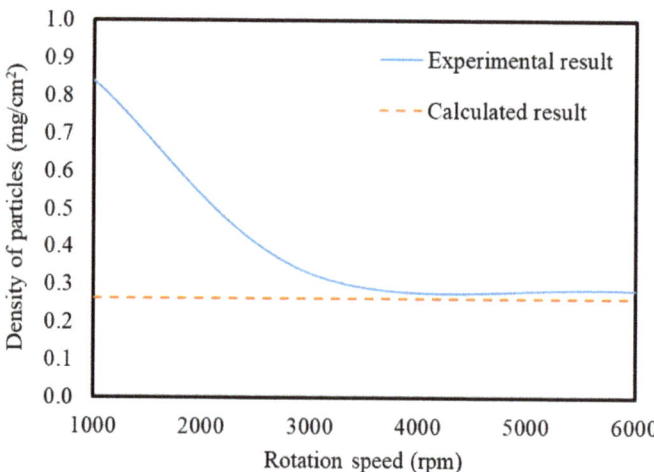

Figure 7. Calculation of the density of the YAG-PVA particle per unit area on the substrate.

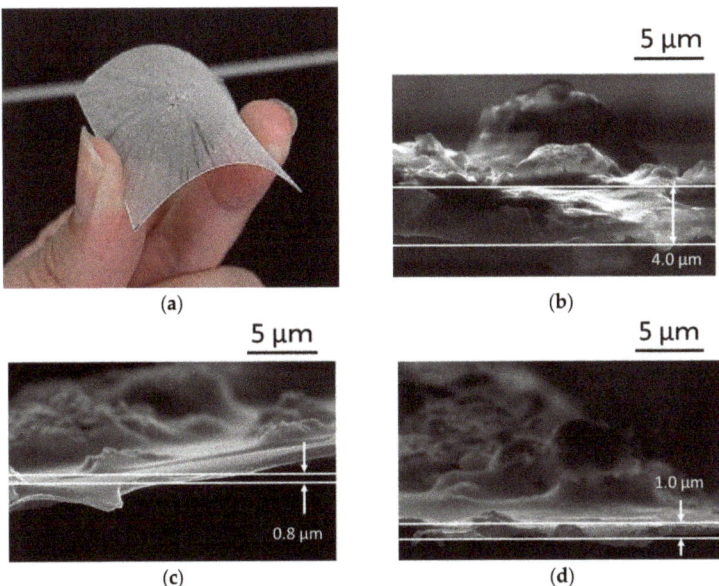

Figure 8. Surface observation of YAG-PVA sediment: (**a**) bending image of flexible YAG-PVA sediment; (**b**) cross-sectional observation of sample 1 by SEM; (**c**) sample 2; (**d**) sample 3.

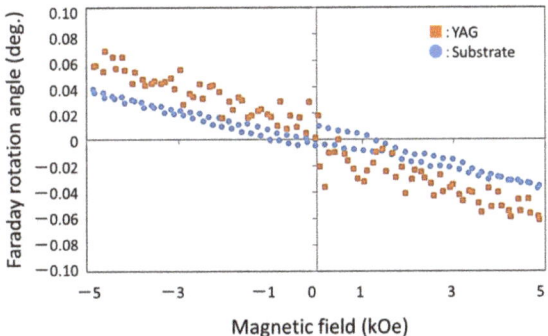

Figure 9. Faraday rotation angle of YAG-PVA sediment.

5. Conclusions

Magnetic garnet sediment that can be flexibly bent was manufactured with a new spin-coating process. We investigated the transmittance, magnetic hysteresis loop and crystallizability of a Bi:YIG-PVA sediment on the flexible substrate deposited by this new spin-coating method. We deposited YAG-PVA sediment, and transmittance was measured. As a result, a translucency of 40% was observed.

The density of a YAG-PVA particle deposited on the substrate became constant at a rotational velocity of the spin coating of 3000 rpm or more, and it was conceivable that the particle deposited densely in one layer. Based on these results, it became clear that void space between particles on the substrate was around 20%, and that the YAG-PVA sediment showed translucency of 20% or more.

In the future, we are planning to develop a hybrid material, in which some of the iron sites of Bi:YIG are substituted with aluminum, for performing MO imaging.

Author Contributions: Conceptualization, R.H.; data curation, R.H., T.I., H.U., Y.F. and S.F.; formal analysis, R.H. and H.U.; investigation, T.I., H.U., Y.F. and S.F. methodology, R.H. writing—original draft preparation, R.H.; writing—review and editing, R.H.; project administration, R.H.; funding acquisition, R.H. All authors have read and agreed to the published version of the manuscript.

Funding: This research was funded by JSPS KAKENHI grant number JP19K14851.

Institutional Review Board Statement: Not applicable.

Informed Consent Statement: Not applicable.

Conflicts of Interest: The authors declare no conflict of interest.

References

1. Fitzpatrick, G.L.; Thome, D.K.; Skaugset, R.L.; Shih, E.Y.C.; Shin, W.C.L. Magneto-optic/eddy current imaging of aging aircraft: A new NDI technique. *Mater. Eval.* **1993**, *51*, 1402–1407.
2. van der Laan, D.C.; van Eck, H.J.N.; Davidson, M.W.; ten Haken, B.; ten Kate, H.H.; Schwartz, J. Magneto-optical imaging study of the crack formation in superconducting tapes caused by applied strain. *Phys. C* **2002**, *372–376*, 1020–1023. [CrossRef]
3. Novotny, P.; Sajdl, P.; Machac, P. A magneto optic imager for NDT applications. *NDT E Int.* **2004**, *37*, 645–649. [CrossRef]
4. Gao, X.; Liu, Y.; You, D. Detection of micro-weld joint by magneto-optical imaging. *Opt. Technol.* **2014**, *62*, 141–151. [CrossRef]
5. Gao, X.; Chen, Y.; You, D.; Xiao, Z.; Chen, X. Detection of micro gap weld joint by using magneto-optical imaging and Kalman filtering compensated with RBF neural network. *Mech. Syst. Signal Process.* **2017**, *84*, 570–583. [CrossRef]
6. Gao, X.; Ma, N.; Du, L. Magneto-optical imaging characteristics of weld defects under alternating magnetic field excitation. *Opt. Express* **2018**, *26*, 9972–9983. [CrossRef] [PubMed]
7. Pershan, P.S. Magneto-optical effects. *J. Appl. Phys.* **1967**, *38*, 1482–1490. [CrossRef]
8. Joubert, P.Y.; Pinassaud, J. Linear magneto-optic imager for non-destructive evaluation. *Sens. Actuat. A* **2006**, *129*, 126–130. [CrossRef]
9. Lee, H.; Kim, T.; Kim, S.; Yoon, Y.; Kim, S.; Babajanyan, A.; Ishibashi, T.; Friedman, B.; Lee, K. Magneto-optical imaging using a garnet indicator film prepared on glass substrates. *J. Magn. Magn. Mater.* **2010**, *322*, 2722–2727. [CrossRef]
10. Bazijievich, M.; Barness, D.; Sinvani, M.; Perel, E.; Shaulov, A.; Yeshurun, Y. Magneto-optical system for high speed real time imaging. *Rev. Sci. Instrum.* **2012**, *83*, 083707. [CrossRef]
11. Murakami, H.; Tonouchi, M. High-sensitive scanning laser magneto-optical imaging system. *Rev. Sci. Instrum.* **2010**, *81*, 013701. [CrossRef]
12. Suzuki, T.; Zaharchuk, G.; Gorman, G.; Sequeda, F.; Labun, P. Magnetic and magneto-optical properties and crystallization kinetics of rapid-thermally crystallized Bi-substituted garnet films. *IEEE Trans. Magn.* **1990**, *26*, 1927–1929. [CrossRef]
13. Takagi, R.H.H.; Yonezawa, T.; Sakaguchi, K.; Inoue, M. Magneto-optical imaging using magnetophotonic crystals. *J. Appl. Phys.* **2014**, *115*, 17A931.
14. Hashimoto, R.; Yonezawa, T.; Takagi, H.; Goto, T.; Endo, H.; Nishimizu, A.; Inoue, M. Defect depth estimation using magneto optical imaging with magnetophotonic crystal. *J. Magn. Soc. Jpn.* **2015**, *39*, 213–215. [CrossRef]
15. Ishibashi, T.; Mizusawa, A.; Nagai, M.; Shimizu, S.; Sato, K. Characterization of epitaxial (Y,Bi)3(Fe,Ga)5O12 thin films grown by metal-organic decomposition method. *J. Appl. Phys.* **2005**, *97*, 013516. [CrossRef]
16. Ishibashi, T.; Kawata, T.; Johansen, T.H.; He, J.; Harada, N.; Sato, K. Magneto-optical indicator garnet films grown by metal-organic decomposition method. *J. Magn. Soc. Jpn.* **2008**, *32*, 150–153. [CrossRef]
17. Ishibashi, T.; Kosaka, T.; Naganuma, M.; Nomura, T. Magneto-optical properties of Bi-substituted yttrium iron garnet films by metal-organic decomposition method. *J. Phys. Conf. Ser.* **2010**, *200*, 112002. [CrossRef]
18. Lee, H.; Yoon, Y.; Kim, S.; Yoo, H.K.; Melikyan, H.; Danielyan, E.; Babajanyan, A.; Ishibashi, T.; Friedman, B.; Lee, K. Preparation of bismuth substituted yttrium iron garnet powder and thin film by the metal-organic decomposition method. *J. Cryst. Growth* **2011**, *329*, 27–32. [CrossRef]
19. Ishibashi, T.; Yoshida, T.; Kobayashi, T.; Ikehara, S.; Nishi, T. Preparation of Y0.5Bi2.5Fe5O12 films on glass substrates using magnetic iron garnet buffer layers by metal-organic decomposition method. *J. Appl. Phys.* **2013**, *113*, 17A926. [CrossRef]
20. Kim, Y.; Bang, D.J.; Kim, Y.; Jung, J.; Hur, N.; You, C.-Y.; Kim, K.H. Magneto-optic property measurement of bismuth substituted yttrium iron garnet films prepared by metal-organic-decomposition method at the 1310-nm and 1550-nm wavelengths. *J. Magn. Magn. Mater.* **2019**, *492*, 165673. [CrossRef]
21. Nadinov, I.; Kovalenko, O.; Rehspringer, J.-L.; Vomir, M.; Mager, L. Limits of the magneto-optical properties of Bi: YIG films prepared on silica by metal organic decomposition. *Ceram. Int.* **2019**, *45*, 21409–21412. [CrossRef]
22. Ishibashi, T. Magneto-optical imaging using bismuth-substituted iron garnet films prepared by metal–organic decomposition. *J. Magn. Soc. Jpn.* **2020**, *44*, 108–116. [CrossRef]
23. Wang, S.-F.; Chorazewicz, K.; Lamichhane, S.; Parrott, R.A.; Cabrini, S.; Fischer, P.; Kent, N.; Turner, J.H.; Ishibashi, T.; Frohock, Z.P.; et al. Sweet Ferromagnetic resonances in single-crystal yttrium iron garnet nanofilms fabricated by metal-organic decomposition. *Appl. Phys. Lett.* **2021**, *119*, 172405. [CrossRef]

24. Matsumoto, K.; Sasaki, S.; Asahara, Y.; Yamaguchi, K.; Fujii, T. Highly bismuth substituted yttrium iron garnet single crystal films prepared by sol-gel method. *J. Magn. Magn. Mater.* **1992**, *104–107*, 451–452. [CrossRef]
25. Arsad, A.Z.; Ibrahim, N.B. Temperature-dependent magnetic properties of YIG thin films with grain size less 12 nm prepared by a sol-gel method. *J. Magn. Magn. Mater.* **2018**, *462*, 70–77. [CrossRef]
26. Leal, L.R.F.; Guerra, Y.; Padrón-Hernández, P.; Rodrigues, A.R.; Santos, F.E.P.; Peña-Garcia, R.P. Structural and magnetic properties of yttrium iron garnet nanoparticles doped with copper obtained by sol gel method. *Mater. Lett.* **2019**, *236*, 547–549. [CrossRef]
27. Elhamali, S.M.; Ibrahim, N.B.; Radiman, S. Structural, optical and magnetic properties of YIG and TbErIG nanofilms prepared using a sol-gel method. *Mater. Res. Bull.* **2019**, *112*, 66–76. [CrossRef]
28. Khan, M.I.; Waqas, M.; Naeem, M.A.; Hasan, M.S.; Iqbal, M.; Mahmood, A.; Ramay, S.M.; Al-Masry, W.A.; Abubshait, S.A.; Abubshait, H.A.; et al. Magnetic behavior of Ga doped yttrium iron garnet ferrite thin films deposited by sol-gel technique. *Ceram. Int.* **2020**, *46*, 27318–27325. [CrossRef]
29. Yao, C.; Hao, A.; Kumar, S.; Huang, T.W.; Qin, N.; Bao, D. Realization of resistive and magnetization switching in sol-gel derived yttrium iron garnet thin films. *Thin Solid Film.* **2020**, *699*, 137889. [CrossRef]
30. SElhamali, M.; Ibrahim, N.B.; Radiman, S. Oxygen vacancy-dependent microstructural, optical and magnetic properties of sol-gel Tb0.2 Er1 Y2.8 Fe5 O12 films. *J. Magn. Magn. Mater.* **2020**, *497*, 166048. [CrossRef]
31. Grosseau, P.; Bachiorrini, A.; Guilhota, B. Preparation of polycrystalline yttrium iron garnet ceramics. *Powder Technol.* **1997**, *93*, 247–251. [CrossRef]
32. Jafelicci, M., Jr.; Godoi, R.H.M. Preparation and characterization of spherical yttrium iron garnet via coprecipitation. *J. Magn. Magn. Mater.* **2001**, *226–230*, 1421–1423. [CrossRef]
33. Kuroda, C.S.; Taniyama, T.; Kitamoto, Y.; Yamazaki, Y. Magneto-optical properties and morphology of particulate film consisting of Bi-YIG coprecipitated particles. *J. Magn. Magn. Mater.* **2002**, *241*, 201–206. [CrossRef]
34. Liu, J.; Jin, Q.; Wang, S.; Yu, P.; Zhang, C.; Luckhardt, C.; Su, Z.; Barua, R.; Harris, V.G. An insight into formation mechanism of rapid chemical Co-precipitation for synthesizing yttrium iron garnet nano powders. *Mater. Chem. Phys.* **2018**, *208*, 169–176. [CrossRef]
35. Vaqueiro, P.; Crosnier-Lopez, M.P.; Lopez-Quintela, M.A. Synthesis and Characterization of Yttrium Iron Garnet Nanoparticles. *J. Solid State Chem.* **1996**, *126*, 161–168. [CrossRef]
36. Rehspringer, J.-L.; Bursik, J.; Niznansky, D.; Klarikova, A. Characterisation of bismuth-doped yttrium iron garnet layers prepared by sol-gel process. *J. Magn. Magn. Mater.* **2000**, *211*, 291–295. [CrossRef]

Article

Morphology Control of Monomer–Polymer Hybrid Electron Acceptor for Bulk-Heterojunction Solar Cell Based on P3HT and Ti-Alkoxide with Ladder Polymer

Yasuyuki Ueda [1,*], Yuki Kurokawa [2], Kei Nishii [1,*], Hideyuki Kanematsu [3], Tadashi Fukumoto [4] and Takehito Kato [1,5,*]

1. Department of Materials Chemistry and Bioengineering, National Institute of Technology, Oyama College, 771 Nakakuki, Oyama 323-0806, Japan
2. Graduate School of Life Science and Systems Engineering, Kyushu Institute of Technology, 2-4 Hibikino, Wakamatsu-ku, Kitakyushu-shi 808-0196, Japan; kurokawa.yuki207@mail.kyutech.jp
3. Department of Materials Science and Engineering, National Institute of Technology, Suzuka College, Shiroko-cho, Suzuka 510-0294, Japan; kanemats@mse.suzuka-ct.ac.jp
4. Technical Research Institute, Nishimatsu Construction Co., Ltd., 1-17-1 Minato-ku, Toranomon, Tokyo 105-6490, Japan; tadashi_fukumoto@nishimatsu.co.jp
5. NPO Energy Education, 6-20-21 Ekiminami, Oyama 323-0822, Japan
* Correspondence: ueda.y.ak@oyama-ct.ac.jp (Y.U.); k.nishii@oyama-ct.ac.jp (K.N.); kato_t@oyama-ct.ac.jp (T.K.)

Abstract: We report the morphology control of a nano-phase-separated structure in the photoactive layer (power generation layer) of organic–inorganic hybrid thin-film solar cells to develop highly functional electronic devices for societal applications. Organic and inorganic–organic hybrid bulk heterojunction solar cells offer several advantages, including low manufacturing costs, light weight, mechanical flexibility, and a potential to be recycled because they can be fabricated by coating them on substrates, such as films. In this study, by incorporating the carrier manager ladder polymer BBL as the third component in a conventional two-component power generation layer consisting of P3HT—the conventional polythiophene derivative and titanium alkoxide—we demonstrate that the phase-separated structure of bulk heterojunction solar cells can be controlled. Accordingly, we developed a discontinuous phase-separated structure suitable for charge transport, obtaining an energy conversion efficiency higher than that of the conventional two-component power generation layer. Titanium alkoxide is an electron acceptor and absorbs light with a wavelength lower than 500 nm. It is highly sensitive to LED light sources, including those used in homes and offices. A conversion efficiency of 4.02% under a 1000 lx LED light source was achieved. Hence, high-performance organic–inorganic hybrid bulk heterojunction solar cells with this three-component system can be used in indoor photovoltaic systems.

Keywords: solar cell; energy conversion; organic–inorganic hybrid material; Ti-alkoxide; P3HT

1. Introduction

In recent years, multiple coated solar cells have been proposed; these include dye-sensitized solar cells [1–4], organic thin-film solar cells, and perovskite solar cells, all of which continue to improve in terms of efficiency and functionality [5–18]. In particular, the power conversion efficiency (PCE) of perovskite solar cells has exceeded 20%, and high-performance electronic devices based on organic–inorganic hybrid materials have been demonstrated [18]. A PCE of 15% has been reported for single-junction organic thin-film solar cells [19], and a further performance improvement is expected as they move towards practical use. The theoretical PCE of organic thin-film solar cells has exceeded 20% [20], and further improvements in efficiency are desirable. The general photoactive layer (power generation layer) of organic thin-film solar cells is composed of p-type semiconductor material as an electron donor and n-type semiconductor material as an electron acceptor. Fullerene derivatives, such as 6,6-phenyl-C61 or C71-butyric acid methyl ester ($PCBM_{60}$ or $PCBM_{70}$), have been used as typical n-type semiconductor materials for electron acceptors. However, fullerene derivatives have low air stability due to oxidation, when the inks of these derivatives are formulated for the fabrication of the power generation layer [21]. Therefore, the development of air-stable electron acceptors is desired. Previous studies from this perspective have reported organic–inorganic hybrid thin-film solar cells that use metal alkoxides and oxides as electron acceptors [22,23]. Among metal oxides and alkoxides, titanium dioxide and alkoxide are light absorbers that can absorb light below 500 nm, making them highly sensitive to LED light sources used indoors and suitable for indoor photovoltaic systems. However, when titanium alkoxide is used as an electron acceptor, the PCE is only 0.03%, and the short-circuit current density (J_{sc}) is 191 µA/cm^2 [22]. Therefore, to achieve the higher efficiency required for practical use, it is necessary to improve the J_{sc}. The thickness of the bulk heteroelectric layer is of the order of approximately 100 nm (97–102 nm), and the ultra-thin film is collectively responsible for light absorption, charge separation, and charge transport [22]. Therefore, it is important to control the morphology, considering all aspects of charge separation, charge transport, and optical absorption [5,6]. Recently, many p-type semiconducting polymers with optical absorption in the long-wavelength range have been developed to obtain a higher J_{sc} [8,9], but their power generation characteristics are inadequate. One of the reasons why sufficient improvements in efficiency are not observed is that it is difficult to maintain solubility in solvent and control the self-assembly (self-aggregation) of a polymer to form a phase-separated structure that is suitable for charge separation and charge transport.

We propose a phase-separated structure for the photoactive layer by utilizing the steric hindrance of the electron-accepting material and demonstrate its effectiveness [22,23]. However, these methods, based on monomers with electron-accepting properties, are insufficient to efficiently suppress the self-assembly of polymers as electron donors. This is because the self-assembly capacity of polymers is higher than the self-assembly capacity of monomers. In contrast, organic thin-film solar cells with bulk heterostructures that use polymers with electron acceptor properties have been reported. However, the formation of dense phase-separated structures that are necessary for the construction of many charge-separated interfaces is not sufficient due to the self-aggregation of the electron-donor and electron-acceptor polymers [24]. This report details our attempt to control the phase-separated structure in the power generation layer by using both monomers and polymers, which have different self-aggregation and molecular hindrance properties, as electron acceptors. In this report, p-type semiconducting polymeric compounds were used as the electron donor and titanium alkoxide was used as the monomer electron acceptor, and poly(benzimidazobenzophenanthroline) (BBL) as the polymer electron acceptor [25]. To create a co-continuous charge transport path in the acceptor, BBL, known to be a ladder polymer with strong self-aggregation, was employed as the electron acceptor [25,26]. It is worth noting that charge separation has been confirmed for each electron acceptor in combination with the p-type semiconducting polymer compound poly(3-hexylthiophene-2,5-diyl) (P3HT) [27]. By controlling the phase-separated structure of the three-component bulk heterostructures using monomer–polymer composite electron acceptors, we achieved high-performance charge management in bulk heterojunction thin-film solar cells.

2. Materials and Methods

General Procedures and Materials

For the photoactive layer, we used P3HT as the electron-donor material, titanium(IV) isopropoxide as the monomeric electron-acceptor material, and BBL as the polymeric electron-acceptor material. These were purchased from Sigma-Aldrich Japan (Tokyo, Japan). Figure 1 shows the molecular structures and energy diagrams of each material.

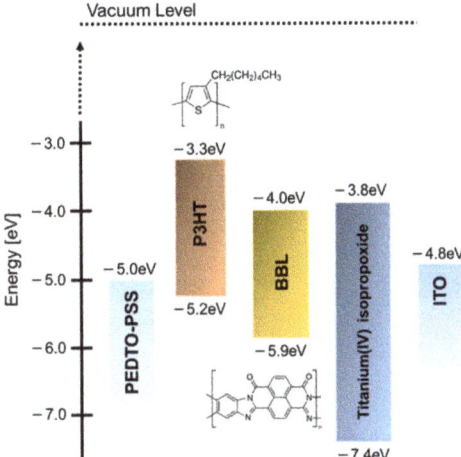

Figure 1. Materials and energy diagrams.

The solar cells were first fabricated by patterning them on an indium tin oxide (ITO)-coated glass substrate (15 Ω per square). These substrates were washed with detergent, acetone, and isopropyl alcohol for 15 min each and then dried in air. In addition, the surface was treated with a UV/O_3 cleaner (Filgen, Model UV253E, Aichi, Japan) for 20 min. For the electron collection layer, a functional thin-film layer was fabricated by dropping 0.50 wt.% titanium(IV) isopropoxide dissolved in isopropyl alcohol onto a substrate and spin coating at 2000 rpm for 30 s on an ITO substrate [23]. Next, a chlorobenzene solution of P3HT (3.0 wt.%) A and chlorobenzene solution of titanium(IV) isopropoxide (6.0 wt.%) B were prepared separately [28].

These solutions were mixed (A+B = 1:1 wt.%) to prepare an ink for forming a charge management layer in the single electron-accepting system. This was spin coated on the functional layer at 2000 to 3500 rpm for 60 s. The functional layers were about 50 nm thick. The charge management layer of the monomer–polymer hybrid electron-accepting system was also fabricated through the same process. For the n-type photoactive layer solution, a solution of P3HT (3.0 wt.%) was added to chlorobenzene (electron-donor solution), and a solution of titanium(IV) alkoxide (6.0 wt.%), methanesulfonic acid (1.2 wt.%), and BBL (0.5, 0.1, 0.15, 2.0, and 3.0 wt.%) dissolved in chlorobenzene was prepared. Each solution was prepared by mixing (1:1 wt.%). These charge management layers were dried in the dark in air and at room temperature for 10 min, after which the thickness of each photoactive layer was about 50 nm. Furthermore, to promote the hydrolysis of titanium(IV) isopropoxide, the substrate containing the photoactive layer was placed in a humidity chamber overnight at 80% humidity. Organic electrodes (50 μm) were applied using a screen printer (Mitani Micronics, MEC-2400, Tokyo, Japan). Poly(3,4-ethylenedioxythiophene)–poly(styrenesulfonic acid) (PEDOT-PSS, Clevios S V3, Leverkusen, Germany) was purchased from Baytron (Tokyo, Japan) and used to form the organic electrodes. Finally, the photoactive layer and organic electrodes were laminated with epoxy resin. Figure 2 shows the structure of the device. The current density–voltage (J–V) characteristics of the solar cells were measured under AM 1.5G quasi-solar radiation (San-Ei Electric, XES-40S1, Osaka, Japan) at 100 mW/cm^2 using a DC voltage and current source/monitor. The incident photon-to-current efficiency (IPCE) of the device was measured using the DC voltage and current source/monitor (Bunko-Keiki Co., Ltd., CEP-2000RS, Tokyo, Japan). The light intensity was corrected using a silicon photodiode reference cell (Bunko-Keiki Co., Ltd., BS-520, Tokyo, Japan). To evaluate the photovoltaic performance, an LED light source (Bunko-Keiki Co., Ltd., BLD-100, Tokyo, Japan) was used at 1000 lx to simulate standard indoor light. The phase-separated structure of the photoactive layer was observed using a scanning electron microscope (SEM, JEOL, JSM-7800, Tokyo, Japan). The UV–Vis/near-IR spectrum of each photoactive layer obtained by spin coating on the glass substrate was measured using a UV spectrophotometer (Shimadzu, UV-1800, Kyoto, Japan). Furthermore, the electrical conductivity of each photoactive layer on the spin-coated glass substrate was measured by the four-point probe method using a resistivity system (Mitsubishi Chemical Analytech Co., Ltd., Lorester-GX MCP-T700, Kanagawa, Japan) [29–33].

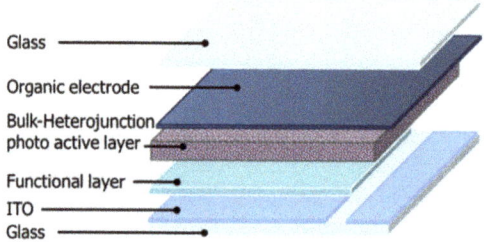

Figure 2. Device structure for solar cells.

3. Results

Figure 3 shows the J–V characteristics of each bulk heterojunction solar cell using titanium-alkoxide and P3HT as the photoactive layer. Table 1 lists the corresponding performance parameters.

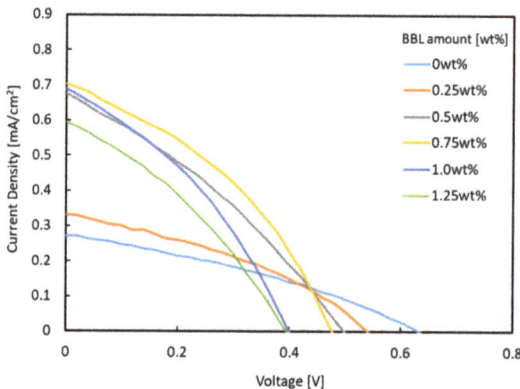

Figure 3. J–V characteristics for solar cells under AM 1.5.

Table 1. Solar cell performance under AM 1.5.

Category	BBL Amount [wt%]	J_{sc} [mA/cm^2]	V_{oc} [V]	Fill Factor	PCE [%]
Single electron acceptor	0	0.272	0.632	0.336	0.057
Hybrid electron acceptor	0.25	0.332	0.540	0.364	0.065
	0.5	0.676	0.498	0.318	0.107
	0.75	0.704	0.476	0.376	0.126
	1.00	0.689	0.399	0.350	0.096
	1.25	0.596	0.395	0.330	0.078

When a monomer–polymer hybrid electron acceptor was used in the charge management layer, the J_{sc} value was higher than that of the device using only titanium alkoxide as a single electron acceptor. In particular, the device with a BBL of 0.75 wt.% showed the highest J_{sc} (0.704 mA/cm^2). The relationship between the amount of BBL added and the J_{sc} value is discussed below. Most of the hybrid electron-accepting devices showed higher fill factor (FF) values than the Ti electron-accepting devices. In contrast, the use of the hybrid electron acceptor reduced the open-circuit voltage (V_{oc}) from 0.632 V to a minimum of 0.451 V. As the lowest unoccupied molecular orbital (LUMO) of BBL is lower than the LUMO of Ti-alkoxide, V_{oc} decreased when BBL was added. As shown in Table 1, the solar cell energy conversion efficiency increased from 0.057% to a maximum of 0.126%, while improving the J_{sc} and FF values. The UV–Vis/near-IR spectra (optical absorption spectra) of each photoactive layer were measured to verify these results. Figure 4 shows the UV–Vis/near-IR spectrum measurement results.

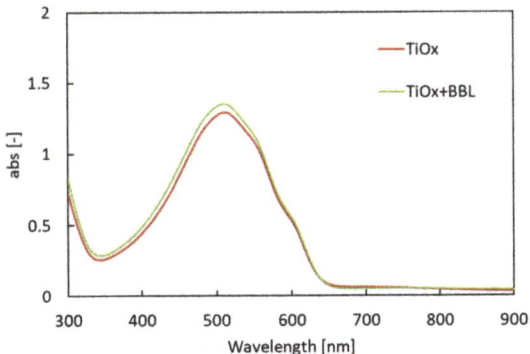

Figure 4. UV–Vis/near-IR spectrum for photoactive layers.

The results show the optical absorption spectrum of each charge management layer to be almost the same, with a peak at 515 nm. Therefore, the improvement in J_{sc} was not affected by the increase in the light absorption of the charge management layer. Furthermore, the addition of BBL does not affect the light absorption of the charge management layer. Next, the IPCE of each solar cell (red: solar cell with single electron acceptor (TiOx only); green: solar cell with hybrid electron acceptor (TiOx + BBL)) was investigated. Figure 5 shows the IPCE spectra [34].

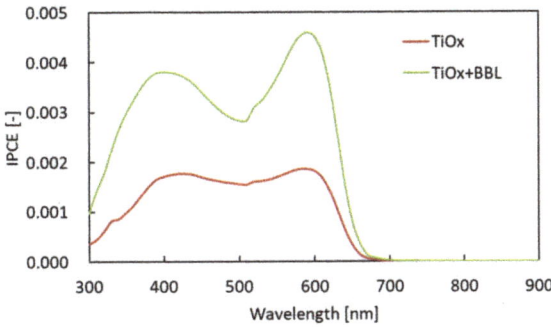

Figure 5. IPCE spectra of each solar cell (red: solar cell with single electron acceptor (TiOx only); green: solar cell with hybrid electron acceptor (TiOx + BBL)).

The IPCE of the solar cell with the hybrid electron acceptor was higher than the IPCE with the Ti-alkoxide single electron acceptor. Therefore, in this case, it was found that the high value of J_{sc} in the cell using the hybrid electron acceptor influences the morphology of the charge management layer. There are three main aspects of the morphology of the charge management layer that influence the achievement of a high J_{sc}. First, it is vital that the excitons generated by photoabsorption diffuse from the electron-donor polymer to the p/n interface. Next, it is important to express charge separation in order to generate a large number of free carriers at the p/n interface. Finally, after charge separation, a charge management layer is required to move the deactivated free carriers from the p/n interface to the electrodes at high speed, without exciting them. In other words, the charge management layer requires the construction of a large number of p/n interfaces for charge separation and a continuous phase-separated structure for the transport of free carriers. Several studies have already reported that control of the phase-separated structure is a critical factor in obtaining high J_{sc} values in bulk heterojunction solar cells [5–9,22–24]. To investigate the reason for the improvement in IPCE and J_{sc}, charge separation and charge transport were studied by SEM for each phase-separated structure of the charge management layer [35–37]. Figure 6 shows an SEM image of the charge management layers. The results show that, in comparison with the conventional single electron-acceptor system, the monomer–polymer hybrid electron acceptor dramatically improves the titanium(IV) isopropoxide network for carrier transport, without reducing the extent of the p/n interface. Preferential self-assembly of the polymer as an electron donor is inhibited by the chemical bulk performance of the monomer as an electron acceptor and the self-assembled polymer. The electron-donor p-type semiconducting polymer has a long chain and is sterically bulky, forming a large structure that affects the photoactive layer. In addition, the conventional electron acceptor was only a monomer. Therefore, the self-assembly of the electron acceptor was not preferential with respect to that of the electron donor. Accordingly, the formation of the electron-donor phase becomes dominant, and the electron-acceptor phase is formed following the electron-donor phase structure.

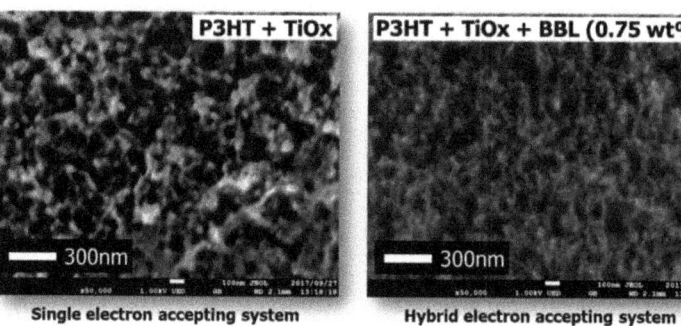

Figure 6. SEM images of photoactive layers.

Therefore, in conventional single electron-acceptor systems, phase formation occurs between polymer phases, and the structure of the electron acceptor becomes coarse and irregular. However, in this monomer–polymer hybrid electron-accepting system, the preferential phase formation of the electron-donor polymer is prevented by the electron acceptor. As a result, the electron donor and acceptor can each form a continuous phase-separated structure due to their individual self-assembly faculties. This phenomenon is also demonstrated by the individual evaluation of the electrical conductivity of the charge management layer in the presence of BBL. Figure 7 shows the measured electrical conductivity of the charge management layer [30,31]. The results show that the addition of BBL significantly improves electrical conductivity and the carrier transport network in the thin films. As the electron-accepting polymers are ladder polymers with high self-assembly and template-forming faculties, the phase of the titanium(IV) isopropoxide monomer followed the ladder structure of BBL [25,26]. Accordingly, the peak increase in J_{sc} occurs at 0.75 wt.% of the BBL addition, which means that the charge separation interface decreases due to the delay in the construction of the template network. Thus, compared with the conventional titanium(IV) isopropoxide system alone, titanium(IV) isopropoxide as an electron acceptor significantly improves the charge transport network, owing to its high electron transport capacity. In other words, the enhanced electronic charge transport efficiency of the titanium(IV) isopropoxide carrier management network resulted in high J_{sc} values, without affecting light absorption. Finally, the performance of the device was measured under a standard room light of 1000 lx. Figure 8 shows the J–V characteristics.

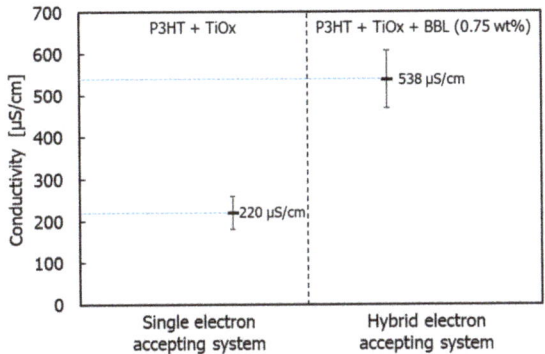

Figure 7. Individual value of electrical conductivity of photoactive layers.

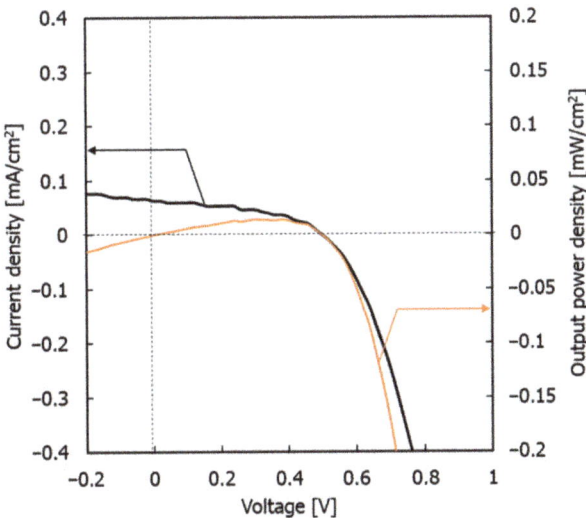

Figure 8. J–V characteristics for solar cell under the LED indoor light at 1000 lx.

The maximum power density (P_{max}), J_{sc}, FF, and PCE were 14.07 µW cm^{-2}, 0.063 mA cm^{-2}, 0.5 V, 44.46%, and 4.02%, respectively. These results demonstrate that the system has the potential to be used as an indoor light-based power generation system [38].

4. Conclusions

In this study, using a hybrid electron acceptor based on the BBL template formation mechanism, the power generation layer was templated into a co-continuous and finely phase-separated structure. This improved the charge transport capacity without decreasing the charge separation efficiency, and increased J_{sc} by approximately 2.2 times. The improvements in the morphology were confirmed by SEM images of the power generation layer (photoactive layer). This confirmed that the hydrolyzed titanium(IV) isopropoxide phase with high electron transport capacity could construct a continuous network supported by the BBL template. The power generation layer has high photosensitivity to LED light sources, and the characteristics under a 1000 lx LED light source showed a high photoelectric conversion efficiency of 4.02%. This demonstrated the potential of using the device as an indoor power source. In contrast, the addition of BBL was found to decrease V_{oc}. This is because the LUMO level of BBL is smaller than that of titanium(IV) isopropoxide, and when using it as a composite material to form the power generation layer, the apparent LUMO of the electron acceptor was lowered. Therefore, to realize higher energy conversion performance and efficiency, it is necessary to search for materials with a high LUMO level and a high template formation function, characterized by the self-assembly capacity.

Author Contributions: Conceptualization, Y.U., K.N. and T.K.; analysis, Y.K. and T.F.; investigation, Y.U., Y.K., H.K. and T.F.; writing—original draft preparation, Y.U. and T.K.; writing—review and editing, Y.U. and T.K.; supervision, K.N. and T.K.; project administration, T.K.; funding acquisition, Y.U. and T.K. All authors have read and agreed to the published version of the manuscript.

Funding: This work was supported by JSPS KAKENHI Grants (No. 19H02662). Y.U. would like to acknowledge the Research Fellowship for Young Scientists from the Tochigi Industrial Promotion Center and K. Matsutani (Executive Director of MANI inc.) for promotion of science (2021) and supported by GEAR 5.0 Project of the National Institute of Technology (KOSEN) in Japan.

Institutional Review Board Statement: Not applicable.

Informed Consent Statement: Not applicable.

Data Availability Statement: The data presented in this study are available in insert article here.

Acknowledgments: We would also like to thank H. Nakata, Y. Negishi, and M. Yamada (NIT, Oyama College) for polymer analysis.

Conflicts of Interest: The authors declare no conflict of interest.

References

1. O'Regan, B.; Grätzel, M. A Low-Cost, High-Efficiency Solar Cell Based on Dye-Sensitized Colloidal TiO$_2$ Films. *Nature* **1991**, *353*, 737–740. [CrossRef]
2. Mathew, S.; Yella, A.; Gao, P.; Humphry-Baker, R.; Curchod, B.F.; Ashari-Astani, N.; Tavernelli, I.; Rothlisberger, U.; Nazeeruddin, M.K.; Grätzel, M. Dye-Sensitized Solar Cells with 13% Efficiency Achieved Through the Molecular Engineering of Porphyrin Sensitizers. *Nat. Chem.* **2014**, *6*, 242–247. [CrossRef] [PubMed]
3. Kakiage, K.; Aoyama, Y.; Yano, T.; Oya, K.; Fujisawa, J.; Hanaya, M. Highly-Efficient Dye-Sensitized Solar Cells with Collaborative Sensitization by Silyl-Anchor and Carboxy-Anchor Dyes. *Chem. Commun.* **2015**, *51*, 15894–15897. [CrossRef] [PubMed]
4. Molla, M.Z.; Baranwal, A.K.; Pandey, S.S.; Ma, T.; Hayase, S. Transparent Conductive Oxide-Less Dye-Sensitized Solar Cells Consisting of Dye-Cocktail and Cobalt Based Redox Electrolyte. *J. Nanosci. Nanotechnol.* **2017**, *17*, 4748–4754. [CrossRef]
5. Shuttle, C.G.; O'Regan, B.; Ballantyne, A.M.; Nelson, J.; Bradley, D.D.C.; de Mello, J.; Durrant, J.R. Experimental Determination of the Rate Law for Charge Carrier Decay in a Polythiophene: Fullerene Solar Cell. *Appl. Phys. Lett.* **2008**, *92*, 183501. [CrossRef]
6. Street, R.A.; Cowan, S.; Heeger, A.J. Experimental Test for Geminate Recombination Applied to Organic Solar Cells. *Phys. Rev. B* **2009**, *82*, 121301. [CrossRef]
7. You, J.; Dou, L.; Yoshimura, K.; Kato, T.; Ohya, K.; Moriarty, T.; Emery, K.; Chen, C.C.; Gao, J.; Li, G.; et al. A Polymer Tandem Solar Cell with 10.6% Power Conversion Efficiency. *Nat. Commun.* **2013**, *4*, 1446. [CrossRef]
8. Gang, L.; Wei, H.C.; Yang, Y. Low-Bandgap Conjugated Polymers Enabling Solution-Processable Tandem Solar Cells. *Nat. Rev. Mater.* **2017**, *2*, 17043.
9. Bae, S.-H.; Zhao, H.; Hsieh, Y.-T.; Zuo, L.; Marco, N.D.; Li, Y.S.; Yang, Y. Second-Generation cycloSal-d4TMP Pronucleotides Bearing Esterase-Cleavable Sites—The "Trapping" Concept. *Eur. J. Org. Chem.* **2006**, *1*, 197.

10. Kojima, A.; Teshima, K.; Shirai, Y.; Miyasaka, T. Organometal Halide Perovskites as Visible-Light Sensitizers for Photovoltaic Cells. *J. Am. Chem. Soc.* **2009**, *131*, 6050–6051. [CrossRef]
11. Lee, M.M.; Teuscher, J.; Miyasaka, T.; Murakami, T.N.; Snaith, H.J. Efficient Hybrid Solar Cells Based on Meso-Superstructured Organometal Halide Perovskites. *Science* **2012**, *338*, 643–647. [CrossRef]
12. Im, J.H.; Jang, I.H.; Pellet, N.; Grätzel, M.; Park, N.G. Growth of CH3NH3PbI3 Cuboids with Controlled Size for High-Efficiency Perovskite Solar Cells. *Nat. Nanotechnol.* **2014**, *9*, 927–932. [CrossRef] [PubMed]
13. Correa Baena, J.P.; Steier, L.; Tress, W.; Saliba, M.; Neutzner, S.; Matsui, T.; Giordano, F.; Jacobsson, T.J.; Srimath Kandada, A.R.; Zakeeruddin, S.M.; et al. Highly Efficient Planar Perovskite Solar Cells Through Band Alignment Engineering. *Energy Environ. Sci.* **2015**, *8*, 2928–2934. [CrossRef]
14. Yang, W.S.; Noh, J.H.; Jeon, N.J.; Kim, Y.C.; Ryu, S.; Seo, J.; Seok, S.I. High-Performance Photovoltaic Perovskite Layers Fabricated Through Intramolecular Exchange. *Science* **2015**, *348*, 1234–1237. [CrossRef] [PubMed]
15. Moriya, M.; Hirotani, D.; Ohta, T.; Ogomi, Y.; Shen, Q.; Ripolles, T.S.; Yoshino, K.; Toyoda, T.; Minemoto, T.; Hayase, S. Architecture of the Interface Between the Perovskite and Hole-Transport Layers in Perovskite Solar Cells. *ChemSusChem* **2016**, *9*, 2634–2639. [CrossRef]
16. Saliba, M.; Matsui, T.; Domanski, K.; Seo, J.-Y.; Ummadisingu, A.; Zakeeruddin, S.M.; Baena, J.-P.C.; Tress, W.R.; Abate, A.; Hagfeldt, A.; et al. Incorporation of rubidium cations into perovskite solar cells improves photovoltaic performance. *Science* **2017**, *354*, 206–209. [CrossRef]
17. Yang, W.S.; Park, B.W.; Jung, E.H.; Jeon, N.J.; Kim, Y.C.; Lee, D.U.; Shin, S.S.; Seo, J.; Kim, E.K.; Noh, J.H.; et al. Iodide Management in Formamidinium-Lead-Halide–Based Perovskite Layers for Efficient Solar Cells. *Science* **2017**, *356*, 1376–1379. [CrossRef]
18. Zuo, L.; Guo, H.; de Quilettes, D.W.; Jariwala, S.; De Marco, N.; Dong, S.; DeBlock, R.; Ginger, D.S.; Dunn, B.; Wang, M.; et al. Polymer-Modified Halide Perovskite Films for Efficient and Stable Planar Heterojunction Solar Cells. *Sci. Adv.* **2017**, *3*, e1700106. [CrossRef]
19. Martin, A.G.; Ewan, D.D.; Jochen, H.-E.; Yoshita, Y.; Nikos, K.; Xiaojing, H. Solar cell efficiency tables(version 56). *Prog. Photovolt.* **2020**, *20*, 629–638.
20. Seki, K.; Furube, A.; Yoshida, Y. Detailed Balance Limit of Power Conversion Efficiency for Organic Photovoltaics. *Appl. Phys. Lett.* **2013**, *103*, 253904. [CrossRef]
21. Krishnan, R.K.R.; Reeves, B.J.; Strauss, S.H.; Boltalina, O.V.; Lüssem, B. C60CF2 Based Organic Field-Effect Transistors with Enhanced Air-Stability. *Org. Electron.* **2020**, *86*, 105898–105905. [CrossRef]
22. Kato, T.; Hagiwara, N.; Suzuki, E.; Nasu, Y.; Izawa, S.; Tanaka, K.; Kato, A. Morphology Control for Highly Efficient Organic–Inorganic Bulk Heterojunction Solar Cell Based on Ti-Alkoxide. *Thin Solid Films* **2016**, *600*, 98–102. [CrossRef]
23. Kato, T.; Oinuma, C.; Otsuka, M.; Hagiwara, N. Morphology Control for Fully Printable Organic–Inorganic Bulk-Heterojunction Solar Cells Based on a Ti-Alkoxide and Semiconducting Polymer. *J. Vis. Exp.* **2017**, 54923. [CrossRef]
24. Benten, H.; Mori, D.; Ohkita, H.; Ito, S. Recent Research Progress of Polymer Donor/Polymer Acceptor Blend Solar Cells. *J. Mater. Chem. A* **2016**, *4*, 5340–5365. [CrossRef]
25. Hirvonen, S.P.; Tenhu, H. Modification of Naphthalenic Unit in B.B.L Main Chain. *Synth. Met.* **2015**, *207*, 87–95. [CrossRef]
26. Hirvonen, S.-P.; Mikko, K.; Erno, K.; Sami, H.; Pasi, L.; Eevakaisa, V.; Sarah, J.B.; Heikki, T. Colloidal properties and gelation of aqueous dispersions of conductive poly(benzimidazobenzophenanthroline) derivatives. *Polymers* **2013**, *54*, 694–701. [CrossRef]
27. Wiatrowski, M.; Dobruchowska, E.; Maniukiewicz, W.; Pietsch, U.; Kowalski, J.; Szamel, Z.; Ulanski, J. Self-Assembly of Perylenediimide Based Semiconductor on Polymer Substrate. *Thin Solid Films* **2010**, *518*, 2266–2270. [CrossRef]
28. Manoj, A.G.; Narayan, K.S. Photovoltaic Properties of Polymer p-n Junctions Made with P3OT/B.B.L Bilayers. *Opt. Mater.* **2003**, *21*, 417–420. [CrossRef]
29. Babel, A.; Jenekhe, S.A. High Electron Mobility in Ladder Polymer Field-Effect Transistors. *J. Am. Chem. Soc.* **2003**, *125*, 13656–13657. [CrossRef]
30. Yamashita, M.; Yamaguchi, S.; Enjoji, H. Resistivity Correction Factor for the Four-Probe Method: Experiment I. *Jpn. J. Appl. Phys.* **1988**, *27*, 869–870. [CrossRef]
31. Yamashita, M.; Yamaguchi, S.; Nishii, T.; Kurihara, H.; Enjoji, H. Resistivity Correction Factor for the Four-Probe Method: Experiment I.I. *Jpn. J. Appl. Phys.* **1989**, *28*, 949–950. [CrossRef]
32. Inoue, M.; Suganuma, K. The Dependence on Thermal History of the Electrical Properties of an Epoxy-Based Isotropic Conductive Adhesive. *J. Electron. Mater.* **2007**, *36*, 669–675. [CrossRef]
33. Shirahata, T.; Kawaharamura, T.; Fujita, S.; Orita, H. Transparent Conductive Zinc–Oxide-Based Films Grown at Low Temperature by Mist Chemical Vapor Deposition. *Thin Solid Films* **2015**, *597*, 30–38. [CrossRef]
34. Liang, Y.; Xu, Z.; Xia, J.; Tsai, S.T.; Wu, Y.; Li, G.; Ray, C.; Yu, L. For the Bright Future. For the Bright Future-Bulk Heterojunction Polymer Solar Cells with Power Conversion Efficiency of 7.4%. *Adv. Mater.* **2010**, *22*, E135–E138. [CrossRef] [PubMed]
35. Shaheen, S.E.; Brabec, C.J.; Sariciftci, N.S.; Padinger, F.; Fromherz, T.; Hummelen, J.C. 2.5% Efficient Organic Plastic Solar Cells. *Appl. Phys. Lett.* **2001**, *78*, 841–843. [CrossRef]
36. Hoppe, H.; Niggemann, M.; Winder, C.; Kraut, J.; Hiesgen, R.; Hinsch, A.; Meissner, D.; Sariciftci, N.S. Nanoscale Morphology of Conjugated Polymer/Fullerene-Based Bulk-Heterojunction Solar Cells. *Adv. Funct. Mater.* **2004**, *14*, 1005–1011. [CrossRef]
37. Lee, J.K.; Ma, W.L.; Brabec, C.J.; Yuen, J.; Moon, J.S.; Kim, J.Y.; Lee, K.; Bazan, G.C.; Heeger, A.J. Processing Additives for Improved Efficiency from Bulk Heterojunction Solar Cells. *J. Am. Chem. Soc.* **2008**, *130*, 3619–3623. [CrossRef]
38. Yang, S.-S.; Hsieh, Z.-C.; Keshtov, M.L.; Sharma, G.D.; Chen, F.-C. Toward High-Performance Polymer Photovoltaic Devices for Low-Power Indoor Applications. *Sol. R.R.L.* **2017**, *1*, 1700174. [CrossRef]

Review

Magnetic Adsorbents for Wastewater Treatment: Advancements in Their Synthesis Methods

Vanpaseuth Phouthavong [1], Ruixin Yan [2], Supinya Nijpanich [1], Takeshi Hagio [1,3], Ryoichi Ichino [1,3,*], Long Kong [2,*] and Liang Li [2]

[1] Department of Chemical Systems Engineering, Graduate School of Engineering, Nagoya University, Furo-cho, Chikusa-ku, Nagoya 464-8603, Japan; phouthavong.vanpaseuth@i.mbox.nagoya-u.ac.jp (V.P.); supinya@slri.or.th (S.N.); hagio@mirai.nagoya-u.ac.jp (T.H.)

[2] School of Environmental Science and Engineering, Shanghai Jiao Tong University, Shanghai 200240, China; risonforever2018@gmail.com (R.Y.); liangli117@sjtu.edu.cn (L.L.)

[3] Institute of Materials Innovation, Institutes for Innovation for Future Society, Nagoya University, Furo-cho, Chikusa-ku, Nagoya 464-8601, Japan

* Correspondence: ichino.ryoichi@material.nagoya-u.ac.jp (R.I.); longmao88@sjtu.edu.cn (L.K.)

Abstract: The remediation of water streams, polluted by various substances, is important for realizing a sustainable future. Magnetic adsorbents are promising materials for wastewater treatment. Although numerous techniques have been developed for the preparation of magnetic adsorbents, with effective adsorption performance, reviews that focus on the synthesis methods of magnetic adsorbents for wastewater treatment and their material structures have not been reported. In this review, advancements in the synthesis methods of magnetic adsorbents for the removal of substances from water streams has been comprehensively summarized and discussed. Generally, the synthesis methods are categorized into five groups, as follows: direct use of magnetic particles as adsorbents, attachment of pre-prepared adsorbents and pre-prepared magnetic particles, synthesis of magnetic particles on pre-prepared adsorbents, synthesis of adsorbents on preprepared magnetic particles, and co-synthesis of adsorbents and magnetic particles. The main improvements in the advanced methods involved making the conventional synthesis a less energy intensive, more efficient, and simpler process, while maintaining or increasing the adsorption performance. The key challenges, such as the enhancement of the adsorption performance of materials and the design of sophisticated material structures, are discussed as well.

Keywords: magnetic adsorbent; synthesis methods; material structure; water treatment

1. Introduction

1.1. Importance and Advantage of Magnetic Adsorbent Technologies for Wastewater Treatment

Natural water sources, apart from soil and air, are the primary medium for pollutant mobility. Over the decades, the release of untreated wastewater into natural water sources, coupled with an ever-increasing population, has resulted in severe environmental problems. Enormous quantities of chemicals, such as pesticides, antibiotics, heavy metals, and dyes are used each year, to drive the global economy. As clean water sources are limited, wastewater should be recycled, by utilizing effective chemical, physical, and biological treatment technologies. Among these techniques, adsorption is a widely exploited physical method because it is simple, environmentally friendly, cost-effective, and reusable [1]. Adsorbents used for the removal of pollutants can be developed to selectively adsorb target adsorbates, with high adsorption performance. The collection of adsorbents after adsorption should be quick and easy. The magnetic separation technique is a good option because magnetic materials can be separated easily from mixtures by applying a magnetic field. The first application of magnetic separation [2] was reported by Robinson et al. [3], in 1973, for the separation of enzymes immobilized on the surface of iron oxide particles. Since

then, the application of magnetic separation in adsorption has been extended in different ways, described below:

(i) Addition of magnetic particles (iron (Fe) or iron oxides (Fe_xO_y)) to assist in the separation of precipitates; that is, coagulation–flocculation during the water treatment process [4,5];
(ii) Direct use of magnetic materials as adsorbents [6];
(iii) Combining magnetic particles with adsorbents to yield magnetic adsorbents [7].

The latter two alternatives have received more interest. Keyword searches in Scopus for "magnetic + adsorbent + separation" and "magnetic + adsorbent + water treatment", reveal a growing trend in the number of research articles (Figure 1).

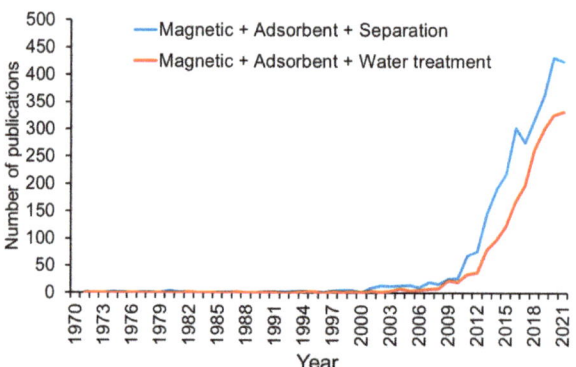

Figure 1. Number of publications that contain the keywords "magnetic + adsorbent + separation" and "magnetic + adsorbent + water treatment" in Scopus. (Accessed on 2 January 2022).

1.2. History and Advancements of Magnetic Adsorbents

Iron-based materials are typically used as magnetic particles because of the strong magnetic moment of unpaired electrons in the 3D orbitals of Fe atoms. The most commonly used materials in the adsorption process are magnetic materials, such as Fe, magnetite (Fe_3O_4), and maghemite (γ-Fe_2O_3). Though Fe and Fe_3O_4 exhibit strong responses to magnetic fields, Fe_3O_4 is used most widely in the literature, owing to its higher stability. Fe_3O_4 particles are superparamagnetic at a size of several nanometers [8]. In aqueous solutions, the abundant hydroxyl groups on the surface of Fe_3O_4 can be protonated or deprotonated by varying the pH, to generate positive or negative charges. Therefore, Fe_3O_4 can be used as an adsorbent to remove ionic species from water via electrostatic interactions [9]. The Fe atom in Fe_3O_4 also acts as an adsorption site for complexation with negatively charged species or the electron-rich functional groups of some organic pollutants. Further, the surface of Fe_3O_4 can be modified to improve functionalization and protect the magnetic core from magnetism degradation [10]. Magnetic adsorbents offer the following advantages:

(i) They can be easily separated from bulk solutions by applying an external magnetic field;
(ii) They are environmentally friendly owing to their biocompatibility [11];
(iii) They are reusable;
(iv) Various organic and inorganic functional groups, to remove diverse target pollutants, can be prepared and modified at the laboratory scale.

In addition to environmental remediation applications, magnetic adsorbents have been used in analytical chemistry in the so-called magnetic solid-phase extraction sample preparation technique. This technique is used to clean up and pre-concentrate the samples, before core analysis improves the detection of anions [12] and organic pollutants [13].

Although magnetic iron oxides are naturally occurring, we can control particle size and achieve better purity by synthesizing them in the laboratory via various routes, such as co-precipitation [14], hydrothermal [15], solvothermal [16], thermal decomposition [17], microemulsion [18], electrochemical [19], and microwave-assisted [20] methods, and greener synthesis methods using biogenic materials [21,22]. Functionalized magnetic particles, or magnetic composites, have been developed or modified using these methods. Among these methods, co-precipitation is the most widely used, due to its simple operation at low temperatures. However, the affinity of hydroxy groups on conventional magnetic iron oxides limits the adsorption performance of some target cationic species. Consequently, they have been coupled with functional adsorbents, such as polymeric ion-exchange resins, used in water treatment processes [23,24]. The magnetic polymer, prepared by the polymerization of γ-Fe_2O_3, is useful for water softening, desalinization, and oil removal from water. In 1996, Chen et al. [25] prepared magnetic composites by simply precipitating ferrihydrite ($Fe(OH)_3$) from Fe_3O_4 and used it to remove heavy metals from water. Since then, the synthesis of magnetic adsorbents has drawn considerable attention, as indicated by the growing number of publications. Diverse effective adsorbents have been reported to incorporate magnetic particles, such as metal oxides [26,27], silica-based materials [28–32], carbon-based materials [33–35], graphene oxide [36], biopolymers, such as lignin [37], cellulose [38], chitosan [39], synthetic polymers [40–42], molecularly imprinted polymers [43], metal organic frameworks [44], and biowaste [45,46]. Magnetic adsorbents can be prepared using physical methods, such as mechanical agitation (stirring, vibration, milling, and ultrasonication) at room or elevated temperatures in some cases, along with the addition of adhesives. Examples of these physical attachments are discussed concretely in this review. Although magnetic adsorbents prepared by physical methods show reasonable adsorption performance, the loose attachment results in a deterioration of their morphology. A low adsorption performance could be due to hidden adsorption sites attaching to the magnetic particles. In contrast, chemically bonded magnetic adsorbents can overcome these drawbacks. Conventional chemical methods utilize specific chemical reactions to form magnetic particles, on pre-synthesized adsorbents or adsorbents on magnetic particles under certain treatment conditions, such as hydrothermal, solvothermal, or reflux.

The adsorption performance of Fe_3O_4 can be improved by doping with sulfur [47,48]. This technology has also been applied to composite adsorbents; namely, converting the Fe_3O_4 priorly embedded onto adsorbents to Fe_3S_4 [49]. The challenge in magnetic adsorbent synthesis is not only to improve the adsorption performance, but also to accomplish good distribution or incorporation of magnetic particles in the adsorbent morphology. Advanced synthesis routes, such as the seed-assisted method [29], pre-milling of magnetic particles with adsorbent seed crystals [30], embedding magnetic particles with a carbon source before carbonization [34], or pre-attaching magnetic particles with precursors as a dry gel before heat treatment [31], help to improve the morphology of magnetic adsorbents. Furthermore, challenges to co-synthesizing magnetic particles and adsorbents in a one-pot synthesis have recently been explored [26].

1.3. Previous Reviews

Many review papers have been published about magnetic particles (not magnetic adsorbents), based on different perspectives, such as the type of surface coating materials, phase of synthesis media, type of reactor, and potential applications. In the area of catalysts and medicine, Lu et al. [10] and Wu et al. [8] classified the synthesis methods of magnetic materials into four popular methods; namely, co-precipitation, thermal decomposition, microemulsion, and hydrothermal synthesis, and also reviewed the surface modification of magnetic particles, based on the type of coating materials. In addition, Teja et al. [50] summarized the synthesis methods of iron oxides and categorized them, according to phase of synthesis media, such as gas phase, liquid phase, and two-phase methods, for the same scientific area. Synthesis methods of magnetic particles, sub-divided using more specific terminology, were found in some review articles [51,52]. Akbarzadeh et al. [53]

focused on magnetic particles for biomedicines. Ali et al. [52] comprehensively summarized and discussed various applications for magnetic materials, for industrial, biomedical, environmental, agricultural, and analytical purposes. Ali's group introduced applications for wastewater treatment briefly, in the section for environmental application. Recently, García-Merino et al. [54] reviewed the synthesis of magnetic particles based on types of reactors. They summarized and compared features of conventional batch synthesis with continuous microfluidic synthesis methods. It was found that most of the above reviews focus on the classification of treatment processes, such as physical, chemical, or biological methods. The relationship between synthesis methods and performance in wastewater treatment has seldom been studied.

As for magnetic adsorbents, Reshadi et al. [55] reviewed magnetic adsorbents used for landfill leachate treatment. Furthermore, magnetic adsorbents for wastewater treatment have been reported by Shukla et al. [56], from the viewpoint of the synthesis methods of magnetic nanoparticles. Abdullah et al. [57] summarized research articles, regarding the synthesis of magnetic adsorbents that used conventional methods, but mainly focused on various types of solid substrates, which were common adsorbents, such as silica-based materials, carbon materials, and graphene, to embed the magnetic particles. Apart from synthesis methods, magnetic adsorbents have been reviewed from another point of view. Abdel Maksoud et al. [58] presented the advantages of using magnetic adsorbents, by classifying them based on types of applied magnetic materials. They additionally discussed their properties and removal efficiency towards organic and inorganic pollutants.

1.4. Objective of This Review

As summarized above, numerous techniques have been developed for the preparation of magnetic adsorbents with effective adsorption performance. To date, magnetic adsorbents have been reviewed from different perspectives; however, reviews that focus on the synthesis methods and the obtained material structures have not been reported. In this review, magnetic adsorbents, mostly used for water pollutant removal, are reviewed and classified by focusing on the synthesis method and the attaching force between magnetic materials and adsorbents, within the obtained material structure. The advantages and drawbacks of these methods, together with the magnetic properties and adsorption performance of the materials, are discussed. Additionally, research papers that illustrate the advancement of each category are summarized.

2. Conventional Synthesis Methods of Magnetic Adsorbents

Conventional synthesis methods can be categorized into four primary groups. Conventional magnetic adsorbents used for the removal of various organic and inorganic pollutants from water, according to their groups with their synthesis methods, adsorption performances, and magnetic performances, are summarized in Table 1.

Table 1. Magnetic adsorbents, their conventional synthesis routes, and performance on adsorption of organic and inorganic pollutants in water.

Adsorbents	Synthesis Method	Magnetic Properties	Pollutant(s)	Adsorption or Removal Performance	Reference
Magnetic materials as adsorbents					
Microsized Fe	Commercial	M_s = 1725 kA/m	Phosphate	q_m = 18.83 mg/g	[59]
Nanosized Fe_3O_4	Sol–gel precipitation and re-crystallization	M_s = 477 kA/m	Phosphate	q_m = 27.15 mg/g	[59]
Fe_3O_4	Ferrite process	Not reported	Phosphate	q_m = 1.9–3.7 mg/g	[15]
Fe_3O_4	Co-precipitation	Not reported	Phosphate	q_m = 15.2 mg/g	[60]
Mixed Fe_3O_4 and γ-Fe_2O_3	Microemulsion	Not reported	Phosphate	Removal efficiency >95%	[18]

Table 1. Cont.

Adsorbents	Synthesis Method	Magnetic Properties	Pollutant(s)	Adsorption or Removal Performance	Reference
Mixed α-Fe_2O_3 and γ-Fe_2O_3	Dispersion-precipitation	M_s = 20 emu/g	Arsenite	q_m = 46.5 mg/g	[61]
Fe_3O_4	Simple mixing and sintering	M_s = 57.4 emu/g	As(V)	q_m = 20.24 mg/g	[48]
Fe_3O_4	Co-precipitation	M_s = 56.86 emu/g	As(V)	q_m = 44.99 mg/g	[62]
Mixed Fe_3O_4 and γ-Fe_2O_3	Co-precipitation	M_s = 67 emu/g	Pb(II) Cr(III) Cd(II) Co(II)	q_m = 617.3 mg/g q_m = 277.0 mg/g q_m = 223.7 mg/g q_m = 27.44 mg/g	[63]
Biogenic Fe_3O_4	Fe^{3+}-reducing bacterial enrichment culture	Not reported	Ni(II) Mn(II) Zn(II)	q_m = 25.22 mg/g q_m = 26.55 mg/g q_m = 77.27 mg/g	[64]
Fe_3O_4	Co-precipitation	M_s = 65.33 emu/g	Pb(II) Cr(VI)	q_m = 53.11 mg/g q_m = 34.87 mg/g	[65]
Attaching of pre-synthesized magnetic particles with adsorbents					
Clinoptilolite–Fe_3O_4	Mechanical method	Not reported	Cd(II) Cu(II) Pb(II)	Removal efficiency =50% Removal efficiency =70% Removal efficiency =90%	[66]
Zeolite–Fe_3O_4	Using organic adhesive	Not reported	Pb(II)	q_m = 133 mg/g	[67]
Faujasite zeolite-$CoFe_2O_4$	Ultrasonication	M_s = 18.93 emu/g	Pb(II)	Removal efficiency = 99%	[68]
Wheat stalk-derived chars–Fe_3O_4	Simple mixing	M_s = 28.6 emu/g	Pb(II)	q_m = 179.85 mg/g	[35]
Rice husk-derived chars–Fe_3O_4	Simple mixing	M_s = 26.1 emu/g	Pb(II)	q_m = 95.44 mg/g	[35]
Activated carbon–Fe_3O_4	Simple mixing	M_s = 27.2 emu/g	Pb(II)	q_m = 43.38 mg/g	[35]
Zeolite–Fe_3O_4	Simple dispersion	Not reported	Zn(II)	q_m = 30 mg/g	[69]
Na-P1 and hydroxysodalite–Fe_3O_4	Mechanical method	Not reported	U(VI)	q_m = 22.4 mg/g	[70]
Kaolin-based zeolite A– Fe_3O_4	Maceration and dispersion	Ms≈37.1 emu/g	Ca(II)	q_m = 54 mg/g	[71]
Kaolin-based zeolite P– Fe_3O_4	Maceration and dispersion	Ms≈37.1 emu/g	Ca(II)	q_m = 51 mg/g	[71]
Zeolite–chitosan–Fe_3O_4	Simple mixing	Not reported	Cr(VI)	Removal efficiency =98%	[72]
Thiol and amine functionalized cellulose–Fe_3O_4	Stirring	Not reported	Pt(IV)	q_m = 40.48 mg/g	[73]
Hydroxyapatite/chitosan cross-linked with green tea derived polyphenol–Fe_3O_4	Stirring	M_s = 53.6 emu/g	Ni(II)	q_m = 112.36 mg/g	[74]
Iminodiacetate functionalized PGMA–Fe_3O_4	High-energy ball milling	M_s = 22.56 emu/g	Uranyl	q_m = 122.9 mg/g	[75]
Iminodiphosphonate functionalized PGMA–Fe_3O_4	High-energy ball milling	M_s = 21.14 emu/g	Uranyl	q_m = 147.0 mg/g	[75]
Phenol modified ZIF-8 functionalized carboxymethyl cellulose–Fe_3O_4	Ultrasonication	Not reported	Rb(I)	q_m = 109 mg/g	[76]
[Epichlorohydrin-co-triethylenetetramine]$_n$-graft-CSSNa–Fe_3O_4 microspheres	Ultrasonication	M_s = 50.51 emu/g	Pb(II) Cd(II) Cu(II) Zn(II)	q_m = 293.38 mg/g q_m = 256.69 mg/g q_m = 277.93 mg/g q_m = 225.07 mg/g	[41]
NiAl LDH–guar gum polymer–Fe_3O_4	Ultrasonication	Not reported	Cr(VI)	q_m = 101 mg/g	[77]
MgAl LDH–Fe_3O_4	Ultrasonication	Not reported	Congo red	q_m = 505 mg/g	[78]
Graphene oxide–Fe_3O_4	Liquid-self assembly	M_s = 18.2 emu/g	Methylene blue	q_m = 172.6 mg/g	[36]
Activated carbon–Fe_3O_4	Ball milling	M_s = 33.8 emu/g	Methylene blue	q_m = 500.5 mg/g	[36]

Table 1. Cont.

Adsorbents	Synthesis Method	Magnetic Properties	Pollutant(s)	Adsorption or Removal Performance	Reference
Larch wood derived lignin hollow microspheres–Fe_3O_4	Mechanical mixing	M_s = 22.7 emu/g	Methylene blue Rhodamine B	q_m = 31.23 mg/g q_m = 17.62 mg/g	[37]
Poplar wood derived lignin hollow microspheres–Fe_3O_4	Mechanical mixing	M_s = 22.7 emu/g	Methylene blue Rhodamine B	q_m = 25.95 mg/g q_m = 15.79 mg/g	[37]
Silica aerogel–Fe_3O_4	Simple stirring	Not reported	Rhodamine B and oil	Removal efficiency =98.5%	[79]
Coffee waste–Fe_3O_4	Dispersion	M_s = 21.5 emu/g	Methylene blue	$q_m \approx$ 128 mg/g	[45]
Zeolite–Fe_3O_4	Simple mixing	Not reported	Reactive orange 16 Indigo carmine	q_m = 1.1 mg/g q_m = 0.58 mg/g	[80]
Polyethylene–Fe_3O_4	Ball milling	M_s = 28.43 emu/g	Pesticides	Recovery = 88–99%	[42]
Synthesis of magnetic particles on adsorbents					
Humic acid–Fe_3O_4	Co-precipitation	Not reported	Phosphate	q_m = 28.9 mg/g	[81]
Activated carbon/MgAl-LDH–Fe_3O_4	Thermal decomposition	M_s = 20.12 emu/g	I-	Adsorption efficiency = 86%	[82]
Calcined orange peel–Fe_3O_4	Co-precipitation and calcination	M_s = 14.6 emu/g	As(III)	q_m = 10.3 mg/g	[46]
NaY zeolite–γFe_2O_3	Co-precipitation	M_s = 18 emu/g	Cr(III) Cu(II) Zn(II) Mn(II)	q_m = 49 mg/g q_m = 87 mg/g q_m = 114 mg/g q_m = 7.97 mg/g	[28]
Polyacrylic acid–$Fe_3O_4/\gamma\text{-}Fe_2O_3$	Co-precipitation	M_s = 50 emu/g	Co(II) Cu(II) Zn(II) Pb(II)	q_m = 12.0 mg/g q_m = 19.2 mg/g q_m = 18.4 mg/g q_m = 29.8 mg/g	[40]
MoS_2–Fe_3O_4	Co-precipitation	M_s = 35.6 emu/g	Cr(VI) Cr(III)	q_m = 218.27 mg/g q_m = 119.38 mg/g	[83]
Lignosulfonate–Fe_3O_4	Co-precipitation	M_s = 43.98 emu/g	Cr(VI)	q_m = 57.14 mg/g	[84]
Guanidinylated chitosan nanobiocomposite–Fe_3O_4	Co-precipitation	M_s = 43.66 emu/g	Pb(II) Cu(II) Cr(VI)	Removal efficiency =98.64% Removal efficiency =100% Removal efficiency =33.76%	[85]
Chitosan–Fe_3O_4	Solvothermal	M_s = 13 emu/g	Pb(II) Cu(II) Zn(II)	q_m = 243 mg/g q_m = 232 mg/g q_m = 131 mg/g	[39]
Carboxymethyl chitosan–Fe_3O_4	Solvothermal	M_s = 15 emu/g	Pb(II) Cu(II) Zn(II)	q_m = 141 mg/g q_m = 123 mg/g q_m = 88 mg/g	[39]
DTPA functionalized chitosan–Fe_3O_4	Co-precipitation	M_s = 35.9 emu/g	U(VI)	$q_m \approx$ 160 mg/g	[86]
Graphene oxide modified with OPO3H2/mesoporous Zr-MOF–Fe_3O_4	Co-precipitation	M_s = 8 emu/g	U(VI)	q_m = 416.7 mg/g	[87]
Cd^{2+} imprinted polymer on carbon nanotube–Fe_3O_4	Solvothermal	Not reported	Cd(II)	q_m = 81 mg/g	[88]
Polystyrene resins on oleic acid–Fe_3O_4	Co-precipitation	Not reported	Cd(II)	q_m = 88.56 mg/g	[89]
Polystyrene–divinylbenzene–Cyanex272–Fe_3O_4	Co-precipitation	M_s = 3.2 emu/g	Cd(II)	q_m = 17.77 mg/g	[90]
Humic acid/L-cystein–Fe_3O_4	Co-precipitation	Not reported	Hg(II)	q_m = 206.5 mg/g	[91]
Holloysite nanotube–Fe_3O_4	Co-precipitation	M_s = 27.91 emu/g	Methylene blue Neutral red Methyl orange	q_m = 18.44 mg/g q_m = 13.62 mg/g q_m = 0.65 mg/g	[92]

Table 1. Cont.

Adsorbents	Synthesis Method	Magnetic Properties	Pollutant(s)	Adsorption or Removal Performance	Reference
Ce-MOF modified activated carbon–Fe_3O_4	Co-precipitation	M_s = 21.39 emu/g	Methylene blue Indigo carmine	Removal efficiency =98–99% Removal efficiency =98–99%	[93]
N-vinylpyrrolidon/chitosan nanocomposite hydrogel–Fe_3O_4	Co-precipitation	M_s = 12 emu/g	Methyl orange	$q_m \approx$ 750 mg/g	[94]
β-cyclodextrin grafted carbon nanotube–Fe_3O_4	Co-precipitation	M_s = 7.15 emu/g	Methylene blue	q_m = 196.5 mg/g	[95]
Activated carbon–Fe_3O_4/γ-Fe_2O_3	Solvothermal	M_s > 30 emu/g	Methylene blue	q_m = 196.5 mg/g	[96]
Mineral derived silica–Fe_2O_3 and plant derived silica–Fe_2O_3	Precipitation, impregnation, and calcination	$M_s \approx$ 0.5–1.3 emu/g	Methylene blue	q_m = 7.0–27.3 mg/g	[97]
Activated sericite clay–Fe_3O_4	Co-precipitation	M_s = 2.17–8.12 emu/g	Methylene blue Crystal violet	Removal efficiency =99% Removal efficiency =99%	[98]
Poly(itaconic acid)/Fe_3O_4–sepiolite	Co-precipitation	M_s = 21.78 emu/g	Methylene blue Methyl violet	q_m = 196.08 mg/g q_m = 175.44 mg/g	[99]
Bentonite/APTMA–Fe_3O_4	Co-precipitation	M_s = 0.7 emu/g	Crystal violet Congo red	q_m = 2286 mg/g q_m = 1210 mg/g	[100]
MoS_2@bentonite–Fe_3O_4	Co-precipitation	M_s = 11.448 emu/g	Crystal violet	q_m = 384.61 mg/g	[101]
Activated carbon–γFe_2O_3/Fe_3O_4/α-FeOOH	Co-precipitation	M_s = 38.5 emu/g	Malachite green	q_m = 486 mg/g	[102]
Lignosulfonate–Fe_3O_4	Co-precipitation	M_s = 43.98 emu/g	Rhodamine B	q_m = 22.47 mg/g	[84]
Reduced graphene oxide–Fe_3O_4	Co-precipitation	M_s = 51.76 emu/g	Rhodamine B	q_m = 432.91 mg/g	[103]
Mesoporous carbon–Fe_3O_4	Solvothermal	M_s = 28.89 emu/g	Ciprofloxacin	q_m = 98.28 mg/g	[104]
Polyacrylonitrile–Fe_3O_4	Solvothermal	Not reported	Tetracycline	q_m = 257.07 mg/g	[105]
Polypyrrole–chitosan–Fe_3O_4	Co-precipitation	M_s = 22.30 emu/g	Carbamazepine	q_m = 121.95 mg/g	[106]
Graphene oxide/cyclodextrin composite–Fe_3O_4	Solvothermal	M_s = 43.96 emu/g	Brivaracetam Eslicarbazepine acetate Carbamazepine	q_m = 36.38 mg/g q_m = 106.86 mg/g q_m = 54.49 mg/g	[107]
Carbon nanofiber aerogels–Fe/Fe_3O_4 core-shell	Co-precipitation	M_s = 102 emu/g	Oil and organic solvents	q_m = 37,000–87,000 mg/g	[108]
Synthesis of adsorbents on magnetic particles					
La(OH)$_3$–Fe_3O_4	Precipitation	M_s = 15–20 emu/g	Phosphate	q_m = 11.77 mg/g	[109]
C18-functionalized silica–Fe_3O_4	Sol–gel	M_s = 39.19 emu/g	Phosphate	q_m = 0.3143 mg/g	[12]
P zeolite–Fe_3O_4	Hydrothermal	M_s = 2.8855 emu/g	K(I)	q_m = 215.1 mg/g	[110]
Mordenite zeolite–Fe_3O_4	Hydrothermal	Magnetic collection rate = 95%	Cs(I)	Removal efficiency >95%	[111]
Amino functionalized silica–Fe_3O_4	Sol–gel	M_s = 60.6 emu/g	Cr(III) Cr(VI)	q_m = 8.22 mg/g q_m = 11.4 mg/g	[112]
Poly(m-phenylenediamine)–Fe_3O_4	Oxidation-polymerization	M_s = 73.78–127.33 emu/g	Cr(VI)	q_m = 125.62–246.09 mg/g	[113]
SDS-PAN functionalized alumina–Fe_3O_4	Direct precipitation	Not reported	Co(II)	Recovery =95.6–98.8%	[114]
ZSM-5 zeolite–Fe_3O_4	Hydrothermal	M_s = 0.8743 emu/g	Pb(II)	q_m = 176.76 mg/g	[115]
Graphene oxide–LDH–Fe_3O_4	Milling and hydrothermal	M_s = 3.5 emu/g	Pb(II)	q_m = 39.7 mg/g	[116]
Amino functionalized silica–Fe_3O_4	Sol–gel	M_s = 29.3 emu/g	Pb(II)	q_m = 238 mg/g	[117]
NaA zeolite–Fe_3O_4	Hydrothermal	χ_ρ = 225–515 m^3/kg	Cu(II) Pb(II)	q_m = 146 mg/g q_m = 477 mg/g	[118]
ZIF-8–Fe_3O_4	Stirring	M_s = 37.26 emu/g	Pb(II) Cu(II)	q_m = 719.42 mg/g q_m = 301.33 mg/g	[119]
Carboxymethylated lignin functionalized silica–Fe_3O_4	Sol–gel	Not reported	Pb(II) Cu(II)	q_m = 150.33 mg/g q_m = 70.69 mg/g	[120]

Table 1. Cont.

Adsorbents	Synthesis Method	Magnetic Properties	Pollutant(s)	Adsorption or Removal Performance	Reference
MnO_2–Fe_3O_4	Hydrothermal	M_s = 14.19 emu/g	Cu(II) Cd(II) Zn(II) Pb(II)	q_m = 498.575 mg/g q_m = 439 mg/g q_m = 416.5 mg/g q_m = 490.5 mg/g	[27]
Siloxydithiocarbamate functionalized silica–Fe_3O_4	Sol–gel	$M_s \approx$ 70 emu/g	Hg(II)	Removal efficiency > 99.8%	[121]
DPTH-functionalized silica–Fe_3O_4	Sol–gel	Not reported	Hg(II)	q_m = 8.39 mg/g	[122]
Sulfur functionalized amide linked organic polymer–MNP-NH_2	Sol–gel	M_s = 15 emu/g	Hg(II)	q_m = 512 mg/g	[123]
Microbial extracellular polymeric substances coated Fe_3O_4	Oxidative copolymerization	M_s = 79.01 emu/g	Ag(I)	q_m = 48 mg/g	[124]
Hydrothermal carbon modified with NaOH–Fe_3O_4	Hydrothermal	Not reported	U(VI)	q_m = 761.20 mg/g	[125]
Amidoxime functionalized flower-like TiO_2 microspheres–Fe_3O_4	Sol–gel	M_s = 15.19 emu/g	U(VI)	q_m = 313.6 mg/g	[126]
Amino-methylene-phosphonic-functionalized silica–Fe_3O_4	Sol–gel	Not reported	Sb(III)	$q_m \approx$ 130 mg/g	[127]
Thiol functionalized silica–Fe_3O_4	Sol–gel	Not reported	$[AuCl_4]^-$	q_m = 115 mg/g	[128]
C18-Silica–Fe_3O_4	Sol–gel	M_s = 41.31 emu/g	Sudan dyes	Recovery = 91–104%	[129]
TiO_2/HKUST-1–Fe_3O_4	Spray-assisted synthesis	M_s = 1.62 emu/g	Methylene blue	q_m > 700 mg/g	[130]
ZIF-8–Fe_3O_4	Mixing and heating	M_s = 14.38 emu/g	Methylene blue	q_m = 20.2 mg/g	[131]
Poly(propylene imine)-functionalized UiO-66–Fe_3O_4	Solvothermal	M_s = 10.5 emu/g	Acid blue 92 Direct red 31	q_m = 122.5 mg/g q_m = 173.7 mg/g	[132]
Chitosan-based adsorbent modified with AO–Fe_3O_4	Sol–gel	M_s = 12.03 emu/g	Orange II Acid red 88 Red amaranth	q_m = 955.0 mg/g q_m = 1075.8 mg/g q_m = 567.5 mg/g	[133]
Polydopamine-coated Fe_3O_4 modified with deep eutectic solvents	Self-polymerization	M_s = 65.71 emu/g	Malachite green Sunset yellow FCF	q_m = 277.78 mg/g q_m = 129.27 mg/g	[134]
Sulfamic acid-functionalized polyamidoamine–Fe_3O_4	Ultrasonication	M_s = 25 emu/g	Malachite green	q_m = 1250 mg/g	[135]
Sulfonic acid functionalized covalent organic polymer–Fe_3O_4	Sol–gel	M_s = 20.2 emu/g	Malachite green	q_m = 333.4 mg/g	[136]
Cationic surfactant functionalized silica–Fe_3O_4	Sol–gel	Not reported	Metal ion-8-hydroxyquinoline complexes	Recovery = 93–113%	[137]
C18-functionalized Fe_3O_4 caged in Ba^{2+}-alginate	Solvothermal	M_s = 49.31 emu/g	PAHs Phthalate esters	Recovery = 72–108%	[138]
C18-modified interior pore wall mesoporous silica–Fe_3O_4	Sol–gel	M_s = 40.8 emu/g	Phthalates	Not reported	[139]
Graphene oxide–LDH–Fe_3O_4	Milling and hydrothermal	M_s = 3.5 emu/g	2,4-dichloro-phenoxyacetic acid	q_m = 173 mg/g	[116]
Agarose coated silica modified with SDS–Fe_3O_4	Sol–gel	M_s = 21.57 emu/g	Phenazopyridine monohydrochloride	q_m = 41 mg/g	[140]
Covalent organic framework–Fe_3O_4	Sol–gel	M_s = 15.8 emu/g	Diclofenac sodium	q_m = 565 mg/g	[141]

M_s: saturation magnetization; χ_ρ = magnetic susceptibility; q_m: maximum adsorption capacity; AO: acryloyloxyethyl dimethylbenzyl ammonium chloride; APTMA: 3-acrylamidopropyltrimethylammonium chloride; CSSNa: sodium dithiocarbamate; DTPA: diethylenetriamine pentaacetic acid; HKUST-1: $Cu_3(1,3,5$-benzenetricarboxylate$)_2$; LDH: layered double hydroxide; MNP-NH_2: Fe_3O_4@SiO_2–NH_2; MOF: metal organic framework; PAHs: polyaromatic hydrocarbons; PAN: 1-(2-pyridylazo)-2-naphthol; PGMA: polyglycidyl methacrylate; SDS: sodium dodecylsulfate; ZIF-8: Zn(2-methylimidazole)$_2$.

2.1. Adsorption Using Magnetic Material Adsorbents

Various magnetic materials, especially iron-based materials, such as nano zerovalent iron (nZVI), maghemite (γ-Fe_2O_3), and magnetite (Fe_3O_4), have received extensive attention for use as magnetic adsorbents. The effective control of particle sizes, crystal structures, and shapes are key issues in the synthesis of magnetic material adsorbents.

2.1.1. nZVI

A wide range of experimental studies have been conducted on magnetic nZVI, owing to its strong reducing properties. Through surface corrosion and precipitation, nZVI can efficiently reduce heavy metal ions, such as hexavalent chromium (Cr(VI)) and some pollutants with low degradability. Compared to other conventional decontamination methods, such as precipitation, ion exchange, and metal complexes, adsorption by nZVI has many advantages. For example, ZVI-based composites could be more cost-effective for in situ environmental applications, and the oxides or hydroxides of iron are natural minerals, which makes the composites eco-friendly.

Chemical reaction, due to its simplicity, is the most frequently used method to synthesize nZVI. Among chemical reactions, the use of sodium borohydride ($NaBH_4$) to reduce the Fe-containing precursor at elevated temperatures has been very popular. Another chemical method to obtain nZVI is the reduction of ferrous oxides in an H_2 atmosphere, at around 500 °C. In addition, some physical methods, such as ball milling and ultrasound assistance, wherein the particle size can be decreased from micrometer to nanometer without using any toxic reagents, can be used to obtain nZVI in a much easier and safer way.

Bare nZVI particles have a large surface area, which is beneficial for increasing the removal capacity. The capacity reached 40 mg/g in a Cd(II) solution, though the capacity varied with different initial concentrations or different initial pH values [142]. However, surface oxidation is a serious problem that limits the further applications of nZVI; the powerful reducing property can act on non-target pollutants, decreasing the selectivity. In air, the formation of a passivation layer inhibits the reactivity of nZVI. Moreover, nZVI particles aggregate much more easily than other non-magnetic, iron-based materials, owing to the surface tension. Surface coating engineering and template confinement can help address these problems (Figure 2).

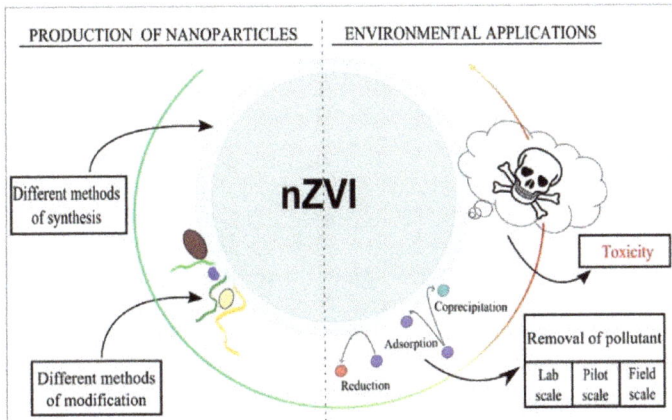

Figure 2. Brief depiction of the synthesis and environmental applications of nZVI (Reprinted from Chem. Eng. J., 287, Stefaniuk, M.; Oleszczuk, P.; Ok, Y.S. Review on nano zerovalent iron (nZVI): From synthesis to environmental applications, 618–632. Ref. [143], Copyright © 2022 with permission from Elsevier).

2.1.2. γ-Fe_2O_3

γ-Fe_2O_3 is mainly obtained by the oxidation of iron, magnetite, or organometallic iron, or the calcination of iron hydroxide. As γ-Fe_2O_3 is a stable iron oxide, its adsorption to heavy metals can often be attributed to physical interactions, such as electrostatic attraction or van der Waals forces. Physical bonding has poor stability in preventing the adsorbents from leaching nanoparticle cores or metal pollutants. Therefore, the construction and design of adsorbents that form more chemical bonds have been investigated. Ahmed et al. [63] studied a mixture of Fe_3O_4 and γ-Fe_2O_3 particles, to remove Pb(II), Cr(III), and Cd(II); the removal capacity reached 617.3, 277.0, and 223.7 mg/g, respectively. For phosphate, this type of mixed adsorbent can also achieve a 95% removal efficiency [18]. Another mixed iron oxide, α-Fe_2O_3 and γ-Fe_2O_3, could uptake 46.5 mg/g As(III), by forming inner-sphere surface complexes [61].

The magnetism and particle sizes of iron oxide nanoparticles are affected by the preparation methods. Hydrothermal methods can achieve smaller nanoparticles and other methods may have aggregation problems. Hence, the stabilizer and surfactant materials should be added in the synthesis process to ensure an even dispersion of the particles. Moreover, the nature of chemical or physical bonding should be further understood, by investigating the mechanisms between the adsorbents and pollutants.

2.1.3. Fe_3O_4

Magnetic Fe_3O_4 has been extensively studied because of its stability, natural abundance, low cost, and environmental friendliness.

Co-precipitation is the conventional method of synthesizing Fe_3O_4. First, Fe(II) and Fe(III) are mixed at a ratio of 1:2, and the resulting black powders are achieved once the solution pH becomes neutral [60]. Although this method is cheap and convenient, the final particle possesses sharp diffraction peaks in the X-ray diffraction (XRD) patterns, revealing that Fe_3O_4 is a type of bulk material. For Fe_3O_4 adsorbents, both physical adsorption and chemical reduction properties have been reported as removal mechanisms. Advanced X-ray photoelectron spectroscopy (XPS) and XRD were used to analyze the lattice structure and elemental valence state, before and after adsorption. For Cr(VI) anions, a Cr(III) oxyhydroxide or hydroxide-phase passive layer was formed on the surface of magnetite, which preferentially reacted with magnetite (111) [144]. In addition, the large surface area and abundant functional groups provided many active sorption sites. Thus, highly dispersed nanoparticles play a key role in the exposure of these sorption sites. To obtain smaller-sized particles, Rajput et al. added tetraethyl ammonium hydroxide (TEAOH) to the Fe_3O_4 product, to prevent agglomeration (Figure 3) [65]. Yusoff et al. [145] found that smaller magnetite particles were easier to obtain when the pH of the co-precipitation solution was greater than 12. After treatment with amino silane groups, the kinetic stability of this magnetic particle can be further developed [59]. Other approaches have also been reported for the preparation of magnetite, including the oxidation of Fe(II) by oxygen, or the reduction of Fe(III) by reducing bacteria [15,64].

The influence of the solution pH on the production of magnetic material adsorbent has been further studied. Gnanaprakash et al. [146] investigated how the initial pH of the salt solution influences the synthesis of magnetite. Additionally, the results showed that 100% spinel iron oxide was formed when the salt solution was below 5; as the pH was increased from 5.7 to 6.7, the percentage of goethite rose from 35% to 78%. The final pH is known to control the nucleation and growth of magnetite nanoparticles and can affect the size and saturation magnetization. In a certain pH range, the protonation/deprotonation of Fe–OH takes place, which can hinder the aggregation of magnetite particles, due to the combined steric and electrostatic stabilization [147,148]. Besides, the solution pH after adsorption treatment also plays a critical role in recycling performance. By adjusting the proton concentration, the surface charge of magnetite changes; thus, the pollutant can be desorbed by electrostatic forces [149].

Magnetite nanoparticles can also be directly synthesized by the thermal decomposition of Fe(III) acetylacetonate (Fe(acac)$_3$) in tris(ethylene glycol), wherein the resultant average crystallite size is approximately 10.7 nm [150]. However, the final magnetite product from ferrous oxalate dihydrate, after heat treatment, is still a bulk material, even at temperatures below 90 °C [151]. Thus, the size distribution and crystallinity are mostly dependent on the Fe-containing precursors. For example, organometallic iron compounds, such as Fe(III)-acetylacetonate and Fe(III)-N-nitroso phenylhydroxylamine, are mixed in organic solvents, such as oleic acid and oleylamine. The stabilizer may inhibit the nucleation and growth process of iron oxide, so that it is well controlled in size and shape.

During the thermal decomposition, the radius of the nanoparticles can be controlled by tuning the reaction temperature and the mass ratios of the solvent reagents. Lassenberger et al. [152] further optimized monodisperse iron oxides using oleic acid as a surfactant, and the results showed that a fast heating rate accelerates the nucleation and growth of iron oxide particles. In contrast, a wide size distribution and larger particle size can also occur when the annealing duration is prolonged.

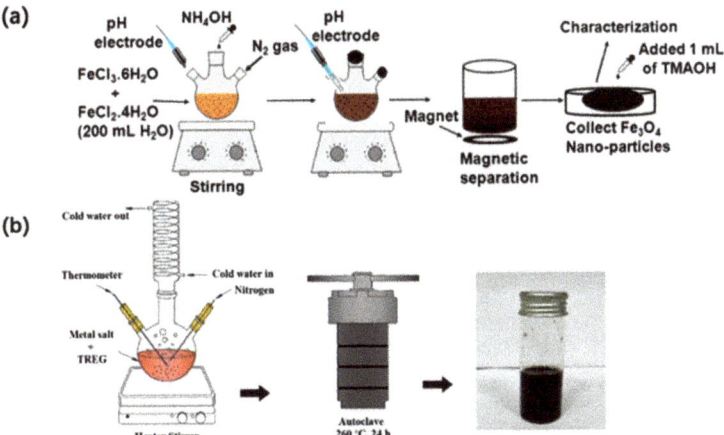

Figure 3. Preparation procedures of Fe$_3$O$_4$ by (**a**) coprecipitation and (**b**) solvothermal synthesis. (Reprinted from J. Colloid. Interface Sci. 468, Rajput, S.; Pittman Jr., C.U.; Mohan, D. Magnetic magnetite (Fe$_3$O$_4$) nanoparticle synthesis and applications for lead (Pb^{2+}) and chromium (Cr^{6+}) removal from water, 334–346. Ref. [65], Copyright © 2022 and J. Alloys Compd. 816, Fotukian, S.M.; Barati, A.; Soleymani, M.; Alizadeh, A.M. Solvothermal synthesis of CuFe$_2$O$_4$ and Fe$_3$O$_4$ nanoparticles with high heating efficiency for magnetic hyperthermia application, 152548 Ref. [153], Copyright © 2022 with permission from Elsevier).

2.2. Attaching Pre-Synthesized Magnetic Particles and Adsorbents

Because the adsorption performance of magnetic materials is rather low, many researchers have sought to embed magnetic properties into well-known adsorbents. The attachment of pre-synthesized adsorbents and pre-synthesized magnetic materials is one such approach. Several methods, including attachment by organic adhesives, electrostatic interactions, self-assembly, crosslinking reactions, and mechanical bonding, can be used in this approach. Pre-synthesized materials include commercial products, apart from the ones developed in laboratories. The attachment methods are summarized in Figure 4.

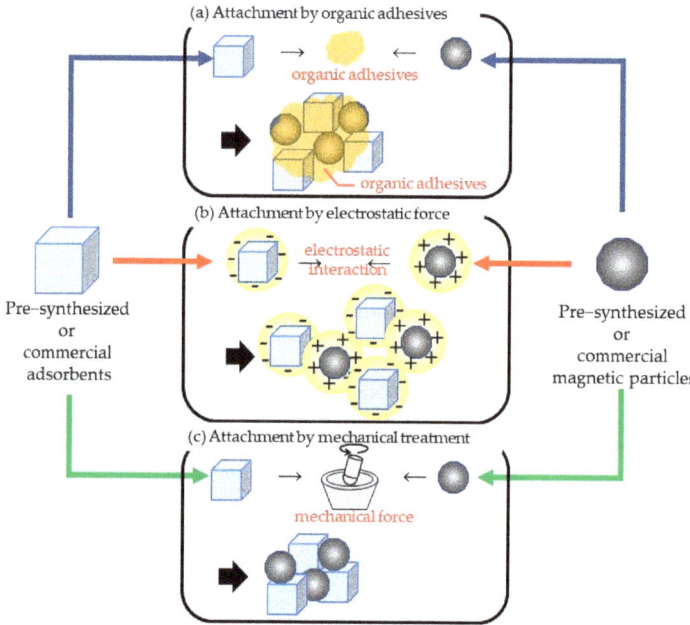

Figure 4. Illustration of attachment methods between adsorbents and magnetic particles.

2.2.1. Attachment Using Organic Adhesives

One of the proposed methods is to simply attach pre-synthesized magnetic particles to pre-synthesized adsorbents. In 2006, Nah et al. [67] bonded commercial zeolites (adsorbents) with commercial Fe_3O_4 (magnetic particles), by adding them into a mixture of urethane and thinner, followed by vacuum drying and ball milling. The urethane functioned as an adhesive and bonded the zeolite and magnetite. The adsorption performance of Pb(II) in magnetic zeolite, tested at pH 5.0, was found to be 133 mg/g. Unfortunately, the adsorption performance was approximately half that of the original zeolite. Polymeric materials have been widely used as adhesives. Minh and Lebedeva [45] reported the attachment of alkali-treated coffee waste to Fe_3O_4, using polyvinyl alcohol (PVA), in 2018. Fe_3O_4 was dispersed in a 2% PVA solution, heated to 80 °C, and coffee waste, pretreated with sodium hydroxide (NaOH), was added, followed by filtering and drying. The obtained coffee waste/Fe_3O_4 composite adsorbent showed a magnetization of 21.5 emu/g and a maximum adsorption of 227 mg/g against methylene blue, at 45 °C. Although this method is very simple and easy, the adverse effect of adhesives attached onto the adsorbent surface must be considered.

2.2.2. Electrostatic Interaction

Another method of attaching adsorbents to magnetic materials is electrostatic interaction. Using this method, Li et al. [37] synthesized magnetic lignin-based hollow microspheres (LHMs). The attached LHMs were prepared by self-assembly, with co-precipitated Fe_3O_4 through electrostatic interactions. Additionally, they stated that some of the Fe_3O_4 nanoparticles could also be immobilized by entering the holes of the LHMs. The adsorption capacity results showed that the magnetic LHM from larch lignin exhibited better adsorption properties for methylene blue and Rhodamine B, which were 31.23 and 17.62 mg/g, than that for poplar lignin, which were 25.95 and 15.79 mg/g, respectively. The saturation magnetization of magnetic LHM was about 22.7 emu/g, which could be easily collected by a magnet. In an experiment on the attachment between cellulose and γ-Fe_2O_3 nanoparticles, Luo et al. [154] reported that iron oxide nanoparticles can be strongly attached through interaction with the electron-rich oxygen atoms of hydroxyl groups. The dependence of the ionic state on pH values implies that the adhesion strength of materials that use electrostatic interactions alone may be questionable.

2.2.3. Liquid Self-Assembly Method

Metin et al. [36,155] prepared graphene oxide (GO)/Fe_3O_4 composites using a liquid self-assembly method. The modified Hummers method and thermal decomposition of iron salt were used to prepare GO and Fe_3O_4, respectively. The composite, developed for Schottky diode applications, has found applications in water treatment. The composite showed a saturation magnetization of 16 emu/g. The adsorption performance was measured using methylene blue (MB) as the adsorbate and showed a maximum capacity of 172.6 mg/g. The adsorbent maintained an adsorption performance of ~80% after five cycles. Guo and Sun [156] reported that the self-assembly method applied to graphene and FePt nanoparticles seemed to stem from the graphene's p-electron polarization, from graphene to FePt, via a possible coordination bonding, due to the close contact between the two materials. Because Metin et al. referred to their report for the self-assembly method, such polarization shall affect the attachment of the magnetic particle and adsorbent; however, further investigation is still needed to understand this mechanism.

2.2.4. Crosslinking Reactions

Another method used for attaching materials is the crosslinking reaction. Hosseini et al. [157] synthesized amine-functionalized magnetic core-coated carboxylated nanochitosan shells, by coupling an amine with carboxylic groups. Chitosan nanoparticles were carboxylated with citric acid (NCS–COOH), Fe_3O_4 nanoparticles were functionalized with amines, using N^1-(3-trimethoxysilylpropyl) diethylenetriamine (NH_2–Fe_3O_4). Fourier transform infrared spectroscopy (FT-IR) results indicated the successful linkage of NCS–COOH on the NH_2–Fe_3O_4 surface, and transmission electron microscope (TEM) images showed that a core shell structure, with NH_2–Fe_3O_4 particles, encapsulated in the NCS–COOH shell was formed. The authors state that high stability over a wide range of pH and dispersity in hydrophilic solvents enables the amphoteric NH_2–Fe_3O_4@NCS–COOH to be a practical candidate for various purposes, including wastewater treatment. The coupling of an amine with a carboxylic acid to form an amide bond is the most popular chemical reaction, which is also used for drug discovery [158]. This technique is effective for the attachment of adsorbent and magnetic particles.

2.2.5. Mechanical Attachment

Others sought the possibility of attaching or embedding magnetic particles to adsorbents by mechanical treatments, such as ball milling. Galhoum [75] prepared magnetic nanocomposites of poly (glycidyl methacrylate) (PGMA) derivatives and nano-sized Fe_3O_4. Fe_3O_4 was prepared by co-precipitation, followed by heating at 80 °C. High-energy ball milling of the two materials was performed at 700 rpm for 60 min (including 1 min stop every 20 min). This treatment was reported to form a core–shell magnetic nanocomposite. The grafting of aminoalkylcarboxylate and aminoalkylphosphonic ligands led to an increase in sorption capacities, due to the specific reactivity of carboxylate and phosphonate. The adsorption against U(VI) for aminoalkylcarboxylate and aminoalkylphosphonic ligand-grafted magnetic PGMA, at the optimum pH of ~4.0, were 122.9 mg/g and 147.0 mg/g, respectively. A similar technique was adopted for magnetized polyethylene composites, by Mohebbi and Farajzadeh, using planetary ball milling [42]. They milled magnetic particles obtained from sand, with polyethylene powder at a ratio of 50:50 (%, w/w), and then subjected them to ball milling for 1.5 h, at a rotational speed of 900 rpm. They insisted that the heat generated by the mechanical collision melted the polyethylene, realizing a strong attachment between the two materials. The prepared composite showed a saturation magnetization of 28.43 emu/g and enabled the extraction and pre-concentration of some pesticides from fruit juices.

2.2.6. Unclear Attachment Methods

Chen et al. [78] combined pre-synthesized nano Fe_3O_4 with pre-synthesized layered double hydroxide (LDH) nanocrystals, by simply mixing suspensions of the two materials, followed by sonication and collection by a magnet. The LDH was prepared by hydrothermal treatment, and Fe_3O_4 was prepared by co-precipitation. This nanohybrid magnetic adsorbent showed a high performance of 505 mg/g against Congo Red (based on LDH). The magnetic adsorbent showed a quick uptake of 96% of the capacity within 5 min and maintained a capacity of over 80% of the initial performance after four cycles. Similarly, Fungaro et al. [69,70,80] prepared a composite of zeolite, synthesized from fly ash and Fe_3O_4, obtained from co-precipitation. The zeolite was slowly added to a suspension of Fe_3O_4 particles and subsequently washed, milled, and dried. The resulting product was found to be easily attracted by a magnet; however, the adhesion force between the two materials was not clarified. Adsorption of Zn(II) [69], U(IV) [70], and two types of dyes, reactive orange 16 and indigo carmine [80], were tested. All adsorbates showed Langmuir-type adsorption, with maximum capacities of 30, 22.4, 1.06, and 0.583 mg/g, respectively. Bessa et al. [71] also synthesized magnetic zeolite by simply mixing Fe_3O_4, prepared by precipitation/partial oxidation, and hydrothermally synthesized zeolites A and P at a 1:3 mass ratio, macerated, and dispersed in distilled water, at 80 °C for 1 h. The saturated magnetization was ~25.5 and ~17.5 emu/g and adsorption performance against Ca(II) was 54 and 51 mg/g for zeolite A and P, respectively. Gaffer et al. [159] further extended this technique and attached chitosan as a secondary adsorbent on a magnetic adsorbent, to form magnetic zeolite–natural polymer composites. The Cr(VI) removal efficiency was 98% at pH 2, when the initial Cr(VI) concentration was 200 mg/L. Many researchers [68–75,80] utilize this technique, but the detailed adhesion mechanism needs further clarification.

2.3. Synthesis of Magnetic Particles on Adsorbents

Many different materials, including active carbon [102], carbon nanofiber aerogels [108], nanotubes [92,95], reduced graphene oxide [103], LDH [159], bentonite [100], and MoS_2 [101], have been studied as potential adsorbents. However, these adsorbents cannot be easily separated after utilization, which restricts their application and development. The introduction of magnetic particles onto adsorbents could be an effective method to facilitate facile separation. Therefore, researchers have attempted to develop a series of methods to synthesize magnetic particles on pre-synthesized adsorbents. A summary of typical magnetic adsorbents with synthesis techniques and properties has been presented in Table 1.

2.3.1. Co-Precipitation

This common method can be used to synthesize magnetic particles on adsorbents. During the co-precipitation process, the magnetic nanoparticles are incorporated with extensively investigated materials, such as active carbon, nanotubes, reduced graphene oxide, zeolite, LDH, and MoS_2. Oliveira et al. [28] reported the introduction of maghemite into commercial NaY zeolite, to produce a novel adsorbent. The bulk magnetization was converted from the original value of 18 to 33 J/T·kg. Uniformly distributed small nanoparticles (3–6 nm) can also be obtained in a cubic zeolite matrix [160,161]. Despite the direct mixing of the precursors with zeolite, 1-butyl-3-methylimidazolium tetrachloroferrate ([bmim]Cl/$FeCl_3$) ionic liquid has been used to immobilize magnetic NaY zeolite. However, the magnetic particles, by occupying the entrance of the zeolite porous structure, prevent the diffusion of pollutants, especially macromolecules or organic dyes [162].

Humic acid, coated on the surface of magnetite as a sorbent, is an environmentally friendly material, and the structure of the magnetite core, with a humic acid shell, has better adsorption performance in an acid condition [81]. Other surface modifiers, such as oleic acid [89], functionalized chitosan [85,86], polyacrylic acid [40], lignosulfonate [84], and activated sericite clay [98], to some extent, could enhance the dispersion of magnetic nanocomposites, because they are widely spaced. In this way, the size of the nanocomposites can be controlled on a small scale, lower than 10 nm.

Recently, more complex hybrid adsorbents have been synthesized using the coprecipitation method. Wan et al. [91] reported the results of a humic acid/L-cysteine-codecorated magnetite that has complex properties, resulting from functional groups, ion exchange, and negatively charged surfaces. In another study, Amini et al. [87] reported a novel composite, GO/Fe_3O_4/OPO_3H_2/PCN-222, that can extract U(VI) in 3 min, with a capacity of 416.7 mg/g. It took three steps to prepare the adsorbents; therefore, large-scale production was unfeasible and not cost-effective. It is worth noting that during this process, some particles were first nucleated into clusters and then aggregated together. The support materials can provide a large surface area for adhesion; however, the adhesion largely involves physical bonding, which can trigger problems, such as leaching and desorption.

2.3.2. Solvothermal

The solvothermal/hydrothermal method provides a simple, direct, and low-temperature method of obtaining nanoparticles with a narrow dispersion and includes an alternative method of calcination at mild temperatures, to promote crystallization. After heating in a Teflon-lined stainless steel autoclave at 200 °C, a crystalline cubic spinel structure of Fe_3O_4 was successfully obtained on the surface of the mesoporous carbon in the composite [104]. In another similar study, Fe_3O_4/γ-Fe_2O_3 was loaded with active carbon, smaller in size [96]. A titanium dioxide-coated magnetic hollow mesoporous silica sphere, with a high surface area, produced by Wu et al. [163], could be used to efficiently and quickly capture phosphopeptides from peptide mixtures.

One-dimensional electrospun nanofibers, with large specific surface areas and high porosities, are connected to each other, making it easy to extract them from water. Liu et al. [105] loaded cubic phase magnetite particles on polyacrylonitrile fibers, through a "two-step" process—electrospinning first and solvothermal next—without changing the morphology and structure of the spinning (Figure 5). Charpentier et al. [39] improved the chitosan-doped Fe_3O_4 adsorbent, using carboxymethyl chitosan. The colloidal magnetic nanoparticles were synthesized via a "one-step" versatile solvothermal method and a simultaneous removal of Pb(II), Cu(II), and Zn(II) was achieved, owing to the chain flexibility and high concentration of chelating groups from carboxymethyl chitosan. Size-controlled magnetic nanoparticles can be synthesized by a solvothermal method, using surfactants. These protective agents can prevent particles from aggregating. However, the size can range from 10 to 200 nm, and the interaction between surfactants and magnetic particles needs further research.

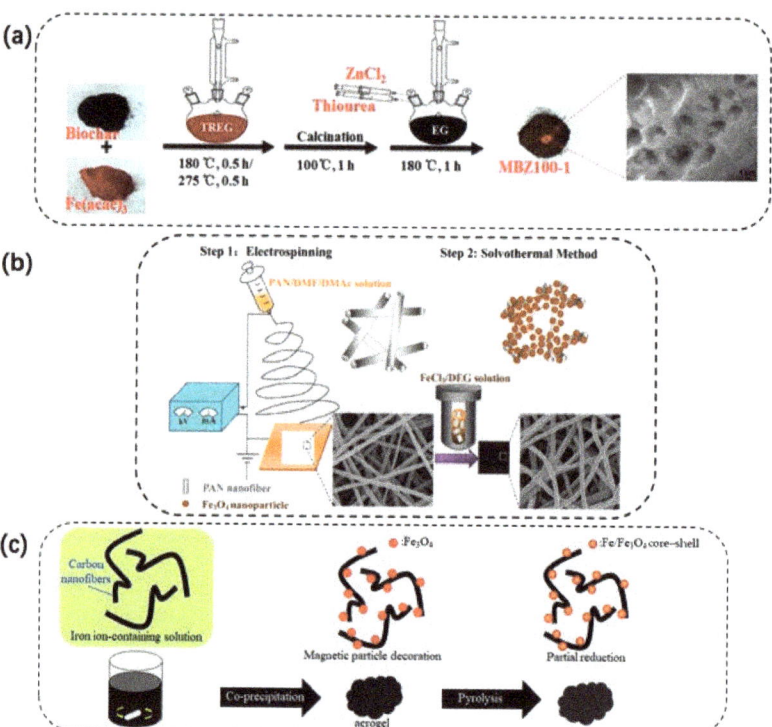

Figure 5. Representative preparation procedures of (**a**) magnetic biochar/ZnS, (**b**) Fe$_3$O$_4$/PAN composite nanofibers (NFs), and (**c**) magnetic carbon fiber aerogels. (Reprinted with permission from (**a**) ACS Sustainable Chem. Eng. 2015, 3, 1, 125−132 Ref. [164] Copyright © 2022 American Chemical Society and (**b**) ACS Appl. Mater. Interfaces 2015, 7, 14573−14583 Ref. [105] Copyright © 2022 American Chemical Society.).

2.3.3. Thermal Decomposition

Thermal decomposition is another facile approach for obtaining efficient adsorbents. Nanoparticles are formed in situ during the heating process and used to absorb directly, without any post-treatment. Recently, a new type of calcined magnetite-activated carbon/MgAl-LDH was prepared and characterized. It effectively removed low-concentration iodide ions (I$^-$), through the "memory effect" [82]. The magnetic adsorbent and layered double oxide were prepared in the same step, making it an efficient and convenient synthesis method. Although thermal decomposition can result in a uniform size distribution and highly crystallized magnetite, it is highly energy intensive. Other essential one-pot methods should be investigated in future studies. The introduction of magnetic particles on adsorbents may occupy the active sites of adsorbents, leading to a decrease in the adsorption properties. These issues should be well addressed for practical applications.

2.4. Synthesis of Adsorbents on Magnetic Particles

Magnetic adsorbents can be prepared in a route opposite to the one described in Section 2.3, by synthesizing the adsorbent onto magnetic materials. In most cases, these are prepared by synthesizing adsorbents in the presence of pre-synthesized magnetic particles. The combination of magnetic particles with inorganic or organic adsorbents/ion exchangers was carried out in different ways, depending on the target adsorbent (Figure 6). Although Fe$_3$O$_4$, pre-synthesized by conventional co-precipitation or solvothermal methods, is the most commonly used magnetic particle, some researchers have used commercialized magnetite for convenience, with an optimal particle size.

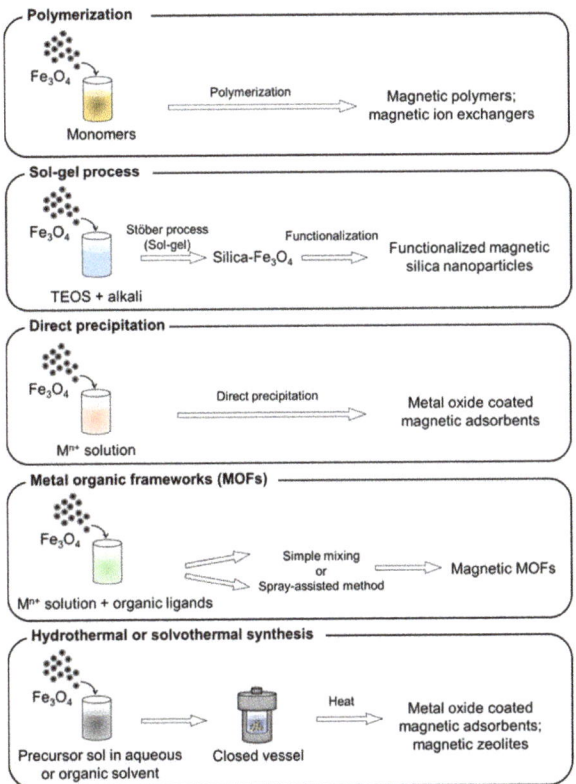

Figure 6. Brief procedures of conventional synthesis methods for adsorbents on magnetic particles.

2.4.1. Polymerization

Polymeric substances, applied as adsorbents or ion exchangers, can be directly grafted onto magnetic iron oxide surfaces. Polymer-coated Fe_3O_4 is prepared by polymerization, depending on the target polymer. A simple protocol, such as suspension of a mixture of Fe_3O_4 and monomers by continuous stirring, was used for grafting at room temperature, under cooling, or at an elevated temperature. For example, Bolto et al. [24] synthesized polyacrylic acid-grafted Fe_3O_4 as an ion exchanger in wastewater treatment; Wei et al. [134] presented a polymer-coated Fe_3O_4 by the self-polymerization of dopamine on Fe_3O_4–COOH at room temperature, for organic dye adsorption. The poly(m-phenylenediamine) layer provided predominant amino groups as adsorption sites for Cr(VI), resulting in a fivefold increase in the maximum Cr(VI) adsorption capacity, compared to that of bare Fe_3O_4 [113]. Another example of grafting polymers at a low temperature is the oxidative copolymerization of microbial extracellular polymeric substances (EPS) on Fe_3O_4, via interaction between Fe and O and C=N on EPS, by continuous stirring at 0 °C [124]. The magnetic polymer was used for the adsorption of Ag(I) and showed a slightly improved maximum adsorption capacity compared to EPS, owing to a higher surface area. Another polymer, sulfonated polystyrene, required an elevated temperature to polymerize on Fe_3O_4, which was then tested for Se adsorption in water samples. Sulfonated polystyrene–Fe_3O_4 was used as a detection tool by magnetic immobilization in a microchannel, to perform on-chip magnetic solid-phase microextraction, with subsequent instrumental analysis [165].

2.4.2. Sol–Gel

The sol–gel process is widely used for coating silica shells on Fe_3O_4 cores, to obtain magnetic core–shell structures. The silica shell can prevent the possible loss of magnetism, due to oxidation of the magnetic oxide core, and can be modified with various functional groups. The Stöber method, a modified sol–gel process for preparing silica, is commonly used [166]. The process is initiated by the hydrolysis of tetraethyl orthosilicate (TEOS) in the presence of ethanol, water, and ammonia, with subsequent co-condensation of silicic acid, to form silica. Sodium silicate can be used for the precipitation of silica in acidic media. In the presence of the hydroxy groups on Fe_3O_4, a silica layer can be formed on the surface of Fe_3O_4 by continuous stirring, under an N_2 atmosphere. Some additives, such as glycerol, were used as porogens, in order to increase the pore size of silica [167], contributing to the occlusion of Fe_3O_4 [122]. Further functionalization of silica with various silane coupling agents has been reported. Octadecyl or C18, a long-chain hydrocarbon group, was grafted onto silica–Fe_3O_4 through alkylation. Jiang et al. [129] used dimethyl octadecyl chlorosilane as a silane coupling agent, to functionalize an ultrafine silica–Fe_3O_4 simple core–shell with C18, for the adsorption of Sudan dyes in water. In another study, Li et al. [139] synthesized C18-functionalized interior pore-wall mesoporous silica, for the adsorption of phthalates in water, by mixing Fe_3O_4 with TEOS and n-octadecyltriethoxysilane. Here, the cationic surfactant cetyl trimethylammonium bromide (CTAB) was employed as the interior wall template. Partial silanol groups of silica were shielded from the C18-functionalized interior by ion pairing with CTAB. Finally, CTAB was removed, and the interior wall structure of C18-functionalized silica was obtained. In this manner, the adsorbents had greater dispersibility in water than the simple core–shell structure. These two types of C18-functionalized silica–Fe_3O_4 showed good magnetic separability, with high adsorption performance, and were utilized as adsorbents in the magnetic solid-phase extraction of organic pollutants. Amino is another widely employed functional group that provides electrostatic interactions with ionic adsorbate species. Huang et al. [112] functionalized silica–Fe_3O_4 with an amino group, by simply refluxing pre-synthesized silica–Fe_3O_4 with 3-aminopropyl trimethoxysilane (APTMS) for 8 h. The obtained product had a microspherical shape and high saturation magnetization (M_s = 60.6 emu/g), which was easily trapped inside a knotted tube, for the online speciation of Cr(VI,III) in water samples. Zhang et al. [117] prepared the same type of magnetic amino adsorbent and found that the amount of APTMS used affected the morphology of the product. An excess amount of APTMS was found to accelerate the hydrolysis rate of TEOS, due to its basicity, resulting in faster precipitation of silica. The faster precipitation led to an incomplete coating of silica on Fe_3O_4. However, well-coated spherical core–shell particles have been used for Pb(II) removal from water. Wang et al. [127] synthesized acid-resistant magnetic adsorbents for the removal of Sb(III) from strong acid solutions. The adsorbent contains an Fe_3O_4 core, protected by a multilayer shell, including silica, and a hydrophobic methyl group, terminated with amino methylene phosphonic acid as the adsorption site. The adsorbents showed high Sb(III) adsorption performance in strong acid media, in the presence of foreign ions. The amino groups can be further modified with various active sites for the removal of different adsorbates. For example, Huang et al. prepared amino-silica-Fe_3O_4 and modified it with various functional groups, such as sulfur-doped amide-linked organic polymers for Hg(II) adsorption [123], covalent organic frameworks (COFs) for diclofenac sodium adsorption [141], and recently sulfonic acid-functionalized covalent organic polymers for the removal of malachite green [136]. These functional adsorbents showed satisfactory adsorption performance and a good magnetic response, with M_s in the range 15−20.2 emu/g. Functionalized silica–Fe_3O_4 adsorbents for the removal of multi-elements have been developed. Examples include amino benzyl EDTA, immobilized on carboxylic-coated silica–Fe_3O_4, for chelating Cu(II), Co(II), Cd(II), and Pb(II) in water [168] and benzyl group-functionalized chitosan, modified on silica–Fe_3O_4, to remove organic dyes [133]. Surfactants have been used for the surface modification of silica–Fe_3O_4 because they possess both hydrophilic and hydrophobic moieties. Karatapanis et al. [137] utilized a cationic surfactant, cetylpyridinium bromide (CPBr), to modify the surface of silica–Fe_3O_4. The positive charge of CPBr attracted a negative charge on silica, in the pH range of 6 to 9, whereas the hydrophobic tails of CPBr served as the adsorption sites. The adsorbents were tested for the adsorption of six heavy metal ions in water samples, after complexation with 8-hydroxyquinoline. Sodium dodecyl sulfate (SDS), an anionic surfactant, is another example of surface modification with surfactants. Adivi et al. [140] prepared a magnetic adsorbent for the removal of phenazopyridine, a cationic drug, from water samples, by initially functionalizing silica–Fe_3O_4 with amino groups, with subsequent electrostatic binding to the negative heads of SDS. The hydrophobic tails of these SDS molecules interacted with the surrounding SDS tails and were finally caged in agarose. Negative charges in SDS and the hydroxy groups on agarose acted as adsorption sites for phenazopyridine. Recently, silica, apart from serving as a protection layer or functionalization bridge, has also been used as a hard template for the crystallization of titanium oxide-based nanosheets. Zhao et al. [126] synthesized amorphous TiO_2 on silica–Fe_3O_4, using the Stöber method. Then, the silica layer was removed by dissolution in a strong alkali solution, followed by recrystallization of the $H_2Ti_2O_5$ nanosheet, by redissolving TiO_2 in diluted hydrochloric acid. After calcination at 400 °C, the crystals became flower-like nanosheets. Finally, amidoxime was grafted onto the magnetic flower-like nanosheet for the removal of U(VI) from seawater samples. The adsorbents, although obtained from a complicated process, showed good magnetic response, with an M_s of 15.19 emu/g.

2.4.3. Direct Precipitation

The precipitation of inorganic compounds on magnetic particle surfaces was employed to prepare magnetic inorganic adsorbents. The adsorbents precipitate and attach to the surface of the magnetic particles, by simple agitation. Wang et al. [114] utilized the Stöber method to precipitate aluminum hydroxide ($Al(OH)_3$) on the surface of Fe_3O_4, by dispersing Fe_3O_4 particles with aluminum isopropoxide, at room temperature. Alumina-coated Fe_3O_4 was obtained by calcination of the precipitate, collected at 500 °C, and subsequently trapped in a microcolumn, using an Nd–Fe–B magnet. It was then used as a solid support for SDS to further bind with 1-(2-pyridylazo)-2-naphthol as the adsorption site for Co(II), in a lab-on-valve sequential injection analysis system. Amorphous calcium silicate ($CaSiO_3$) and manganese dioxide (MnO_2)-coated Fe_3O_4 were prepared separately by Briso et al. [169], who evaluated it for the removal of multi-elements from acidic mine water. They found that approximately 90% of the heavy metal ions were removed by using only $CaSiO_3$-coated Fe_3O_4 in the first step, whereas the MnO_2-coated Fe_3O_4 decreased the concentration of toxic metal ions to below the permissible contamination levels, in the second step. In a more recent study, the precipitation of lanthanum hydroxide ($La(OH)_3$) on the surface of Fe_3O_4 was accomplished on a kilogram-scale, by Ahmed et al. [109]. The preparation was performed in a tank, with continuous stirring, and using sodium hydroxide as the precipitating agent, at 50 °C. After collection, the adsorbents showed a magnetic separation rate of 98% within 3 min and were evaluated for phosphate adsorption at batch scale and pilot plant scale, using river water samples. Phosphate adsorption occurs via electrostatic interaction with protonated hydroxide groups and complexation with La. A removal efficiency of 40% was achieved at the pilot plant scale, which is approximately half that of the batch scale.

2.4.4. Metal Organic Frameworks (MOFs)

MOFs are utilized as adsorbents and combined with magnetic particles to improve separation performance. A zeolitic imidazole framework (ZIF-8) was prepared on the surface of Fe_3O_4, pre-synthesized via a solvothermal method, in the presence of citric acid. Fe_3O_4 was dispersed in a mixture of zinc nitrate and imidazole at 70 °C [131] or pretreated with polystyrene sulfonate, before mixing with the ZIF-8 precursors at room temperature [119]. MOF formation is initiated by the interaction between Zn(II) and carboxylate groups in Fe_3O_4, followed by coordination of Zn(II) with imidazole. Morphology studies of ZIF-8 modified Fe_3O_4 prepared by these two synthesis processes revealed an obvious core–shell structure. Their adsorption performance was evaluated by the adsorption of methylene blue [131], Pb(II), and Cu(II) [119] in water. Another type of magnetic MOF was developed for methylene blue adsorption. Fe_3O_4 and TiO_2, as photocatalysts, were incorporated into $Cu_3(1,3,5$-benzenetricarboxylate$)_2$ or HKUST-1, via a spray-assisted synthetic process [130]. This method is one of the fast MOF synthesis methods [170]. Fe_3O_4 and TiO_2 particles were pre-mixed with the HKUST-1 mother solution, before spraying through two heated nozzles. The product was collected, washed, and dried before use. The process is fast, but the crystallinity of the product obtained by the spray-assisted process was found to be lower than that of the batch process, and the incorporation of Fe_3O_4 particles was quite non-uniform. The adsorbents had much lower magnetic saturation than that of bare Fe_3O_4, due to the thickness of HKUST-1, but were separable from solutions using a simple magnet. However, the incorporated TiO_2 did not significantly contribute to methylene blue degradation.

2.4.5. Hydrothermal and Solvothermal

Hydrothermal and solvothermal methods have been widely used for the preparation of nanoparticles, especially metal oxides. When water is used as the solvent, the process is termed hydrothermal; when other solvents are used, it is termed solvothermal. The synthesis is performed in a closed reaction vessel, above the critical temperature of solvents, in the range of 130–250 °C, and high pressure (0.3–4.0 Mpa) [8,171]. Zeolites, a group of aluminosilicate compounds that are widely used as adsorbents in environmental remediation, have been combined with magnetic particles, to improve the performance of adsorbents. The preparation of magnetic zeolites using conventional hydrothermal methods with different frameworks has been reported, including P zeolite [110], ZSM-5 [115], NaA [118], and mordenite [111]. The synthesis begins by mixing Fe_3O_4 with a zeolite precursor solution, before transferring it to a stainless steel autoclave, followed by crystallization at specific temperatures, for a determined amount of time. In most cases, Fe_3O_4 and zeolite particles are attached to each other on the surface. The ability of the synthesized magnetic zeolites to adsorb monovalent and divalent metal ions in water was evaluated. Another metal oxide, MnO_2, was also prepared by a hydrothermal process, around the Fe_3O_4 core, to obtain a core–shell structure, for the removal of Cu(II) Cd(II), Zn(II), and Pb(II) [27]. The hydrothermal method was applied to synthesize magnetic graphene oxide with an LDH composite, for the first time, by Zhang et al. [116]. Before being subjected to the hydrothermal process, hydroxides of magnesium and aluminum were pre-milled, using ball milling, and were mixed with graphene oxide and Fe_3O_4. This process is called the mechano-hydrothermal method. During the hydrothermal process, LDHs were formed with the subsequent formation of the magnetic composite. LDHs can induce the precipitation of other metal ions. Carboxylic groups and aromatic rings in graphene oxide are attractive to metal ions and organic compounds, respectively. Therefore, the composite showed simultaneous adsorption of Pb(II) and 2,4-dichlorophenoxy acetic acid, a type of pesticide, in water. Magnetic carbon-based adsorbents were prepared using a conventional hydrothermal method. Lai et al. [125] pre-mixed sucrose solution as the carbon source with Fe_3O_4, before the hydrothermal treatment. The collected product was then refluxed with NaOH. The FT-IR analysis showed the partial carbonization of sucrose, to obtain carbon, whereas carboxylate and f-lactonic groups were mostly found on the surface. These functional groups were able to coordinate with U(VI); therefore, magnetic adsorbents were used to remove U(VI) from water. The solvothermal method was used to directly functionalize C18 on Fe_3O_4 without surface modification, using silica. Zhang et al. [138] prepared C18–Fe_3O_4 by dispersing Fe_3O_4 in ethanol and toluene, before adding octadecyltriethoxysilane. The solvothermal process was performed at 120 °C for 12 h. After collection, the C18–Fe_3O_4 was caged in a hydrophilic barium–alignate polymer, for magnetic solid-phase extraction of polycyclic aromatic hydrocarbons and phthalate esters from water. This is an example application of magnetic adsorbents in sample preparation, prior to instrumental analysis. MOFs can also be produced by a solvothermal process in the presence of Fe_3O_4. Far et al. [132] modified Fe_3O_4, prepared by the co-precipitation method, with zirconium-based MOFs (coded UiO-66). Fe_3O_4 was dispersed in N,N-dimethylformamide, in the presence of UiO-66 precursors, before the solvothermal process. The obtained particles were then modified with a polypropylene imine dendrimer by continuous stirring. The final product was found to aggregate with a non-uniform shape and size and showed soft ferromagnetism, with an M_s of 10.5 emu/g. The polypropylene imine dendrimer was used as an adsorption site for the evaluation of anionic dye removal in wastewater treatment plants.

3. Recent Advancements in Synthesis Methods of Magnetic Adsorbents

Although conventional methods continue to be used in a variety of studies, advanced techniques are being developed, some of which are based on conventional techniques (Figure 7). In this section, recent advancements in synthesis techniques are discussed. Advanced magnetic adsorbents in the removal of various organic and inorganic pollutants from water, according to their categorized groups, with their synthesis methods, adsorption performances, and magnetic performances are summarized in Table 2.

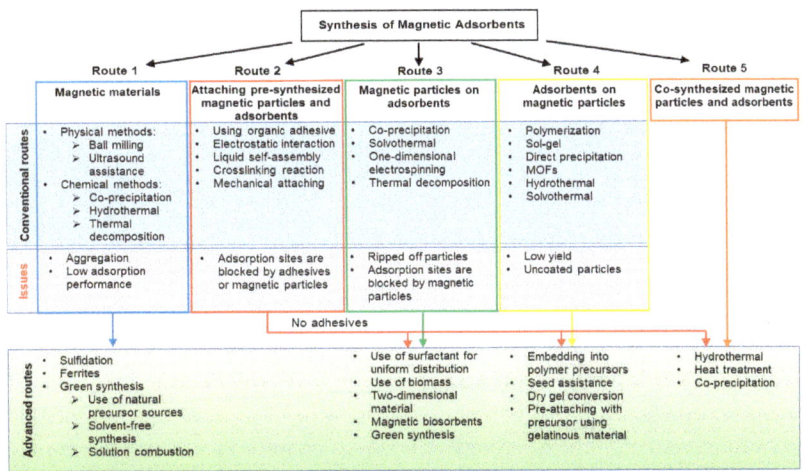

Figure 7. Schematic classification of the conventional synthesis methods of magnetic adsorbents covered in this review with their issues and advancements.

The improvements of these new methods make conventional synthesis less energy intensive, more efficient, and simpler, while maintaining or increasing the adsorption performance in most cases. These approaches are also important from the viewpoint of cost. The addition of magnetic properties to adsorbents increases the cost of their synthesis. However, previous reports show that the economic performance of magnetic adsorbents is already competitive in some limited cases. Oladipo et al. [172] reported that magnetic hybrids have a comparative advantage, regarding operational cost (including energy input and sludge formation), separation after spent, regeneration, and reuse in the removal of boron. Reshadi et al. [55] also claimed that rather expensive, but reusable, magnetic adsorbents have become more cost-effective than low-cost, single-use, conventional adsorbents. Therefore, magnetic adsorbents developed by the aforementioned advanced methods are expected to be competitive in wider applications.

Table 2. Advanced synthesis methods of magnetic adsorbents, and their performance on adsorption of organic and inorganic pollutants in water.

Adsorbents	Synthesis Method	Magnetic Properties	Pollutant(s)	Adsorption or Removal Performance	Reference
Advanced synthesis of magnetic materials as adsorbents					
Sulfur-doped Fe_3O_4	Simple mixing and sintering	M_s = 37.1 emu/g	As(V)	q_m = 58.38 mg/g	[48]
Dendrimerlike biosorbent–Fe_3O_4/Fe_2O_3 based on orange peel waste	Co-precipitation	Not reported	As(V)	q_m = 81.3 mg/g	[173]
Sulfur-doped Fe_3O_4	Simple mixing and sintering	M_s = 32.97 emu/g	Pb(II)	q_m = 500 mg/g	[47]
β-cyclodextrin-stabilized Fe_3S_4	Thermal decomposition	M_s = 37.1 emu/g	Pb(II)	q_m = 256.41 mg/g	[174]
Fe_3S_4-reduced graphene oxide	Thermal decomposition and sulfuration	M_s = 20.67 emu/g	Pb(II)	q_m = 285.71 mg/g	[49]
Fe/FeS	Sulfidation	M_s = 78.0 emu/g	Cr(VI)	q_m = 69.7 mg/g	[175]
$MgFe_2O_4$	Sol–gel	M_s = 9.4 emu/g	Indigo carmine dye	q_m = 46 mg/g	[176]
$CuFe_2O_4$	Solution combustion	M_s = 18.1 emu/g	Malachite green	q_m = 22 mg/g	[177]
Bio-synthesized Fe_3O_4	Simple precipitation using microalgae extract	M_s = 0.2705 emu/g	Crystal violet Methyl orange	q_m = 256.41 mg/g q_m = 270.27 mg/g	[21]
Bio-synthesized Fe_3O_4	Simple precipitation using microalgae extract	M_s = 0.2705 emu/g	Methylene blue	q_m = 312.5 mg/g	[22]
Starch-coated Fe_3O_4	Green co-precipitation	M_s = 46.8 emu/g	Optilan blue	Removal efficiency =72–89%	[178]
S-nZVI	Sulfidation	Not reported	Florfenicol	Removal efficiency >98%	[179]
S-nZVI	Sulfidation	Not reported	Diclofenac	Removal efficiency >85.9%	[180]
Advanced synthesis of magnetic particles on adsorbents					
Pinewood-derived biochar–$MnFe_2O_4$	Direct pyrolysis	Not reported	As(V)	q_m = 3.44 g/kg	[181]
Pinewood-derived biochar–γ-Fe_2O_3	Direct pyrolysis	Not reported	As(V)	q_m = 428.7 mg/kg	[182]
Sodium alginate-dispersed nZVI	Sulfidation	Not reported	Cr(VI)	Removal efficiency =96.4%	[183]
Fe-coated bamboo charcoal	Impregnation and microwave heating	Not reported	Pb(II)	q_m = 200.38 mg/g	[184]
Bagasse-derived biochar	Co-precipitation	M_s = 0.49–1.17 emu/g	17β-estradiol	q_m = 34.06–50.24 mg/g	[185]
Biotemplate-fabricated $ZnFe_2O_4$/MgAl LDH	Thermal decomposition	M_s = 31.8 emu/g	Congo red	q_m = 294.12 mg/g	[186]
Ag–C–Fe_3O_4	Solution combustion	M_s = 2.6 emu/g	Methylene blue Acid orange 7 Rhodamine 6G	q_m = 152.62 mg/g q_m = 154.57 mg/g q_m = 168.68 mg/g	[187]
Activated carbon–Fe_3O_4	Solution combustion	M_s = 4.82–13.5 emu/g	Acid yellow 42 Acid red 213	q_m = 62.36 mg/g q_m = 77.99 mg/g	[188]
C–Fe_3O_4	Solution combustion	M_s = 2.43 emu/g	Acid orange 7 Acid blue 129 Methylene blue Rhodamine 6G	q_m = 126.19 mg/g q_m = 83.42 mg/g q_m = 118.15 mg/g q_m = 131.80 mg/g	[189]

Table 2. Cont.

Adsorbents	Synthesis Method	Magnetic Properties	Pollutant(s)	Adsorption or Removal Performance	Reference
Advanced synthesis of adsorbents on magnetic particles					
Faujasite-type zeolite–Fe_3O_4	Seed-assisted hydrothermal with seed crystal/Fe_3O_4 mixture	Not reported	Methylene blue	q_m = 35.7 mg/g	[29]
Activated carbon–Fe_3O_4	Carbonization of Fe_3O_4 embedded polymer precursor	Not reported	Methylene blue	q_m = 650 mg/g	[34]
BEA-type zeolite–Fe_3O_4	Dry-gel conversion of Fe_3O_4 pre-mixed precursor gel	Not reported	Methylene blue	q_m = 133 mg/g	[31]
Zn-based zeolitic Imidazolate MOF-basil seed mucilage nanocomposite	Ultrasonication	M_s = 2.22 emu/g	Methylene blue Eriochrome black T	q_m = 9.09 mg/g q_m = 13.21 mg/g	[190]
MOR-type zeolite–Fe_3O_4	Seed-assisted hydrothermal with seed crystal/Fe_3O_4 mixture	Not reported	Benzene	q_m = 6.9 mg/g	[30]
Co-synthesis of magnetic particles and adsorbents					
ZrO_2–Fe_3O_4	Co-precipitation	M_s > 23.65 emu/g	Phosphate	q_m = 27.93–69.44 mg/g	[26]
Ma/Al/La–Fe_3O_4	Co-precipitation and calcination	Not reported	F-	q_m = 65.75 mg/g	[191]
Triethylene tetramine functionalized chitosan resin–Fe_3O_4	Precipitation and crosslinking	M_s = 30 emu/g	Uranyl	q_m = 166.6 mg/g	[192]
MgAl LDH on carbon–Fe_3O_4	Hydrothermal self-assembly and Sol-gel	M_s = 5.84 emu/g	Cr(VI)	q_m = 152.0 mg/g	[193]
Rice husk-derived carbonaceous material–Fe_3O_4	Carbon-thermal	M_s = 77.8 emu/g	Cr(VI)	q_m = 157.7 mg/g	[194]
Sludge biochar–Fe_3O_4	Hydrothermal	M_s = 29.94 emu/g	Pb(II)	q_m = 174.216 mg/g	[195]
Biochar–Fe_3O_4	Electromagnetization and pyrolysis	M_s = 26.79 emu/g	Acid orange 7	q_m = 382.01 mg/g	[33]
Fullerene–Fe_3O_4	Solvent-free catalytic thermal decomposition	M_s = 7.002 emu/g	Acid blue 25 Methylene blue	q_m = 806.5 mg/g q_m = 833.3 mg/g	[196]
Polyvinylpyrrolidone–Fe_3O_4	Modified hydrothermal	Not reported	Crude oil	Removal efficiency ≈100%	[197]

M_s: saturation magnetization; q_m: maximum adsorption capacity; LDH: layered double hydroxide; MOF: metal organic framework; S-nZVI: sulfidized nano zerovalent iron.

3.1. Advancements of Magnetic Material Adsorbents

Recently, the development of more effective adsorbents, based on magnetic materials, has attracted much attention. In this section, we focus on the advancements in the synthesis methods for magnetic material adsorbents. A summary of the recent advancements in the synthesis of magnetic adsorbents has been shown in Table 2.

3.1.1. S-nZVI

Although nZVI technology, from laboratory synthesis to environmental application, has made progress through a comprehensive amount of research, there are still two main technical obstacles to overcome, outlined here: (i) Unstable nZVI often has high surface energy and intrinsic magnetic interactions, meaning that when coupled with the van der Waals force between the particles, nZVI is easier to aggregate. The aggregated nZVI may not be as active as expected. (ii) The reducing property of nZVI may be consumed by non-target pollutants, which leads to a decrease in selectivity [198]. In the research to solve these two problems with nZVI, the sulfidation of nZVI (S-nZVI), a method of chemical modification of nZVI particles by adding sulfur compounds, recently showed that the technology is simple, inexpensive, and environmentally acceptable.

The common precipitation synthesis method of S-nZVI can be roughly divided into the following two categories: the one-step synthesis method and the two-step synthesis method. Sodium dithionite ($Na_2S_2O_4$) is the most commonly used sulfidation reagent for one-step synthesis of n-ZVI, where iron salts are precipitated with $NaBH_4$ and $Na_2S_2O_4$, in one aqueous solution. Results have shown that core–shell structure nanoparticles are formed during the one-step method process, with a wide [S/Fe] dose range (0.07–0.4) [199]. The mixed shell structure is mainly composed of nonuniformly distributed iron hydroxides and iron sulfides (Fe_xS_y). Some studies have also reported the effect of sulfidation on the aggregation of nZVI and the resultant several-fold greater sequestration performance, compared with bare nZVI [142,143]. Song et al. used dithionite as a sulfur source to control the S/Fe molar ratio of S-nZVI. Their XANES results showed that the sulfidation process can effectively inhibit the formation of FeOOH on the nZVI surface and promote the crystallization of the Fe core. FeS formed on the surface can not only activate oxygen molecules into reactive oxygen species (ROS), improving the reactivity, but also suppress aggregation, due to the increased electrostatic and steric repulsion and the decreased magnetic attraction [180].

Two-step synthesis is conducted by depositing sulfur species on the surface of pre-synthesized nZVI with Na_2S (or other sulfur-containing reagents). Unlike the one-step method that produces rough and irregular particles, the two-step method can generate nanoparticles with similar core–shell morphology and smaller sizes. Smooth spheres are usually arranged in a typical chain-like shape, with needles and plates around the core. Mangayayam et al. [200] reported that the surface layer of S-nZVI synthesized by two steps is about 5 nm thick, and the surface has defects and heterogeneous crystal orientations. In addition, post-sulfidation, using dithionite, can greatly improve the selectivity of CMC-nZVI on trichloroethylene and eliminate the reaction with water [201]. At the same time, compared to using nZVI alone, S-nZVI usually maintains a longer reaction lifespan. Dithionite dissociates to form sulfur dioxide free radicals and produces sulfite and thiosulfate, which can be used as a scavenger for electron-accepting compounds (such as O_2) in an aqueous solution, thereby maintaining the zerovalent state of nZVI for a longer period of time [202].

3.1.2. Ferrite (Mfe_2O_4)

In recent years, spinel ferrite has become an important magnetic nanoparticle for water treatment. The general formula of ferrite is Mfe_2O_4, where M could be Zn, Mg, Co, Ni, etc. Depending on the position of M(II) and Fe(III) in the crystallographic sites, there are three different types of ferrite, as follows: normal, inverse, and mixed [203]. It is also easy to separate ferrites from wastewater using an external magnetic field.

"Bottom-up" synthesis methods, including sol–gel, hydrothermal, co-precipitation, and solvothermal, have been the most frequently used methods in recent years. The adsorbent morphology, especially the capacity, is greatly affected by the different raw materials and techniques used. For example, Adel et al. [176] obtained $MgFe_2O_4$, with both microporous and mesoporous structures, by a simple sol–gel method, followed by calcination at 500 °C. Its adsorption capacity for indigo carmine dye reached 46 mg/g with spherical particles. For another magnesium–zinc ferrite composite, the increase in zinc content improved the removal efficiency of both Cr(VI) and Ni(II). The auto-combustion method enhanced the porous structure of the adsorbent [204]. In a study comparing different green synthesis methods, samples obtained by co-precipitation had better removal performance than combustion and microwave-assisted methods [63].

Fe_3O_4 is one of the most common ferrites and the mechanism study for the formation of Fe_3O_4 has been widely studied. The reaction temperature and solvent mass ratio are two factors that have a significant effect on the production of Fe_3O_4, not only for the adsorbent morphology but also the nanoparticle magnetism [205]. Firstly, the size of Fe_3O_4 particles will increase, with the increase in reaction temperatures, which has been confirmed by TEM images and calculated by the Debye–Scherrer equation, using XRD data [206]. It is believed that magnetic nanoparticles exhibit a size effect or a high surface area to volume ratio, which results in a higher metal removal adsorption capacity [207]. Secondly, the solvent used in the synthesis process is sometimes used as a multitask agent, to help produce precursors, decorate the particles or change the aggregation state of the target product. For example, in order to obtain the monodisperse nanoparticles, Xu et al. reported a simplified method of monodisperse Fe_3O_4, through the decomposition of iron acetylacetonate in benzyl ether and oleylamine [208]. The size of Fe_3O_4 can be controlled from 14 to 100 nm, by varying the heating conditions and ratios of oleylamine and oleic acid. The experiment also proved that excessive oleylamine can provide a sufficient reducing environment for the Fe precursor and promote the formation of Fe_3O_4 nanoparticles, at a relatively low temperature. In another study, Mohapatra et al. confirmed that the oxidation state of Fe in the prepared nanoparticles affects the superparamagnetic or nonmagnetic state, so as to influence the purity of the Fe_3O_4 phase [209]. Thus, both reaction temperature and solvent mass ratios are key factors for magnetite's properties.

3.1.3. Surfactant Modification

The functional modification of the surface is a common method of preventing the agglomeration of magnetic nanoparticles. Due to the steric hindrance or electrostatic repulsion, the influence from magnetic forces and van der Waals forces can be effectively suppressed. According to the surface characteristics and application scenarios, surfactant-functionalized magnetic nanocomposites can be easily classified into oil-soluble, water-soluble, and amphiphilic [210]. Chin et al. has reported a controllable method to obtain magnetite nanoparticles, using environmentally benign and non-toxic polyethylene oxide (PEO) as the solvent and surfactant simultaneously [211]. His study confirmed that the spherical-shaped Fe_3O_4 particles were more easily obtained when carboxylic acid (-COOH) existed during the iron acetylacetonate hydrolysis. As a common cationic surfactant, cetyltrimethylammonium bromide (CTAB) has a long, apolar chain that was used to modify palygorskite–Fe_3O_4. The treated palygorskite–Fe_3O_4 was positively charged and changed from partially hydrophobic to hydrophilic, which was favorable for the removal of anionic dyes [212]. In contrast, anionic surfactants, such as sodium dodecyl sulfate (SDS), has exhibited the hydrophobic effect and good electrostatic attraction to metal cations, such as $Ni(II)$, $Cu(II)$, $Zn(II)$, and other contaminants, including norfloxacin [213,214]. Nonionic surfactants, including silica, carbon, and precious metal, can be formed as the coatings or outer shells [215–217]. Cendrowski et al. compared the magnetites coated with solid silica, mesoporous shell, and pristine nanoparticles. The results showed great differences in thermal and chemical stability. Due to the lack of diffusion of oxygen and hydrochloric acid through the silica structure, the thermal stability and acid resistance were both enhanced [218]. For an amphoteric surfactant, Al_2O_3 is an example that can react both as an acid and a base. In a study of fluoride adsorption, Chai et al. found that sulfate-doped Fe_3O_4/Al_2O_3 exhibited a high capacity, over a wide pH range. In acidic solutions, the equilibrium pH is much higher than the initial pH; in alkaline conditions it will drop to a lower value [219]. It has shown a good amphoteric property, which is favorable to the application of adsorbents in natural water environments.

3.1.4. Green Synthesis

In recent years, a number of magnetic nanoparticle synthesis methods have been developed. The particle size of nanoparticles can be controlled by great thermal decomposition of the toxic and expensive precursors and surfactant organic solvents. The high thermal energy consumption and large amount of organic waste solvents present significant environmental challenges. Thus, more solvent-less or solvent-free green synthesis methods are being investigated.

Some natural biomass has abundant functional groups and can promote the formation of magnetic particles. Coconut husks have been reported to contain phenolic substances with carboxyl groups on the surface, such as benzoic acid and caffeic acid, which could stabilize the magnetite dispersions, over a wide pH range [220]. In a similar study, iron oxide nanoparticles were mixed with tangerine peel extract. When the pH value was 4 and the adsorbent dosage was 4 g/100 mL, the maximum removal rate of Cd(II) ions was 90% [221]. The magnetic material modified by starch also has a good crystal form and maintains a good removal effect on textile dyes, other than heavy metals [178]. However, a pre-synthesized magnetic nanoparticle and aqueous solution are required to obtain the nanocomposite.

Using waste ferrous sulfate as the main iron source, magnetite ($FeFe_2O_4$) nanoparticles were synthesized through solvent-free reduction reactions [222]. At room temperature, the porous magnetite nanoparticles have an M_s of approximately 77 emu/g, which is sufficient for separation from wastewater, using an external magnetic field. Using industrial waste as a raw material to synthesize porous magnetite nanoparticles not only reduces production costs but also ensures clean production and eases environmental pressure. Sulfur dioxide produced in the reduction reaction is recycled with water to produce sulfuric acid [222]. In addition, magnetic carbonaceous adsorbents produced by ball milling biochar or activated carbon and Fe_3O_4 nanoparticles were obtained using a solvent-free method, and their ability to adsorb methylene blue from water was evaluated and compared [223] (Figure 8a). A possible multiple adsorption mechanism includes electrostatic interaction bonding, ion exchange, and π electronic interaction, resulting in a maximum capacity of 500.5 mg/g. Although green synthesis has many advantages, the removal efficiency and the solvent-free reaction mechanism need to be improved.

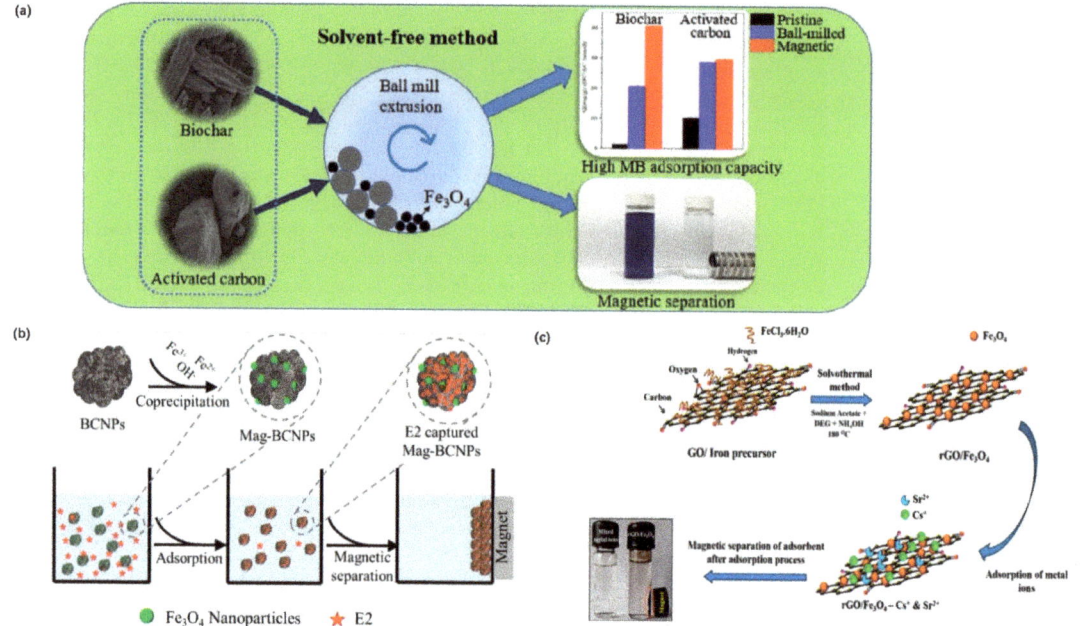

Figure 8. Preparation procedures of Fe_3O_4 loaded on (**a**) magnetic biochar and activated carbon, (**b**) nanosized biochar particles and (**c**) RGO. (Reprinted from (**a**) Sci. Total Environ. 722, Li, Y.; Zimmerman, A.R.; He F.; Chen, J.; Han, L.; Chen, H.; Hu, X.; Gao, B. Solvent-free synthesis of magnetic biochar and activated carbon through ball-mill extrusion with Fe_3O_4 nanoparticles for enhancing adsorption of methylene blue. 137972. Ref. [223], Copyright © 2022 and (**b**) Chem. Eng. J. 352, Dong, X.; He, L.; Hu, H.; Liu, N.; Gao, S.; Piao, Y. Removal of 17β-estradiol by using highly adsorptive magnetic bio-char nanoparticles from aqueous solution. 371–379. Ref. [185], Copyright © 2022 with permission from Elsevier and (**c**) Ind. Eng. Chem. Res. 2018, 57, 4, 1225–1232 Ref. [224] Copyright © 2022 American Chemical Society).

Recently, solution combustion synthesis (SCS) has become a widely adopted technique for fabricating nanomaterials, especially nano oxides, due to its attractive advantages of being simple, non-polluting, energy saving, and highly efficient. Typically, metal hydrazinecarboxylate hydrates are solid at room temperature and have a complex crystal structure. They will produce ultrafine solid oxides of corresponding metals and a large quantity of gaseous products, at relatively low temperatures (125–250 °C). Due to their exothermic and self-sustained properties, the reaction can be carried out until complete conversion, without any additional external energy input. Therefore, it is an energy-saving and sustainable synthesis method [225,226]. Xuanli Wang et al. developed a one-step SCS method to prepare Fe_3O_4 nanoparticles and the obtained samples have a high saturation magnetization of 89.17 emu/g and a small grain size of 57.3 nm. The flame temperature and the quantity of gaseous products released, depend on the nature of the fuel and the ratio of fuel to oxidant (φ). As the molar ratio of glycine (fuel) increases, the combustion mode changes from self-propagating combustion to smoldering combustion, and the oxide phase of SCS products sequentially changes from α-Fe_2O_3 → Fe_3O_4 → FeO [227]. The mechanisms for the formation of magnetite were further discussed by Aali et al. [228]. Glycine, urea and citric acid were used as fuels, and an overview of the results revealed that porous single-phase magnetite nano powder, with high crystallinity and saturation magnetization, was only achieved when φ = 0.95; the highest saturation magnetization reached 99 emu/g. Magnetite and hematite were produced simultaneously in the case of urea, while in the case of glycine and citric acid, first magnetite was prepared and then oxidized to hematite. SCS also allows for the effective doping of materials and mixing with other supporting materials, to achieve large-scale production, for a wider range of applications.

3.2. Advancements in Synthesis of Magnetic Particles on Adsorbents

The synthesis of adsorbents with magnetic particles and high adsorption performance through conventional synthesis methods is still challenging. Therefore, an effective and green synthesis process should be developed. To maximize the efficiency of magnetic particles on adsorbents, methods that facilitate the formation of particles with uniform size, monodispersity, and non-aggregation can be applied.

3.2.1. Uniform Distribution of Nanoparticles

The diffusion, transfer, fate, and environmental risks of engineered nanoparticles (ENPs) dispersed in liquid are significantly different from those of aggregation. Zhang et al. [229] showed that when the concentration of ENPs is lower than the critical micelle concentration (CMC), steric hindrance and/or electrostatic repulsion play a major role in their dispersion. When the dispersant is higher than its CMC, the dispersion of ENPs can be regarded as the "dissolution" process of the dispersant in the micelles. To overcome these agglomeration problems, associated with nanometer size, nanoparticles with high dispersion and high thermodynamic stability have been investigated.

In a recent study, sodium alginate was used to disperse nZVI to develop a new type of nZVI material [183]. Polymers chelate and coordinate ferrous ions before in situ redox, which results in an artificial "concentration" of iron into the polymer area; thus, aggregation of polymer-modified nanoparticles may be lower in an aqueous solution [230]. In addition, hydroxyl group-containing molecules that have a surface passivation effect can also inhibit subsequent crystallization, by preventing further oxidation bond formation. The increase in the number of hydroxyl groups results in a large surface energy and enhances the stability of the intermediate, consequently restraining the transformation into a crystalline form.

Functionalized, high-salt-tolerant magnetite nanoparticles were synthesized by a one-step synthesis, with sulfonated phenolic resin. Even in high-salt environments, strong anionic sulfonate groups can promote sufficient electrostatic repulsion between particles to separate them [231]. The deprotonated anionic coating can maintain the high negative charge on the surface of nanoparticles and can provide strong electrostatic repulsion, in the typical pH range of the natural aquatic environment, to maintain a highly dispersed state. However, pollutants that also have a negatively charged surface will affect adsorption and cause instability in the colloidal solution. The influence of pH, cations, anions, and humic acid on the aggregation behavior of citric acid-coated magnetite nanoparticles in the aqueous phase has been reported by Liu et al. [232].

3.2.2. Bio-Derived Magnetic Nanocomposite

Many conversion schemes have been developed to use the characteristics of biomass feed to prepare adsorbents. Both biological (anaerobic digestion, hydrolysis, and fermentation) and thermal (combustion, pyrolysis, liquefaction, baking, and gasification) methods are used to convert biomass into adsorbents. Biochar from heat treatment usually has a high energy density (usually >28 kJ/g) [233]. Magnetic biochar is mainly prepared by the following three methods: calcination, co-precipitation, and pyrolysis [234]. Generally, the typical pyrolysis process can be subdivided into fast pyrolysis, slow pyrolysis, and flash pyrolysis. Co-precipitation is a bottom-up ion precipitation reaction. Calcination is a heat treatment process in which small molecules, such as H_2O, CO_2, and SO_2, are removed [234]. The factors affecting the adsorption of pollutants by the magnetic biochar are summarized below.

There have been attempts to develop a magnetic biochar adsorbent to ensure a better and more effective separation of biochar particles, after the wastewater treatment process [181,233]. Further, several studies have been conducted, wherein biochar is magnetized, characterized, and applied to wastewater treatment [182,184,185,235] (Figure 8). Park et al. [236] used sesame straw raw biochar to adsorb multiple metal ions (Pb, Cr, Cd, Cu, and Zn) from wastewater. The results showed that its adsorption behavior for multi-metals was different from single-metal adsorption; in multi-metals, the adsorption of Cd was particularly impeded due to competition. However, to ensure the applicability of magnetic biochar for wastewater treatment, it is necessary to use physical and chemical conditions to simulate polluted water or use actual polluted water.

The high surface area produces rapid adsorption kinetics and, thus, a relatively short contact time. Furthermore, these nanocomposites are magnetic and can be separated from aqueous streams using an external magnetic field. Overall, magnetic nanocomposites have become a revolutionary tool for wastewater treatment, due to their unique properties and the overlap with current technologies.

3.2.3. Matrice-Confined NPs

The physical confinement of nanoparticles within the structure is mainly based on weak interactions, generated by hydrogen bonds, π–π bonds, or covalent grafting, which can be achieved by in situ chemistry or post-processing methods. A physical confinement strategy is attractive as the specific reactivity of the nanoparticles is retained, and the performance can be easily adjusted by changing the size and shape of the pores, to promote the regional growth of the nanocrystals. Among them, porous carbon, mesoporous silicon oxide, aluminum oxide, and montmorillonite have been widely studied.

Silica has many advantages, such as low cost, easy preparation, good liquid dispersion, easy modification, pH resistance, and thermal stability, making it highly suitable for use in magnetic composites. Generally, the magnetic particles may be arranged on different areas of the mesoporous silicon oxides, for example, as a core, scattered distribution, or fixed/grafted on the surface or in the siliceous framework, by functionalization in the mesopores. For instance, Fe_3O_4 nanoparticles coated with silica and naphthoyl chloride (Fe_3O_4@SiO_2@Nap) [237], porphyrin copolymer (Fe_3O_4 @SiO_2-TCPP) [238], lignin (Fe_3O_4@SiO_2-NH-MFL) [120], and porous aromatic frameworks (Fe_3O_4@SiO_2@PAF-6) [239] have been applied to water samples, to achieve ultrafast absorption. Improved Stöber and template removal methods are used to prepare magnetic mesoporous silica nanoparticles (MSNs), which contain a Fe_3O_4 core and a mesoporous silica shell [240].

Compared with one-dimensional materials with lower adsorption capacity, advanced two-dimensional materials have a higher surface area and better adsorption potential. Graphene is a typical two-dimensional material. As a carbon derivative, graphene exhibits high porosity. Graphene-based adsorbents have been widely studied for water treatment applications. In recent studies, magnetic iron sulfide (Fe_3S_4) was synthesized via a solvothermal process. The synthesized samples were analyzed, using various characterization techniques, to understand the adsorption mechanism of Pb(II) [49]. In another study, the authors compared the removal effect of a magnetic composite consisting of one-dimensional nanofibers and two-dimensional graphene, for Pb(II) and Cr(VI), respectively. The adsorption capacities of graphene and nanofibers for Pb(II) ions were 131.40 and 42.90 mg/g, respectively. In the case of Cr(VI) ions, the adsorption capacities were 68.85 and 51.07 mg/g, respectively [241]. In addition, two-dimensional, ultrathin nanosheet-like LDHs were modified with magnetic particles [242,243]. In this work, the improved adsorption performance was due to the enhanced exposure of active sites to pollutants.

Magnetic materials based on biosorbents are considered to be highly efficient and environmentally friendly. These synergistic biomaterials can be used to modify magnetic nanoparticles with various adsorption mechanisms, for use in simple applications and to achieve reusability [244]. For instance, cholesterol improves the stability of magnetic phosphatidylcholine, by increasing the accumulation of phospholipid molecules, which may improve the reusability of the adsorbent (>8 times) for the extraction of Organochlorine pesticides (OCPs). Cholesterol-functionalized magnetic nanoparticles can be obtained using APTES and cholesteryl chloroformate modification [245]. Verma et al. [246] synthesized magnetic biosorbents from citrus (peel and pulp) biomass waste, for wastewater treatment. Samples were synthesized at 500 °C, and both As(III) and As(V) from groundwater were removed. Many new adsorbents are still in the laboratory stage. They have demonstrated the ability to adsorb pollutants from water, under different pH values, different ionic strengths, and mixed with a large amount of organic matter. The adsorption capacity and removal mechanisms are gradually being understood. In future studies, the main challenges faced by wastewater treatment may result in the inclination to investigate environmental application risk, life cycle, and long-term evolution mechanisms.

3.3. Advancements in Synthesis of Adsorbents on Magnetic Particles

Efforts to synthesize well-incorporated magnetic adsorbents with uniform size and good adsorption performance are still challenging. The incorporation of magnetic particles into the adsorbents can be achieved by limiting the adsorbent formation in the vicinity of the magnetic particles. The advanced synthesis of adsorbents on magnetic particles can be accomplished by controlling or directing the growth of the adsorbent on magnetic particles, or by using pre-attached magnetic particles with adsorbent precursors. The synthesis procedures introduced in research articles, representing recent advancements in this category, are briefly depicted in Figure 9.

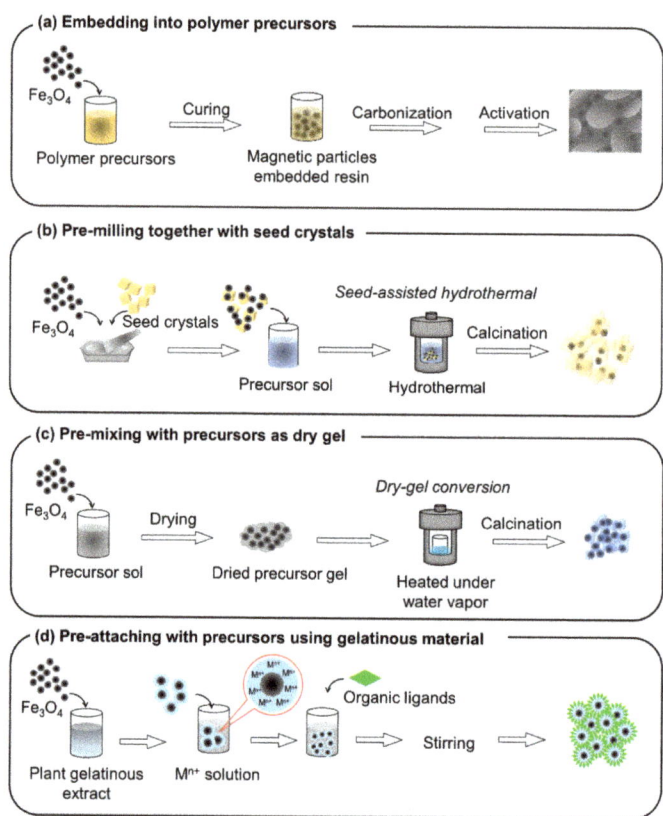

Figure 9. Brief procedures of the advanced synthesis of adsorbents on magnetic particles. (**a**) Embedding Fe_3O_4 into polymer precursors to yield magnetic activated carbon. (**b**) Pre-milling of Fe_3O_4 with zeolite seed crystals together to produce magnetic zeolites. (**c**) Pre-mixing Fe_3O_4 with zeolite precursors to form dried precursor gel before dry gel conversion to magnetic zeolite. (**d**) Pre-attaching gelatinous material coated Fe_3O_4 with metal ion precursor with subsequent formation of MOFs.

3.3.1. Embedding into Polymer Precursors

The rigid nature of natural carbon sources limits their ability to combine with magnetic particles. Nijpanich et al. [34] produced the first study to successfully incorporate magnetic particles into small-sized activated carbon, by embedding Fe_3O_4 particles into epoxy resin as a carbon source. The starting materials of the epoxy resin were mixed with Fe_3O_4 particles, before curing at room temperature (Figure 9a). This process allowed the incorporation of magnetic particles into the carbon source, prior to the carbonization and activation process. The as-prepared magnetic-activated carbon showed no significant difference in methylene blue adsorption performance, compared to non-magnetic-activated carbon prepared by the same route. The performance, however, was higher than that of magnetic-activated carbon derived from other carbon sources. There was no negative effect due to the embedded magnetic particles on the porosity and adsorption performance of the activated carbon. Although the Fe phase was found in the final product by XRD analysis due to the reduction of Fe_3O_4 under the activation conditions, the adsorbents were more easily collected using an external magnet, owing to a higher magnetic susceptibility.

3.3.2. Application of Seed Crystals

Recent studies have shown advancements in the synthesis of magnetic zeolites. Hagio et al. applied seed-assisted hydrothermal synthesis, an alternative but selective technique for preparing zeolites, to synthesize magnetic FAU-type [29] and MOR-type zeolites [30] for the first time (Figure 9b). In principle, with hydrothermal treatment, the addition of the target zeolite crystals as the seed into precursors can induce and accelerate the crystallization of zeolites. Therefore, a high product yield was obtained. Seed crystals act as nucleation centers for crystal growth. According to this principle, if the seed crystals are well mixed with magnetic particles, the target zeolites crystallize near the magnetic particles. Hagio et al. discovered that pre-milling seed crystals and Fe_3O_4 particles together, before adding to the precursors, allowed for good pre-attachment among them [30]. The growth of zeolite crystals occurred in the tiny gaps between the seed and Fe_3O_4 particles. Consequently, the Fe_3O_4 particles were completely incorporated inside the zeolite crystals. Furthermore, the amount of magnetic zeolite produced by the seed-assisted technique was found to be almost two times higher than that prepared without seed crystals. Both magnetic FAU-type and MOR-type zeolites showed good separability from solutions, using an external magnet. The magnetic FAU-type zeolite showed similar performance of methylene blue adsorption to the non-magnetic one, which revealed no adverse effects from the incorporated Fe_3O_4, inside the FAU frameworks.

3.3.3. Pre-Mixing into Precursor Gel

Another advancement of magnetic zeolite synthesis is the utilization of the dry-gel conversion (DGC) method, first presented by Phouthavong et al. [31]. Some researchers have already introduced this technique, which crystallizes a precursor gel under heated vapor to synthesize well-crystallized non-magnetic zeolites [247,248]. In this route, Fe_3O_4 particles were homogenously pre-mixed into dry gel precursors before placing under heat water vapor (Figure 9c). Because the Fe_3O_4 particles were closely attached to the precursors in the dry gel, they were subsequently incorporated into the BEA-type zeolite during crystal growth. After the DGC, the obtained particles were collected without washing and subsequently calcinated to remove the organic template. Low waste generation is another advantage of DGC. The obtained magnetic BEA-type zeolite had a homogenous and uniform shape, size, and magnetic response. Their adsorption ability to remove methylene blue from water was similar to that of non-magnetic BEA. This indicates that the incorporated Fe_3O_4 particles did not clog the pores of the zeolite.

3.3.4. Pre-Attaching with Precursor Using Gelatinous Material

Recently, basil seed mucilage, a plant-based product, was also used in magnetic composite synthesis technology. Mahmoodi and Javanbakht [190] prepared magnetic MOF nano-biocomposites by depositing gelatinous parts, extracted from basil seed mucilage, on Fe_3O_4 particles and further modifying them with ZIF-8 MOFs. To synthesize ZIF-8, a Zn(II)-methanol solution was stirred in the presence of mucilage-coated Fe_3O_4 particles. The Zn(II) ions were absorbed into the mucilage via electrostatic interactions (Figure 9d). Therefore, the Zn(II)-loaded, mucilage-coated Fe_3O_4 particles acted as nucleation centers for the growth of ZIF-8, after the addition of the organic precursor, 2-methylimidazole.

3.4. Co-Synthesis of Magnetic Particles and Adsorbents

In some cases, adsorbents and magnetic materials can be synthesized using a similar process. Methods to simultaneously synthesize adsorbents and magnetic materials have recently been proposed. To the best of our knowledge, this one-pot synthesis method was first introduced in the past decade. The co-synthesis methods are briefly introduced in Figure 10.

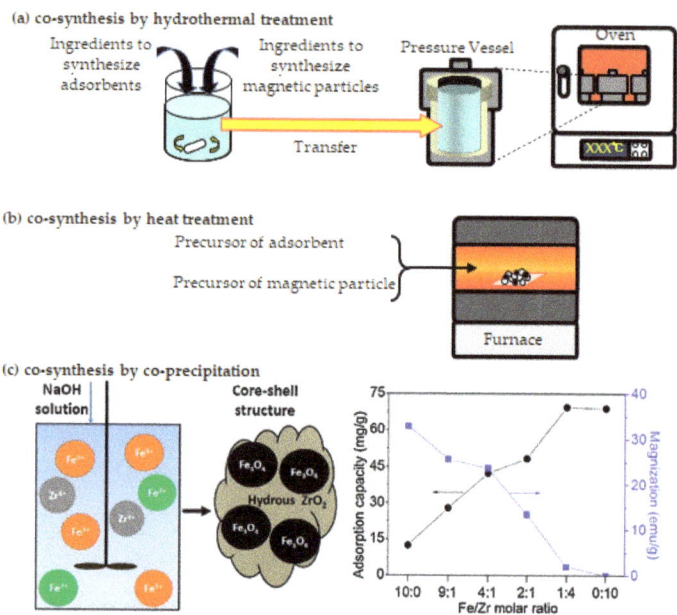

Figure 10. Illustrations of procedures for co-synthesis methods. (**a**) Hydrothermal, (**b**) heat treatment, and (**c**) co-synthesis methods. ((**c**) Reprinted from Appl. Surf. Sci. 366, Wang, Z.; Xing, M.; Fang, W.; Wu, D. One-step synthesis of magnetite core/zirconia shell nanocomposite for high efficiency removal of phosphate from water, 67–77. Ref. [26], Copyright © 2022 with permission from Elsevier).

3.4.1. Co-Synthesis via Hydrothermal Synthesis

Because Fe_3O_4 (magnetic particles) can be synthesized from alkaline solutions containing Fe(II) and Fe(III), adsorbents that crystallize from alkaline solutions have been considered for the co-synthesis of adsorbents and magnetic particles. In 2013, Aono et al. [249] prepared a composite material, consisting of Na-P1-type zeolite, and nanosized magnetite, by alkali processing from a mixed solution of fly ash and $FeCl_2$ and $FeCl_3$, followed by hydrothermal synthesis at 100 °C for 24 h. From TEM observations, Fe_3O_4 nanoparticles were observed at the grain boundaries between the polycrystalline zeolites. This was expected, as the formation of zeolite crystals is slower than the formation of nanosized Fe_3O_4. The resulting magnetic zeolite could achieve a 61% decontamination of soil containing radioactive Cs. Chen et al. [195] prepared magnetic $MnFe_2O_4$-sludge biochar via a one-pot hydrothermal synthesis for Pb(II) removal. Dried sludge obtained from sewage was added to deionized water and mixed with $MnCl_2 \cdot 4H_2O$ and $FeCl_3 \cdot 6H_2O$. NaOH (5 M) was added dropwise, until the pH reached 10–10.5. This was followed by hydrothermal synthesis at 180 °C for 10 h. The washed and dried product was tested with a Pb(II)-containing solution and showed a maximum adsorption amount of 174.216 mg/g.

3.4.2. Co-Synthesis via Heat Treatment

Another approach is to embed the iron ions into organic materials, followed by heat treatment in a vacuum or inert gas atmosphere. In this method, the polymeric material turns into active carbon adsorbents, and iron ions crystallize into Fe_3O_4 particles. Fan et al. [194] used rice husks, pretreated with a 2 M NaOH solution, and added them to ethanol, containing $FeCl_3 \cdot 6H_2O$. After drying, heat treatment at 800 °C was conducted for 2 h, under a N_2 atmosphere. Fe_3O_4 particles were deposited on carbonaceous material. Adsorption capacities against Cr(VI) of this magnetic carbonaceous material in 50 and 100 mg/L Cr(VI) solutions were 49.913 and 99.158 mg/g, respectively. The saturated magnetization was 77.8 emu/g. Jung et al. [33] prepared magnetic biochar/Fe_3O_4 nanocomposites, using brown marine macroalgae as the precursor for biochar. An electrochemical technique was used to supply the iron ions, and heat treatment was carried out at 600 °C for 1 h. The magnetic biochar possessed superparamagnetic characteristics, with a saturation magnetization of approximately 26.79 emu/g, and showed a maximum adsorption capacity of 382.01 mg/g for acid orange 7, a type of dye, at 30 °C. Although the process is one-pot synthesis, these methods require some pretreatments. Elessawy et al. [196] synthesized functionalized magnetic fullerene nanocomposites in a one-pot process, using cut PET bottle waste and ferrocene. The two materials were introduced into a stainless steel autoclave reactor, which was heated to 800 °C for 20 h, resulting in a black product. This was a facile, one-step, green synthesis route, using catalytic thermal decomposition of PET bottle waste as a precursor and ferrocene as a catalyst and precursor for magnetic nanoparticles in the composite. The nanocomposite showed saturated magnetization of 7.002 emu/g and a maximum adsorption of 833.3 and 806.5 mg/g against methylene blue and acid blue 25, respectively, at 25 °C.

3.4.3. Co-Synthesis via Co-Precipitation

Co-precipitation is also extended to realize the one-pot synthesis of magnetic adsorbents. Wang et al. [26] demonstrated the one-step synthesis of Fe_3O_4 core/zirconia shell nanocomposites, using the co-precipitation method. Particular amounts of $FeSO_4 \cdot 7H_2O$, $FeCl_3 \cdot 6H_2O$, and $ZrOCl_2 \cdot 8H_2O$ were dissolved in doubly distilled water. Precipitation was caused by the dropwise addition of a 6 M NaOH solution, under stirring at 400 rpm, until the pH reached 7.6, and was aged without stirring at 60 °C for 18 h. The washed, collected, and dried product consisted of ball-like or hexagonal particles of Fe_3O_4 and cloud-like zirconia. The adsorption performance was evaluated through phosphate removal, which exhibited maximum adsorption capacities of 27.93–69.44 mg/g, when the Fe/Zr molar ratio was varied between 9:1 to 1:4, as shown in Figure 10c.

Zhao et al. [191] synthesized a magnetic Mg–Al–La composite using co-precipitation, followed by calcination for fluoride removal. Similar to the aforementioned study, a solution containing $FeSO_4 \cdot 7H_2O$ and $FeCl_3 \cdot 6H_2O$ was mixed with another solution containing $Mg(NO_3)_2 \cdot 6H_2O$, $Al(NO_3)_3 \cdot 9H_2O$, and $La(NO_3)_3 \cdot 6H_2O$, and then a NaOH solution was added dropwise to carry out co-precipitation. The Mg–Al–La hydroxide material exhibited a hydrotalcite-like structure, and iron was converted to Fe_3O_4 by calcination. The calcined magnetic Mg–Al–La composite showed a maximum adsorption capacity of 65.75 mg/g against fluoride.

Although the co-synthesis of adsorbents and magnetic particles is rather simple, it should be noted that it does not solve all the difficulties in the preparation of magnetic adsorbents, such as the prevention of adsorption sites by magnetic particles. Control of the structure during the synthesis and the combination of appropriate adsorbents and magnetic materials may overcome these issues.

4. Summary and Future View

In conclusion, the advancement in the synthesis methods of magnetic adsorbents, for the removal of substances from water streams, has been comprehensively summarized and discussed. Although numerous techniques have been developed for the preparation of magnetic adsorbents with effective adsorption performance, reviews that focus on the synthesis methods of magnetic adsorbents for wastewater treatment and the obtained material structures have not been reported, which are important for the future development of such materials. Generally, the typical synthesis methods are categorized into the following five groups: direct use of magnetic particles, attachment of pre-prepared adsorbents and pre-prepared magnetic particles, synthesis of magnetic particles on pre-prepared adsorbents, synthesis of adsorbents on pre-prepared magnetic particles, and co-synthesis of adsorbents and magnetic particles. The improvements in these new methods make conventional synthesis more energy saving, more efficient, and simpler in process while maintaining or increasing the adsorption performance. Advanced methods have overcome the negative aspects of compositing, such as the coverage of adsorption sites by the magnetic substance, through control of material structures and/or enhancing the adsorption ability of magnetic particles.

Despite the long history of magnetic adsorbents, there are still issues with their synthesis methods that need to be addressed. Primarily, the improvement of the magnetic adsorbent synthesis method to realize effective and uniform compositing is still necessary. The yields of the magnetic adsorbents are not always stated, which has a significant impact on the process simplicity and material cost. In addition, a thorough investigation is needed to enable the design and control of magnetic adsorbent structures to achieve the intensive and selective removal of specific pollutants. Moreover, the development of simulations and observations to understand the synthesis process and adsorption process is still challenging; however, it is expected that this will provide greater insight into the design of ideal magnetic adsorbents, through information on the relationship between the composite structure and performance. Finally, further improvement in reusability is essential for realizing an efficient, economical, and environmentally friendly synthesis method for magnetic adsorbents, which is required to expand applicable cases in social implementation.

Author Contributions: Conceptualization, R.I. and L.L.; methodology, V.P., R.Y., S.N., T.H., R.I., L.K. and L.L.; resources, V.P., R.Y., S.N., T.H., R.I., L.K. and L.L.; writing—original draft preparation, V.P., R.Y., T.H. and L.K.; writing—review and editing, R.I. and L.L.; visualization, V.P., T.H. and L.K.; supervision, R.I. and L.L. All authors have read and agreed to the published version of the manuscript.

Funding: This research was funded by JST SICORP (Grant Number JPMJSC18H1) and the National Key Research and Development Program of China (No. 2017YFE0127100).

Institutional Review Board Statement: Not applicable.

Informed Consent Statement: Not applicable.

Data Availability Statement: No new data were created or analyzed in this study. Data sharing is not applicable.

Acknowledgments: The authors appreciate the support from JST SICORP (Grant Number JPMJSC18H1), Japan and from the National Key Research and Development Program of China (No. 2017YFE0127100).

Conflicts of Interest: The authors declare no conflict of interest.

References

1. Rashid, R.; Shafiq, I.; Akhter, P.; Iqbal, M.J.; Hussain, M. A state-of-the-art review on wastewater treatment techniques: The effectiveness of adsorption method. *Environ. Sci. Pollut. Res.* **2021**, *28*, 9050–9066. [CrossRef] [PubMed]
2. Giakisikli, G.; Anthemidis, A.N. Magnetic materials as sorbents for metal/metalloid preconcentration and/or separation. A review. *Anal. Chim. Acta* **2013**, *789*, 1–16. [CrossRef] [PubMed]
3. Robinson, P.J.; Dunnill, P.; Lilly, M.D. The properties of magnetic supports in relation to immobilized enzyme reactors. *Biotechnol. Bioeng.* **1973**, *15*, 603–606. [CrossRef]
4. de Latour, C.; Kolm, H.H. High gradient magnetic separation A water-treatment alternative. *J. Am. Water Work. Assoc.* **1976**, *78*, 325–327. [CrossRef]
5. Booker, N.A.; Keir, D.; Priestley, A.J.; Ritchie, C.B.; Sudarmana, D.L.; Woods, M.A. Sewage clarification with magnetite particles. *Water Sci. Technol.* **1991**, *23*, 1703–1712. [CrossRef]
6. Matei, E.; Predescu, A.; Vasile, E.; Predescu, A. Properties of magnetic iron oxides used as materials for wastewater treatment. *J. Phys. Conf. Ser.* **2011**, *304*, 012022. [CrossRef]
7. Gutierrez, A.M.; Dziubla, T.D.; Hilt, J.Z. Recent advances on iron oxide magnetic nanoparticles as sorbents of organic pollutants in water and wastewater treatment. *Rev. Environ. Health* **2017**, *32*, 111–117. [CrossRef]
8. Wu, W.; He, Q.; Jiang, C. Magnetic iron oxide nanoparticles: Synthesis and surface functionalization strategies. *Nanoscale Res. Lett.* **2008**, *3*, 397–415. [CrossRef]
9. Zhang, J.; Lin, S.; Han, M.; Su, Q.; Xia, L.; Hui, Z. Adsorption properties of magnetic magnetite nanoparticle for coexistent Cr(VI) and Cu(II) in mixed solution. *Water* **2020**, *12*, 446. [CrossRef]

10. Lu, A.H.; Salabas, E.L.; Schüth, F. Magnetic nanoparticles: Synthesis, protection, functionalization, and application. *Angew. Chem. Int. Ed.* **2007**, *46*, 1222–1244. [CrossRef]
11. de Dios, A.S.; Díaz-García, M.E. Multifunctional nanoparticles: Analytical prospects. *Anal. Chim. Acta* **2010**, *666*, 1–22. [CrossRef] [PubMed]
12. Phouthavong, V.; Manakasettharn, S.; Viboonratanasri, D.; Buajarern, S.; Prompinit, P.; Sereenonchai, K. Colorimetric determination of trace orthophosphate in water by using C18-functionalized silica coated magnetite. *Sci. Rep.* **2021**, *11*, 23073. [CrossRef] [PubMed]
13. Chen, L.; Wang, T.; Tong, J. Application of derivatized magnetic materials to the separation and the preconcentration of pollutants in water samples. *TrAC Trends Anal. Chem.* **2011**, *30*, 1095–1108. [CrossRef]
14. Petcharoen, K.; Sirivat, A. Synthesis and characterization of magnetite nanoparticles via the chemical co-precipitation method. *Mater. Sci. Eng. B* **2012**, *177*, 421–427. [CrossRef]
15. Tu, Y.-J.; You, C.F.; Chang, C.K.; Chen, M.H. Application of magnetic nano-particles for phosphorus removal/recovery in aqueous solution. *J. Taiwan Inst. Chem. Eng.* **2015**, *46*, 148–154. [CrossRef]
16. Hou, Y.; Yu, J.; Gao, S. Solvothermal reduction synthesis and characterization of superparamagnetic magnetite nanoparticles. *J. Mater. Chem.* **2003**, *13*, 1983–1987. [CrossRef]
17. Zhao, F.; Zhang, B.; Feng, L. Preparation and magnetic properties of magnetite nanoparticles. *Mater. Lett.* **2012**, *68*, 112–114. [CrossRef]
18. Lakshmanan, R.; Okoli, C.; Boutonnet, M.; Järås, S.; Rajarao, G.K. Microemulsion prepared mangetic nanoparticles for phosphate removal: Time efficient studies. *J. Environ. Chem. Eng.* **2014**, *2*, 185–189. [CrossRef]
19. Es'haghzade, Z.; Pajootan, E.; Bahrami, H.; Arami, M. Facile synthesis of Fe_3O_4 nanoparticles via aqueous based electrochemical route for heterogeneous electro-Fenton removal of azo dyes. *J. Taiwan Inst. Chem. Eng.* **2017**, *71*, 91–105. [CrossRef]
20. Miao, F.; Hua, W.; Hu, L.; Huang, K. Magnetic Fe_3O_4 nanoparticles prepared by a facile and green microwave-assisted approach. *Mater. Lett.* **2011**, *65*, 1031–1033. [CrossRef]
21. Shalaby, M.; Madkour, F.F.; El-Kassas, H.Y.; Mohamed, A.A.; Elgarahyet, A.M. Green synthesis of recyclable iron oxide nanoparticles using *Spirulina platensis* microalgae for adsorptive removal of cationic and anionic dyes. *Environ. Sci. Pollut. Res.* **2021**, *28*, 65549–65572. [CrossRef] [PubMed]
22. Shalaby, M.; Madkour, F.F.; El-Kassas, H.Y.; Mohamed, A.A.; Elgarahyet, A.M. Microwave enhanced sorption of methylene blue dye onto bio-synthesized iron oxide nanoparticles: Kinetics, isotherms, and thermodynamics studies. *Int. J. Phytoremediation* **2021**, in press. [CrossRef] [PubMed]
23. Bolto, B.A.; Dixon, D.R.; Eldridge, R.J.; Swinton, E.A.; Weiss, D.E.; Willis, D. The use of magnetic polymers in water treatment. *J. Polym. Sci.* **1975**, *49*, 211–219. [CrossRef]
24. Bolto, B.A.; Dixon, D.R.; Eldridge, R.J. Graft polymerization on magnetic polymer substrates. *J. Appl. Polym. Sci.* **1978**, *22*, 1977–1982. [CrossRef]
25. Chen, W.Y.; Anderson, P.R.; Holsen, T.M. Recovery and recycle of metals from wastewater with a magnetite-based adsorption process. *Res. J. Water Pollut. Control Fed.* **1996**, *63*, 958–964.
26. Wang, Z.; Xing, M.; Fang, W.; Wu, D. One-Step synthesis of magnetite core/zirconia shell nanocomposite for high efficiency removal of phosphate from water. *Appl. Surf. Sci.* **2016**, *366*, 67–77. [CrossRef]
27. Li, Q.; Yang, F.; Zhang, J.; Zhou, C. Magnetic Fe_3O_4/MnO_2 core–shell nano-composite for removal of heavy metals from wastewater. *SN Appl. Sci.* **2020**, *2*, 1375. [CrossRef]
28. Oliveira, L.C.A.; Petkowicz, D.I.; Smaniotto, A.; Pergher, S.B.C. Magnetic zeolites: A new adsorbent for removal of metallic contaminants from water. *Water Res.* **2004**, *38*, 3699–3704. [CrossRef]
29. Hagio, T.; Kunishi, H.; Yamaoka, K.; Kamimoto, Y.; Ichino, R. Seed-Assisted synthesis of magnetic faujasite-type zeolite and its adsorption performance. *Nanosci. Nanotechnol. Lett.* **2018**, *10*, 862–867. [CrossRef]
30. Hagio, T.; Nijpanich, S.; Kunishi, H.; Yamaoka, K.; Phouthavong, V.; Kamimoto, Y.; Ichino, R.; Iwai, K. Synthesis of MOR zeolite/magnetite composite via seed assisted method. *J. Nanosci. Nanotechnol.* **2019**, *19*, 6841–6848. [CrossRef]
31. Phouthavong, V.; Hiraiwa, M.; Hagio, T.; Nijpanich, S.; Chounlamany, V.; Nishihama, T.; Kamimoto, Y.; Ichino, R. Magnetic BEA-type zeolites: Preparation by dry-gel conversion method and assessment of dye removal performance. *J. Mater. Cycles Waste Manag.* **2020**, *22*, 375–382. [CrossRef]
32. Fadillah, G.; Yudha, S.P.; Sagadevan, S.; Fatimah, I.; Muraza, O. Magnetic iron oxide/clay nanocomposites for adsorption and catalytic oxidation in water treatment applications. *Open Chem.* **2020**, *18*, 1148–1166. [CrossRef]
33. Jung, K.-W.; Choi, B.H.; Jeong, T.-U.; Ahn, K.-H. Facile synthesis of magnetic biochar/Fe_3O_4 nanocomposites using electro-magnetization technique and its application on the removal of acid orange 7 from aqueous media. *Bioresour. Technol.* **2016**, *220*, 672–676. [CrossRef]
34. Nijpanich, S.; Morihashi, R.; Hagio, T.; Kamimoto, Y.; Ichino, R. Synthesis of magnetic activated carbon based on a magnetite/butyl glycidyl ether-diluted bisphenol A/diethylenetriamine epoxy resin system. *Nanosci. Nanotechnol. Lett.* **2018**, *10*, 843–848. [CrossRef]
35. Li, Y.; Zhang, X.; Zhang, P.; Liu, X.; Han, L. Facile fabrication of magnetic bio-derived chars by co-mixing with Fe_3O_4 nanoparticles for effective Pb^{2+} adsorption: Properties and mechanism. *J. Clean. Prod.* **2020**, *262*, 121350. [CrossRef]

36. Meral, K.; Metın, Ö. Graphene oxide-magnetite nanocomposite as an efficient and magnetically separable adsorbent for methylene blue removal from aqueous solution. *Turk. J. Chem.* **2014**, *38*, 775–782. [CrossRef]
37. Li, Y.; Wu, M.; Wang, B.; Wu, Y.; Ma, M.; Zhang, X. Synthesis of magnetic lignin-based hollow microspheres: A highly adsorptive and reusable adsorbent derived from renewable resources. *ACS Sustain. Chem. Eng.* **2016**, *4*, 5523–5532. [CrossRef]
38. Nata, I.F.; Wicakso, D.R.; Mirwan, A.; Irawan, C.; Ramadhani, D.; Ursulla. Selective adsorption of Pb(II) ion on amine-rich functionalized rice husk magnetic nanoparticles biocomposites in aqueous solution. *J. Environ. Chem. Eng.* **2020**, *8*, 104339. [CrossRef]
39. Charpentier, T.V.J.; Neville, A.; Lanigan, J.L.; Barker, R.; Smith, M.J.; Richardson, T. Preparation of magnetic carboxymethylchitosan nanoparticles for adsorption of heavy metal ions. *ACS Omega* **2016**, *1*, 77–83. [CrossRef]
40. Lee, P.L.; Sun, Y.C.; Ling, Y.C. Magnetic nano-adsorbent integrated with lab-on-valve system for trace analysis of multiple heavy metals. *J. Anal. At. Spectrom.* **2009**, *24*, 320–327. [CrossRef]
41. Liu, L.; Liu, S.; Zhao, L.; Su, G.; Liu, X.; Peng, H.; Xue, J.; Tang, A. Fabrication of novel magnetic core–shell chelating adsorbent for rapid and highly efficient adsorption of heavy metal ions from aqueous solution. *J. Mol. Liq.* **2020**, *313*, 113593. [CrossRef]
42. Mohebbi, A.; Farajzadeh, M.A. Chemical synthesis-free and facile preparation of magnetized polyethylene composite and its application as an efficient magnetic sorbent for some pesticides. *J. Chromatogr. A* **2020**, *1625*, 461340. [CrossRef] [PubMed]
43. Huang, S.; Xu, J.; Zheng, J.; Zhu, F.; Xie, L.; Ouyang, G. Synthesis and application of magnetic molecularly imprinted polymers in sample preparation. *Anal. Bioanal. Chem.* **2018**, *410*, 3991–4014. [CrossRef] [PubMed]
44. Meteku, B.E.; Huang, J.; Zeng, J.; Subhan, F.; Feng, Z.; Zhang, Y.; Qiu, Z.; Aslam, S.; Li, G.; Yan, Z. Magnetic metal-organic framework composites for environmental monitoring and remediation. *Coord. Chem. Rev.* **2020**, *413*, 213261. [CrossRef]
45. Minh, P.T.; Lebedeva, O.E. Adsorption properties of a magnetite composite with coffee waste. *Russ. J. Phys. Chem. A* **2018**, *92*, 2044–2047. [CrossRef]
46. Shehzad, K.; Xie, C.; He, J.; Cai, X.; Xu, W.; Liu, J. Facile synthesis of novel calcined magnetic orange peel composites for efficient removal of arsenite through simultaneous oxidation and adsorption. *J. Colloid Interface Sci.* **2018**, *511*, 155–164. [CrossRef]
47. Huang, X.; Kong, L.; Huang, S.; Liu, M.; Li, L. Synthesis of novel magnetic sulfur-doped Fe_3O_4 nanoparticles for efficient removal of Pb(II). *Sci. China Chem.* **2018**, *61*, 164–171. [CrossRef]
48. Liu, J.; Kong, L.; Huang, X.; Liu, M.; Li, L. Removal of arsenic(V) from aqueous solutions using sulfur-doped Fe_3O_4 nanoparticles. *RSC Adv.* **2018**, *8*, 40804. [CrossRef]
49. Kong, L.; Li, Z.; Huang, X.; Huang, S.; Sun, H.; Liu, M.; Li, L. Efficient removal of Pb(II) from water using magnetic Fe_3S_4/reduced graphene oxide composites. *J. Mater. Chem. A* **2017**, *5*, 19333–19342. [CrossRef]
50. Teja, A.S.; Koh, P.-Y. Synthesis, properties, and applications of magnetic iron oxide nanoparticles. *Prog. Cryst. Growth Charact. Mater.* **2009**, *55*, 22–45. [CrossRef]
51. Majidi, S.; Sehrig, F.Z.; Farkhani, S.M.; Goloujeh, M.S.; Akbarzadeh, A. Current methods for synthesis of magnetic nanoparticles. *Artif. Cells Nanomed. Biotechnol.* **2016**, *44*, 722–734. [CrossRef] [PubMed]
52. Ali, A.; Shah, T.; Ullah, R.; Zhou, P.; Guo, M.; Ovais, M.; Tan, Z.; Rui, Y.K. Review on recent progress in magnetic nanoparticles: Synthesis, characterization, and diverse applications. *Front. Chem.* **2021**, *9*, 629054. [CrossRef] [PubMed]
53. Akbarzadeh, A.; Samiei, M.; Davaran, S. Magnetic nanoparticles: Preparation, physical properties, and applications in biomedicine. *Nanoscale Res. Lett.* **2012**, *7*, 144. [CrossRef] [PubMed]
54. García-Merino, B.; Bringas, E.; Ortiz, I. Synthesis and applications of surface-modified magnetic nanoparticles: Progress and future prospects. *Rev. Chem. Eng.* **2021**, 000010151520200072. [CrossRef]
55. Reshadi, M.A.M.; Bazargan, A.; McKay, G. A review of the application of adsorbents for landfill leachate treatment: Focus on magnetic adsorption. *Sci. Total Environ.* **2020**, *731*, 138863. [CrossRef]
56. Shukla, S.; Khan, R.; Daverey, A. Synthesis and characterization of magnetic nanoparticles, and their applications in wastewater treatment: A review. *Environ. Technol. Innov.* **2021**, *24*, 101924. [CrossRef]
57. Abdullah, N.H.; Shameli, K.; Abdullah, E.C.; Abdullah, L.C. Solid matrices for fabrication of magnetic iron oxide nanocomposites: Synthesis, properties, and application for the adsorption of heavy metal ions and dyes. *Compos. B Eng.* **2019**, *162*, 538–568. [CrossRef]
58. Abdel Maksoud, M.I.A.; Elgarahy, A.M.; Farrell, C.; Al-Muhtaseb, A.H.; Rooney, D.W.; Osman, A.I. Insight on water remediation application using magnetic nanomaterials and biosorbents. *Coord. Chem. Rev.* **2020**, *403*, 213096. [CrossRef]
59. de Vicente, I.; Merino-Martos, A.; Cruz-Pizarro, L.; de Vicente, J. On the use of magnetic nano and microparticles for lake restoration. *J. Hazard. Mater.* **2010**, *181*, 375–381. [CrossRef]
60. Choi, J.; Chung, J.; Lee, W.; Kim, J.-O. Phosphorous adsorption on synthesized magnetite in wastewater. *J. Ind. Eng. Chem.* **2016**, *34*, 198–203. [CrossRef]
61. Cheng, W.; Xu, J.; Wang, Y.; Wu, F.; Xu, X.; Li, J. Dispersion–Precipitation synthesis of nanosized magnetic iron oxide for efficient removal of arsenite in water. *J. Colloid Interface Sci.* **2015**, *445*, 93–101. [CrossRef] [PubMed]
62. Nikraftar, N.; Ghorbani, F. Adsorption of As(V) using modified magnetic nanoparticles with ascorbic acid: Optimization by response surface methodology. *Water Air Soil Pollut.* **2016**, *227*, 178. [CrossRef]
63. Ahmed, M.A.; Ali, S.M.; El-Deka, S.I.; Galal, A. Magnetite–Hematite nanoparticles prepared by green methods for heavy metal ions removal from water. *Mater. Sci. Eng. B* **2013**, *178*, 744–751. [CrossRef]

64. Iwahori, K.; Watanabe, J.; Tani, Y.; Seyama, H.; Miyata, N. Removal of heavy metal cations by biogenic magnetite nanoparticles produced in Fe(III)-reducing microbial enrichment cultures. *J. Biosci. Bioeng.* **2014**, *117*, 333–335. [CrossRef] [PubMed]
65. Rajput, S.; Pittman, C.U., Jr.; Mohan, D. Magnetic magnetite (Fe_3O_4) nanoparticle synthesis and applications for lead (Pb^{2+}) and chromium (Cr^{6+}) removal from water. *J. Colloid Interface Sci.* **2016**, *468*, 334–346. [CrossRef] [PubMed]
66. Feng, D.; Aldrich, C.; Tan, H. Removal of heavy metal ions by carrier magnetic separation of adsorptive particulates. *Hydrometallurgy* **2000**, *56*, 359–368. [CrossRef]
67. Nah, I.W.; Hwang, K.-Y.; Jeon, C.; Choi, H.B. Removal of Pb ion from water by magnetically modified zeolite. *Miner. Eng.* **2006**, *19*, 1452–1455. [CrossRef]
68. Paris, E.C.; Malafatti, J.O.D.; Musetti, H.C.; Manzoli, A.; Zenatti, A.; Escote, M.T. Faujasite zeolite decorated with cobalt ferrite nanoparticles for improving removal and reuse in Pb^{2+} ions adsorption. *Chin. J. Chem. Eng.* **2020**, *28*, 1884–1890. [CrossRef]
69. Fungaro, D.A.; Graciano, J.E.A. Adsorption of zinc ions from water using zeolite/iron oxide composites. *Adsorp. Sci. Technol.* **2007**, *25*, 729–740. [CrossRef]
70. Fungaro, D.A.; Yamaura, M.; Craesmeyer, G.R. Uranium removal from aqueous solution by zeolite from fly ash-iron oxide magnetic nanocomposite. *Int. Rev. Chem. Eng.* **2012**, *4*, 353–358.
71. Bessa, R.A.; Costa, L.S.; Oliveira, C.P.; Bohn, F.; do Nascimento, R.F.; Sasaki, J.M.; Loiola, A.R. Kaolin-Based magnetic zeolites A and P as water softeners. *Microporous Mesoporous Mater.* **2017**, *245*, 64–72. [CrossRef]
72. Gaffer, A.; Kahlawy, A.A.A.; Aman, D. Magnetic zeolite-natural polymer composite for adsorption of chromium (VI). *Egypt. J. Pet.* **2017**, *26*, 995–999. [CrossRef]
73. Anbia, M.; Rahimi, F. Adsorption of platinum(IV) from an aqueous solution with magnetic cellulose functionalized with thiol and amine as a nano-active adsorbent. *J. Appl. Polym. Sci.* **2017**, *134*, 45361. [CrossRef]
74. Le, V.T.; Doan, V.D.; Nguyen, D.D.; Nguyen, H.T.; Ngo, Q.P.; Tran, T.K.N.; Le, H.S. A novel cross-linked magnetic hydroxyapatite/chitosan composite: Preparation, characterization, and application for Ni(II) ion removal from aqueous solution. *Water Air Soil Pollut.* **2018**, *229*, 101. [CrossRef]
75. Galhoum, A.A. Facile synthesis of functionalized polyglycidyl methacrylate-magnetic nanocomposites for enhanced uranium sorption. *RSC Adv.* **2019**, *9*, 38783. [CrossRef]
76. Tian, N.; Wu, J.; Wang, J.; Dai, W. Development of a novel core–shell magnetic Fe_3O_4@CMC@ZIF-8-OH composite with outstanding rubidium-ion capacity. *J. Chem. Eng. Data* **2019**, *64*, 5716–5724. [CrossRef]
77. Dinari, M.; Shirani, M.A.; Maleki, M.H.; Tabatabaeian, R. Green cross-linked bionanocomposite of magnetic layered double hydroxide/guar gum polymer as an efficient adsorbent of Cr(VI) from aqueous solution. *Carbohydr. Polym.* **2020**, *236*, 116070. [CrossRef] [PubMed]
78. Chen, C.; Gunawan, P.; Xu, R. Self-Assembled Fe_3O_4-layered double hydroxide colloidal nanohybrids with excellent performance for treatment of organic dyes in water. *J. Mater. Chem.* **2011**, *21*, 1218–1225. [CrossRef]
79. Hu, S.C.; Shi, F.; Liu, J.X.; Yu, L.; Liu, S.H. Magnetic mesoporous iron oxide/silica composite aerogels with high adsorption ability for organic pollutant removal. *J. Porous Mater.* **2016**, *23*, 655–661. [CrossRef]
80. Fungaro, D.A.; Yamaura, M.; Carvalho, T.E.M. Adsorption of anionic dyes from aqueous solution on zeolite from fly ash-iron oxide magnetic nanocomposite. *J. At. Mol. Sci.* **2011**, *2*, 305–316. [CrossRef]
81. Rashid, M.; Price, N.T.; Pinilla, M.Á.G.; O'Shea, K.E. Effective removal of phosphate from aqueous solution using humic acid coated magnetite nanoparticles. *Water Res.* **2017**, *123*, 353–360. [CrossRef] [PubMed]
82. Wu, H.; Xia, T.; Yin, L.; Ji, Y. Adsorption of iodide from an aqueous solution via calcined magnetite-activated carbon/MgAl-layered double hydroxide. *Chem. Phys. Lett.* **2021**, *774*, 138612. [CrossRef]
83. Kumar, A.S.K.; Jiang, S.J.; Warchoł, J.K. Synthesis and characterization of two-dimensional transition metal dichalcogenide magnetic MoS_2@Fe_3O_4 Nanoparticles for Adsorption of Cr(VI)/Cr(III). *ACS Omega* **2017**, *2*, 6187–6200. [CrossRef] [PubMed]
84. Geng, J.; Gu, F.; Chang, J. Fabrication of magnetic lignosulfonate using ultrasonic-assisted in situ synthesis for efficient removal of Cr(VI) and Rhodamine B from wastewater. *J. Hazard. Mater.* **2019**, *375*, 174–181. [CrossRef]
85. Eivazzadeh-Keihan, R.; Radinekiyan, F.; Asgharnasl, S.; Maleki, A.; Bahreinizad, H. A natural and eco-friendly magnetic nanobiocomposite based on activated chitosan for heavy metals adsorption and the in-vitro hyperthermia of cancer therapy. *J. Mater. Res. Technol.* **2020**, *9*, 12244–12259. [CrossRef]
86. Sheng, L.; Zhou, L.; Huang, Z.; Liu, Z.; Chen, Q.; Huang, G.; Adesina, A.A. Facile synthesis of magnetic chitosan nano-particles functionalized with N/O-containing groups for efficient adsorption of U(VI) from aqueous solution. *J. Radioanal. Nucl. Chem.* **2016**, *310*, 1361–1371. [CrossRef]
87. Amini, A.; Khajeh, M.; Oveisi, A.R.; Daliran, S.; Ghaffari-Moghaddam, M.; Delarami, H.S. A porous multifunctional and magnetic layered graphene oxide/3D mesoporous MOF nanocomposite for rapid adsorption of uranium(VI) from aqueous solutions. *J. Ind. Eng. Chem.* **2021**, *93*, 322–332. [CrossRef]
88. Zargar, B.; Khazaeifar, A. Synthesis of an ion-imprinted sorbent by surface imprinting of magnetized carbon nanotubes for determination of trace amounts of cadmium ions. *Microchim. Acta* **2017**, *184*, 4521–4529. [CrossRef]
89. Wang, Z.; Ding, S.; Li, Z.; Li, F.; Zhao, T.; Li, J.; Lin, H.; Chen, C. Synthesis of a magnetic polystyrene-based cation-exchange resin and its utilization for the efficient removal of cadmium (II). *Water Sci. Technol.* **2018**, *2017*, 770–781. [CrossRef]
90. Dong, T.; Xing, H.; Wu, H.; Lv, Y.; Wu, L.; Mi, S.; Yang, L. Preparation of magnetic Levextrel resin for cadmium(II) removal. *Environ. Technol. Innov.* **2021**, *23*, 101657. [CrossRef]

91. Wan, K.; Wang, G.; Xue, S.; Xiao, Y.; Fan, J.; Li, L.; Miao, Z. Preparation of humic acid/L-cysteine-codecorated magnetic Fe_3O_4 nanoparticles for selective and highly efficient adsorption of mercury. *ACS Omega* **2021**, *6*, 7941–7950. [CrossRef] [PubMed]
92. Xie, Y.; Qian, D.; Wu, D.; Ma, X. Magnetic halloysite nanotubes/iron oxide composites for the adsorption of dyes. *Chem. Eng. J.* **2011**, *168*, 959–963. [CrossRef]
93. Paz, R.; Viltres, H.; Gupta, N.K.; Leyva, C. Fabrication of magnetic cerium-organic framework-activated carbon composite for charged dye removal from aqueous solutions. *J. Mol. Liq.* **2021**, *337*, 116578. [CrossRef]
94. Abou Taleb, M.F.; Abou El Fadl, F.I.; Albalwi, H. Adsorption of toxic dye in wastewater onto magnetic NVP/CS nanocomposite hydrogels synthesized using gamma radiation. *Sep. Purif. Technol.* **2021**, *266*, 118551. [CrossRef]
95. Cheng, J.; Chang, P.R.; Zheng, P.; Ma, X. Characterization of magnetic carbon nanotube–cyclodextrin composite and its adsorption of dye. *Ind. Eng. Chem. Res.* **2014**, *53*, 1415–1421. [CrossRef]
96. Stoia, M.; Păcurariu, C.; Istratie, R.; Nižňansky, D. Solvothermal synthesis of magnetic Fe_xO_y/C nanocomposites used as adsorbents for the removal of methylene blue from wastewater. *J. Therm. Anal. Calorim.* **2015**, *121*, 989–1001. [CrossRef]
97. Panasenko, A.; Pirogovskaya, P.; Tkachenko, I.; Ivannikov, S.; Arefieva, O.; Marchenko, Y. Synthesis and characterization of magnetic silica/iron oxide composite as a sorbent for the removal of methylene blue. *Mater. Chem. Phys.* **2020**, *245*, 122759. [CrossRef]
98. Abdullah, N.H.; Shameli, K.; Abdullah, E.C.; Abdullah, L.C. Low cost and efficient synthesis of magnetic iron oxide/activated sericite nanocomposites for rapid removal of methylene blue and crystal violet dyes. *Mater. Charact.* **2020**, *163*, 110275. [CrossRef]
99. Ge, H.; Zhang, Z.; Zhao, X.; Li, H.; Sun, J.; Jv, X. Adsorption performance of organic dyes in single and binary systems onto poly(itaconic acid)/magnetite sepiolite composite prepared via the green synthetic methods. *Can. J. Chem. Eng.* **2021**, *99*, S157–S167. [CrossRef]
100. Ain, Q.U.; Rasheed, U.; Yaseen, M.; Zhang, H.; He, R.; Tong, Z. Fabrication of magnetically separable 3-acrylamidopropyltrimethylammo chloride intercalated bentonite composite for the efficient adsorption of cationic and anionic dyes. *Appl. Surf. Sci.* **2020**, *514*, 145929. [CrossRef]
101. Uddin, M.K.; Mashkoor, F.; AlArifi, I.M.; Nasar, A. Simple one-step synthesis process of novel MoS_2@bentonite magnetic nanocomposite for efficient adsorption of crystal violet from aqueous solution. *Mater. Res. Bull.* **2021**, *139*, 111279. [CrossRef]
102. Zhu, X.; Liu, Y.; Zhou, C.; Zhang, S.; Chen, J. Novel and high-performance magnetic carbon composite prepared from waste hydrochar for dye removal. *ACS Sustain. Chem. Eng.* **2014**, *2*, 969–977. [CrossRef]
103. Qin, Y.; Long, M.; Tan, B.; Zhou, B. RhB adsorption performance of magnetic adsorbent Fe_3O_4/RGO composite and its regeneration through a fenton-like reaction. *Nano-Micro Lett.* **2014**, *6*, 125–135. [CrossRef]
104. Shi, S.; Fan, Y.; Huang, Y. Facile low temperature hydrothermal synthesis of magnetic mesoporous carbon nanocomposite for adsorption removal of ciprofloxacin antibiotics. *Ind. Eng. Chem. Res.* **2013**, *52*, 2604–2612. [CrossRef]
105. Liu, Q.; Zhong, L.-B.; Zhao, Q.-B.; Frear, C.; Zheng, Y.-M. Synthesis of Fe_3O_4/polyacrylonitrile composite electrospun nanofiber mat for effective adsorption of tetracycline. *ACS Appl. Mater. Interfaces* **2015**, *7*, 14573–14583. [CrossRef] [PubMed]
106. Nezhadali, A.; Koushali, S.E.; Divsar, F. Synthesis of polypyrrole—Chitosan magnetic nanocomposite for the removal of carbamazepine from wastewater: Adsorption isotherm and kinetic study. *J. Environ. Chem. Eng.* **2021**, *9*, 105648. [CrossRef]
107. Palakeeti, B.; Reddy, K.V.; Gobi, K.V.; Rao, P.N.; Chinta, J.P. Simple and efficient method for the quantification of antiepileptic drugs in human plasma by using magnetic graphene oxide-β-cyclodextrin composite as a sorbent. *Futur. J. Pharm. Sci.* **2021**, *7*, 93. [CrossRef]
108. Ieamviteevanich, P.; Palaporn, D.; Chanlek, N.; Poo-arporn, Y.; Mongkolthanaruk, W.; Eichhorn, S.J.; Pinitsoontorn, S. Carbon nanofiber aerogel/magnetic core–shell nanoparticle composites as recyclable oil sorbents. *ACS Appl. Nano Mater.* **2020**, *3*, 3939–3950. [CrossRef]
109. Ahmed, S.; Zhang, Y.; Wu, B.; Zheng, Z.; Leung, C.F.; Choy, T.; Kwok, Y.; Lo, I.M.C. Scaled-Up development of magnetically recyclable Fe_3O_4/La(OH)$_3$ composite for river water phosphate removal: From bench-scale to pilot-scale study. *Sci. Total Environ.* **2021**, *791*, 148281. [CrossRef]
110. Cao, J.; Liu, X.W.; Fu, R.; Tan, Z. Magnetic P zeolites: Synthesis, characterization and the behavior in potassium extraction from seawater. *Sep. Purif. Technol.* **2008**, *63*, 92–100. [CrossRef]
111. Aono, H.; Kaji, N.; Itagaki, Y.; Johan, E.; Matsue, N. Synthesis of mordenite and its composite material using chemical reagents for Cs decontamination. *J. Ceram. Soc. Jpn.* **2016**, *124*, 617–623. [CrossRef]
112. Huang, Y.F.; Li, Y.; Jiang, Y.; Yan, X.P. Magnetic immobilization of amine-functionalized magnetite microspheres in a knotted reactor for on-line solid-phase extraction coupled with ICP-MS for speciation analysis of trace chromium. *J. Anal. At. Spectrom.* **2010**, *25*, 1467–1474. [CrossRef]
113. Wang, T.; Zhang, L.; Li, Y.; Yang, W.; Song, T.; Tang, C.; Meng, Y.; Dai, S.; Wang, H.; Chai, L.; et al. Synthesis of core–shell magnetic Fe_3O_4@poly(m-phenylenediamine) particles for chromium reduction and adsorption. *Environ. Sci. Technol.* **2015**, *49*, 5654–5662. [CrossRef] [PubMed]
114. Wang, Y.; Luo, X.; Tang, J.; Hu, X.; Xu, Q.; Yang, C. Extraction and preconcentration of trace levels of cobalt using functionalized magnetic nanoparticles in a sequential injection lab-on-valve system with detection by electrothermal atomic adsorption spectrometry. *Anal. Chim. Acta* **2012**, *713*, 92–96. [CrossRef] [PubMed]
115. Jilin, C.; Guihuan, C.; Hongfei, G.; Jianxin, C. Synthesis and characterization of magnetic ZSM-5 zeolite. *Trans. Tianjin Univ.* **2013**, *19*, 326–331.

116. Zhang, F.; Song, Y.; Song, S.; Zhang, R.; Hou, W. Synthesis of magnetite-graphene oxide-layered double hydroxide composites and applications for the removal of Pb(II) and 2,4-dichlorophenoxyacetic acid from aqueous solutions. *ACS Appl. Mater. Interfaces* **2015**, *7*, 7251–7263. [CrossRef]
117. Zhang, J.; Zhai, S.; Li, S.; Xiao, Z.; Song, Y.; An, Q.; Tian, G. Pb(II) removal of Fe_3O_4@SiO_2-NH_2 core–shell nanomaterials prepared via a controllable sol–gel process. *Chem. Eng. J.* **2013**, *215–216*, 461–471. [CrossRef]
118. Liu, H.; Peng, S.; Shu, L.; Chen, T.; Bao, T.; Frost, R.L. Magnetic zeolite NaA: Synthesis, characterization based on metakaolin and its application for the removal of Cu^{2+}, Pb^{2+}. *Chemosphere* **2013**, *91*, 1539–1546. [CrossRef]
119. Jiang, X.; Su, S.; Rao, J.; Li, S.; Lei, T.; Bai, H.; Wang, S.; Yang, X. Magnetic metal-organic framework (Fe_3O_4@ZIF-8) core–shell composite for the efficient removal of Pb(II) and Cu(II) from water. *J. Environ. Chem. Eng.* **2021**, *9*, 105959. [CrossRef]
120. Zhang, Y.; Ni, S.; Wang, X.; Zhang, W.; Lagerquist, L.; Qin, M.; Willför, S.; Xu, C.; Fatehi, P. Ultrafast adsorption of heavy metal ions onto functionalized lignin-based hybrid magnetic nanoparticles. *Chem. Eng. J.* **2019**, *372*, 82–91. [CrossRef]
121. Tavares, D.S.; Daniel-da-Silva, A.L.; Lopes, C.B.; Silva, N.J.O.; Amaral, V.S.; Rocha, J.; Pereira, E.; Trindade, T. Efficient sorbents based on magnetite coated with siliceous hybrid shells for removal of mercury ions. *J. Mater. Chem. A* **2013**, *1*, 8134–8143. [CrossRef]
122. Alonso, E.V.; Guerrero, M.M.L.; Cueto, P.C.; Benítez, J.B.; Pavón, J.M.C.; de Torres, A.G. Development of an on-line solid phase extraction method based on new functionalized magnetic nanoparticles. Use in the determination of mercury in biological and sea-water samples. *Talanta* **2016**, *153*, 228–239. [CrossRef] [PubMed]
123. Huang, L.; Shuai, Q. Facile approach to prepare sulfur-functionalized magnetic amide-linked organic polymers for enhanced Hg(II) removal from water. *ACS Sustain. Chem. Eng.* **2019**, *7*, 9957–9965. [CrossRef]
124. Wei, W.; Li, A.; Pi, S.; Wang, Q.; Zhou, L.; Yang, J.; Ma, F.; Ni, B.J. Synthesis of core–shell magnetic nanocomposite Fe_3O_4@microbial extracellular polymeric substances for simultaneous redox sorption and recovery of silver ions as silver nanoparticles. *ACS Sustain. Chem. Eng.* **2018**, *6*, 749–756. [CrossRef]
125. Lai, Z.; Xuan, Z.; Yu, S.; Zhang, Z.; Cao, Y.; Zhao, Y.; Li, Y.; Luo, J.; Li, X. Synthesis of magnetic-carbon sorbent for removal of U(VI) from aqueous solution. *J. Radioanal. Nucl. Chem.* **2019**, *322*, 2079–2089. [CrossRef]
126. Zhao, M.; Cui, Z.; Pan, D.; Fan, F.; Tang, J.; Hu, Y.; Zhang, P.; Li, P.; Kong, X.Y.; et al. An efficient uranium adsorption magnetic platform based on amidoxime-functionalized flower-like Fe_3O_4@TiO_2 core–shell microspheres. *ACS Appl. Mater. Interfaces* **2021**, *13*, 17931–17939. [CrossRef] [PubMed]
127. Wang, D.; Guan, K.; Bai, Z.; Liu, F. Facile preparation of acid-resistant magnetite particles for removal of Sb(III) from strong acidic solution. *Sci. Technol. Adv. Mater.* **2016**, *17*, 80–88. [CrossRef] [PubMed]
128. Roto, R.; Yusran, Y.; Kuncaka, A. Magnetic adsorbent of Fe_3O_4@SiO_2 core–shell nanoparticles modified with thiol group for chloroauric ion adsorption. *Appl. Surf. Sci.* **2016**, *377*, 30–36. [CrossRef]
129. Jiang, C.; Sun, Y.; Yu, X.; Zhang, L.; Sun, X.; Gao, Y.; Zhang, H.; Song, D. Removal of sudan dyes from water with C18-functional ultrafine magnetic silica nanoparticles. *Talanta* **2012**, *89*, 38–46. [CrossRef] [PubMed]
130. Kubo, M.; Moriyama, R.; Shimada, M. Facile fabrication of HKUST-1 nanocomposites incorporating Fe_3O_4 and TiO_2 nanoparticles by a spray-assisted synthetic process and their dye adsorption performances. *Microporous Mesoporous Mater.* **2019**, *280*, 227–235. [CrossRef]
131. Zheng, J.; Cheng, C.; Fang, W.J.; Chen, C.; Yan, R.W.; Huai, H.X.; Wang, C.C. Surfactant-Free synthesis of a Fe_3O_4@ZIF-8 core–shell heterostructure for adsorption of methylene blue. *Cryst. Eng. Commun.* **2014**, *16*, 3960–3964. [CrossRef]
132. Far, H.S.; Hasanzadeh, M.; Nashtaei, M.S.; Rabbani, M.; Haji, A.; Moghadam, B.H. PPI-Dendrimer-Functionalized magnetic metal-organic framework (Fe_3O_4@MOF@PPI) with high adsorption capacity for sustainable wastewater treatment. *ACS Appl. Mater. Interfaces* **2020**, *12*, 25294–25303. [CrossRef]
133. Liu, Y.; Zheng, H.; Han, Y.; Wu, Y.; Wang, Y.; Liu, Y.; Feng, L. Amphiphilic magnetic copolymer for enhanced removal of anionic dyes: Fabrication, application and adsorption mechanism. *Colloids Surf. A Physicochem. Eng. Asp.* **2021**, *623*, 126674. [CrossRef]
134. Wei, X.; Wang, Y.; Chen, J.; Liu, Z.; Xu, F.; He, X.; Li, H.; Zhou, Y. Fabrication of di-selective adsorption platform based on deep eutectic solvent stabilized magnetic polydopamine: Achieving di-selectivity conversion through adding $CaCl_2$. *Chem. Eng. J.* **2021**, *421*, 127815. [CrossRef]
135. Karbasaki, S.S.; Bagherzade, G.; Maleki, B.; Ghani, M. Fabrication of sulfamic acid functionalized magnetic nanoparticles with denderimeric linkers and its application for microextraction purposes, one-pot preparation of pyrans pigments and removal of malachite green. *J. Taiwan Inst. Chem. Eng.* **2021**, *118*, 342–354. [CrossRef]
136. Liu, K.; Huang, L.; Suai, Q. Solvent and catalyst free preparation of sulfonic acid functionalized magnetic covalent organic polymer as efficient adsorbent for malachite green removal. *J. Water Process. Eng.* **2021**, *43*, 102306. [CrossRef]
137. Karatapanis, A.E.; Fiamegos, Y.; Stalikas, C.D. Silica-Modified magnetic nanoparticles functionalized with cetylpyridinium bromide for the preconcentration of metals after complexation with 8-hydroxyquinoline. *Talanta* **2011**, *84*, 834–839. [CrossRef]
138. Zhang, S.; Niu, H.; Cai, Y.; Shi, Y. Barium alginate caged Fe_3O_4@C18 magnetic nanoparticles for the pre-concentration of polycyclic aromatic hydrocarbons and phthalate esters from environmental water samples. *Anal. Chim. Acta* **2010**, *665*, 167–175. [CrossRef] [PubMed]
139. Li, Z.; Huang, D.; Fu, C.; Wei, B.; Yu, W.; Deng, C.; Zhang, X. Preparation of magnetic core mesoporous shell microspheres with C18-modified interior pore-walls for fast extraction and analysis of phthalates in water samples. *J. Chromatogr. A* **2011**, *1218*, 6232–6239. [CrossRef] [PubMed]

140. Adivi, F.G.; Hashemi, P.; Tehrani, A.D. Agarose-Coated Fe$_3$O$_4$@SiO$_2$ magnetic nanoparticles modified with sodium dodecyl sulfate, a new promising sorbent for fast adsorption/desorption of cationic drugs. *Polym. Bull.* **2019**, *76*, 1239–1256. [CrossRef]
141. Huang, L.; Mao, N.; Yan, Q.; Zhang, D.; Shuai, Q. Magnetic covalent organic frameworks for the removal of diclofenac sodium from water. *ACS Appl. Nano Mater.* **2020**, *3*, 319–326. [CrossRef]
142. Su, Y.; Adeleye, A.S.; Huang, Y.; Sun, X.; Dai, C.; Zhou, X.; Zhang, Y.; Keller, A.A. Simultaneous removal of cadmium and nitrate in aqueous media by nanoscale zerovalent iron (nZVI) and Au doped nZVI particles. *Water Res.* **2014**, *63*, 102–111. [CrossRef] [PubMed]
143. Stefaniuk, M.; Oleszczuk, P.; Ok, Y.S. Review on nano zerovalent iron (nZVI): From synthesis to environmental applications. *Chem. Eng. J.* **2016**, *287*, 618–632. [CrossRef]
144. Kendelewicz, T.; Liu, P.; Doyle, C.S.; Brown, G.E., Jr. Spectroscopic study of the reaction of aqueous Cr(VI) with Fe$_3$O$_4$ (111) surfaces. *Surf. Sci.* **2000**, *469*, 144–163. [CrossRef]
145. Yusoff, A.H.M.; Salimi, M.N.; Jamlos, M.F. Critical parametric study on final size of magnetite nanoparticles. *IOP Conf. Ser. Mater. Sci. Eng.* **2018**, *318*, 012020. [CrossRef]
146. Gnanaprakash, G.; Mahadevan, S.; Jayakumar, T.; Kalyanasundaram, P.; Philip, J.; Raj, B. Effect of initial pH and temperature of iron salt solutions on formation of magnetite nanoparticles. *Mater. Chem. Phys.* **2007**, *103*, 168–175. [CrossRef]
147. Mascolo, M.C.; Pei, Y.; Ring, T.A. Room temperature co-precipitation synthesis of magnetite nanoparticles in a large pH window with different bases. *Materials* **2013**, *6*, 5549–5567. [CrossRef]
148. Hajdú, A.; Illés, E.; Tombácz, E.; Borbáth, I. Surface charging, polyanionic coating and colloid stability of magnetite nanoparticles. *Colloids Surf. A Physicochem. Eng. Asp.* **2009**, *347*, 104–108. [CrossRef]
149. Chang, M.; Shih, Y. Synthesis and application of magnetic iron oxide nanoparticles on the removal of Reactive Black 5: Reaction mechanism, temperature and pH effects. *J. Environ. Manage.* **2018**, *224*, 235–242. [CrossRef]
150. Maity, D.; Kale, S.N.; Kaul-Ghanekar, R.; Xue, J.-M.; Ding, J. Studies of magnetite nanoparticles synthesized by thermal decomposition of iron (III) acetylacetonate in tri(ethylene glycol). *J. Magn. Magn. Mater.* **2009**, *321*, 3093–3098. [CrossRef]
151. Angermann, A.; Töpfer, J. Synthesis of magnetite nanoparticles by thermal decomposition of ferrous oxalate dihydrate. *J. Mater. Sci.* **2008**, *43*, 5123–5130. [CrossRef]
152. Lassenberger, A.; Grünewald, T.A.; van Oostrum, P.D.J.; Rennhofer, H.; Amenitsch, H.; Zirbs, R.; Lichtenegger, H.C.; Reimhult, E. Monodisperse iron oxide nanoparticles by thermal decomposition: Elucidating particle formation by second-resolved in situ small-angle X-ray scattering. *Chem. Mater.* **2017**, *29*, 4511–4522. [CrossRef] [PubMed]
153. Fotukian, S.M.; Barati, A.; Soleymani, M.; Alizadeh, A.M. Solvothermal synthesis of CuFe$_2$O$_4$ and Fe$_3$O$_4$ nanoparticles with high heating efficiency for magnetic hyperthermia application. *J. Alloys Compd.* **2020**, *816*, 152548. [CrossRef]
154. Luo, X.; Zhang, L. High effective adsorption of organic dyes on magnetic cellulose beads entrapping activated carbon. *J. Hazard. Mater.* **2009**, *171*, 340–347. [CrossRef]
155. Metin, Ö.; Aydoğan, Ş.; Meral, K. A new route for the synthesis of graphene oxide–Fe$_3$O$_4$ (GO–Fe$_3$O$_4$) nanocomposites and their Schottky diode applications. *J. Alloys Compd.* **2014**, *585*, 681–688. [CrossRef]
156. Guo, S.; Sun, S. FePt nanoparticles assembled on graphene as enhanced catalyst for oxygen reduction reaction. *J. Am. Chem. Soc.* **2012**, *134*, 2492–2495. [CrossRef] [PubMed]
157. Hosseini, S.M.; Younesi, H.; Bahramifar, N.; Mehraban, Z. A novel facile synthesis of the amine-functionalized magnetic core coated carboxylated nanochitosan shells as an amphoteric nanobiosupport. *Carbohydr. Polym.* **2019**, *221*, 174–185. [CrossRef] [PubMed]
158. Mahjour, B.; Shen, Y.; Liu, W.; Cernak, T. A map of the amine–carboxylic acid coupling system. *Nature* **2020**, *580*, 71–75. [CrossRef] [PubMed]
159. Shou, J.; Jiang, C.; Wang, F.; Qiu, M.; Xu, Q. Fabrication of Fe$_3$O$_4$/MgAl-layered double hydroxide magnetic composites for the effective decontamination of Co(II) from synthetic wastewater. *J. Mol. Liq.* **2015**, *207*, 216–223. [CrossRef]
160. Salah El-Din, T.A.; Elzatahry, A.A.; Aldhayan, D.M.; Al-Enizi, A.M.; Al-Deyab, S.S. Synthesis and characterization of magnetite zeolite nano composite. *Int. J. Electrochem. Sci.* **2011**, *6*, 6177–6183.
161. Jahangirian, H.; Ismail, M.H.S.; Haron, M.J.; Rafiee-Moghaddam, R.; Shameli, K.; Hosseini, S.; Kalantari, K.; Khandanlou, R.; Gharibshahi, E.; Soltaninejad, S. Synthesis and characterization of Zeolite/Fe$_3$O$_4$ nanocomposite by green quick precipitation method. *Dig. J. Nanomater. Biostructures* **2013**, *8*, 1405–1413.
162. Shirani, M.; Semnani, A.; Habibollahi, S.; Haddadi, H.; Narimani, M. Synthesis and application of magnetic NaY zeolite composite immobilized with ionic liquid for adsorption desulfurization of fuel using response surface methodology. *J. Porous Mater.* **2016**, *23*, 701–712. [CrossRef]
163. Wu, J.H.; Li, X.-S.; Zhao, Y.; Gao, Q.; Guo, L.; Feng, Y.-Q. Titania coated magnetic mesoporous hollow silica microspheres: Fabrication and application to selective enrichment of phosphopeptides. *Chem. Commun.* **2010**, *46*, 9031–9033. [CrossRef] [PubMed]
164. Yan, L.; Kong, L.; Qu, Z.; Li, L.; Shen, G. Magnetic biochar decorated with ZnS nanocrystals for Pb(II) removal. *ACS Sustain. Chem. Eng.* **2015**, *3*, 125–132. [CrossRef]
165. Chen, B.; Hu, B.; He, M.; Huang, Q.; Zhang, Y.; Zhang, X. Speciation of selenium in cells by HPLC-ICP-MS after (on-chip) magnetic solid phase extraction. *J. Anal. At. Spectrom.* **2013**, *28*, 334–343. [CrossRef]

166. Stöber, W.; Fink, A.; Bohn, E. Controlled growth of monodisperse silica spheres in the micron size range. *J. Colloid Interface Sci.* **1968**, *26*, 62–69. [CrossRef]
167. Vacassy, R.; Flatt, R.J.; Hofmann, H.; Choi, K.S.; Singh, R.K. Synthesis of microporous silica spheres. *J. Colloid Interface Sci.* **2000**, *227*, 302–315. [CrossRef]
168. Jang, J.H.; Lim, H.B. Characterization and analytical application of surface modified magnetic nanoparticles. *Microchem. J.* **2010**, *94*, 148–158. [CrossRef]
169. Briso, A.; Quintana, G.; Ide, V.; Basualto, C.; Molina, L.; Montes, G.; Valenzuela, F. Integrated use of magnetic nanostructured calcium silicate hydrate and magnetic manganese dioxide adsorbents for remediation of an acidic mine water. *J. Water. Process. Eng.* **2018**, *25*, 247–257. [CrossRef]
170. Carné-Sánchez, A.; Imaz, I.; Cano-Sarabia, M.; Maspoch, D. A spray-drying strategy for synthesis of nanoscale metal–organic frameworks and their assembly into hollow superstructures. *Nat. Chem.* **2013**, *5*, 203–211. [CrossRef]
171. Qiao, K.; Tian, W.; Bai, J.; Wang, L.; Zhao, J.; Du, Z.; Gong, X. Application of magnetic adsorbents based on iron oxide nanoparticles for oil spill remediation: A review. *J. Taiwan Inst. Chem. Eng.* **2019**, *97*, 227–236. [CrossRef]
172. Oladipo, A.A.; Gazi, M. Efficient boron abstraction using honeycomb-like porous magnetic hybrids: Assessment of techno-economic recovery of boric acid. *J. Environ. Manage.* **2016**, *183*, 917–924. [CrossRef] [PubMed]
173. Meng, F.; Yang, B.; Wang, B.; Duan, S.; Chen, Z.; Ma, W. Novel dendrimerlike magnetic biosorbent based on modified orange peel waste: Adsorption–reduction behavior of arsenic. *ACS Sustain. Chem. Eng.* **2017**, *5*, 9692–9700. [CrossRef]
174. Kong, L.; Yan, L.; Qu, Z.; Yana, N.; Li, L. β-Cyclodextrin stabilized magnetic Fe_3S_4 nanoparticles for efficient removal of Pb(II). *J. Mater. Chem. A* **2015**, *3*, 15755–15763. [CrossRef]
175. Gong, Y.; Gai, L.; Tang, J.; Fu, J.; Wang, Q.; Zeng, E.Y. Reduction of Cr(VI) in simulated groundwater by FeS-coated ironmagnetic nanoparticles. *Sci. Total Environ.* **2017**, *595*, 743–751. [CrossRef]
176. Adel, M.; Ahmed, M.A.; Mohamed, A.A. Effective removal of indigo carmine dye from wastewaters by adsorption onto mesoporous magnesium nanoparticles. *Environ. Nanotechnol. Monit. Manag.* **2021**, *16*, 100550. [CrossRef]
177. Vergis, B.R.; Hari Krishna, R.; Kottam, N.; Nagabhushana, B.M.; Sharath, R.; Darukaprasad, B. Removal of malachite green from aqueous solution by magnetic $CuFe_2O_4$ nano-adsorbent synthesized by one pot solution combustion method. *J. Nanostruct. Chem.* **2018**, *8*, 1–12. [CrossRef]
178. Stan, M.; Lung, I.; Soran, M.-L.; Opris, O.; Leostean, C.; Popa, A.; Copaciu, F.; Lazar, M.D.; Kacso, I.; Silipas, T.-D.; et al. Starch-Coated green synthesized magnetite nanoparticles for removal of textile dye Optilan blue from aqueous media. *J. Taiwan Inst. Chem. Eng.* **2019**, *100*, 65–73. [CrossRef]
179. Cao, Z.; Liu, X.; Xu, J.; Zhang, J.; Yang, Y.; Zhou, J.; Xu, X.; Lowry, G.V. Removal of antibiotic florfenicol by sulfide-modified nanoscale zero-valent iron. *Environ. Sci. Technol.* **2017**, *51*, 11269–11277. [CrossRef]
180. Song, S.; Su, Y.; Adeleye, A.S.; Zhang, Y.; Zhou, X. Optimal design and characterization of sulfide-modified nanoscale zerovalent iron for diclofenac removal. *Appl. Catal. B Environ.* **2017**, *201*, 211–220. [CrossRef]
181. Wang, S.; Gao, B.; Li, Y.; Wan, Y.; Creamer, A.E. Sorption of arsenate onto magnetic iron–manganese (Fe–Mn) biochar composites. *RSC Adv.* **2015**, *5*, 67971–67978. [CrossRef]
182. Wang, S.; Gao, B.; Zimmerman, A.R.; Li, Y.; Ma, L.; Harris, W.G.; Migliaccio, K.W. Removal of arsenic by magnetic biochar prepared from pinewood and natural hematite. *Bioresour. Technol.* **2015**, *175*, 391–395. [CrossRef] [PubMed]
183. Li, Z.; Xu, S.; Xiao, F.; Qian, L.; Song, Y. Removal of hexavalent chromium from groundwater using sodium alginate dispersed nano zero-valent iron. *J. Environ. Manage.* **2019**, *244*, 33–39. [CrossRef] [PubMed]
184. Zhang, Z.; Wang, X.; Wang, Y.; Xia, S.; Chen, L.; Zhang, Y.; Zhao, J. Pb(II) removal from water using Fe-coated bamboo charcoal with the assistance of microwaves. *J. Environ. Sci.* **2013**, *25*, 1044–1053. [CrossRef]
185. Dong, X.; He, L.; Hu, H.; Liu, N.; Gao, S.; Piao, Y. Removal of 17β-estradiol by using highly adsorptive magnetic biochar nanoparticles from aqueous solution. *Chem. Eng. J.* **2018**, *352*, 371–379. [CrossRef]
186. Sun, Q.; Tang, M.; Hendriksen, P.V.; Chen, B. Biotemplated fabrication of a 3D hierarchical structure of magnetic $ZnFe_2O_4$/MgAl-LDH for efficient elimination of dye from water. *J. Alloys Compd.* **2020**, *829*, 154552. [CrossRef]
187. Muntean, S.G.; Nistor, M.A.; Ianoş, R.; Păcurariu, C.; Căpraru, A.; Surdu, V.A. Combustion synthesis of Fe_3O_4/Ag/C nanocomposite and application for dyes removal from multicomponent systems. *Appl. Surf. Sci.* **2019**, *481*, 825–837. [CrossRef]
188. Nistor, M.A.; Muntean, S.G.; Ianoş, R.; Racoviceanu, R.; Ianaşi, C.; Cseh, L. Adsorption of anionic dyes from wastewater onto magnetic nanocomposite powders synthesized by combustion method. *Appl. Sci.* **2021**, *11*, 9236. [CrossRef]
189. Ianoş, R.; Păcurariu, C.; Muntean, S.G.; Muntean, E.; Nistor, M.A.; Nižňanský, D. Combustion synthesis of iron oxide/carbon nanocomposites, efficient adsorbents for anionic and cationic dyes removal from wastewaters. *J. Alloys Compd.* **2018**, *741*, 1235–1246. [CrossRef]
190. Mahmoodi, M.; Javanbakht, V. Fabrication of Zn-based magnetic zeolitic imidazolate framework bionanocomposite using basil seed mucilage for removal of azo cationic and anionic dyes from aqueous solution. *Int. J. Biol. Macromol.* **2021**, *167*, 1076–1090. [CrossRef]
191. Zhao, W.; Chen, Y.; Zhang, W. Rapid and convenient removal of fluoride by magnetic magnesium–aluminum–lanthanum composite: Synthesis, performance and mechanism. *Asia-Pac. J. Chem. Eng.* **2017**, *12*, 640–650. [CrossRef]
192. Jin, J.; Huang, X.; Zhou, L.; Peng, J.; Wang, Y. In situ preparation of magnetic chitosan resins functionalized with triethylenetetramine for the adsorption of uranyl(II) ions. *J. Radioanal. Nucl. Chem.* **2015**, *303*, 797–806. [CrossRef]

193. Zhang, H.; Huang, F.; Liu, D.L.; Shi, P. Highly efficient removal of Cr(VI) from wastewater via adsorption with novel magnetic Fe_3O_4@C@MgAl-layered double-hydroxide. *Chin. Chem. Lett.* **2015**, *26*, 1137–1143. [CrossRef]
194. Fan, Y.; Yang, R.; Lei, Z.; Liu, N.; Lv, J.; Zhai, S.; Zhai, B.; Wang, L. Removal of Cr(VI) from aqueous solution by rice husk derived magnetic sorbents. *Korean J. Chem. Eng.* **2016**, *33*, 1416–1424. [CrossRef]
195. Chen, Y.; Xu, F.; Li, H.; Li, Y.; Liu, Y.; Chen, Y.; Li, M.; Li, L.; Jiang, H.; Chen, L. Simple hydrothermal synthesis of magnetic $MnFe_2O_4$-sludge biochar composites for removal of aqueous Pb^{2+}. *J. Anal. Appl. Pyrolysis* **2021**, *156*, 105173. [CrossRef]
196. Elessawy, N.A.; El-Sayed, E.M.; Ali, S.; Elkady, M.F.; Elnouby, M.; Hamad, H.A. One-Pot green synthesis of magnetic fullerene nanocomposite for adsorption characteristics. *J. Water Process. Eng.* **2020**, *34*, 101047. [CrossRef]
197. Mirshahghassemi, S.; Lead, J.R. Oil recovery from water under environmentally relevant conditions using magnetic nanoparticles. *Environ. Sci. Technol.* **2015**, *49*, 11729–11736. [CrossRef]
198. Yan, W.; Lien, H.-L.; Koel, B.E.; Zhang, W. Iron nanoparticles for environmental clean-up: Recent developments and future outlook. *Environ. Sci. Processes Impacts* **2013**, *15*, 63–77. [CrossRef]
199. Su, Y.; Jassby, D.; Zhang, Y.; Keller, A.A.; Adeleye, A.S. Comparison of the colloidal stability, mobility, and performance of nanoscale zerovalent iron and sulfidated derivatives. *J. Hazard. Mater.* **2020**, *396*, 122691. [CrossRef]
200. Mangayayam, M.C.; Perez, J.P.H.; Dideriksen, K.; Freeman, H.M.; Bovet, N.; Benning, L.G.; Tobler, D.J. Structural transformation of sulfidized zerovalent iron and its impact on long-term reactivity. *Environ. Sci. Nano* **2019**, *6*, 3422–3430. [CrossRef]
201. Fan, D.; Johnson, G.O.; Tratnyek, P.G.; Johnson, R.L. Sulfidation of nano zerovalent iron (nZVI) for improved selectivity during iin-situ chemical reduction (ISCR). *Environ. Sci. Technol.* **2016**, *50*, 9558–9565. [CrossRef]
202. Garcia, A.N.; Boparai, H.K.; O'Carroll, D.M. Enhanced ichlorination of 1,2-dichloroethane by coupled nanoIron-dithionite treatment. *Environ. Sci. Technol.* **2016**, *50*, 5243–5251. [CrossRef] [PubMed]
203. Kefeni, K.K.; Mamba, B.B.; Msagati, T.A.M. Application of Spinel ferrite nanoparticles in water and wastewater treatment: A review. *Sep. Purif. Technol.* **2017**, *118*, 92–100. [CrossRef]
204. Tatarchuk, T.; Myslin, M.; Lapchuk, I.; Shyichuk, A.; Murthy, A.P.; Gargula, R.; Kurzydło, P.; Bogacz, B.F.; Pędzwiatr, A.T. Magnesium-Zinc ferrites as magnetic adsorbents for Cr(VI) and Ni(II) ion removal: Cation distribution and antistructure modeling. *Chemosphere* **2021**, *270*, 129414. [CrossRef] [PubMed]
205. Yan, H.; Zhang, J.; You, C.; Song, Z.; Yu, B.; Shen, Y. Influences of different synthesis conditions on properties of Fe_3O_4 nanoparticles. *Mater. Chem. Phys.* **2009**, *113*, 46–52. [CrossRef]
206. Vuong, T.K.O.; Tran, D.L.; Le, T.L.; Pham, D.V.; Pham, H.N.; Ngo, T.H.L.; Do, H.M.; Nguyen, X.P. Synthesis of high-magnetization and monodisperse Fe_3O_4 nanoparticles via thermal decomposition. *Mater. Chem. Phys.* **2015**, *163*, 537–544. [CrossRef]
207. Shen, Y.F.; Tang, J.; Nie, Z.H.; Wang, D.; Ren, Y.; Zuo, L. Preparation and application of magnetic Fe_3O_4 nanoparticles for wastewater purification. *Sep. Purif. Technol.* **2009**, *68*, 312–319. [CrossRef]
208. Xu, Z.; Shen, C.; Hou, Y.; Gao, H.; Sun, S. Oleylamine as both reducing agent and stabilizer in a facile synthesis of magnetite nanoparticles. *Chem. Mater.* **2009**, *21*, 1778–1780. [CrossRef]
209. Mohapatra, J.; Zeng, F.; Elkins, K.; Xing, M.; Ghimire, M.; Yoon, S.; Mishra, S.R.; Liu, J.P. Size-Dependent magnetic and inductive heating properties of Fe_3O_4 nanoparticles: Scaling laws across the superparamagnetic size. *Phys. Chem. Chem. Phys.* **2018**, *20*, 12879–12887. [CrossRef] [PubMed]
210. Liu, S.; Yu, B.; Wang, S.; Shen, Y.; Cong, H. Preparation, surface functionalization and application of Fe_3O_4 magnetic nanoparticles. *Adv. Colloid Interface Sci.* **2020**, *281*, 102165. [CrossRef]
211. Chin, S.F.; Pang, S.C.; Tan, C.H. Green synthesis of magnetite nanoparticles (via thermal decomposition method) with controllable size and shape. *J. Mater. Environ. Sci.* **2011**, *2*, 299–302.
212. Middea, A.; Spinelli, L.S.; Souza, F.G., Jr.; Neumann, R.; Fernandes, T.L.A.P.; Gomes, O.F.M. Preparation and characterization of an organo-palygorskite-Fe_3O_4 nanomaterial for removal of anionic dyes from wastewater. *Appl. Clay Sci.* **2017**, *139*, 45–53. [CrossRef]
213. Adeli, M.; Yamini, Y.; Faraji, M. Removal of copper, nickel and zinc by sodium dodecyl sulphate coated magnetite nanoparticles from water and wastewater samples. *Arab. J. Chem.* **2017**, *10*, S514–S521. [CrossRef]
214. Li, C.; Gao, Y.; Li, A.; Zhang, L.; Ji, G.; Zhu, K.; Wang, X.; Zhang, Y. Synergistic effects of anionic surfactants on adsorption of norfloxacin by magnetic biochar derived from furfural residue. *Environ. Pollut.* **2019**, *254*, 113005. [CrossRef] [PubMed]
215. Sotiriou, G.A.; Hirt, A.M.; Lozach, P.Y.; Teleki, A.; Krumeich, F.; Pratsinis, S.E. Hybrid, silica-coated, janus-like plasmonic-magnetic nanoparticles. *Chem. Mater.* **2011**, *23*, 1985–1992. [CrossRef] [PubMed]
216. Robinson, I.; Tung, L.D.; Maenosono, S.; Wältid, C.; Thanh, N.T.K. Synthesis of core–shell gold coated magnetic nanoparticles and their interaction with thiolated DNA. *Nanoscale* **2010**, *2*, 2624–2630. [CrossRef] [PubMed]
217. Kralj, S.; Drofenik, M.; Makovec, D. Controlled surface functionalization of silica-coated magnetic nanoparticles with terminal amino and carboxyl groups. *J. Nanopart. Res.* **2011**, *13*, 2829–2841. [CrossRef]
218. Cendrowski, K.; Sikora, P.; Zielinska, B.; Horszczaruk, E.; Mijowska, E. Chemical and thermal stability of core-shelled magnetite nanoparticles and solid silica. *Appl. Surf. Sci.* **2017**, *407*, 391–397. [CrossRef]
219. Chai, L.; Wang, Y.; Zhao, N.; Yang, W.; You, X. Sulfate-Doped Fe_3O_4/Al_2O_3 nanoparticles as a novel adsorbent for fluoride removal from drinking water. *Water Res.* **2013**, *47*, 4040–4049. [CrossRef]

220. Sebastian, A.; Nangia, A.; Prasad, M.N.V. A green synthetic route to phenolics fabricated magnetite nanoparticles from coconut husk extract implications to treat metal contaminated water and heavy metal stress in Oryza sativa L. *J. Clean. Prod.* **2018**, *174*, 355–366. [CrossRef]
221. Ehrampoush, M.H.; Miria, M.; Salmani, M.H.; Mahvi, A.H. Cadmium removal from aqueous solution by green synthesis iron oxide nanoparticles with tangerine peel extract. *J. Environ. Health Sci. Eng.* **2015**, *13*, 84–366. [CrossRef] [PubMed]
222. Ren, G.; Yang, L.; Zhang, Z.; Zhong, B.; Yang, X.; Wang, X. A new green synthesis of porous magnetite nanoparticles from waste ferrous sulfate by solid-phase reduction reaction. *J. Alloys Compd.* **2017**, *710*, 875–879. [CrossRef]
223. Li, Y.; Zimmerman, A.R.; He, F.; Chen, J.; Han, L.; Chen, H.; Hu, X.; Gao, B. Solvent-Free synthesis of magnetic biochar and activated carbon through ball-mill extrusion with Fe_3O_4 nanoparticles for enhancing adsorption of methylene blue. *Sci. Total Environ.* **2020**, *722*, 137972. [CrossRef]
224. Minitha, C.R.; Suresh, R.; Maity, U.K.; Holdorai, Y.; Subramaniam, V.; Manoravi, P.; Joseph, M.; Kumar, R.T.R. Magnetite nanoparticle decorated reduced graphene oxide composite as an efficient and recoverable adsorbent for the removal of cesium and strontium ions. *Ind. Eng. Chem. Res.* **2018**, *57*, 1225–1232. [CrossRef]
225. Wen, W.; Wu, J.M. Nanomaterials via solution combustion synthesis: A step nearer to controllability. *RSC Adv.* **2014**, *4*, 58090–58100. [CrossRef]
226. Varma, A.; Mukasyan, A.S.; Rogachev, A.S.; Manukyan, K.V. Solution combustion synthesis of nanoscale materials. *Chem. Rev.* **2016**, *116*, 14493–14586. [CrossRef]
227. Wang, X.; Qin, M.; Fang, F.; Jia, B.; Wu, H.; Qu, X.; Volinsky, A.A. Effect of glycine on one-step solution combustion synthesis of magnetite nanoparticles. *J. Alloys Compd.* **2017**, *719*, 288–295. [CrossRef]
228. Aali, H.; Mollazadeh, S.; Vahdati Khaki, J. Single-phase magnetite with high saturation magnetization synthesized via modified solution combustion synthesis procedure. *Ceram. Int.* **2018**, *44*, 20267–20274. [CrossRef]
229. Zhang, D.; Qiu, J.; Shi, L.; Liu, Y.; Pan, B.; Xing, B. The mechanisms and environmental implications of engineered nanoparticles dispersion. *Sci. Total Environ.* **2020**, *722*, 137781. [CrossRef]
230. Cirtiu, C.M.; Raychoudhury, T.; Ghoshal, S.; Moores, A. Systematic comparison of the size, surface characteristics and colloidal stability of zero valent iron nanoparticles pre- and post-grafted with common polymers. *Colloids Surf. A Physicochem. Eng. Asp.* **2011**, *390*, 95–104. [CrossRef]
231. Park, Y.; Huh, C.; Ok, J.; Cho, H. One-Step synthesis and functionalization of high-salinity-tolerant magnetite nanoparticles with sulfonated phenolic resin. *Langmuir* **2019**, *35*, 8769–8775. [CrossRef]
232. Liu, J.; Dai, C.; Hu, Y. Aqueous aggregation behavior of citric acid coated magnetite nanoparticles: Effects of pH, cations, anions, and humic acid. *Environ. Res.* **2018**, *161*, 49–60. [CrossRef]
233. Mohan, D.; Sarswat, A.; Ok, Y.S.; Pittman, C.U., Jr. Organic and inorganic contaminants removal from water with biochar, a renewable, low cost and sustainable adsorbent—A critical review. *Bioresour. Technol.* **2014**, *160*, 191–202. [CrossRef]
234. Thines, K.R.; Abdullah, E.C.; Mubarak, N.M.; Ruthiraan, M. Synthesis of magnetic biochar from agricultural waste biomass to enhancing route for waste water and polymer application: A review. *Renew. Sustain. Energ. Rev.* **2017**, *67*, 257–276. [CrossRef]
235. Wang, S.; Tang, Y.; Chen, C.; Wu, J.; Huang, Z.; Mo, Y.; Zhang, K.; Chen, J. Regeneration of magnetic biochar derived from eucalyptus leaf residue for lead(II) removal. *Bioresour. Technol.* **2015**, *186*, 360–364. [CrossRef]
236. Park, J.-H.; Ok, Y.S.; Kim, S.-H.; Cho, J.-H.; Heo, J.-S.; Delaune, R.D.; Seo, D.-C. Competitive adsorption of heavy metals onto sesame straw biochar in aqueous solutions. *Chemosphere* **2016**, *142*, 77–83. [CrossRef]
237. Cai, Y.; Yan, Z.-H.; Wang, N.-Y.; Cai, Q.-Y.; Yao, S.-Z. Preparation of naphthyl functionalized magnetic nanoparticles for extraction of polycyclic aromatic hydrocarbons from river waters. *RSC Adv.* **2015**, *5*, 56189–56197. [CrossRef]
238. Yu, J.; Zhu, S.; Pang, L.; Chen, P.; Zhu, G.-T. Porphyrin-Based magnetic nanocomposites for efficient extraction of polycyclic aromatic hydrocarbons from water samples. *J. Chromatogr. A* **2018**, *1540*, 1–10. [CrossRef]
239. Chen, Y.; Zhang, W.; Zhang, Y.; Deng, Z.; Zhao, W.; Du, H.; Ma, Z.; Yin, D.; Xie, F.; Chen, Y.; et al. In situ preparation of core–shell magnetic porous aromatic framework nanoparticles for mixed–mode solid–phase extraction of trace multitarget analytes. *J. Chromatogr. A* **2018**, *1556*, 1–9. [CrossRef]
240. Meng, C.; Zhikun, Z.; Qiang, L.; Chunling, L.; Shuangqing, S.; Songqing, H. Preparation of amino-functionalized Fe_3O_4@$mSiO_2$ core–shell magnetic nanoparticles and their application for aqueous Fe^{3+} removal. *J. Hazard. Mater.* **2018**, *341*, 198–206. [CrossRef]
241. Santhosh, C.; Nivetha, R.; Kollu, P.; Srivastava, V.; Sillanpää, M.; Grace, A.N.; Bhatnagar, A. Removal of cationic and anionic heavy metals from water by 1D and 2D-carbon structures decorated with magnetic nanoparticles. *Sci. Rep.* **2017**, *7*, 14107. [CrossRef]
242. Zhou, Q.; Lei, M.; Li, J.; Zhao, K.; Liu, Y. Determination of 1-naphthol and 2-naphthol from environmental waters by magnetic solid phase extraction with Fe@MgAl-layered double hydroxides nanoparticles as the adsorbents prior to high performance liquid chromatography. *J. Chromatogr. A* **2016**, *1441*, 1–7. [CrossRef]
243. Yan, R.; Feng, X.; Kong, L.; Wan, Q.; Zheng, W.; Hagio, T.; Ichino, R.; Cao, X.; Li, L. Evenly distribution of amorphous iron sulfides on reconstructed Mg-Al hydrotalcites for improving Cr(VI) removal efficiency. *Chem. Eng. J.* **2021**, *417*, 129228. [CrossRef]
244. Naing, N.N.; Li, S.F.Y.; Lee, H.K. Magnetic micro-solid-phase-extraction of polycyclic aromatic hydrocarbons in water. *J. Chromatogr. A* **2016**, *1441*, 23–30. [CrossRef]
245. Cai, Y.; Yan, Z.; Wang, L.; Van, M.N.; Cai, Q. Magnetic solid phase extraction and static headspace gas chromatography–mass spectrometry method for the analysis of polycyclic aromatic hydrocarbons. *J. Chromatogr. A* **2016**, *1429*, 97–106. [CrossRef]

246. Verma, L.; Siddique, M.A.; Singh, J.; Bharagava, R.N. As(III) and As(V) removal by using iron impregnated biosorbents derived from waste biomass of Citrus limmeta (peel and pulp) from the aqueous solution and ground water. *J. Environ. Manag.* **2019**, *250*, 109452. [CrossRef]
247. Hari Prasad Rao, P.R.; Matsukata, M. Dry-Gel conversion technique for synthesis of zeolite BEA. *Chem. Commun.* **1996**, *12*, 1441–1442.
248. Xu, W.; Dong, J.; Li, J.; Li, J.; Wu, F. A novel method for the preparation of zeolite ZSM-5. *J. Chem. Soc. Chem. Commun.* **1990**, *10*, 755–756. [CrossRef]
249. Aono, H.; Tamura, K.; Johan, E.; Yamauchi, R.; Yamamoto, T.; Matsue, N.; Henmi, T. Preparation of composite material of Na-P1-type zeolite and magnetite for Cs decontamination. *Chem. Lett.* **2013**, *42*, 589591. [CrossRef]

Article

Self-Pierce Riveting of Three Thin Sheets of Aluminum Alloy A5052 and 980 MPa Steel

Satoshi Achira *, Yohei Abe and Ken-ichiro Mori

Department of Mechanical Engineering, Toyohashi University of Technology, Toyohashi 441-8580, Japan; abe@plast.me.tut.ac.jp (Y.A.); mori@plast.me.tut.ac.jp (K.-i.M.)
* Correspondence: achira@plast.me.tut.ac.jp

Abstract: Self-pierce riveting of three thin sheets of 980 MPa steel and 5052 aluminum alloy was performed to investigate the effect of sheet configuration on the deforming behaviors of the sheets and the rivet and joint strength. When the lower sheet was aluminum alloy, the joining range was relatively wide, i.e., the interlock hooking the rivet leg tended to be large. In the sheet configuration in which the upper and lower sheets were A5052 and the middle sheet was 980 MPa steel, the rivet leg spread out moderately and the joint without defects was obtained. In the lower 980 MPa steel sheet, fracture tended to occur due to the low ductility of the lower sheet, and the joining range was narrow with the small interlock although the three sheets were joined by an appropriate die shape. In joint strength of joined three sheets, fracture occurred in the lower-strength aluminum alloy sheet if interlocks of about 300 μm and 150 μm could be formed in the lower aluminum alloy sheet and 980 MPa steel sheet, respectively.

Keywords: joining; self-pierce riveting; ultra-high strength steel sheets; aluminum alloy sheets; three sheets; plastic deformation; joint strength

Citation: Achira, S.; Abe, Y.; Mori, K.-i. Self-Pierce Riveting of Three Thin Sheets of Aluminum Alloy A5052 and 980 MPa Steel. *Materials* **2022**, *15*, 1010. https://doi.org/10.3390/ma15031010

Academic Editors: Bolv Xiao and Adam Grajcar

Received: 30 November 2021
Accepted: 24 January 2022
Published: 28 January 2022

Publisher's Note: MDPI stays neutral with regard to jurisdictional claims in published maps and institutional affiliations.

Copyright: © 2022 by the authors. Licensee MDPI, Basel, Switzerland. This article is an open access article distributed under the terms and conditions of the Creative Commons Attribution (CC BY) license (https://creativecommons.org/licenses/by/4.0/).

1. Introduction

The concentration of carbon dioxide, which is the cause of global warming, is increasing every year. Reduction in car body weight is one of the most important methods to reduce carbon dioxide emissions. In addition to the development of new powertrains, a significant reduction in vehicle weight is necessary to meet the stringent carbon dioxide emission regulations. Various lightweight materials such as advanced high strength steel sheets, aluminum alloy sheets, aluminum castings, and carbon fibre reinforced plastics are often introduced into car body structures. Among them, the use of aluminum alloy sheet and high strength steel sheet is increasing [1]. Reducing the weight of car body often conflicts with other missions that require increased mass, such as crashworthiness and comfort. The use of high strength steel is the most competitive way to achieve weight reduction and is often used for car body parts that require less deformation due to impact load. However, the formability and weldability of high strength steel sheet decrease as the tensile strength increases. Aluminum alloy sheet has one-third the density of steel. It also is characterized by high thermal conductivity, a natural oxide film on the surface, and a low melting point. Estimates of the amount of aluminum alloy sheet used for automobile are increasing every year, and the presence of aluminum alloy sheet in automotive materials is expected to increase in the future. The sheet configuration including these various materials are not easy to join by conventional resistance spot welding due to the difference in melting points and the problem of brittle intermetallic layer formed at the interface [2]. In addition, although two sheets are usually joined together, it is desirable to join three or more sheets to improve the design flexibility of car body panels. Lei et al. [3] conducted finite element simulations of the resistance spot welding process in joining three sheets of mild steel to investigate the transient thermal characteristics of the resistance spot welding process. Ma et al. [4] investigated the nugget formation process for three high strength steel sheets by the experiments and simulations. However, welding three sheets of dissimilar materials is even more difficult. The method of joining sheets using plastic deformation is very advantageous for dissimilar materials because it avoids difficulties in joining due to differences in melting points. Various joining methods such as friction stir welding, mechanical clinching and self-pierce riveting have been developed and used in place of the welding processes [5,6].

Fereiduni et al. [7] performed friction stir spot welding on the aluminum alloy and carbon steel to investigate the effects of tool rotation speed and dwell time on the microstructure in the joint interface and joint strength. Sato et al. [8], Feng et al. [9] and Yamamoto et al. [10] performed friction stir spot welding on aluminum alloy and galvanized steel sheets. In these studies, the two overlapping specimens, and only the upper aluminum alloy sheet was stirred, i.e., it is difficult to apply this method to multiple overlapping joints of three or more sheets.

Mechanical clinching is a method of joining sheets by locally hemming them with a punch and die without rivets. Peng et al. [11] reviewed the recent development of the clinching process. He et al. [12] investigated the clinching process using an extensible die by experiments and simulations. Chen et al. [13] joined aluminum alloy sheets by the two-steps clinching method using clinch rivets to improve the neck thickness and interlock. Réjane et al. [14] investigated the effect of the shape of the lower hole on the mechanical properties of the shear-clinched joint for aluminum alloy and steel sheets. Lee et al. [15] used a hole clinching process to joined dissimilar materials of aluminum alloy sheet and carbon fibre reinforced plastic. Abe et al. [16] prevented fracture due to concentration of deformation at the punch corner by changing the die shape in mechanical clinching of high tensile steel and aluminum alloy sheets. Not only two sheets, but also three sheets were joined by mechanical clinching, e.g., Kaðèák et al. [17] joined three steel sheets. However, the joint strength of mechanical clinched sheets obtained from these studies is not high, and it is desirable to make a joint with high joint loads.

Self-pierce riveting, is a method that allows joining without drilled holes and overlapping of sheets with gaps, is an alternative method to resistance spot welding in the joining process of car body [18]. Mori et al. [19] reviewed the mechanical joining process of dissimilar materials including self-pierce riveting. Self-pierce riveting, which mechanically joined sheets without metallic bonding, is possible to join high-tensile steel and aluminum alloy sheets at room temperature. This riveting is characterized by higher joint strength than resistance spot welding and mechanical clinching [20]. Abe et al. [21] evaluated the joining process of aluminum alloy sheet and steel sheet by self-pierce riveting using the experiments and simulations. Atzeni et al. [22] simulated the self-pierce riveting of two aluminum alloy sheets and calculated the tension-shearing strength of the joint and compared it with experimental results. Wood et al. [23] developed a finite element model of a test and measurement system at automotive crash speeds to investigate the joinability of self-pierce riveted joints of aluminum alloy sheets. Zhang [24] performed self-pierce riveting of a 1 mm thick high-strength steel and aluminum alloy sheets and investigated the effect of the sequence of riveting on the dimensional stability of the product. Jeong et al. [25] performed self-pierce riveting of high strength steel and aluminum alloy sheets and investigated the effect of sheet constraint conditions on the cross-sectional properties and joint strength.

The interlock of self-pierce riveting depends on the thickness of the sheet, the flow stress and ductility of the sheet and rivet, and the shape of the rivet and die [19]. A lot of studies were conducted to obtain the optimum joining quality for each combination of sheets. Porcaro et al. [26] conducted the tension-shearing and peel tests on the aluminum alloy sheets joined by self-pierce riveting to investigate the mechanical properties of sheets and riveted sheet shape. Ma et al. [27] investigated the effects of the rivet properties and the die shape on the deforming behaviors for the aluminum alloy and mild steel sheets. Xu [28] investigated the effect of rivet length on the cross-sectional shapes of the joint.

The studies about the optimization process of self-pierce riveting using finite element simulation were also performed. Moraes et al. [29] simulated the self-pierce riveting process of the magnesium and aluminum alloy sheets. Abe et al. [30] performed the joining of high tensile strength steel and aluminum alloy sheets and optimised die shape. Mori et al. [31] performed self-pierce riveting of ultra-high strength steel and aluminum alloy sheets and the optimum joining conditions were evaluated. However, the optimization process shown in these studies is for joining two sheets. The optimization process for severe sheet configurations such three thin sheets with large strength differences has not been clarified, i.e., it is desired to show the fundamental joinabilities.

Self-pierce riveting can be applied to joining three or more sheets [32]. Abe et al. [33] and Mori et al. [34] performed self-pierce riveting on three sheets containing high strength steel and aluminum alloy sheets. However, it was limited to conditions where the lower sheet thickness ratio was large. Mori et al. [35] conducted the joining experiments under conditions where the lower sheet thickness ratio was small, but the joining by self-pierce riveting is difficult because of the risk of cracking and the large difference in flow stress of the sheets [19,36]. In addition, self-pierce riveting has the problem that different sheet strengths affect the joinabilities, and the effect of different combinations of sheets on the deforming behaviors has not been clarified.

In this study, self-pierce riveting of thin three sheets of 980 MPa ultra-high strength steel and aluminum alloy A5052 and finite element simulation were performed to investigate the effect of sheet configuration on the joinabilities and joint strength.

2. Three Sheets of Self-Pierce Riveting

2.1. Process and Conditions for Self-Pierce Riveting of Three Sheets

The joining process of three sheets of self-pierce riveting is shown in Figure 1. First, the three sheets placed on top of the die are fixed with a blankholder. The rivet is then placed on top of the sheets and pushed into the sheets. After the rivet is punched through the upper and middle sheets, the rivet leg spreads out and enters the lower sheet and then three sheets are joined by an interlock.

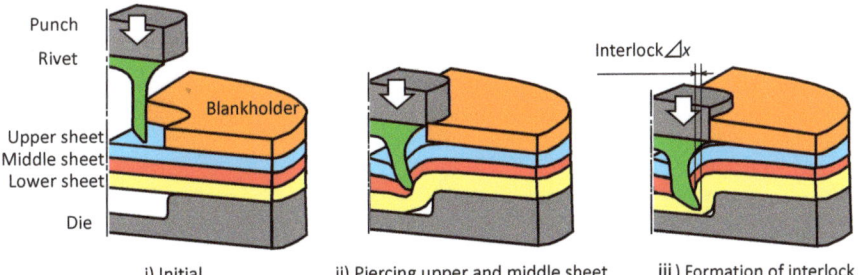

Figure 1. Self-pierce riveting process of three sheets.

A schematic illustration of the self-pierce riveting equipment is shown in Figure 2. Four coil springs in the upper part of the blankholder were used to apply 5 kN in blank holding force to the sheets. The die diameter D and the die depth H were changed by the die and the counter punch, respectively. The punch was pressed by a 250 kN screw driven type universal testing instrument (Autograph AGS-J, SHIMADZU Co., Kyoto, Japan). The riveting speed is 50 mm/min.

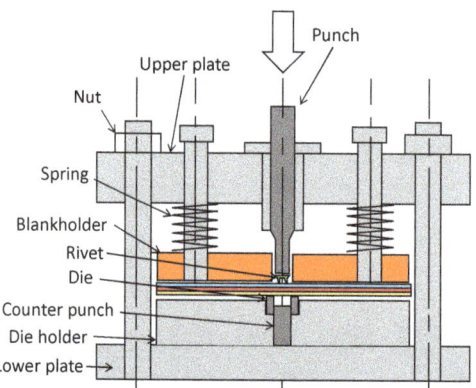

Figure 2. Schematic illustration of self-pierce riveting equipment.

The riveting conditions used in the experiments are shown in Figure 3. Aluminum alloy A5052 sheet and 980 MPa steel sheet as ultra-high strength steel sheet for car body panels were used as specimens. The material properties of the sheets obtained by the uniaxial tensile test are shown in Table 1. Three tensile test specimens were given, and the average value are shown. The sheet configuration for three-sheet joining is shown in Table 2.

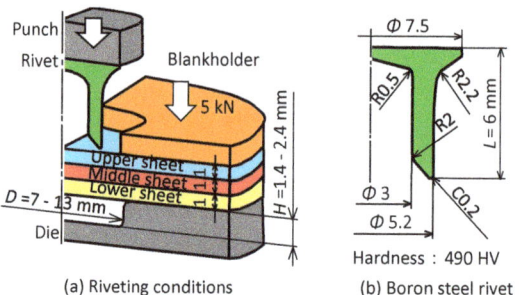

(a) Riveting conditions   (b) Boron steel rivet

Figure 3. Riveting conditions of die and rivet.

Table 1. Material properties of sheets.

Sheet	Sheet Thickness [mm]	Tensile Strength [MPa]	Elongation [%]	Reduction in Area [%]
A5052	1.05	275	25	65
980 MPa steel	1.05	1002	14	35

Table 2. Configuration of sheets used in joining experiment.

Sheet	1	2	3	4	5	6	7	8
Upper	A5052	A5052	980 MPa steel	980 MPa steel	A5052	A5052	980 MPa steel	980 MPa steel
Middle	A5052	980 MPa steel	A5052	980 MPa steel	A5052	980 MPa steel	A5052	980 MPa steel
Lower	A5052	A5052	A5052	A5052	980 MPa steel	980 MPa steel	980 MPa steel	980 MPa steel

The effect of the die shape on the deforming behaviors of the sheets and the rivet is shown in Figure 4. When the die depth is small, the rivet is compressed, and the rivet leg does not spread. When the die depth is large, fracture tends to occur because the rivet leg penetrates the lower sheet. When the die diameter is small, the deformation concentrates at the die corner, causing the lower sheet to fracture. When the die diameter is large, the interlock cannot be formed because the reaction force from the die wall cannot be obtained.

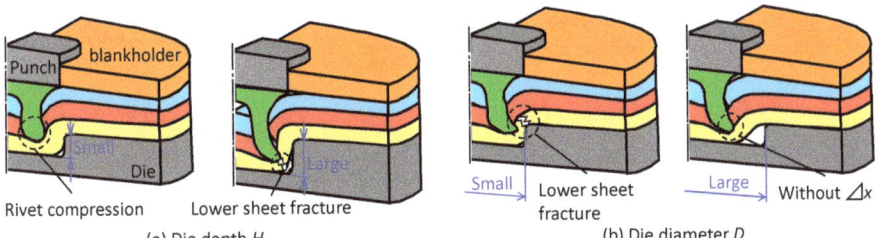

(a) Die depth *H*

(b) Die diameter *D*

Figure 4. Effects of die shape on deforming behaviors of sheets and rivet.

2.2. Simulation Conditions and Results

The commercial finite element code LS-DYNA was used for the simulation to predict the deforming behaviors of the sheets and rivet. The conditions of simulation are shown in Table 3. A dynamic explicit method was used. The symmetric conditions were applied. The punch, blankholder and die were rigid bodies, and the sheets were isotropic elasto-plastic materials, taking into account the strain effect using the power hardening law. The rivet and the sheet were divided by a quadrilateral solid element, and the element size was 0.1 mm × 0.1 mm. In the simulation, remeshing of the elements was automatically applied to increase the calculating accuracy. When the size of element in the sheet fell below a certain value, the element was removed. The certain value of the element size was set to the minimum thickness before fracture in the experiment, which was 0.1 mm for aluminum alloy sheet and 0.35 mm for 980 MPa steel sheet, respectively. The coulomb friction was assumed for all interfaces among sheets, tools and rivets, and the coefficient of friction was 0.20. The material properties of the sheets in the simulation are shown in Table 4.

Table 3. Conditions of simulation.

Solver	LS-DYNA
Simulation method	Dynamic explicit
Model	Axial symmetry
Plastic deformation	Isotropy
Yield criterion	Von Mises
Hardening equation	$\bar{\sigma} = K\bar{\varepsilon}^n$
Coefficient of friction	0.2

Table 4. Material properties of sheets in simulation.

Sheet	Young's Modulus [GPa]	K-Value * [MPa]	n-Value * [-]
A5052	70.3	550	0.32
980 MPa steel	210	1406	0.13

* $\bar{\sigma} = K\bar{\varepsilon}^n$.

The simulated and experimental cross-sectional shapes are shown in Figure 5. The simulated cross-sectional shape is similar to the experimental one, although the experimental interlock is larger than the simulated one.

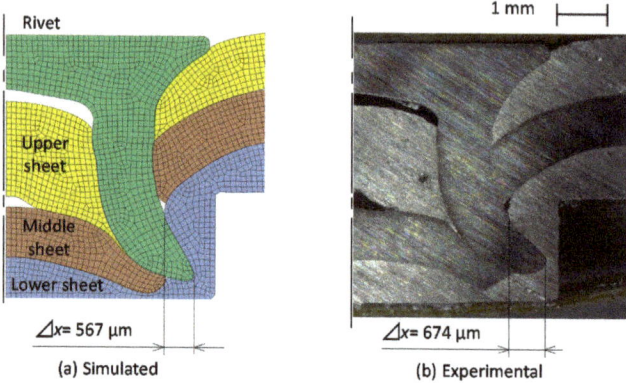

Figure 5. Comparison of simulated cross-sectional shapes with experimental one.

2.3. Joining Requirements

The cross-sectional shapes of sheets in the experiment are shown in Figure 6. In this study, a joint without defects was defined as the formation of 50 μm or more in length of the interlock without fracture in the lower sheet as shown in Figure 6a. The conditions that did not satisfy these requirements were defined as defects: for example, no interlock and the lower sheet fracture as shown in Figure 6b.

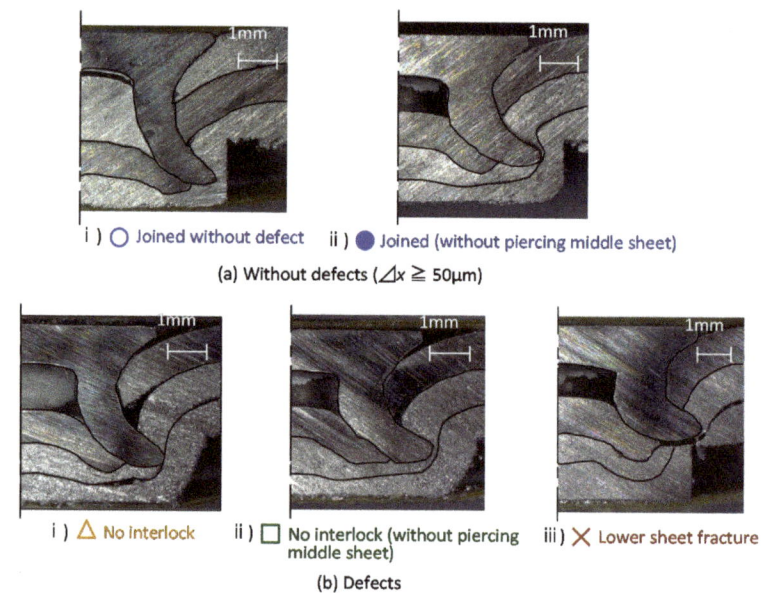

Figure 6. Cross-sectional shapes in experiment.

3. Three Sheets Joining of 980 MPa Steel and Aluminum Alloy
3.1. Three Sheets Joining in Lower Aluminum Alloy

The punch-load-stroke curves for $D = 9$ mm and $H = 1.8$ mm obtained from the experiment are shown in Figure 7. The punch load increased as the stroke increased. The punch load was higher as the number of steel sheets increased. The punch load decreased when the rivet punched through the sheet.

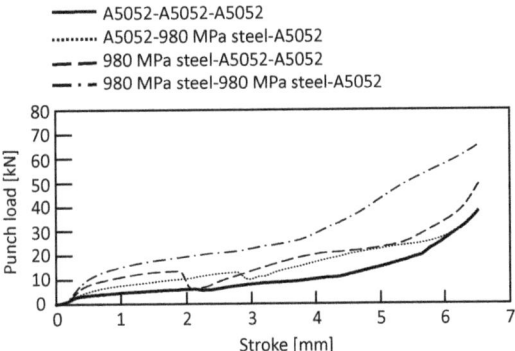

Figure 7. Punch-load-stroke curves in lower aluminum alloy sheet for D = 9 mm and H = 1.8 mm obtained from experiment.

The joining results in the lower aluminum alloy sheet are shown in Table 5, and the joining range is shown in Figure 8. A relatively wide joining range was obtained for the lower aluminum alloy sheet, and the widest range was obtained for A5052-980 MPa steel-A5052. The joining range in the configuration of the upper 980 MPa steel sheet tended to be narrow. The joining range of the three aluminum alloy sheets was limited.

Table 5. Joining results in lower aluminum alloy sheets.

Upper Sheet	Middle Sheet	Die Depth H [mm]	Die Diameter D [mm]	Joining Results
A5052	A5052	1.4	9.0	×
A5052	A5052	1.6	9.0	×
A5052	A5052	1.6	10.0	×
A5052	A5052	1.6	11.0	×
A5052	A5052	1.6	12.0	×
A5052	A5052	1.8	9.0	×
A5052	A5052	1.8	10.0	×
A5052	A5052	1.8	11.0	×
A5052	A5052	2.0	9.0	×
A5052	A5052	2.0	10.0	×
A5052	A5052	2.0	11.0	○
A5052	A5052	2.4	9.0	×
A5052	A5052	2.4	10.0	×
A5052	A5052	2.4	11.0	×
A5052	980 MPa Steel	1.6	8.0	○
A5052	980 MPa Steel	1.6	9.0	○
A5052	980 MPa Steel	1.6	10.0	○
A5052	980 MPa Steel	1.6	11.0	□
A5052	980 MPa Steel	1.8	8.0	○
A5052	980 MPa Steel	1.8	9.0	○
A5052	980 MPa Steel	1.8	10.0	○
A5052	980 MPa Steel	1.8	11.0	○
A5052	980 MPa Steel	2.0	8.0	○
A5052	980 MPa Steel	2.0	9.0	○
A5052	980 MPa Steel	2.0	10.0	○
A5052	980 MPa Steel	2.0	11.0	○
A5052	980 MPa Steel	2.4	8.0	×
A5052	980 MPa Steel	2.4	9.0	×
A5052	980 MPa Steel	2.4	10.0	○
A5052	980 MPa Steel	2.4	11.0	○
980 MPa Steel	A5052	1.2	9.0	□
980 MPa Steel	A5052	1.2	10.0	●

Table 5. Cont.

Upper Sheet	Middle Sheet	Die Depth H [mm]	Die Diameter D [mm]	Joining Results
980 MPa Steel	A5052	1.2	11.0	□
980 MPa Steel	A5052	1.4	9.0	×
980 MPa Steel	A5052	1.4	10.0	●
980 MPa Steel	A5052	1.4	11.0	□
980 MPa Steel	A5052	1.6	8.0	×
980 MPa Steel	A5052	1.6	9.0	○
980 MPa Steel	A5052	1.6	10.0	●
980 MPa Steel	A5052	1.6	11.0	□
980 MPa Steel	A5052	1.8	8.0	×
980 MPa Steel	A5052	1.8	9.0	○
980 MPa Steel	A5052	1.8	10.0	●
980 MPa Steel	A5052	1.8	11.0	△
980 MPa Steel	A5052	2.0	8.0	×
980 MPa Steel	A5052	2.0	9.0	○
980 MPa Steel	A5052	2.0	10.0	○
980 MPa Steel	A5052	2.0	11.0	△
980 MPa Steel	980 MPa Steel	1.6	9.0	□
980 MPa Steel	980 MPa Steel	1.8	9.0	△
980 MPa Steel	980 MPa Steel	1.8	9.5	□
980 MPa Steel	980 MPa Steel	2.0	9.0	○
980 MPa Steel	980 MPa Steel	2.0	9.5	○
980 MPa Steel	980 MPa Steel	2.0	10.0	□
980 MPa Steel	980 MPa Steel	2.0	11.0	□
980 MPa Steel	980 MPa Steel	2.2	9.0	×
980 MPa Steel	980 MPa Steel	2.2	9.5	○
980 MPa Steel	980 MPa Steel	2.2	10.0	○
980 MPa Steel	980 MPa Steel	2.2	11.0	○
980 MPa Steel	980 MPa Steel	2.2	12.0	□
980 MPa Steel	980 MPa Steel	2.4	9.5	○
980 MPa Steel	980 MPa Steel	2.4	10.0	○
980 MPa Steel	980 MPa Steel	2.4	11.0	○
980 MPa Steel	980 MPa Steel	2.4	12.0	△

○ Joined without defect   ● Joined (without piercing middle sheet)
□ No interlock (without piercing middle sheet)   △ No interlock   × Lower sheet fracture

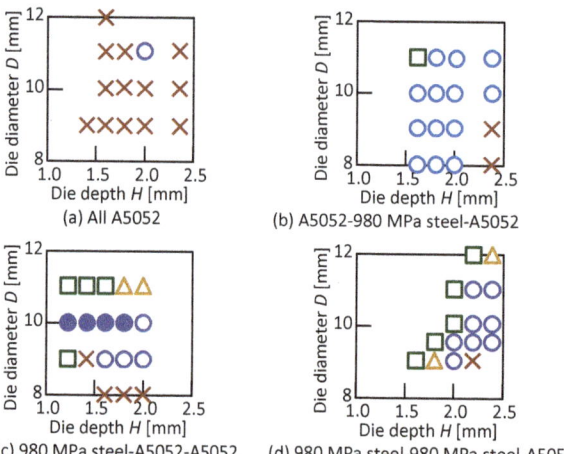

(a) All A5052
(b) A5052-980 MPa steel-A5052
(c) 980 MPa steel-A5052-A5052
(d) 980 MPa steel-980 MPa steel-A5052

Figure 8. Joining range in lower aluminum alloy sheet.

The relationship between the die shape and the interlock in the joined three sheets is shown in Figure 9. In A5052-980 MPa steel-A5052, which showed the widest joining range, the interlock of 250 μm or more was obtained in all conditions. The interlock decreased in both the 980 MPa steel-A5052-A5052 and 980 MPa steel-980 MPa steel-A5052 than that of A5052-980 MPa steel-A5052 under the same die shape.

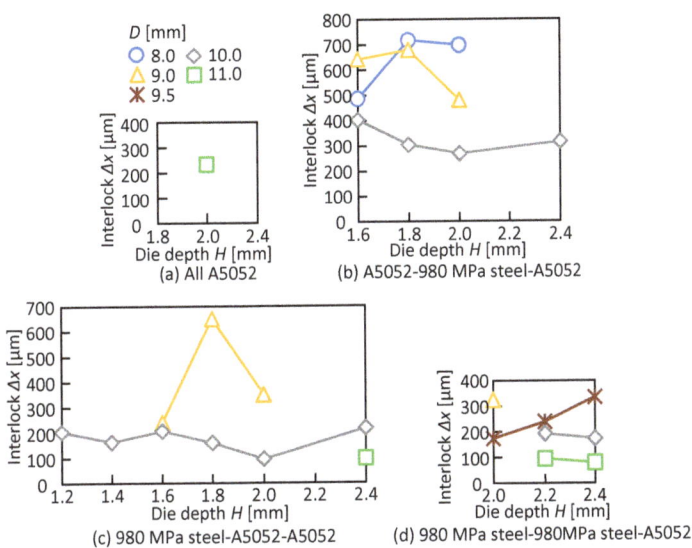

Figure 9. Relationship between die shape and interlock.

The simulated cross-sectional shapes in the lower aluminum alloy sheet for A5052-980 MPa steel-A5052 are shown in Figure 10. The amount of rivet compression was small even when the rivet entered the lower sheet because of the low strength of the lower sheet, and the rivet tended to easily form the interlock. Thus, it is considered that a relatively wide joining range was obtained in the lower aluminum alloy sheet.

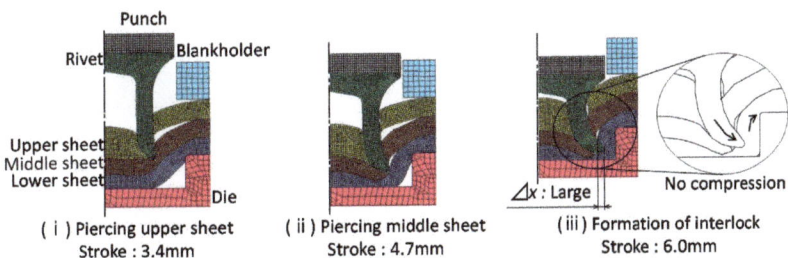

Figure 10. Cross-sectional shapes in simulation for $D = 9$ mm and $H = 1.8$ mm and A5052-980 MPa steel-A5052.

The effects of the upper and middle sheets on the deforming behaviors are shown in Figure 11. In Figure 11a, the rivet leg did not spread in the soft three A5052 sheets, and the rivet tip pierced the lower sheet as shown in Figure 8a. The rivet leg tended to spread moderately upon the penetration of the middle sheet due to the high strength of the middle sheet in Figure 11b. In the experiments as shown in Figures 8b and 9b, the wide joint range and the large interlock were obtained. In Figure 11c,d, since the upper sheet was of high strength, the rivet leg tended to be highly compressed, resulting in unpenetrated middle sheets. The result in the experiment as shown in Figure 8 was similar. For the sheet configurations in Figure 11c,d, the joining range can be obtained by setting the die shape parameters *D* and *H* appropriately.

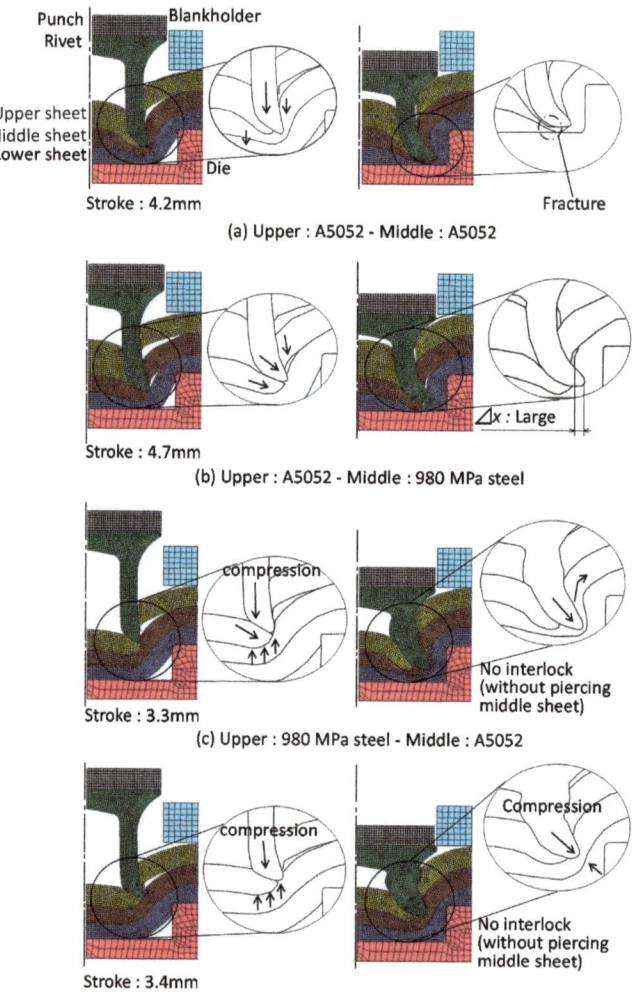

Figure 11. Effect of strength of upper and middle sheets on deforming behaviors of sheets and rivet for *D* = 9 mm, *H* = 1.8 mm and lower A5052 sheet.

3.2. The Tension-Shearing Test in Lower Aluminum Alloy

The joining strength was measured by a tension-shearing test. The measuring conditions of the joint strength are shown in Figure 12. The tension-shearing test was performed based on the test for the resistance spot welded joints [37]. The specimens were selected according to the JIS Z3136. A single joint was made at the center of the overlapped sheets as shown in Figure 12a. A 50 kN screw driven type universal testing instrument (Autograph AGS-J, SHIMADZU Co., Kyoto, Japan) was used for this test, and the tensile speed was 10 mm/min. Since the specimens were in triplicate, there were three different loading combinations: the upper and lower, the upper and middle, and the middle and lower sheets as shown in Figure 12b. The tension-shearing load F_s in fracture by shearing in the area of the rivet outer diameter to the sheet edge is expressed by the following Equation (1).

$$F_s = Ak\sigma_a \tag{1}$$

where A is the cross-sectional area from the rivet outer diameter to the sheet edge, σ_a is the tensile strength of the aluminum alloy sheet, and k is the factor of shear strength. The values in Table 1, $k = \frac{1}{\sqrt{3}}$ and $A = 30$ mm^2 were substituted in the Equation (1);

$$F_s = 30 \times \frac{1}{\sqrt{3}} \times 275 = 4.76 \text{ [kN]}$$

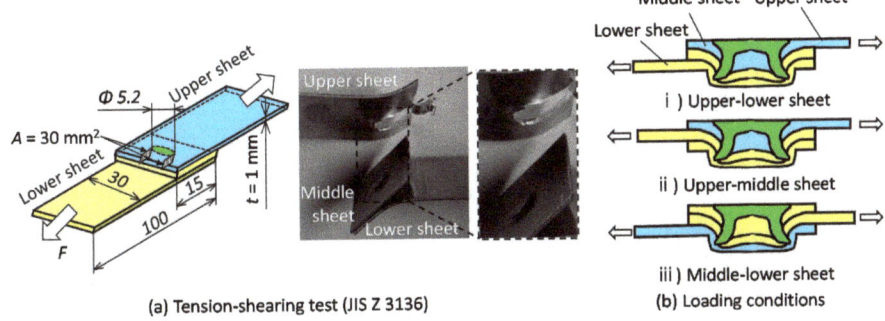

Figure 12. Measuring conditions of joint strength.

The tension-shearing load F_r at which the rivet leg is fractured by shearing is expressed by the following equation.

$$F_r = \frac{\pi}{4}\left(d_o^2 - d_i^2\right)k\sigma_r \tag{2}$$

where, the tensile strength of a rivet is σ_r and the external and internal diameters of the rivet leg are d_o and d_i, respectively. The values in Figure 3b and $\sigma_r = 1600$ MPa were substituted in the Equation (2).

$$F_r = \frac{\pi}{4}(5.2^2 - 3^2) \times \frac{1}{\sqrt{3}} \times 1600 = 13.1 \text{ [kN]}$$

thus, the tension-shearing load F_s, assuming fracture of the aluminum alloy sheet, is smaller than the load F_r.

The tension-shearing load in the lower aluminum alloy sheet is shown in Figure 13. In the case of configurations containing low strength aluminum alloy sheets, fracture occurred mostly in the aluminum alloy sheet. In the upper-lower sheet tensile test shown in Figure 13a,b, pulling of the rivet occurred before fracture in the aluminum alloy sheet. The maximum tension-shearing load under the conditions where fracture occurred in the aluminum alloy sheet was about 3.0 kN, which was about 63% of the Equation (1). However, this was the load at which the rivet was pulled out before the aluminum alloy sheet was completely fractured, as shown in the deformed joint. It was considered that the measured load was lower than the Equation (1). In the upper-middle sheet in Figure 13c, the aluminum alloy sheet completely fractured, and the load was 4.7 kN, which was equivalent to the Equation (1). In the upper-middle sheet in Figure 13d, the load was as high as 7.0 kN even when the pulling rivet occurred, because the tension-shearing load was applied between the steel sheets.

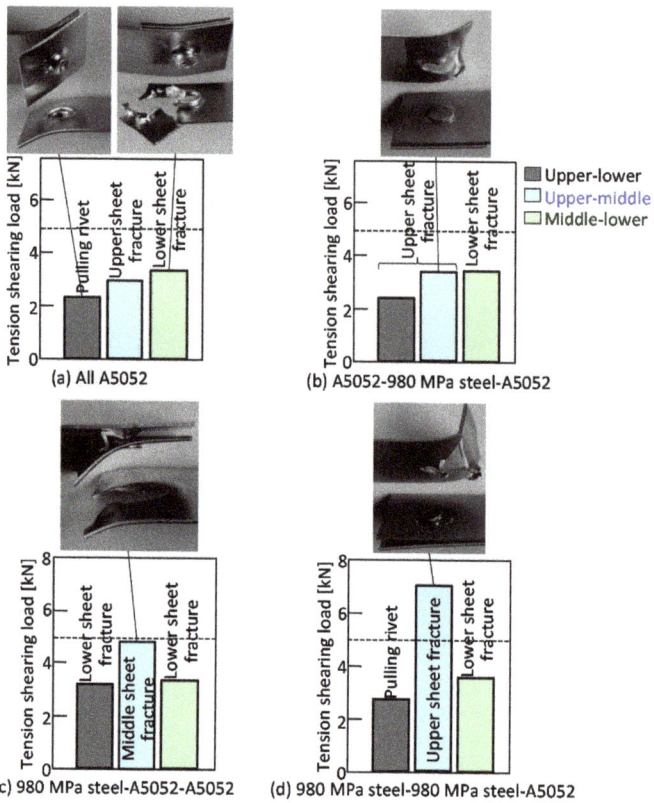

Figure 13. Tension-shearing load in lower aluminum alloy sheet.

The relationship between the configuration and the interlock in the lower aluminum alloy sheet is shown in the Figure 14. In the three aluminum alloy sheets, the rivet was pulled out from the lower sheet. Of the three sheet configurations in which the lower aluminum alloy sheet was fractured, 980 MPa steel-980 MPa steel-A5052 had the lowest interlock, i.e., about 300 μm in interlock was sufficient in the lower aluminum alloy sheet.

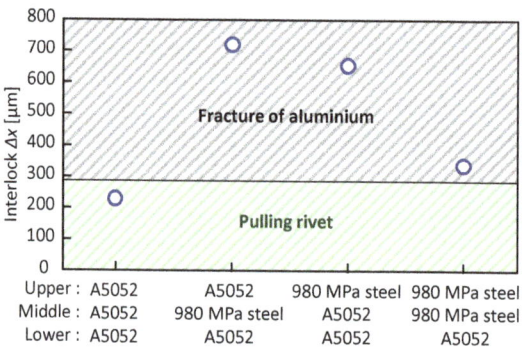

Figure 14. Relationship between configuration and interlock in lower aluminum alloy sheet.

3.3. Three Sheets Joining in Lower 980 MPa Steel

The punch-load-stroke curves in the lower 980 MPa steel sheet for $D = 9$ mm and $H = 1.8$ mm obtained from the experiment are shown in Figure 15. Compared with the lower aluminum alloy sheet, the maximum punch load was larger for all sheet configurations.

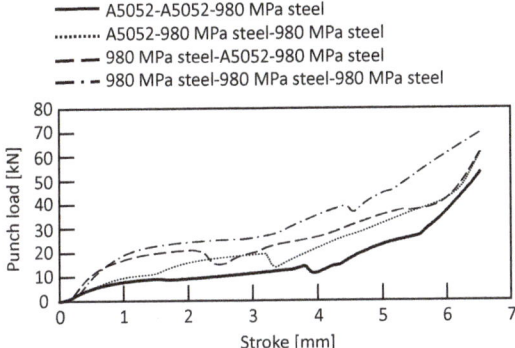

Figure 15. Punch-load-stroke curves in lower 980 MPa steel sheet for $D = 9$ mm and $H = 1.8$ mm obtained from experiment.

The joining results in the lower 980 MPa steel sheet are shown in Table 6, and then the joining range is shown in Figure 16. In Figure 16a,b, the joining conditions without defect were obtained, but the joining range was narrower than for the lower aluminum alloy sheet with the same configuration of upper and middle. The sheet configuration in Figure 16c was not joined because the rivet leg tended to spread to the die corner, causing the lower sheet to fracture. In Figure 16d, the lower sheet fractured due to low ductility of the sheets in all conditions. Among the eight types of sheet configurations including Figure 14, A5052-980 MPa steel-A5052 was the best configuration to join because it provided the widest joining range by moderately spreading the rivet leg so that the large interlock was obtained.

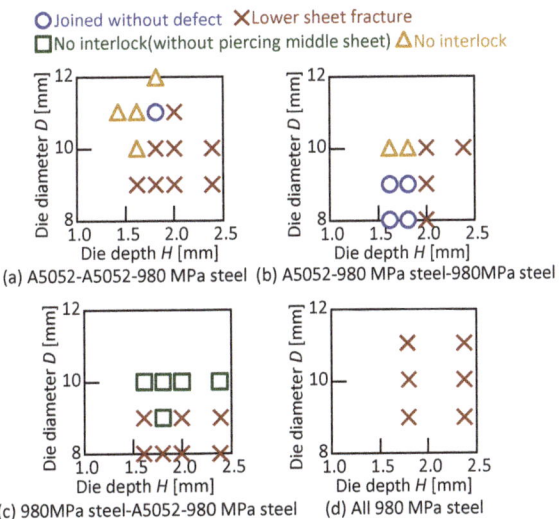

Figure 16. Joining range in lower 980 MPa steel sheet.

Table 6. Joining results in lower 980 MPa steel sheet.

Upper Sheet	Middle Sheet	Die Depth H [mm]	Die Diameter D [mm]	Joining Results
A5052	A5052	1.4	11.0	△
A5052	A5052	1.6	9.0	×
A5052	A5052	1.6	10.0	△
A5052	A5052	1.6	11.0	△
A5052	A5052	1.8	9.0	×
A5052	A5052	1.8	10.0	×
A5052	A5052	1.8	11.0	○
A5052	A5052	1.8	12.0	△
A5052	A5052	2.0	9.0	×
A5052	A5052	2.0	10.0	×
A5052	A5052	2.0	11.0	×
A5052	A5052	2.4	9.0	×
A5052	A5052	2.4	10.0	×
A5052	980 MPa Steel	1.6	8.0	○
A5052	980 MPa Steel	1.6	9.0	○
A5052	980 MPa Steel	1.6	10.0	△
A5052	980 MPa Steel	1.8	8.0	○
A5052	980 MPa Steel	1.8	9.0	○
A5052	980 MPa Steel	1.8	10.0	△
A5052	980 MPa Steel	2.0	8.0	×
A5052	980 MPa Steel	2.0	9.0	×
A5052	980 MPa Steel	2.0	10.0	×
A5052	980 MPa Steel	2.4	10.0	×
980 MPa Steel	A5052	1.6	8.0	×
980 MPa Steel	A5052	1.6	9.0	×
980 MPa Steel	A5052	1.6	10.0	□
980 MPa Steel	A5052	1.8	8.0	×
980 MPa Steel	A5052	1.8	9.0	□
980 MPa Steel	A5052	1.8	10.0	□
980 MPa Steel	A5052	2.0	8.0	×
980 MPa Steel	A5052	2.0	9.0	×
980 MPa Steel	A5052	2.0	10.0	□
980 MPa Steel	A5052	2.4	8.0	×
980 MPa Steel	A5052	2.4	9.0	×
980 MPa Steel	A5052	2.4	10.0	□
980 MPa Steel	980 MPa Steel	1.6	9.0	×
980 MPa Steel	980 MPa Steel	1.6	10.0	×
980 MPa Steel	980 MPa Steel	1.6	11.0	×
980 MPa Steel	980 MPa Steel	2.4	9.0	×
980 MPa Steel	980 MPa Steel	2.4	10.0	×
980 MPa Steel	980 MPa Steel	2.4	11.0	×

The cross-sectional shapes of the sheets and rivet in the simulation for A5052-980 MPa steel-980 MPa steel are shown in Figure 17. The rivet leg tended to be compressed because the middle and lower sheets were high strength whereas the rivet penetrated the upper sheet easily. The deformations of the middle and lower sheets became small, and the interlock tended to be small.

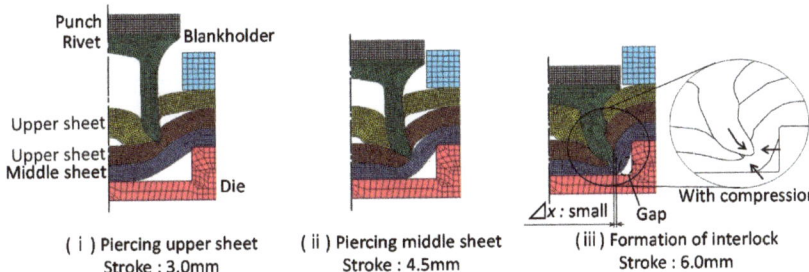

(i) Piercing upper sheet Stroke : 3.0mm
(ii) Piercing middle sheet Stroke : 4.5mm
(iii) Formation of interlock Stroke : 6.0mm

Figure 17. Cross-sectional shapes of sheets and rivet in simulation for D = 9 mm, H = 1.8 mm and A5052-980 MPa steel-980 MPa steel.

3.4. The Tension-Shearing Test in Lower 980 MPa Steel

The tension-shearing load in the lower 980 MPa steel sheet is shown in Figure 18. In Figure 18a, the pulling rivet occurred before the fracture aluminum alloy sheet in all conditions. The maximum tension-shearing load was about 2 to 3 kN, which was about 42 to 63% of the Equation (1). In Figure 18b, fracture occurred in the low strength aluminum alloy sheet. The maximum tension-shearing load was about 3.0 kN, it was similar to the load in the lower aluminum alloy sheet. In the middle-lower sheet, the tension-shearing load was as high as 5.3 kN, even when the pulling rivet occurred, because the tension-shearing load was applied between the steel sheets.

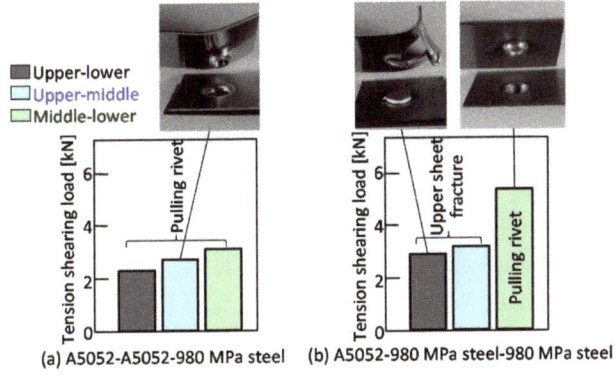

(a) A5052-A5052-980 MPa steel   (b) A5052-980 MPa steel-980 MPa steel

Figure 18. Tension-shearing load in lower 980 MPa sheet.

The relationship between the sheet configuration and the interlock in the lower 980 MPa steel sheet is shown in Figure 19. The interlock tended to be smaller in all configurations than that in the lower aluminum alloy sheet in Figure 14. The interlock in A5052-980 MPa steel-980 MPa steel, for which without defect conditions were obtained, was 198 μm. Fracture occurred in the aluminum alloy sheet, except for the condition where the steel sheets were given a tension. The formation of about 150 μm in interlock is sufficient because fracture occurs in the low strength aluminum alloy sheet in the lower 980 MPa steel sheet.

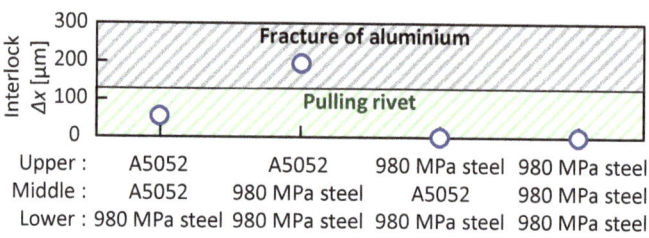

Figure 19. Relationship between configuration and interlock in lower 980 MPa steel sheet.

4. Conclusions

In this study, the deforming behaviors of three thin sheets and a rivet—including the ultra-high strength steel and aluminum alloy sheets—were investigated by self-pierce riveting and finite element simulation. The results obtained are shown below;

(1) When the lower sheet was the aluminum alloy, the joint range was relatively wide, and the interlock tended to be large;

(2) In the lower 980 MPa steel sheet, A5052-A5052-980 MPa steel and A5052-980 MPa steel-980 MPa steel were joined by selecting an appropriate die shape. Due to the low ductility and high flow stress of the lower sheet, fracture tended to occur, resulting in a narrower joining range where the interlock was not large;

(3) Among the eight types of sheet configurations, A5052-980 MPa steel-A5052 was the best configuration to join because it provided the widest joining range by moderately spreading the rivet leg so that a large interlock was obtained;

(4) In the tension-shearing test, fracture occurred in the lower-strength aluminum alloy sheet if interlocks of about 300 μm and 150 μm could be formed in the lower aluminum alloy sheet and 980 MPa steel sheet, respectively.

Author Contributions: Conceptualization, Y.A. and K.-i.M.; methodology and investigation, S.A. and Y.A.; writing—original draft preparation, S.A. and Y.A.; visualization, S.A. and Y.A.; supervision, K.-i.M. All authors have read and agreed to the published version of the manuscript.

Funding: This research received no external funding.

Institutional Review Board Statement: Not applicable.

Informed Consent Statement: Not applicable.

Data Availability Statement: All data are available from the corresponding author upon reasonable request.

Acknowledgments: The authors would like to thank Daiki Yoshioka for their great support and fruitful discussions in the experiment.

Conflicts of Interest: The authors declare no conflict of interest.

References

1. Lai, M.; Brun, R. Latest developments in sheet metal forming technology and materials for automotive application: The use of ultra high strength steels at Fiat to reach weight reduction at sustainable costs. *Key Eng. Mater.* **2007**, *344*, 1–8.
2. Haque, R.; Beynon, J.H.; Durandet, Y.; Kirstein, O.; Blacket, S. Feasibility of measuring residual stress profile in different self-pierce riveted joints. *Sci. Technol. Weld. Join.* **2013**, *17*, 60–68. [CrossRef]
3. Lei, Z.; Kang, H.; Liu, Y. Finite Element Analysis for Transient Thermal Characteristics of Resistance Spot Welding Process with Three Sheets Assemblies. *Procedia Eng.* **2011**, *16*, 622–631. [CrossRef]
4. Ma, N.; Murakawa, H. Numerical and experimental study on nugget formation in resistance spot welding for three pieces of high strength steel sheets. *J. Mater. Proc. Technol.* **2010**, *210*, 2045–2052. [CrossRef]
5. Sakiyama, T.; Naito, Y.; Miyazaki, Y.; Nose, T.; Murayama, G.; Saita, K.; Oikawa, H. Dissimilar Metal Joining Technologies for Steel Sheet and Aluminium Alloy Sheet in Auto Body. *Nippon Steel Tech. Rep.* **2013**, *103*, 91–98. Available online: https://www.nipponsteel.com/en/tech/report/nsc/pdf/103-14.pdf (accessed on 1 September 2021).

6. Meschut, G.; Janzen, V.; Olfermann, T. Innovative and Highly Productive Joining Technologies for Multi-Material Lightweight Car Body Structures. *J. Mater. Eng. Perform.* **2014**, *23*, 1515–1523. [CrossRef]
7. Fereiduni, E.; Movahedi, M.; Kokabi, A.H. Aluminium/steel joints made by an alternative friction stir spot welding process. *J. Mater. Process. Technol.* **2015**, *224*, 1–10. [CrossRef]
8. Sato, Y.; Tada, M.; Shiota, A.; Kokawa, H.; Nakagawa, S.; Miyamoto, K. Effect of Zn coating on tensile shear strength in dissimilar friction stir spot welds of Al alloy and Zn-coated steel. *Prepr. Natl. Meet. JWS* **2009**, *84*, 46–47. (In Japanese) [CrossRef]
9. Feng, K.; Watanabe, M.; Kumai, S. Microstructure and Joint Strength of Friction Stir Spot Welded 6022 Aluminium Alloy Sheets and Plated Steel Sheets. *Mater. Trans.* **2011**, *52*, 48–1425. [CrossRef]
10. Yamamoto, M.; Ogura, T.; Ohashi, R.; Fujimoto, M.; Hirose, A. Effects of interfacial microstructures on joint strength in friction stir spot welded 6061 aluminium alloy/zinc coated steel joints. *J. Light Met. Weld.* **2013**, *51*, 223–232. (In Japanese) [CrossRef]
11. Peng, H.; Chen, C.; Zhang, H.; Ran, X. Recent Development of Improved Clinching Process. *Int. J. Adv. Manuf. Technol.* **2020**, *110*, 3169–3199. [CrossRef]
12. He, X.; Liu, F.; Xing, B.; Yang, H.; Wang, Y.; Gu, F.; Ball, A. Numerical and experimental investigations of extensible die clinching. *Int. J. Adv. Manuf. Technol.* **2014**, *74*, 1229–1236. [CrossRef]
13. Chen, C.; Zhao, S.; Cui, M.; Han, X.; Fan, S. Mechanical properties of the two-steps clinched joint with a clinch-rivet. *J. Mater. Process. Technol.* **2016**, *237*, 361–370. [CrossRef]
14. Hörhold, R.; Müller, M.; Merklein, M.; Meschut, G. Mechanical properties of an innovative shear-clinching technology for ultra-high-strength steel and aluminium in lightweight car body structures. *Weld. World* **2016**, *60*, 613–620. [CrossRef]
15. Lee, C.J.; Lee, S.H.; Lee, J.M.; Kim, B.H.; Kim, B.M.; Ko, D.C. Design of Hole-Clinching Process for Joining CFRP and Aluminium Alloy Sheet. *Int. J. Precis. Eng. Manuf.* **2014**, *15*, 1151–1157. [CrossRef]
16. Abe, Y.; Mori, K.; Kato, T. Joining of high strength steel and aluminium alloy sheets by mechanical clinching with dies for control of metal flow. *J. Mater. Process. Technol.* **2012**, *212*, 884–889. [CrossRef]
17. Kaďèák, L.; Spiďák, E.; Kubík, R.; Mucha, J. Finite Element Calculation of Clinching with Rigid Die of Three Steel Sheets. *Strength Mater.* **2017**, *49*, 488–499. [CrossRef]
18. Barnes, T.A.; Pashby, I.R. Joining Techniques for Aluminium Spaceframes Used in Automobiles: Part II-Adhesive Bonding and Mechanical Fasteners. *J. Mater. Process. Technol.* **2000**, *99*, 72–79. [CrossRef]
19. Mori, K.; Bay, N.; Fratini, L.; Micari, F.; Tekkaya, A.E. Joining by plastic deformation. *CIRP Ann.–Manuf. Technol.* **2013**, *62*, 673–694. [CrossRef]
20. Mori, K.; Abe, Y.; Kato, T. Mechanism of Superiority of Fatigue Strength for Aluminium Alloy Sheets Joined by Mechanical Clinching and Self-Pierce Riveting. *J. Mater. Process. Technol.* **2012**, *212*, 1900–1905. [CrossRef]
21. Abe, Y.; Kato, T.; Mori, K. Joinability of Aluminium Alloy and Mild Steel Sheets by Self Piercing Rivet. *J. Mater. Process. Technol.* **2006**, *177*, 417–421. [CrossRef]
22. Atzeni, E.; Ippolito, R.; Settineril, L. FEM Modeling of Self Piercing Riveted Joint. *Key Eng. Mater.* **2007**, *344*, 655–662.
23. Wood, P.K.C.; Schley, C.A.; Williams, M.A.; Rusinek, A. Model to Describe the High Rate Performance of Self-Piercing Riveted Joints in Sheet Aluminium. *Mater. Des.* **2011**, *32*, 2246–2259. [CrossRef]
24. Zhang, H. Influence of riveting sequence/direction on distortion of steel and aluminum sheets. *J. Manuf. Processes.* **2020**, *53*, 304–309. [CrossRef]
25. Jeong, T.-E.; Kam, D.-H.; Kim, C. Parametric investigation of effect of abnormal process conditions on self-piercing riveting. *Appl. Sci.* **2020**, *10*, 2520. [CrossRef]
26. Porcaro, R.; Hanssen, A.G.; Langseth, M.; Aalberg, A. The behaviour of a self-piercing riveted connection under quasi-static loading conditions. *Int. J. Solids Struct.* **2016**, *43*, 5110–5131. [CrossRef]
27. Ma, Y.M.; Lou, M.; Li, Y.B.; Lin, Z.Q. Effect of Rivet and Die on Self-Piercing Rivetability of AA6061-T6 and Mild Steel CR4 of Different Gauges. *J. Mater. Process. Technol.* **2018**, *251*, 282–294. [CrossRef]
28. Xu, Y. Effects of factors on physical attributes of self-piercing riveted joints. *Sci. Technol. Weld. Join.* **2006**, *11*, 666–671. [CrossRef]
29. Moraes, J.; Jordon, J.; Su, X.; Brewer, L.; Fay, B.; Bunn, J.; Sochalski-Kolbus, L.; Barkey, M. Residual stresses and plastic deformation in self-pierce riveting of dissimilar aluminum-to-magnesium alloys. *SAE Int. J. Mater. Manf.* **2018**, *11*, 139–150. [CrossRef]
30. Abe, Y.; Kato, T.; Mori, K. Self-piercing riveting of high tensile strength steel and aluminium alloy sheets using conventional rivet and die. *J. Mater. Process. Technol.* **2009**, *209*, 3914–3922. [CrossRef]
31. Mori, K.; Kato, T.; Abe, Y.; Ravshanbek, Y. Plastic joining of ultra high strength steel and aluminium alloy sheets by self piercing rivet. *Ann. CIRP* **2006**, *55*, 283–286. [CrossRef]
32. Han, L.; Chrysanthou, A.; Young, K.W. Mechanical Behaviour of Self-Piercing Riveted Multi-Layer Joints under Different Specimen Configurations. *Mater. Des.* **2007**, *28*, 2024–2033. [CrossRef]
33. Abe, Y.; Kato, T.; Mori, K. Self-Pierce Riveting of Three High Strength Steel and Aluminium Alloy Sheets. *Int. J. Mater. Form.* **2008**, *1*, 1271–1274. [CrossRef]
34. Mori, K.; Abe, Y.; Kato, T. Self-pierce riveting of multiple steel and aluminium alloy sheets. *J. Mater. Process. Technol.* **2014**, *214*, 2002–2008. [CrossRef]
35. Mori, K.; Abe, Y.; Kato, T.; Sakai, S. Self-Pierce Riveting of Three Aluminium Alloy and Mild Steel Sheets. *AIP Conf. Proc.* **2010**, *1252*, 673–680. [CrossRef]

36. Abe, Y.; Maeda, T.; Yoshioka, D.; Mori, K. Mechanical clinching and self-pierce riveting of thin three sheets of 5000 series aluminium alloy and 980 MPa grade cold rolled ultra-high strength steel. *Materials* **2020**, *13*, 4741. [CrossRef]
37. Japanese Industrial Standards. *Specimen Dimensions and Procedure for Shear Testing Resistance Spot and Embossed Projection Welds*; JIS Z3136; Japanese Industrial Standards (JIS): Tokyo, Japan, 1999. (In Japanese)

Article

Effect of Sb and Zn Addition on the Microstructures and Tensile Properties of Sn–Bi-Based Alloys

Akira Yamauchi * and Masashi Kurose

National Institute of Technology, Gunma College, 580 Toriba-machi, Maebashi 371-8530, Japan; kurose@gunma-ct.ac.jp
* Correspondence: ayana@gunma-ct.ac.jp

Abstract: The tensile behavior of Sn–Bi–Cu and Sn–Bi–Ni alloys has been widely investigated. Reportedly, the addition of small amounts of a third element can refine the microstructures of the eutectic Sn-58mass% Bi solder and improve its ductility. However, the superplasticity mechanism of Sn-based alloys has not been clearly established. Therefore, in this study, the effects of Sb and Zn addition on the microstructures and tensile properties of Sn–Bi-based alloys were investigated. The alloys were subjected to tensile tests under various strain rates and temperatures. We found that Zn- and Sb-added Sn–Bi-based alloys demonstrated superplastic deformation at high temperatures and low strain rates. Sb addition significantly affected the elongation of the Sn–Bi–Sb alloys because the metal dissolves in both the primary Sn phase and the eutectic Sn–Bi matrix. The segregation of Zn and formation of needle-like Zn particles at the eutectic Sn–Bi phase boundary affected the superplastic deformation of the alloys. The deformation of the Sn–40Bi-based alloys at high temperatures and low strain rates led to dynamic recovery, dynamic recrystallization, and/or grain boundary slip because of the accumulation of voids.

Keywords: Sn–Bi-based alloy; superplasticity; tensile property; deformation; low-melting-point solder

Citation: Yamauchi, A.; Kurose, M. Effect of Sb and Zn Addition on the Microstructures and Tensile Properties of Sn–Bi-Based Alloys. *Materials* **2022**, *15*, 884. https://doi.org/10.3390/ma15030884

Academic Editors: Gábor Harsányi, Yoshikazu Todaka, Hideyuki Kanematsu and Takaya Sato

Received: 2 December 2021
Accepted: 19 January 2022
Published: 24 January 2022

Publisher's Note: MDPI stays neutral with regard to jurisdictional claims in published maps and institutional affiliations.

Copyright: © 2022 by the authors. Licensee MDPI, Basel, Switzerland. This article is an open access article distributed under the terms and conditions of the Creative Commons Attribution (CC BY) license (https://creativecommons.org/licenses/by/4.0/).

1. Introduction

The use of Pb and Pb-containing products has been banned in many countries because of their harmful effects on the human body and environment [1,2]. Therefore, several types of Pb-free solder alloys, including Sn-58mass%Bi (412 K) [3], Sn-9mass%Zn (471 K) [4], Sn-0.7mass%Cu (500 K) [5], Sn-3.5mass%Ag (494 K) [6], Sn-70mass%Au (553 K) [7], Sn-35mass%Bi-1mass%Ag (460 K) [8], Sn-8mass%Zn-3mass%Bi (461 K) [9], and Sn-3mass%–0.5mass%Cu (490 K) [10] systems, have been developed to replace Sn-37mass%Pb (456 K) systems in the electronic packaging industries. For example, Sn-3.0mass% Ag-0.5mass% Cu (SAC305), a Pb-free solder, was developed as a substitute for Sn–Pb solder because of its high strength and joint reliability. However, SAC305 is unsuitable for some low-heat-resistance components because its melting temperature, at approximately 494 K, is higher than that of the Sn–Pb eutectic solder (456 K).

Eutectic Sn-58mass% Bi alloy, which has a low melting temperature of 412 K, presents several advantages over other types of solders in low-temperature soldering applications, because it can protect electronic devices from thermal damage under high reflow temperatures. The Sn–Bi Pb-free solder alloys are characterized by a relatively high tensile strength and good creep resistance [11–13]. However, because Sn–Bi-based solders also show frangibility and poor ductility, their applications in the packaging industry are limited [14,15].

Recently, several research groups have focused on manufacturing new alloys doped with micro-/nanometer-sized particles, e.g., Ni, Fe, Zn, Ag, Al, ZrO_2, Al_2O_3, TiO_2, etc., to enhance the mechanical properties and wettability of Pb-free solders for green electronic devices [16–22]. Takao et al. [23], for instance, reported that Sn–Bi and Sn–Bi–Cu alloys exhibit superplasticity. However, the superplasticity mechanism of Sn-based alloys has not been clearly established. In previous studies, the tensile behavior of Sn–Bi–Cu and Sn–Bi–Ni alloys has been investigated, and the corresponding results indicated that these alloys show superplasticity at high temperatures and low strain rates [24,25]. Cu and Ni occupy a small solid-solution region in the Sn phase diagram and easily form an intermetallic compound with the metal. However, the effect of elements occupying large solid-solution regions in the phase diagrams of Sn or Bi on the superplastic deformation of the resultant alloy has not yet been elucidated. The present study was conducted to investigate the effect of Sb and Zn addition on the microstructures and tensile behavior of Sn-40mass% Bi alloys. Tensile tests were then performed under various temperatures and strain rates to examine the superplastic behavior of the modified alloys. To confirm the superplastic deformation, the concentration of the added element and evolution of microstructures were also evaluated.

2. Materials and Methods

Sn, Sn-57mass% Bi, Sn-5mass% Sb, and Sn-9mass% Zn (purity, 99.5 mass%) ingot bars were used to synthesize Sn-40mass% Bi-X mass% Sb (X = 0.1, 0.5, and 1.0) and Sn-40mass% Bi-Y mass% Zn (Y = 0.1, 1.0, and 3.0) (the compositional unit "mass%" is omitted hereafter for convenience). Appropriate amounts of the initial ingots were weighed, placed in an Al_2O_3 crucible, fused in an electric furnace at 673 K, and then left to solidify for 24 h to achieve a homogeneous composition. The fused ingots were then melted at 653 K, cast in an Al mold, and cooled at a rate of approximately 15 K·min^{-1} to form a cylindrical ingot. Finally, dog-bone-type specimens (Figure 1) were machined from the cylindrical Sn-40Bi-based alloy ingots for the tensile tests.

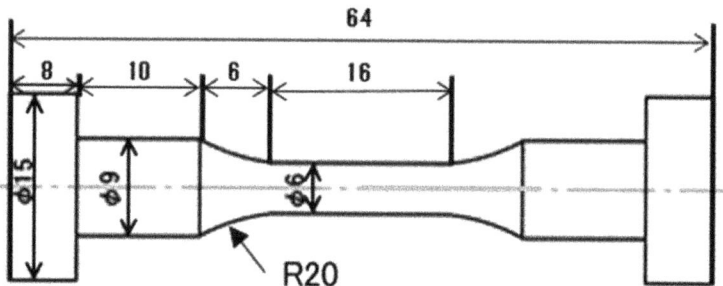

Figure 1. Schematic of the tensile test specimen (unit: mm), adapted from [25].

The tensile tests were performed on a universal material testing machine at strain rates of 5.25×10^{-2}, 5.25×10^{-3}, and 5.25×10^{-4} s^{-1}. Each test piece was exposed to various temperatures of 298, 313, 333, and 353 K in a controlled-atmosphere furnace until complete fracture occurred. In this paper, measurements obtained over three trials under each set of conditions were statistically averaged and reported as the tensile test results.

Specimens were embedded in resin and cut for microstructural observation before and after the tensile testing. The surfaces of the specimens were polished first with SiC papers of up to 1500 grit and then with a 1 μm diamond abrasive and colloidal silica to achieve a mirror-like finish. The cross-sectional and vertical-sectional microstructures and fracture surfaces of the specimens were observed using an optical microscope and a scanning electron microscope (SEM). The crystal orientations were determined by electron backscatter diffraction analysis, and the elemental distributions were evaluated by electron probe microanalysis (EPMA). The solidus and liquidus temperatures of the solder alloys were measured using differential scanning calorimetry (DSC).

3. Results and Discussion

3.1. Microstructures

Figure 2 shows the backscattered electron (BSE) images of the Sn–Bi alloys containing various amounts of Sb. In the images, the dark regions represent the solidified β-Sn matrix, and the bright regions reflect the Bi phase dispersed in the β-Sn matrix. The Sn-40Bi-Sb alloys show a hypoeutectic structure composed of primary Sn dendrites with an average diameter of ~20 μm and eutectic Sn–Bi phases with an average diameter of ~5 μm. The addition of Sb to the Sn-40Bi alloys only slightly affected the grain size of the primary Sn phase and the overall alloy microstructure. EPMA of the alloys (Figure 3) revealed that the Sb atoms preferentially exist in the primary Sn phase.

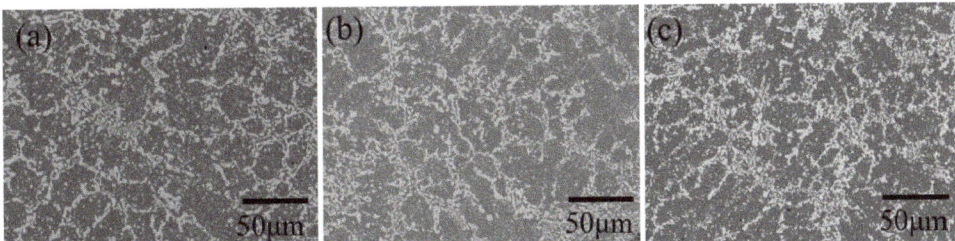

Figure 2. BSE images of (**a**) Sn-40Bi-0.1Sb, (**b**) Sn-40Bi-0.5Sb, and (**c**) Sn-40Bi-1.0Sb.

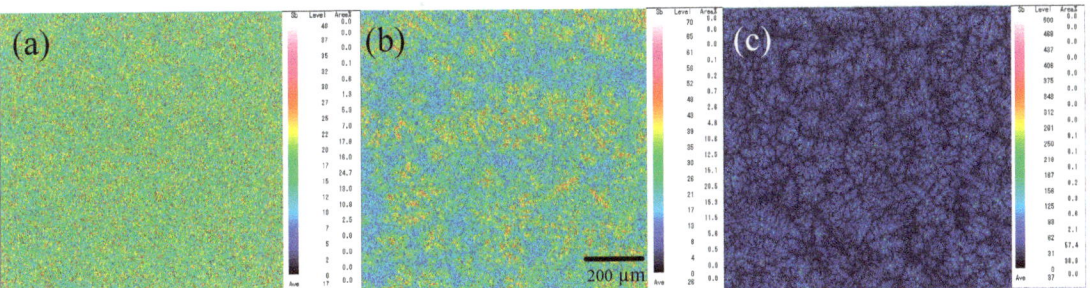

Figure 3. EPMA mapping images of Sb in (**a**) Sn-40Bi-0.1Sb, (**b**) Sn-40Bi-0.5Sb, and (**c**) Sn-40Bi-1.0Sb.

The results of the quantitative EPMA are summarized in Table 1. Sn-40Bi-1.0Sb showed a relatively high Sb concentration (2.4 mass%), as well as a preferential segregation of the Sb atoms in the primary Sn phase (Figure 3c). Further, an increase in the Sb concentration also led to the dissolution of Sb atoms in the eutectic Sn–Bi matrix.

Table 1. EPMA results of the Sn-40Bi-Sb alloys.

Specimen	Sb Concentration (Mass %)	
	Primary Sn Phase	Eutectic Sn–Bi Matrix
Sn-40Bi-0.1Sb	0.09	0
Sn-40Bi-0.5Sb	1.0	0.16
Sn-40Bi-1.0Sb	2.4	0.41

Figure 4 shows the microstructures of the Sn-40Bi-Zn alloys. The EPMA results presented in Figure 5 reveal that the alloys consist of a gray β-Sn phase, a eutectic Sn–Bi phase, and a needle-like Zn-rich phase. A significant increase in the Zn-rich phases with increasing Zn concentration was observed. The EPMA further confirmed that the Zn atoms preferentially existed in the Bi phase of the eutectic Sn–Bi matrix. This result agrees with those of Mokhtari et al. [26] and Hirata et al. [27]. The grain size of the primary Sn phase of Sn-40Bi-1.0Zn, shown in Figure 4b, was larger than that of the primary Sn phase of the other Sn–Bi–Zn alloys. The EPMA of the Zn-rich phase showed that the phases were mainly composed of Zn, and the Zn mass percentage increased to more than 50%, as shown in Figure 5b,c.

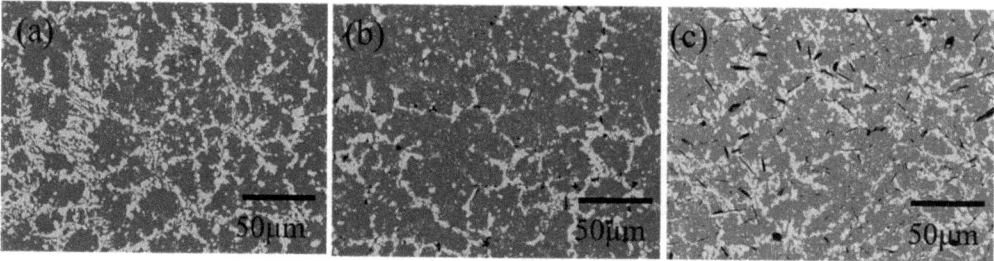

Figure 4. BSE images of (**a**) Sn-40Bi-0.1Zn, (**b**) Sn-40Bi-1.0Zn, and (**c**) Sn-40Bi-3.0Zn.

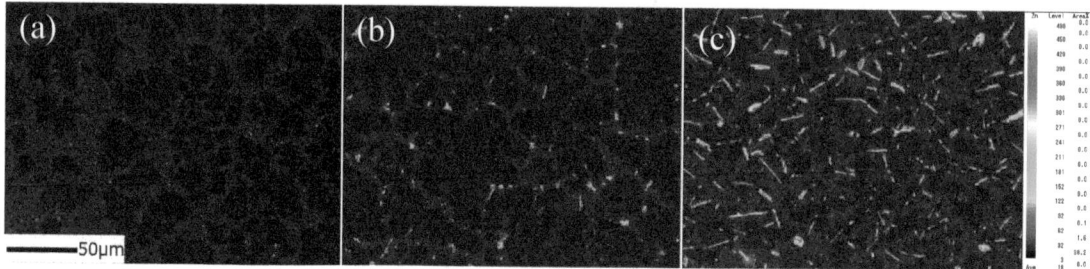

Figure 5. EPMA mapping images of Zn in (**a**) Sn-40Bi-0.1Zn, (**b**) Sn-40Bi-1.0Zn, and (**c**) Sn-40Bi-3.0Zn.

3.2. Tensile Properties

Figure 6 shows the stress–strain curves of Sn-40Bi-0.1Sb at various strain rates and temperatures (298 and 353 K). At 298 K, the alloy exhibited a nearly steady-state flow as the stress approached its yield strength. The tensile strength of the specimen decreased, whereas its elongation increased with the decreasing strain rate. The tensile behavior of Sn-40Bi-0.1Sb did not reflect superplasticity at 298 K. At 353 K, the alloy exhibited a gradual decrease in tensile strength under strain rates of 5.25×10^{-2} and 5.25×10^{-3} s^{-1}. The specimen fractured when the stress reached its yield strength. Under a strain rate of 5.25×10^{-4} s^{-1}, the yield strength of the alloy exhibited an initial sharp decrease, followed by a more gradual decrease until rupture occurred. The greatest elongation of this specimen observed under the conditions of 353 K and 5.25×10^{-4} s^{-1} was 206%. Superplastic behavior is defined as the ability to withstand a strain of >200% [28,29]. Sn-40Bi-0.1Sb demonstrated superplasticity at a temperature of 353 K and a strain rate of 5.25×10^{-4} s^{-1}.

In general, the stress–strain curves of the Sn-40Bi-Sb alloys are similar to those of Sn-40Bi-0.1Cu and Sn-40Bi-0.01Ni [24,25]. Figure 7 shows the stress–strain curves of the Sn-40Bi-Sb alloys at 353 K under a strain rate of 5.25×10^{-4} s^{-1}. The highest tensile strength (45 MPa) under these conditions is observed in Sn-40Bi-0.1Sb. The tensile strength of the alloys decreases with increasing Sb concentration. Indeed, the tensile strength of Sn-40Bi-1.0Sb, at 23 MPa, is only half of that of Sn-40Bi-0.1Sb at 353 K under a strain rate of 5.25×10^{-4} s^{-1}. Conversely, the elongation increases with the increasing Sb concentration. The maximum elongation of Sn-40Bi-1.0Sb under the conditions of 353 K and 5.25×10^{-4} s^{-1} is no less than 900%, as illustrated in Figure 8. The specimens generally exhibited longer elongation, without necking, and chisel-point fractures as their fracture mode. This result is obtained possibly because Sb is solid-solution not only in the primary Sn phase but also in the Bi phase. Therefore, in contrast to Cu and Ni, which are typically added in low amounts, a larger amount of Sb must be added to the alloy to achieve superplastic deformation with long elongation. The microstructural observations further indicated that the superplastic deformation is likely to occur in the Sn-40Bi-Sb alloys, because the structures of the primary Sn phase and eutectic Sn–Bi matrix become finer as the Sb concentration increases. Figure 9 shows the fractograph of Sn-40Bi-1.0Sb after the tensile test under a strain rate of 5.25×10^{-4} s^{-1} at 298 and 353 K. At 298 K, the alloy displays a ductile fracture mode, whereas, at 353 K, it displays a chisel-point fracture mode. The area of the Sn-40Bi-1.0Sb alloy at 353 K was reduced by approximately 98%. Thus, the reduction in the area and elongation of the Sn-40Bi-Sb alloys increases with an increasing concentration of Sb.

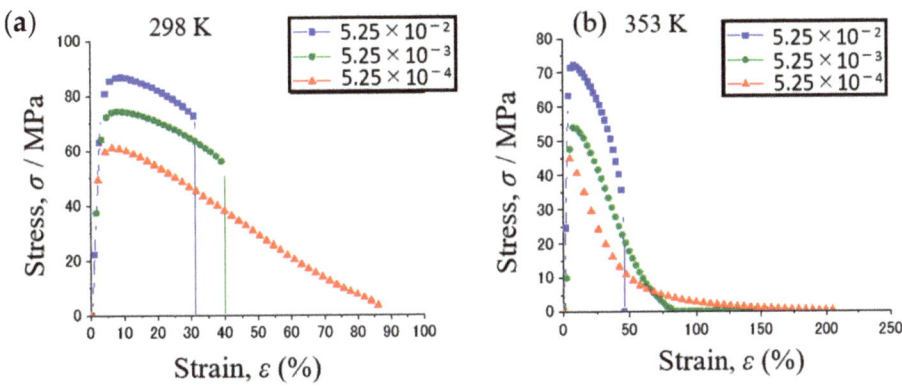

Figure 6. Stress–strain curves of Sn-40Bi-0.1Sb under various strain rates at (**a**) 298 and (**b**) 353 K.

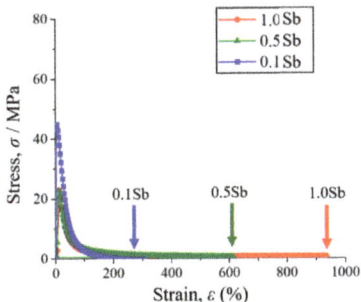

Figure 7. Stress–strain curves of various Sn-40Bi-Sb alloys at 353 K under a strain rate of 5.25×10^{-4} s^{-1}.

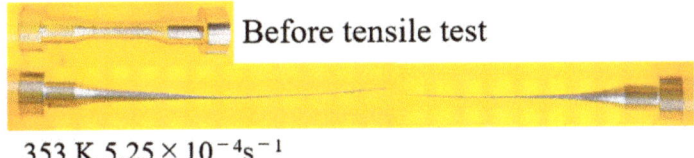

Figure 8. Photographs of Sn-40Bi-1.0Sb before and after the tensile testing at 353 K under a strain rate of 5.25×10^{-4} s^{-1}.

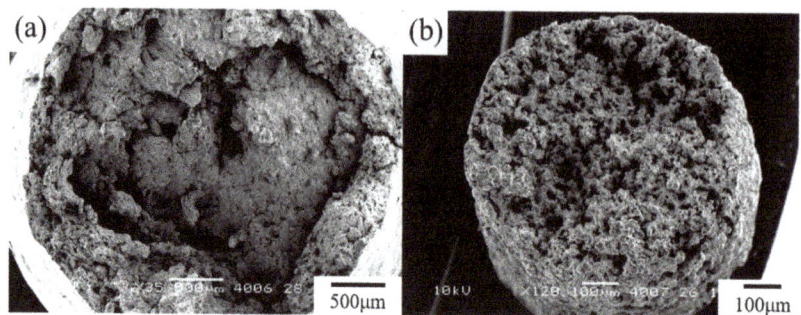

Figure 9. Fractured surface of the Sn-40Bi-1.0Sb alloy after the tensile test under a strain rate of 5.25×10^{-4} s^{-1} at (**a**) 298 K and (**b**) 353 K.

Figure 10 shows the stress–strain curves of Sn-40Bi-0.1Zn at various temperatures under a strain rate of 5.25×10^{-4} s^{-1}. The ductility of Sn-40Bi-0.1Zn exhibited gradual improvements as the temperature increased. The specimen showed superplastic deformation at temperatures above 333 K and elongation of up to 630%. Furthermore, specimens with different Zn concentrations demonstrated large elongations at 333 and 353 K. The Sn-40Bi-Zn alloys show the same temperature and strain-rate dependences as the Sn-40Bi-Sb alloys (Figure 6); thus, the stress–strain curves in various conditions are omitted for this alloy in this paper.

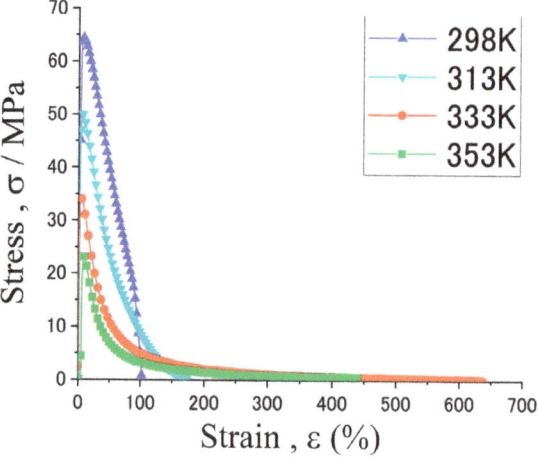

Figure 10. Stress–strain curves of Sn-40Bi-0.1Zn under a strain rate of 5.25×10^{-4} s^{-1} at various temperatures.

Figure 11 shows the stress–strain curves of the various Sn-40Bi-Zn specimens at 333 K, under a strain rate of 5.25×10^{-4} s^{-1}. The elongation of the alloys increased in the order of 1Zn < 3Zn < 0.1Zn, likely because their microstructures were modified by Zn addition (Figure 4). Interestingly, the Sn-40Bi-Zn alloys showed extensive crystallization of the needle-like Zn particles when the metal was added at a rate of 1 mass% or higher. Therefore, the elongations of Sn-40Bi-1Zn and Sn-40Bi-3Zn were lower than that of Sn-40Bi-Zn. The fractured surfaces of the Sn-40Bi-Zn alloys after the tensile test are almost the same as those of the Sn-40Bi-Sb alloys (Figure 9). Thus, there is no difference between the additional elements on the fractured photographs.

Figure 12 shows the vertical and cross-sectional images of various Sn-40Bi-Zn alloys after the tensile tests. In these tests, the tensile stress was applied to the specimens in the vertical direction. All the alloys displayed chisel-point fractures, which indicate superplastic deformation. The area of the Sn-40Bi-0.1Zn alloy at 353 K was reduced by approximately 99.5%, indicating that the reduction in the area and elongation of the Sn-40Bi-Zn alloys at 353 K decreased with the increasing Zn concentration. The number of voids in the specimens clearly increased with the increasing Zn concentration. Whether there is a correlation between the needle-like Zn particles and the number of voids is unknown, but the tendency of the elongation to decrease with the increasing Zn concentration is considered to be related to the number of voids in the specimens. At high temperatures and low strain rates, which are conducive to superplastic behavior, specimens do not show remarkable elongation of the primary Sn phase in the tensile direction; cracks and voids are observed to be segregated at the grain boundaries around the intermetallic compounds [25]. These results indicate that void accumulation begins from the intermetallic compounds or metal particles, and the final structure of the alloy may be different from its initial dendritic structure. Therefore, dynamic recovery as well as dynamic recrystallization occur after the ultimate tensile strength is achieved, and grain boundary slip deformation may be accompanied by diffusion creep due to void accumulation. Thus, the deformation of the primary Sn phase is the dominant deformation mechanism of the Sn-40Bi-Zn alloys at low temperatures and high strain rates. At high temperatures and low strain rates, the deformation of these alloys leads to dynamic recovery, dynamic recrystallization, and/or grain boundary slip.

Slip deformation may be speculated to be the dominant deformation mechanism of the eutectic Sn–Bi–Zn matrix. Table 2 summarizes the solidus and liquidus temperatures of the Sn-40Bi and Sn-40Bi-Zn alloys obtained from DSC measurements. The melting point temperature of the alloys decreased when the amount of Zn added was equal to or exceeded 1 mass%, and it was 406 K when the added Zn amount was 3 mass%. The liquidus temperature increased from 440 K in the alloy without Zn to 452 K in the alloy with 1 mass% Zn, but it decreased sharply to 427 K in the alloy with 3 mass% Zn, likely because the ternary eutectic composition of the alloy is close to that of Sn-40Bi-3Zn. Because the tensile test temperature (353 K) is only 87% of the solidus temperature of Sn-40Bi-3Zn, the elongation of this alloy may be estimated to be larger than that of Sn-40Bi-1Zn. This is because superplastic deformation occurring at high temperatures and low liquidus temperatures tends to promote plastic flow.

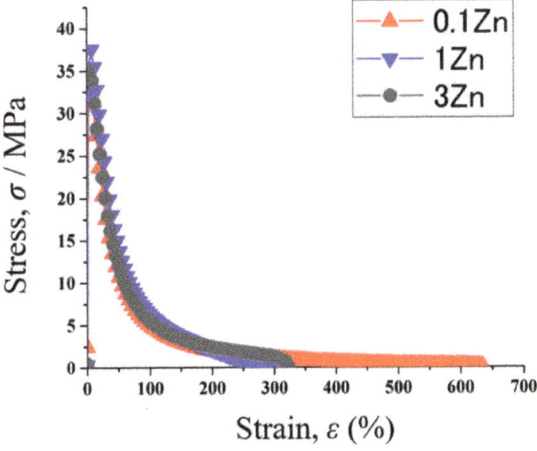

Figure 11. Stress–strain curves of various Sn-40Bi-Zn alloys at 333 K under a strain rate of 5.25×10^{-4} s^{-1}.

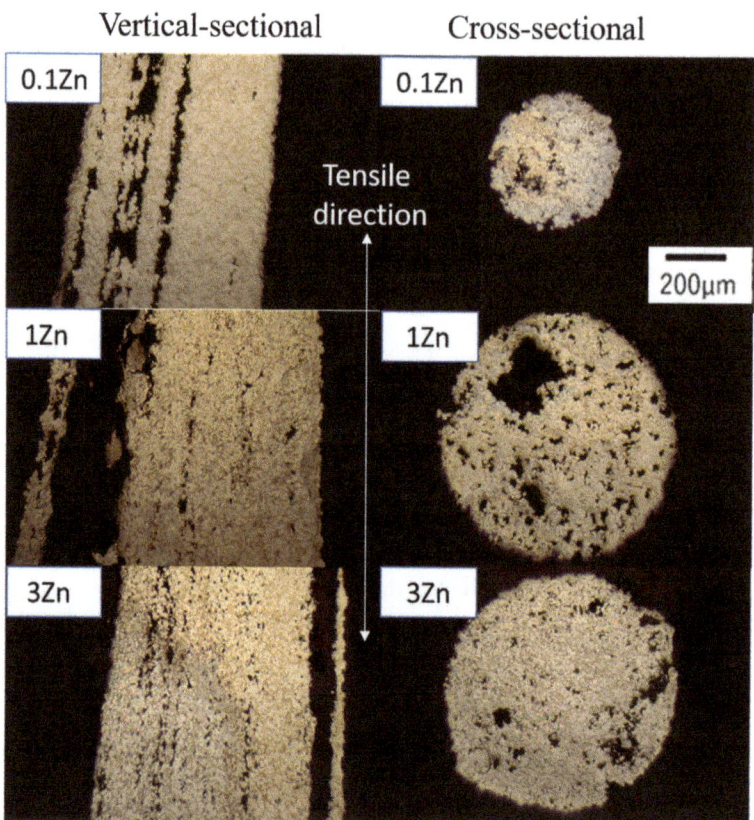

Figure 12. Vertical and cross-sectional optical microscopic images of various Sn-40Bi-Zn alloys after the tensile tests.

Table 2. Solidus and liquidus temperatures of the Sn-40Bi-Zn alloys.

Specimen	Solidus Temperature (K)	Liquidus Temperature (K)
Sn-40Bi	411	440
Sn-40Bi-0.1Zn	411	445
Sn-40Bi-1Zn	406	452
Sn-40Bi-3Zn	406	427

3.3. Strain Rate Sensitivity Index (m)

Figure 13 shows the strain rate sensitivity index (*m*) of various Sn-40Bi-Sb alloys at 353 K. In this study, *m* was generally defined by Equation (1):

$$m = \frac{d \ln \sigma}{d \ln \dot{\varepsilon}}, \tag{1}$$

where σ is the stress, and $\dot{\varepsilon}$ is the strain rate. The *m* value is an important parameter that describes the high-temperature tensile ductility of an alloy material; it is calculated from the slopes of the lines obtained at various temperatures and increases with increasing temperature. For materials without superplasticity, $m < 0.2$, whereas, for general superplastic materials with fine grains, the *m* value is usually very large, i.e., $m = 0.3–1.0$. A large

m value indicates that large superplastic elongation can be obtained [30–32]. The m value calculated from the slopes of the lines shown in Figure 13 is lower than the critical value of 0.3, indicating that the alloys are characterized by fine-grained superplasticity. This result is similar to previous findings reported for Sn-40Bi-0.1Cu and Sn-40Bi-0.01Ni [24,25]. Because the microstructures and average grain sizes of the Sn-40Bi-Sb alloys are similar to those of Sn-40Bi-0.1Cu, the m value of the Sn-40Bi-Sb alloys also increases with the increasing Sb concentration. Evidently, increasing the Sb concentration influences the superplastic-like deformation of the resultant alloys. These results suggest that although the Sn-40Bi-Sb alloys show superplastic-like deformation, their deformation mechanism is not based on fine-grained superplasticity. Thus, it can be considered that the deformation mechanism of the Sn-40Bi-X alloy might be caused by the grain boundary slip, diffusion creep, and dislocation creep.

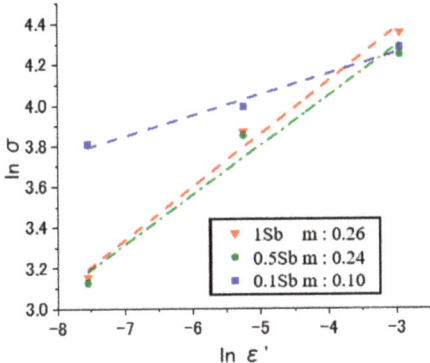

Figure 13. Strain rate sensitivity index (m) of the Sn-40Bi-Sb alloys at 353 K (σ, stress; $\dot{\varepsilon}$, strain rate).

4. Conclusions

In this study, the microstructures and tensile behaviors of the Sn-40Bi-X% Sb (X = 0.1, 0.5, 1.0) and Sn-40Bi-Y% Zn (Y = 0.1, 1.0, 3.0) alloys at various temperatures and strain rates were investigated using SEM and tensile tests. The following results were obtained.

(1) The tensile strength decreased and the elongation increased with the increasing temperature or decreasing strain rate.
(2) The Sn-40Bi-Sb and Sn-40Bi-Zn alloys demonstrated superplasticity at high temperatures (>333 K) and low strain rates (<5.25 × 10^{-3} s^{-1}).
(3) The m value of the alloys increased with the increasing temperature. Moreover, the m value of the Sn-based alloys was lower than the critical value of 0.3, implying that these alloys did not exhibit fine-grained superplasticity but superplastic-like deformation with grain boundary slip and diffusion creep. The maximum m value of Sn-40Bi-1Sb was 0.26.
(4) The Sb atoms were dissolved in both the primary Sn phase and the eutectic Sn–Bi matrix. In contrast, the Zn atoms were dissolved in the eutectic Sn–Bi matrix, and needle-like Zn crystals were formed when the Zn concentration exceeded 1 mass%. The deformation of the primary Sn phases is the dominant deformation mechanism at low temperatures and high strain rates. Moreover, deformation at high temperatures and low strain rates leads to recovery, dynamic recrystallization, and/or grain boundary slip.

Further research on complex Sn–Bi-based solder alloys will be conducted in the future to obtain more low-temperature solder candidates that could be applied to electronic devices.

Author Contributions: A.Y.: Conceptualization, project administration, methodology, formal analysis, investigation, writing—original draft and review, and editing. M.K.: Investigation, review, and supervision. All authors have read and agreed to the published version of the manuscript.

Funding: This research was funded by JSPS KAKENHI (Grant Number 18K03859), and the Inter-University Cooperative Research Program (Proposal Nos. 19G0023, 20G0041, and 202112-CRKEQ-0012) of the Cooperative Research and Development Center for Advanced Materials, Institute for Materials Research, Tohoku University.

Institutional Review Board Statement: Not applicable.

Informed Consent Statement: Not applicable.

Data Availability Statement: All data are available from the corresponding author upon reasonable request.

Acknowledgments: This work was supported by JSPS KAKENHI Grant Number 18K03859. This work was partly performed under the Inter-University Cooperative Research Program (Proposal Nos. 19G0023 and 20G0041) of the Cooperative Research and Development Center for Advanced Materials, Institute for Materials Research, Tohoku University. We would like to thank our former students, C. Nagata and K. Tsukamoto, for their technical support during the tensile testing of the Sn–Bi-based alloys.

Conflicts of Interest: The authors declare no conflict of interest.

References

1. Suganuma, K. Advances in lead-free electronics soldering. *Curr. Opin. Solid State Mater. Sci.* **2001**, *5*, 55–64. [CrossRef]
2. Synedlina, J.J.; Nurmib, S.T.; Lepisto, T.K.; Ristolainen, E.O. Mechanical and microstructural properties of SnAgCu solder joints. *Mater. Sci. Eng. A* **2006**, *420*, 55–62.
3. Gain, A.K.; Zhang, L. Growth mechanism of intermetallic compound and mechanical properties of nickel (Ni) nanoparticle doped low melting temperature tin-bismuth (Sn-Bi) solder. *J. Mater. Sci. Mater. Electron.* **2016**, *27*, 781–794. [CrossRef]
4. Wu, M.L.; Yu, D.Q.; Law, C.M.T.; Wang, L. The properties of Sn-9Zn lead-free solder alloys doped with trace rare earth elements. *J. Electron. Mater.* **2002**, *31*, 921–927. [CrossRef]
5. Gain, A.K.; Zhang, L.; Quadir, M.Z. Thermal aging effects on microstructures and mechanical properties of an environmentally friendly eutectic tin-copper solder alloy. *Mater. Des.* **2016**, *110*, 275–283. [CrossRef]
6. Sahin, M.; Cadirli, E. The effects of temperature gradient and growth rate on the microstructure of directionally solidified Sn-3.5Ag eutectic solder. *J. Mater. Sci. Mater. Electron.* **2012**, *23*, 484–492. [CrossRef]
7. Matijasevic, S.; Lee, C.C.; Wang, C.Y. Au-Sn alloy phase diagram and properties related to its use as a bonding medium. *Thin Solid Films* **1993**, *223*, 276–287. [CrossRef]
8. Gain, A.K.; Zhang, L. Interfacial microstructure, wettability and material properties of nickel (Ni) nanoparticle doped tin-bismuth-silver (Sn-Bi-Ag) solder on copper (Cu) substrate. *J. Mater. Sci. Mater. Electron.* **2016**, *27*, 3982–3994. [CrossRef]
9. Hsi, S.; Lin, C.T.; Chang, T.C.; Wang, M.C.; Liang, M.K. Interfacial Reactions, Microstructure, and Strength of Sn-8Zn-3Bi and Sn-9Zn-Al Solder on Cu and Au/Ni (P) Pads. *Metall. Mater. Trans. A* **2010**, *41*, 275–284. [CrossRef]
10. Gain, A.K.; Zhang, L. Harsh service environment effects on the microstructure and mechanical properties of Sn–Ag–Cu-1 wt% nano-Al solder alloy. *J. Mater. Sci. Mater. Electron.* **2016**, *27*, 11273–11283. [CrossRef]
11. Tomlinson, W.J.; Collier, I. The mechanical properties and microstructures of copper and brass joints soldered with eutectic tin-bismuth solder. *J. Mater. Sci.* **1987**, *22*, 1835–1839. [CrossRef]
12. Mei, Z.; Morris, J.W. Characterization of eutectic Sn-Bi solder joints. *J. Electron. Mater.* **1992**, *21*, 599–607. [CrossRef]
13. Yebisuya, T.; Kawakubo, T. Creep and tensile properties of cast Bi-Sn, Bi-Pb and Bi-Sn-Pb solders. *J. Jpn. Inst. Met. Mater.* **1993**, *57*, 455–462. [CrossRef]
14. Watanabe, H. The lead-free solder of addition micro-elements in industrial products. *J. Jpn. Inst. Electron. Packag.* **2005**, *8*, 183–187. [CrossRef]
15. Nagano, M.; Hidaka, N.; Watanabe, H.; Shimoda, M.; Ono, M. Effect of additional elements on creep properties of the Sn-Ag-Cu lead free solder. *J. Jpn. Inst. Electron. Packag.* **2006**, *9*, 171–179. [CrossRef]
16. Mccormack, M.; Chen, H.S.; Kammlott, G.W.; Jin, S. Significantly improved mechanical properties of Bi-Sn solder alloys by Ag-doping. *J. Electron. Mater.* **1997**, *26*, 954–958. [CrossRef]
17. Sakuyama, S.; Akamatsu, T.; Uenishi, K.; Sato, T. Effects of a third element on microstructure and mechanical properties of eutectic Sn-Bi solder. *Trans. Jpn. Inst. Electron. Packag.* **2009**, *2*, 98–103.
18. Okamoto, K.; Nomura, K.; Doi, S.; Akamatsu, T.; Sakuyama, S.; Uenishi, K. Effect of Sb and Zn addition on impact resistance improvement of Sn-Bi solder joints. *Int. Symp. Microelectron.* **2013**, *2013*, 000104–000108. [CrossRef]
19. Zhang, L.; Tu, K.N. Structure and properties of lead-free solders bearing micro and nano particles. *Mater. Sci. Eng. R* **2014**, *82*, 1–32. [CrossRef]

20. Gain, A.K.; Zhang, L. Effects of Ni nanoparticles addition on the microstructure, electrical and mechanical properties of Sn-Ag-Cu alloy. *Materialia* **2019**, *5*, 100234. [CrossRef]
21. Tsao, L.C.; Chang, S.Y. Effects of Nano-TiO_2 additions on thermal analysis, microstructure and tensile properties of Sn3.5Ag0.25Cu solder. *Mater. Des.* **2010**, *31*, 990–993. [CrossRef]
22. Hirata, A.; Shoji, I.; Tsuchida, T.; Ookubo, T. Effect of electrode material on joint strength of soldered joints with Sn-Bi and Sn-Bi-Sb lead-free solder balls. In Proceedings of the ASME 2013 International Technical Conference and Exhibition on Packaging and Integration of Electronic and Photonic Microsystems, Burlingame, CA, USA, 16–18 July 2013.
23. Takao, H.; Yamada, A.; Hasegawa, H.; Matsui, M. Mechanical properties and solder joint reliability of low-melting Sn-Bi-Cu lead free solder alloy. *J. Jpn. Inst. Electron. Packag.* **2002**, *5*, 152–158. [CrossRef]
24. Yamauchi, A.; Ida, K.; Fukuda, M.; Yamaguchi, T. Tensile properties of Sn-Bi lead-free solder alloys. *Solid State Phenom.* **2018**, *273*, 72–76. [CrossRef]
25. Umeyama, J.; Yamauchi, A. Tensile behavior and superplastic deformation of Sn-Bi-Cu Alloy. *Mater. Trans.* **2019**, *60*, 882–887. [CrossRef]
26. Mokhtari, O.; Zhou, S.; Chan, Y.C.; Nishikawa, H. Effect of Zn addition on interfacial reactions between Sn-Bi solder and Cu substrate. *Mater. Trans.* **2016**, *57*, 1272–1276. [CrossRef]
27. Hirata, Y.; Yang, C.; Lin, S.; Nishikawa, H. Improvements in mechanical properties of Sn-Bi alloys with addition of Zn and In. *Mater. Sci. Eng. A* **2021**, *813*, 141131. [CrossRef]
28. Ohsawa, H.; Nishimura, H. Manufacturing method of superplastic materials and commercial applications. *J. Jpn. Inst. Light Met.* **1989**, *39*, 765–775. [CrossRef]
29. JIS H 7007:2002; Glossary of Terms Used in Metallic Superplastic Materials. Japanese Standards Association: Tokyo, Japan, 2002.
30. Maruyama, K.; Nakashima, H. *Materials Science for High Temperature Strength*; Uchida-Rokakuho: Tokyo, Japan, 1997.
31. Nieh, T.G.; Wadsworth, J.; Sherby, O.D. *Superplasticity in Metals and Ceramics*; Cambridge University Press: Cambridge, UK, 1997; pp. 22–31.
32. Ridley, N. Metals for superplastic forming. In *Superplastic Forming of Advanced Metallic Materials: Methods and Applications*; Giuliano, G., Ed.; Woodhead Publishing Limited: Cambridge, UK, 2011; pp. 3–33.

Article

Copper Surface Treatment Method with Antibacterial Performance Using "Super-Spread Wetting" Properties

Beomdeok Seo [1,*], Hideyuki Kanematsu [2], Masashi Nakamoto [1], Yoshitsugu Miyabayashi [3] and Toshihiro Tanaka [1]

1. Division of Materials and Manufacturing Science, Graduate School of Engineering, Osaka University, 2-1 Yamadaoka, Suita 565-0871, Osaka, Japan; nakamoto@mat.eng.osaka-u.ac.jp (M.N.); tanaka@mat.eng.osaka-u.ac.jp (T.T.)
2. Department of Materials Science and Engineering, National Institute of Technology (KOSEN), Suzuka College, Suzuka 510-0294, Mie, Japan; kanemats@mse.suzuka-ct.ac.jp
3. Graduate School of Engineering, Osaka University, 2-8 Yamadaoka, Suita 565-0871, Osaka, Japan; miyabayashi@jrl.eng.osaka-u.ac.jp
* Correspondence: seobeomdeok@mat.eng.osaka-u.ac.jp

Abstract: In this work, a copper coating is developed on a carbon steel substrate by exploiting the superwetting properties of liquid copper. We characterize the surface morphology, chemical composition, roughness, wettability, ability to release a copper ion from surfaces, and antibacterial efficacy (against *Escherichia coli* and *Staphylococcus aureus*). The coating shows a dense microstructure and good adhesion, with thicknesses of approximately 20–40 μm. X-ray diffraction (XRD) analysis reveals that the coated surface structure is composed of Cu, Cu_2O, and CuO. The surface roughness and contact angle measurements suggest that the copper coating is rougher and more hydrophobic than the substrate. Inductively coupled plasma atomic emission spectroscopy (ICP-AES) measurements reveal a dissolution of copper ions in chloride-containing environments. The antibacterial test shows that the copper coating achieves a 99.99% reduction of *E. coli* and *S. aureus*. This study suggests that the characteristics of the copper-coated surface, including the chemical composition, high surface roughness, good wettability, and ability for copper ion release, may result in surfaces with antibacterial properties.

Keywords: antibacterial properties; coating; copper; fine crevice structure; super-spread wetting properties

Citation: Seo, B.; Kanematsu, H.; Nakamoto, M.; Miyabayashi, Y.; Tanaka, T. Copper Surface Treatment Method with Antibacterial Performance Using "Super-Spread Wetting" Properties. *Materials* 2022, 15, 392. https://doi.org/10.3390/ma15010392

Academic Editor: Fernão D. Magalhães

Received: 26 November 2021
Accepted: 3 January 2022
Published: 5 January 2022

Publisher's Note: MDPI stays neutral with regard to jurisdictional claims in published maps and institutional affiliations.

Copyright: © 2022 by the authors. Licensee MDPI, Basel, Switzerland. This article is an open access article distributed under the terms and conditions of the Creative Commons Attribution (CC BY) license (https://creativecommons.org/licenses/by/4.0/).

1. Introduction

The Centers for Disease Control and Prevention reported that healthcare-associated infections (HAIs) cause or contribute to 99,000 deaths and add approximately $40 billion to healthcare coasts each year [1]. As a possible cause of infection, bacterial contamination on the surfaces of materials, especially in hospitals and public places, is proposed as a serious threat [2]. On surfaces, many types of bacteria can survive for long periods, with some even able to survive for more than a month [2]. Various efforts, such as hand washing, disinfection, and antibacterial surfaces, have been developed to control infection, but the problem has not been resolved [2–4]. A recent trend in risk management of the transfer of bacteria from surface to surface is the use of copper in the manufacture of public and hospital materials [5,6]. Some of these studies reported that the use of copper alloys in intensive care unit rooms can significantly reduce HAIs compared with a standard room [5,7].

Although the antibacterial mechanism of the solid copper surface has yet to be clearly understood, several studies have investigated the result of the so-called contact killing [8,9]. When bacteria are directly in contact with metallic copper, copper ions accumulate inside the cell because the bacteria recognize the copper ions as essential nutrients [10,11]. The cell and DNA are then damaged and destroyed by the depolarization effect and reactive oxygen species (ROS) [12]. Surface properties, including roughness, wetting behavior, and contact angle, significantly influence contact killing [12–20]. Together with the contact-killing effect, the direct release of copper ions from metallic copper plays a decisive role in the bacterial killing process [21–23]. Copper ions prompt the generation of ROS and cause bacterial cell damage or death. It is also reported that there are differences depending on the type of copper oxide. The antibacterial performance can be further improved by applying a copper oxide surface because of the extensive release of copper ions from the copper oxide surface [23–25]. Based on these observations, the use of copper is a promising strategy to prevent HAIs.

When applying copper to materials in public and hospital settings, it is the preferred method to coat copper on an inexpensive metal such as carbon steel in consideration of the economic aspects [26]. However, there are problems associated with the copper coating process. The widely used copper-plating process uses cyanide ions, which can cause serious environmental pollution problems [27,28]. Other methods, such as plasma treatment with oxygen, chemical vapor deposition, and ammonia plasma, require complex equipment and procedures [29–32]. Therefore, to widely apply the antibacterial properties of copper, a new method to solve these problems is important.

In this study, for copper surface coating, the super-spread wettability properties of liquid copper are exploited. The literature reports that liquid copper is not able to wet a solid oxide [33,34]. However, our previous works have shown that liquid copper unusually penetrates and spreads on a surface with fine crevice structures formed by capillary action [35]. The resulting phenomenon, named "super-spread wetting", caused by the capillary characteristics of the liquid metal and metal surface with a fine crevice structure, differs from an ordinary occurrence [36–38]. In addition, using the super-spread wetting property, the liquid copper is able to flow to the desired target point [39]. Although some research has been conducted to understand this unusual phenomenon, studies on the coating technology have not been conducted [40,41].

We provide a new method of copper coating with antibacterial properties on carbon steel using the super-spread wetting properties of liquid copper. In addition, our research suggests a method for coating copper that does not use complex manufacturing equipment and processes and does not emit pollutants. The surface characterizations, including morphology, chemical composition, phase, roughness, wettability, and the ability for a copper ion to be released from the surfaces, have been systematically investigated. Furthermore, the antibacterial properties are determined by ISO 22196:2011 method against *Escherichia coli* and *Staphylococcus aureus*. In this work, we will discuss the antibacterial mechanism for copper-coated surfaces.

2. Materials and Methods

2.1. Materials and Fabrication of the Test Samples

The substrate was cut from JIS-SS400 carbon steel plate with the chemical composition: C 0.148; Si 0.213; Mn 0.458; S 0.018; P 0.012 (wt.%), Fe balance. The specimens were machined into rectangular shapes, with dimensions of 10 mm × 10 mm × 2 mm, then sequentially ground by emery papers up to 1200 grit and degreased in acetone using an ultrasonic bath. Copper powder (99 purity, Sigma Aldrich, St Louis, MO, USA) was applied for the coating process.

Figure 1 shows the different steps involved with fabricating the test samples. First, a fine structure was formed on the surface to allow liquid copper to spread on the surface. Our previous research confirmed that surfaces with a fine crevice structure can be created by laser irradiation [38]. As shown in Figure 1a, a continuous Nd: YAG laser (ML-7062A, Miyachi Corporation, Tokyo, Japan) was used to fabricate the fine crevice structure with two types of patterns: covering all (10 × 10 mm square) and 48% (0.8 mm × 9.9 mm × 6 pcs rectangle arranged at intervals of 1.0 mm) of the substrate, respectively. Laser irradiation was performed on the substrate positioned 110 mm under the scanning lens with an average power of 30 W at a frequency of 6.0 kHz, a spot diameter of 0.1 mm, a pitch of 0.01 mm and a scan speed of 9.0 mm/s under air atmosphere. Then, the prepared substrate with a crevice structure was fed with 6 mg/cm^2 copper and heated to 1100 °C, which is slightly above the melting point of copper, to coat it with liquid copper in an electric furnace (see Figure 1b) [36]. The temperature profile for the coating process is shown in Figure 1c. To prevent oxidation during the heating process, Ar gas (60 mL/min) and H$_2$ gas (15 mL/min) were supplied, and the oxygen partial pressure was maintained at approximately 10^{-3} atm in the furnace. After a prescribed heating time, the specimens were cooled in the furnace at a rate of 400 °C per hour to 600 °C under an Ar gas environment and then air-cooled. The cooling condition was determined by focusing on the formation of copper and copper compounds with antibacterial effects by using a thermodynamic calculation with the FactSage software (version 7.1).

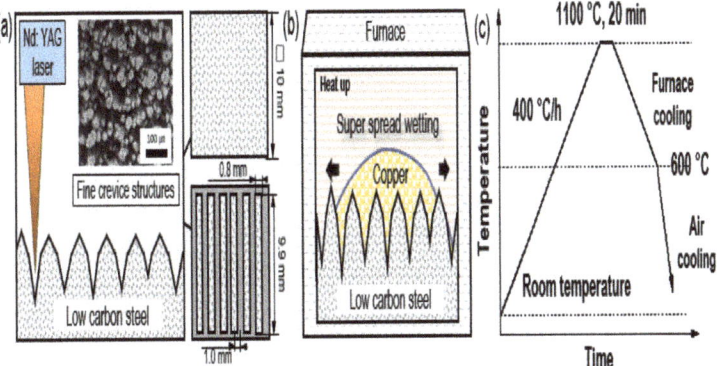

Figure 1. Schematic diagram of the different steps of fabricating procedure: (**a**) formation of fine crevice structure and pattern shape using Nd: YAG laser; (**b**) process for copper coating by super-spread wetting properties; (**c**) heat temperature profile for the coating process.

2.2. Strains and Culture Conditions

We used two different bacterial strains to determine the antibacterial effect, *Escherichia coli* (*E. coli*, ATCC 25922) and *Staphylococcus aureus* (*S. aureus*, ATCC12228), the most frequently used gram-negative and gram-positive organisms, respectively. All the investigations were conducted in Luria Broth (LB, Nacalai tesque, Kyoto, Japan) consisting of 10 g of bactotrypton, 5 g of yeast extract and 10 g of NaCl per liter. All solutions were sterilized by autoclaving at 120 °C for 15 min before use. Both bacteria were cultured in 10 mL of LB broth on a swing bed at 35 °C overnight. This solution was then diluted in sterile LB broth to 10^5 CFU/mL.

2.3. Surface Characterization

The surface morphology and composition were investigated using scanning electron microscopy (SEM, Miniscope TM-1000, Hitachi, Japan) and energy-dispersive spectroscopy (EDS, JSM-6500F, JEOL, Tokyo, Japan). The surface phase was analyzed with X-ray diffraction analysis (XRD, SmartLab, Rigaku, Tokyo, Japan) using CuKα radiation in the range of 2θ from 20°–80°. The surface roughness and profile were measured by three-dimensional (3D) laser scanning microscopy (VK-9700, Keyence, Osaka, Japan) and analyzed with the VK-H1XP software. Results for five random areas were presented as the average roughness (R_a), peak-to-valley roughness (R_z) and root-mean-square roughness (R_q). The contact angle measurement was used to characterize the wettability by LB broth. The contact angle was measured at room temperature using the sessile drop method from a contact angle meter (CA, DMo-501, Kyowa Interface Science, Saitama, Japan). The measurements were repeated three times for each specimen. The "Standard test methods for measuring adhesion by tape test (ASTM D3359-02)" were performed to investigate the adhesion of the copper coating on the specimens [42].

2.4. Anti-Bacterial Activity Test

We used a modified ISO 22196:2011 (Measurement of Antibacterial Activity on Plastics and Other Non-Porous Surfaces) test to characterize the antibacterial properties of the specimens [43,44]. Prior to the experiments, the specimens were sterilized with 75 vol.% ethyl alcohol and UV-light for 30 min. The prepared bacterial suspension (16 µL) was applied to the specimen surfaces. A piece of polymer film, 10 × 10 mm, sterilized with 70% ethanol and dried, was placed on the surface to spread the suspension and to reduce evaporation. The samples were then incubated for 24 h at 35 °C. After the incubation, the specimen and polymer film were vortexed for 1 min with 1 mL of sterilized water containing Tween 80 (20 µL) to remove attached bacteria from the surfaces. The bacterial solution was diluted with fresh LB broth by a factor of 10–10^4. Subsequently, 100 µL of diluted solution was evenly distributed over the surface of the LB agar in petri dishes, followed by incubation at 35 °C for 24 h. Afterwards, the numbers of bacterial colonies were counted to determine the bacterial cell concentration. Each sample type was tested in triplicate. Log reduction was determined by the following equation (Equation (1)) [45,46]:

$$\text{Log reduction} = \log_{10} \frac{A}{B} \quad (1)$$

where A and B refer to the number of bacterial colonies on the control sample and test sample, respectively, after a designated contact time.

2.5. Measurement of Copper Ion Release

The copper ion release from the coating was measured using inductively coupled plasma-atomic emission spectrometry (ICP–AES, Optima 8300, Norwalk, Connecticut, USA). The specimens were immersed in 100 mL of sterilized LB broth, and then, the solution was extracted from samples after 1, 4, 8, and 24 h to analyze for copper release. The copper ion concentration was quantified with a standard ICP ionic solution with different concentrations (from 0.02 to 0.5 ppm) that was used to plot the calibration curve. The element copper was analyzed using an emission wavelength of 325 nm.

3. Results

3.1. Characteristics of Copper Coating by Super-Wetting Properties

3.1.1. Surface Morphology and Cross-Sectional Analysis

Figure 2 indicates the scanning electron microscopy (SEM) images before and after coating. The substrate had a surface with a smooth and even structure (Figure 2a). Our previous work confirmed that laser irradiation melts the metal and causes swelling and spattering, and as a result, the liquid metal accumulates and forms a fine crevice structure (Figure 2b) [38]. Before the coating process, the fine crevice structure formed on all and 48% of the substrate by laser irradiation (Figure 2c,d). After the coating process by the super-spread wetting properties, the copper was coated on the surface (Figure 2e,f). It clearly indicates that the coating film was formed only on the fine crevice structure fabricated by laser irradiation (Figure 2f). This means that we are able to coat copper on the required area through the process of modifying the surface.

Figure 3 shows the cross-sectional images and the energy-dispersive X-ray spectroscopy (EDS) results for copper coating by the superspreading properties. The cross-sectional images reveal that the coating, with approximately 20–40 µm thickness and homogeneous microstructure, was composed of a substrate outer layer. Furthermore, the copper-coating surface had the highest adhesion strength grade of 5B (no detachment of the squares of the lattice), according to ASTM D3359-02, as shown in Table 1 [42]. Furthermore, it clearly shows that copper was coated along a complex surface structure, as shown in Figure 3e,f. This phenomenon results from liquid copper penetrating and spreading through the complex fine crevice structure by capillary action [35,36]. These results provide evidence that a coating can be controlled by the fine crevice surface structure and super-spread wetting properties. Furthermore, additional research is needed for the possibility of controlling the coating thickness according to the surface structure and the copper supplied.

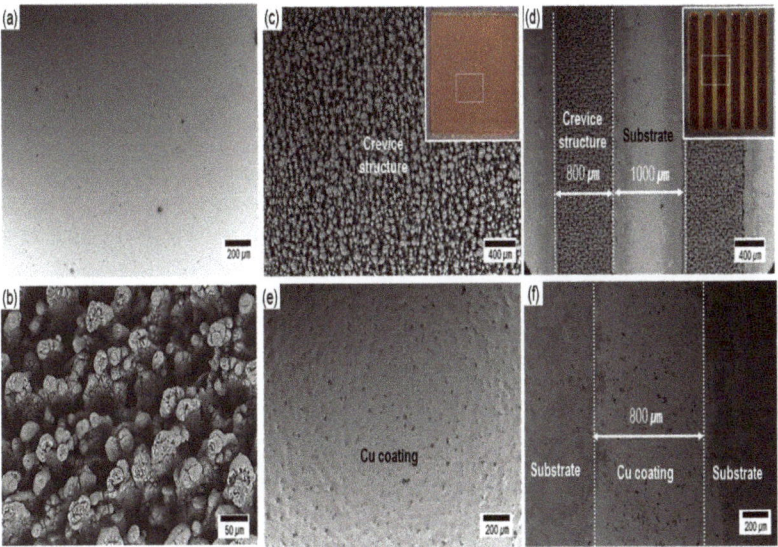

Figure 2. SEM images before (**a**–**d**) and after (**e**,**f**) copper coating. (**a**) as-received carbon steel substrate. (**b**) fine crevice structure by laser irradiation. Uncoated (**c**) all and (**d**) 48% of the substrate. Coated (**e**) all (**f**) 48% of the substrate.

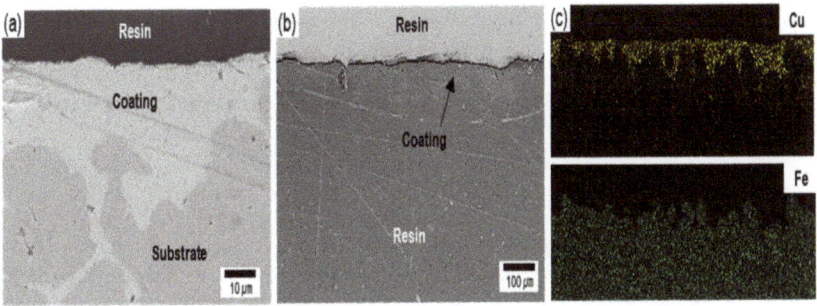

Figure 3. The cross-sectional images at (**a**) high- and (**b**) low-magnification with (**c**) EDS mapping.

Table 1. Adhesion tests according to ASTM D-3359-02 on the copper coating surface.

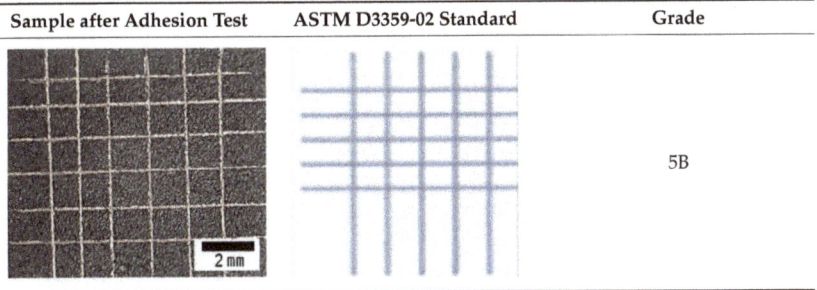

Sample after Adhesion Test	ASTM D3359-02 Standard	Grade
		5B

3.1.2. Phase Analysis

Figure 4 shows the XRD patterns of the substrate and the copper-coated specimens by the super-spread wetting properties in the range of 20°–80°. According to the XRD analysis results of the all- and 48%-copper-coated surfaces, diffraction peaks were observed at 2θ values of 43.39°, 50.49°, and 74.18°, which corresponded to the (111), (200) and (200) planes of the metallic Cu (JCPDS No. 04-0836). In addition to the peaks of metallic Cu, the diffraction peaks at 36.4° and 42.3° corresponded to the (111) and (200) planes of the Cu_2O (JCPDS No. 05-0667), and those at 33.17°, 35.4°, and 38.7° corresponded to (100), (002), and (111) planes of CuO (JCPDS No. 48-1548). Of course, peaks from the crystal phases of α-Fe and Fe_3O_4 related to the substrate also appeared. These results indicate that the substrate with a surface crevice structure was coated with Cu by the super-spread wetting properties. In addition, as predicted by the thermodynamics calculation, the surface structure formed from Cu to Cu, Cu_2O and CuO by oxidation during the fabrication and cooling processes.

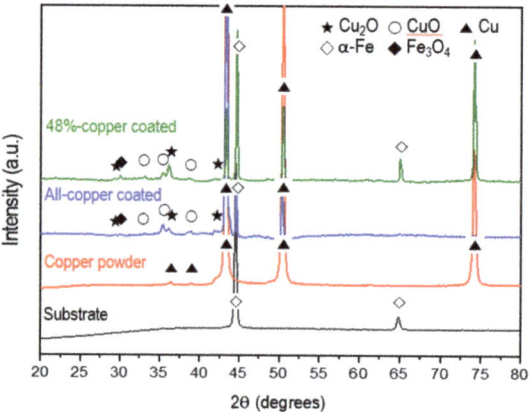

Figure 4. XRD patterns for specimens.

3.1.3. Measurement of Surface Roughness and Wettability

To further identify the surface topology after the coating process by the super-spread wetting properties, the surface roughness was assessed by a 3D laser scanning microscope, as shown in Figure 5. The substrate showed a homogeneous and regular topology, with low peaks (green) and low valleys (yellow) (Figure 5a,c). In contrast, the mountains (red) and valleys (blue) identified on the coated surface revealed a heterogeneous and irregular topology (Figure 5b,d). Table 2 offers, in addition, some objective parameters to determine the surface characteristics of the samples, i.e., average roughness (R_a), peak-to-valley roughness (R_z) and root-mean-square roughness (R_q). Based on statistical analysis, the difference before and after the coating process was obvious. For example, R_a was 0.24 and 6.35 μm before and after coating, respectively. The R_z and R_q parameters, reflecting the local height variations in a surface area, were 9.55 and 0.33 μm before and 80.57 and 7.89 μm after coating, respectively. This result is caused by the super-spread wetting properties of liquid copper through capillary action into the surface with crevice structure fabricated by a laser.

Table 2. Average roughness (R_a), peak-to-valley roughness (R_z) and root-mean-square roughness (R_q) determined by 3D microscope.

	Substrate	Coated Surface
Ra (μm)	0.24 ± 0.03	6.35 ± 0.16
Rz (μm)	9.55 ± 2.35	80.57 ± 3.10
Rq (μm)	0.33 ± 0.03	7.89 ± 0.18

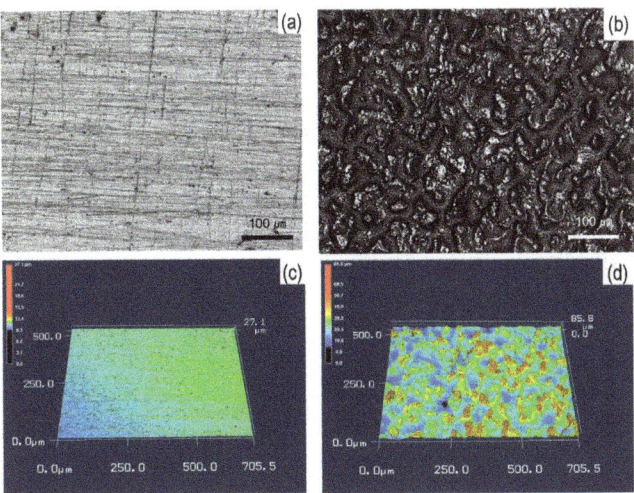

Figure 5. Surface and 3D images for surface topography before and after the coating process; (**a**–**c**) substrate and (**b**–**d**) coated surface.

Contact angle measurements characterized the degree of wettability for the specimens in the LB broth, and representative images and average values are shown in Figure 6. The substrate before the coating process had a contact angle of 61.3° (Figure 6a). After the coating process, the specimen became more wettable, which was indicated by a reduction of the contact angle to 56.5° (Figure 6b). Therefore, the coating process by the super- spread wetting properties led to changes in the wettability (Figure 6c).

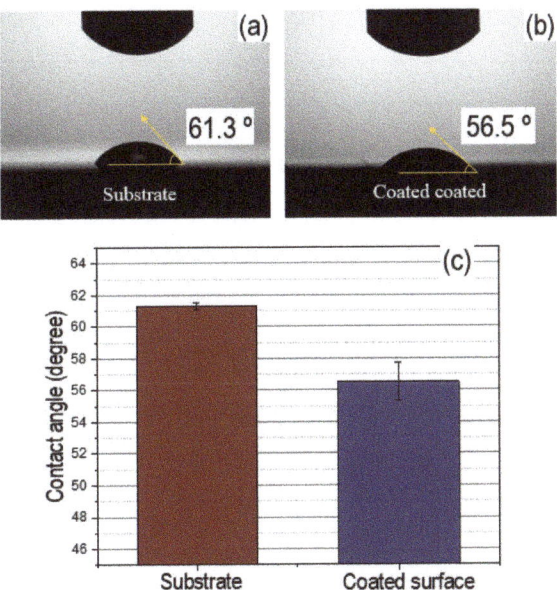

Figure 6. Representative images and average values for contact angle in Luria broth medium; substrate (**a**) coated surface (**b**) average values for the contact angle (**c**).

3.2. Anti-Bacterial Nature of the Cu Coating by Super-Spread Wetting Properties

The antibacterial nature of the specimens coated by the super-spread wetting properties was evaluated by ISO 22196:2011 against *E. coli* and *S. aureus*. Figure 7a shows representative images of bacteria colonies after incubation for *E. coli* and *S. aureus* of the solution collected after contact with the specimen for 24 h. Large amounts of bacterial colonies appeared on the substrate (control) for both types of bacteria. In contrast, in the dish cases, for the all copper-coated samples that were fabricated with the super-spread properties, there were no colonies for both types of bacteria and a similar aspect as the copper plate was even shown. As shown in Figure 7b, the number of bacteria inoculated on the copper-coated surface was reduced to less than 10 CFU/mL after a 24-h incubation period. Therefore, the log reduction value for all copper-coated samples against *E. coli* and *S. aureus* was 4.08 and 4.08, respectively, by Equation (1) (see Figure 7c). The copper-coated surface supported less than 0.01% of both types of bacterial grown on the copper coating by super-spread wetting properties. These results demonstrate that the copper-coated specimens by super-spread wettability have antibacterial properties to both gram-positive and gram-negative bacteria. In addition, on the 48%-copper-coated samples, only a few bacterial colonies were found. The number of bacteria inoculated on the 48%-copper-coated surface was reduced to $(2.1 \pm 0.1) \times 10^3$ CFU/mL of *E. coli* and $(2.6 \pm 0.1) \times 10^3$ CFU/mL of *S. aureus* after a 24-h incubation period (see Figure 7b).

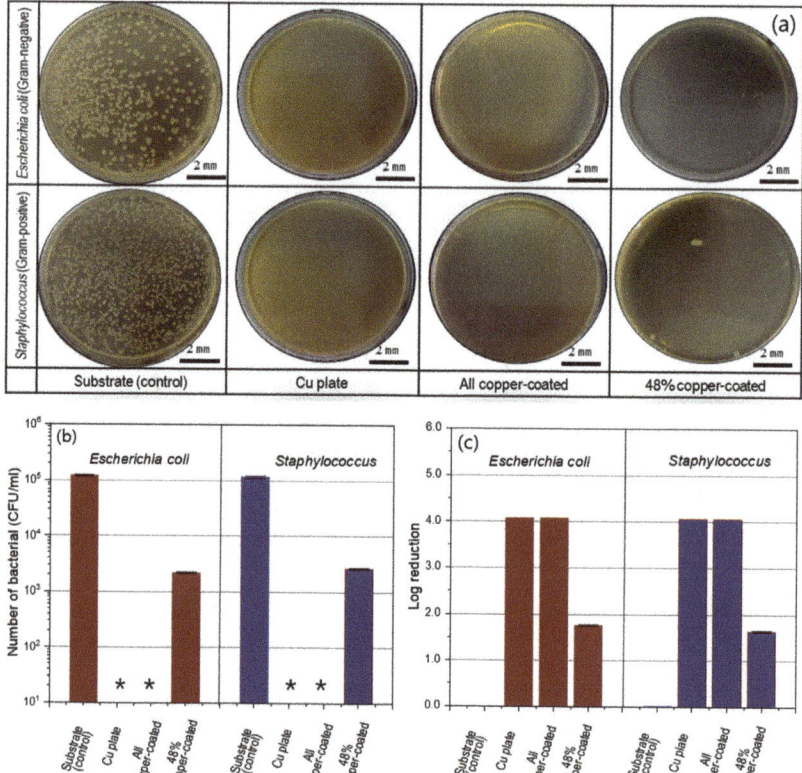

Figure 7. Antibacterial activity of the specimens determined by ISO 22196:2011. (**a**) Images of colonies after 24 h of incubation for *Escherichia coli* and *Staphylococcus* (dilution factor of 10^{-2}), (**b**) quantified number of bacterial (* indicates that no colony was observed, limit of detection < 10 CFU/mL) and (**c**) log reduction in the different groups. The data are expressed as mean ± S.D. of triplicates.

3.3. Copper-Ion Release

Figure 8 summarizes the release of copper ions from the all- and 48%-copper-coated samples versus exposure time in uninoculated LB broth. The copper ion concentration for the all-copper-coated sample increased with the extension of immersion duration time. However, copper emission from the 48%-copper-coated sample proceeded rapidly in the early immersion stage but was relatively slower over time. This was proposed to arise from the difference of the chemical state depending on the exposed area of the copper and the oxide type. After 24 h, the concentration of copper ions was 295 ppb and 100 ppb for the all- and 48%-copper-coated samples, respectively. These data clearly showed that copper ions were released from both specimens.

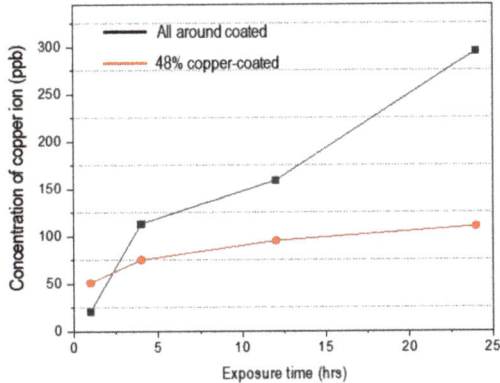

Figure 8. The variation in copper ion release from all-around and 48%-copper-coated in un-inoculated Luria broth medium vs. exposure time.

4. Discussion

Figure 9 presents why the copper coating with the super-spread wetting property is antibacterial, based on the experimental results so far. The coated specimen has an irregular surface structure because liquid copper is wetted to the crevice structure by capillary action using the superwetting property. Several studies found that these micro-sized rough surfaces enhance bacterial adhesion to the surface, described as an anchoring effect [20,47]. Bacteria would contact the rough surface easily composed of copper and copper oxides that have antibacterial performance formed through the coating process. Additionally, the coated surfaces are 56.5° through contact angle measurement, which means that the surface has a hydrophilic character with good wettability. It is known through many studies that when the surface has hydrophilic properties, bacteria can easily attach to the surface [15,16]. These wetting properties can further improve the antibacterial properties of copper compounds known as contact-killing. The influx of copper ions into the cytoplasm would be the key to antibacterial performance in contact-killing [48,49]. In addition, the antimicrobial efficacy of a copper coating is dependent on the number of copper ions released from the surface to the electrolyte. The LB broth used in this study contains chloride, and copper has the property of being dissolved in the form of complex ions in such an environment [50–52]. This phenomenon is related to the breaking of the equilibrium state of the copper surface into a polarized state by chloride. The elution phenomenon of copper from the coating surface can be explained as follows. When the coated surface is exposed to these environments, transitional products ($CuCl_{ads}$) are formed according to the interaction between copper atoms on the coating surface and Cl. Because this product is not stable, it combines with more Cl^- ions and oxidizes into soluble oxidation products $CuCl_2^-$, as shown in Equations (2)–(4) [50–52].

$$Cu+ \rightarrow CuCl^-_{ads} \tag{2}$$

$$CuCl^-_{ads} + Cl^- \rightarrow CuCl_2^- \quad (3)$$

$$Cu + 2Cl^- \rightarrow CuCl_2^- + 2e^- \quad (4)$$

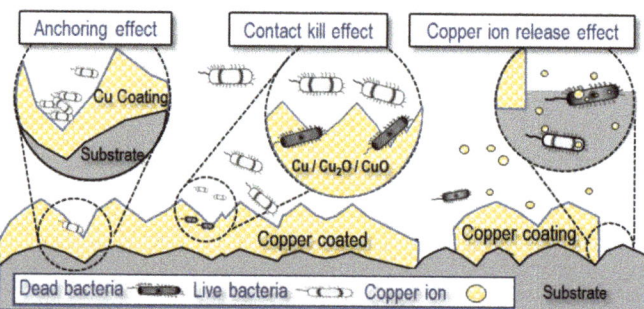

Figure 9. Schematic of the antimicrobial activity mechanism of copper coating specimens with super-spread wetting properties.

It has been reported that when the copper ion concentration in the immersion solution is higher than 0.036 mg/L, the antibacterial rate is more than 99% [21,22]. Our results also indicate that the copper ions concentration in the immersion solution is more than 295 and 110 ppb in the all- and 48%-copper-coated specimens, respectively. For that reason, it is considered that antibacterial properties are shown not only in the specimen coated with copper over the entire area, but also in the specimen in which the base material is partially exposed. The toxicity of copper ions is still unclear but is usually owing to their ability to catalyze Fenton chemistry according to Equation (5) [8,53]. Combined with Equation (6), these reactions can provide a reactive oxygen species that can destroy bacterial cells [53].

$$Cu^+ + H_2O_2 \rightarrow Cu^{2+} + OH^- + OH^\bullet \quad (5)$$

$$H_2O_2 + OH^\bullet \rightarrow H_2O + O_2^- + H^+ \quad (6)$$

5. Conclusions

Using the superwetting property of liquid copper, we have coated copper on carbon steel surfaces and also only on a desired area. In addition, we have discussed the properties of the coating surface and the correlation between antibacterial properties. The results demonstrated that a coating without visible cracks or voids between two metals can be manufactured using the superwetting property of liquid copper. In addition, we confirm that coating using the superwetting property has excellent antibacterial performance. Additionally, it is also interesting that it has excellent antibacterial performance even when coated over only 48% of the area. It is considered that the properties of the copper coating, which include the chemical composition of the surface, high surface roughness, good wettability, and ability for copper ion release, influence the antibacterial properties. Our research presents a new method for copper coating with antibacterial properties in a simple to produce and environmentally friendly way. In addition, we expect that this study can be applied to various base materials requiring antibacterial properties.

Author Contributions: Methodology, B.S., H.K. and M.N.; investigation, B.S. and M.N.; writing—original draft preparation, B.S.; writing—review and editing, B.S., T.T. and H.K.; supervision, T.T.; funding acquisition, Y.M. and T.T. All authors have read and agreed to the published version of the manuscript.

Funding: This research received no external funding.

Institutional Review Board Statement: Not applicable.

Informed Consent Statement: Not applicable.

Data Availability Statement: Data will be made available upon reasonable request.

Acknowledgments: The authors would like to thank Nobumitsu Hirai and Hyojin Kim for their great support and fruitful discussions in the experiment.

Conflicts of Interest: The authors declare no conflict of interest.

References

1. Klevens, R.M.; Edwards, J.R.; Richards, C.L., Jr.; Horan, T.C.; Gaynes, R.P.; Pollock, D.A.; Cardo, D.M. Estimating Health Care-Associated Infections and Deaths in US Hospitals, 2002. *Public Health Rep.* **2007**, *122*, 160–166. [CrossRef]
2. Kramer, A.; Schwebke, I.; Kampf, G. How Long Do Nosocomial Pathogens Persist on Inanimate Surfaces? A Systematic Review. *BMC Infect. Dis.* **2006**, *6*, 1–8. [CrossRef] [PubMed]
3. Saint, S.; Kowalski, C.P.; Banaszak-Holl, J.; Forman, J.; Damschroder, L.; Krein, S.L. The Importance of Leadership in Preventing Healthcare-Associated Infection: Results of a Multisite Qualitative Study. *Infect. Control Hosp. Epidemiol.* **2010**, *31*, 901–907. [CrossRef]
4. Huskins, W.C.; Huckabee, C.M.; O'Grady, N.P.; Murray, P.; Kopetskie, H.; Zimmer, L.; Walker, M.E.; Sinkowitz-Cochran, R.L.; Jernigan, J.A.; Samore, M. Intervention to Reduce Transmission of Resistant Bacteria in Intensive Care. *N. Engl. J. Med.* **2011**, *364*, 1407–1418. [CrossRef] [PubMed]
5. Schmidt, M.G.; von Dessauer, B.; Benavente, C.; Benadof, D.; Cifuentes, P.; Elgueta, A.; Duran, C.; Navarrete, M.S. Copper Surfaces Are Associated with Significantly Lower Concentrations of Bacteria on Selected Surfaces within a Pediatric Intensive Care Unit. *Am. J. Infect. Control* **2016**, *44*, 203–209. [CrossRef]
6. Schmidt, M.G.; Attaway, H.H.; Sharpe, P.A.; John Jr, J.; Sepkowitz, K.A.; Morgan, A.; Fairey, S.E.; Singh, S.; Steed, L.L.; Cantey, J.R. Sustained Reduction of Microbial Burden on Common Hospital Surfaces through Introduction of Copper. *J. Clin. Microbiol.* **2012**, *50*(7), 2217–2223. [CrossRef]
7. Salgado, C.D.; Sepkowitz, K.A.; John, J.F.; Cantey, J.R.; Attaway, H.H.; Freeman, K.D.; Sharpe, P.A.; Michels, H.T.; Schmidt, M.G. Copper Surfaces Reduce the Rate of Healthcare-Acquired Infections in the Intensive Care Unit. *Infect. Control Hosp. Epidemiol.* **2013**, *34*, 479–486. [CrossRef]
8. Mathews, S.; Hans, M.; Mücklich, F.; Solioz, M. Contact Killing of Bacteria on Copper Is Suppressed If Bacterial-Metal Contact Is Prevented and Is Induced on Iron by Copper Ions. *Appl. Environ. Microbiol.* **2013**, *79*, 2605–2611. [CrossRef] [PubMed]
9. Santo, C.E.; Quaranta, D.; Grass, G. Antimicrobial Metallic Copper Surfaces Kill Staphylococcus Haemolyticus via Membrane Damage. *Microbiologyopen* **2012**, *1*, 46–52. [CrossRef] [PubMed]
10. Warnes, S.L.; Caves, V.; Keevil, C.W. Mechanism of Copper Surface Toxicity in Escherichia Coli O157: H7 and Salmonella Involves Immediate Membrane Depolarization Followed by Slower Rate of DNA Destruction Which Differs from That Observed for Gram-positive Bacteria. *Environ. Microbiol.* **2012**, *14*, 1730–1743. [CrossRef] [PubMed]
11. do Espírito Santo, A.P.; Perego, P.; Converti, A.; Oliveira, M.N. Influence of Milk Type and Addition of Passion Fruit Peel Powder on Fermentation Kinetics, Texture Profile and Bacterial Viability in Probiotic Yoghurts. *LWT* **2012**, *47*, 393–399. [CrossRef]
12. Dalecki, A.G.; Crawford, C.L.; Wolschendorf, F. Copper and Antibiotics: Discovery, Modes of Action, and Opportunities for Medicinal Applications. *Adv. Microb. Physiol.* **2017**, *70*, 193–260.
13. Chen, S.; Li, Y.; Cheng, Y.F. Nanopatterning of Steel by One-Step Anodization for Anti-Adhesion of Bacteria. *Sci. Rep.* **2017**, *7*, 1–9. [CrossRef]
14. Zhang, X.; Yang, C.; Xi, T.; Zhao, J.; Yang, K. Surface Roughness of Cu-Bearing Stainless Steel Affects Its Contact-Killing Efficiency by Mediating the Interfacial Interaction with Bacteria. *ACS Appl. Mater. Interfaces* **2021**, *13*, 2303–2315. [CrossRef]
15. Xu, J.; Sun, T.T.; Jiang, S.; Munroe, P.; Xie, Z.H. Antimicrobial and Biocorrosion-Resistant MoO_3-SiO_2 Nanocomposite Coating Prepared by Double Cathode Glow Discharge Technique. *Appl. Surf. Sci.* **2018**, *447*, 500–511. [CrossRef]
16. Wang, G.; Weng, D.; Chen, C.; Chen, L.; Wang, J. Influence of TiO_2 Nanostructure Size and Surface Modification on Surface Wettability and Bacterial Adhesion. *Colloids Interface Sci. Commun.* **2020**, *34*, 100220. [CrossRef]
17. Dürr, H. Influence of Surface Roughness and Wettability of Stainless Steel on Soil Adhesion, Cleanability and Microbial Inactivation. *Food Bioprod. Process.* **2007**, *85*, 49–56. [CrossRef]
18. Katsikogianni, M.; Missirlis, Y.F. Concise Review of Mechanisms of Bacterial Adhesion to Biomaterials and of Techniques Used in Estimating Bacteria-Material Interactions. *Eur Cell Mater* **2004**, *8*, 37–57. [CrossRef] [PubMed]
19. Medilanski, E.; Kaufmann, K.; Wick, L.Y.; Wanner, O.; Harms, H. Influence of the Surface Topography of Stainless Steel on Bacterial Adhesion. *Biofouling* **2002**, *18*, 193–203. [CrossRef]
20. Jia, Z.; Xiu, P.; Li, M.; Xu, X.; Shi, Y.; Cheng, Y.; Wei, S.; Zheng, Y.; Xi, T.; Cai, H. Bioinspired Anchoring AgNPs onto Micro-Nanoporous TiO_2 Orthopedic Coatings: Trap-Killing of Bacteria, Surface-Regulated Osteoblast Functions and Host Responses. *Biomaterials* **2016**, *75*, 203–222. [CrossRef]
21. Wang, X.; Dong, H.; Liu, J.; Qin, G.; Chen, D.; Zhang, E. In Vivo Antibacterial Property of Ti-Cu Sintered Alloy Implant. *Mater. Sci. Eng. C* **2019**, *100*, 38–47. [CrossRef] [PubMed]
22. Liu, J.; Li, F.; Liu, C.; Wang, H.; Ren, B.; Yang, K.; Zhang, E. Effect of Cu Content on the Antibacterial Activity of Titanium–Copper Sintered Alloys. *Mater. Sci. Eng. C* **2014**, *35*, 392–400. [CrossRef]

23. Horton, D.J.; Ha, H.; Foster, L.L.; Bindig, H.J.; Scully, J.R. Tarnishing and Cu Ion Release in Selected Copper-Base Alloys: Implications towards Antimicrobial Functionality. *Electrochim. Acta* **2015**, *169*, 351–366. [CrossRef]
24. Anita, S.; Ramachandran, T.; Rajendran, R.; Koushik, C.V.; Mahalakshmi, M. A Study of the Antimicrobial Property of Encapsulated Copper Oxide Nanoparticles on Cotton Fabric. *Text. Res. J.* **2011**, *81*, 1081–1088. [CrossRef]
25. Gabbay, J.; Borkow, G.; Mishal, J.; Magen, E.; Zatcoff, R.; Shemer-Avni, Y. Copper Oxide Impregnated Textiles with Potent Biocidal Activities. *J. Ind. Text.* **2006**, *35*, 323–335. [CrossRef]
26. Mitra, D.; Kang, E.T.; Neoh, K.G. Antimicrobial Copper-Based Materials and Coatings: Potential Multifaceted Biomedical Applications. *ACS Appl. Mater. Interfaces* **2020**, *12*, 21159–21182. [CrossRef] [PubMed]
27. Saeki, I.; Harada, T.; Tanaka, I.; Ando, T.; Gan, L.; Murakami, H. Electroplating of Copper on Low Carbon Steel from Alkaline Citrate Complex Baths. *ISIJ Int.* **2020**, *60*, 2031–2037. [CrossRef]
28. Dash, R.R.; Gaur, A.; Balomajumder, C. Cyanide in Industrial Wastewaters and Its Removal: A Review on Biotreatment. *J. Hazard. Mater.* **2009**, *163*, 1–11. [CrossRef]
29. Varghese, S.; ElFakhri, S.O.; Sheel, D.W.; Sheel, P.; Eric Bolton, F.J.; Foster, H.A. Antimicrobial Activity of Novel Nanostructured Cu-SiO2 Coatings Prepared by Chemical Vapour Deposition against Hospital Related Pathogens. *AMB Express* **2013**, *3*, 1–8. [CrossRef] [PubMed]
30. Bharadishettar, N.; Bhat, K.U.; Panemangalore, D.B. Coating Technologies for Copper Based Antimicrobial Active Surfaces: A Perspective Review. *Metals* **2021**, *11*, 711. [CrossRef]
31. Tian, J.; Xu, K.; Hu, J.; Zhang, S.; Cao, G.; Shao, G. Durable Self-Polishing Antifouling Cu-Ti Coating by a Micron-Scale Cu/Ti Laminated Microstructure Design. *J. Mater. Sci. Technol.* **2021**, *79*, 62–74. [CrossRef]
32. Hannula, P.M.; Masquelier, N.; Lassila, S.; Aromaa, J.; Janas, D.; Forsén, O.; Lundström, M. Corrosion Behaviour of Cast and Deformed Copper-Carbon Nanotube Composite Wires in Chloride Media. *J. Alloys Compd.* **2018**, *746*, 218–226. [CrossRef]
33. Daehn, K.E.; Serrenho, A.C.; Allwood, J. Preventing Wetting Between Liquid Copper and Solid Steel: A Simple Extraction Technique. *Metall. Mater. Trans. B* **2019**, *50*, 1637–1651. [CrossRef]
34. Eustathopoulos, N.; Nicholas, M.G.; Drevet, B. *Wettability at High Temperatures*; Elsevier: Amsterdam, The Netherlands, 1999.
35. Takahira, N.; Tanaka, T.; Hara, S.; Lee, J. Unusual Wetting of Liquid Metals on Iron Substrate with Oxidized Surface in Reduced Atmosphere. *Mater. Trans.* **2005**, *46*, 3008–3014. [CrossRef]
36. Fukuda, A.; Matsukawa, H.; Goto, H.; Suzuki, M.; Nakamoto, M.; Matsumoto, R.; Utsunomiya, H.; Tanaka, T. Metal–Metal Joining by Unusual Wetting on Surface Fine Crevice Structure Formed by Laser Treatment. *Mater. Trans.* **2015**, *56*, 1852–1856. [CrossRef]
37. Takahira, N.; Yoshikawa, T.; Tanaka, T.; Holappa, L. Wettability of Liquid in and Bi on Flat and Porous Solid Iron Substrate. *Mater. Trans.* **2007**, *48*, 2708–2711. [CrossRef]
38. Yeon, J.; Ni, P.; Nakamoto, M.; Tanaka, T. In Situ Observations of the Formation of Surface Fine Crevice Structures Created by Laser Irradiation. *Mater. Trans.* **2021**, *62*, 261–270. [CrossRef]
39. Siboniso, V.; Yeon, J.; Grozescu, C.; Goto, H.; Nakamoto, M.; Matsumoto, R.; Utsunomiya, H.; Tanaka, T. Mechanism of the Unusual Wetting of a Surface Fine Crevice Structure Created by Laser Irradiation. *Mater. Trans.* **2017**, *58*, 1227–1230. [CrossRef]
40. Nakamoto, M.; Fukuda, A.; Pinkham, J.; Vilakazi, S.; Goto, H.; Matsumoto, R.; Utsunomiya, H.; Tanaka, T. Joining of Copper Plates by Unusual Wetting with Liquid Tin and Tin–Lead Solder on "Surface Fine Crevice Structure". *Mater. Trans.* **2016**, *57*, 973–977. [CrossRef]
41. Yeon, J.; Ishida, Y.; Nakamoto, M.; Tanaka, T. Joining of Metals by Super-Spread Wetting on Surface Fine Crevice Structure Created by Reduction-Sintering Copper Oxide Powder. *Mater. Trans.* **2018**, *59*, 1192–1197. [CrossRef]
42. Astm, D. *3359-02: Standard Test Methods for Measuring Adhesion by Tape Test*; ASTM Int.: West Conshohocken, PA, USA, 2002.
43. Uhm, S.-H.; Song, D.-H.; Kwon, J.-S.; Lee, S.-B.; Han, J.-G.; Kim, K.-M.; Kim, K.-N. E-Beam Fabrication of Antibacterial Silver Nanoparticles on Diameter-Controlled TiO$_2$ Nanotubes for Bio-Implants. *Surf. Coatings Technol.* **2013**, *228*, S360–S366. [CrossRef]
44. International Organization for Standardization. *Measurement of Antibacterial Activity on Plastics and Other Non-Porous Surfaces*; International Organization for Standardization: Geneva, Switzerland, 2011.
45. Ando, Y.; Miyamoto, H.; Noda, I.; Sakurai, N.; Akiyama, T.; Yonekura, Y.; Shimazaki, T.; Miyazaki, M.; Mawatari, M.; Hotokebuchi, T. Calcium Phosphate Coating Containing Silver Shows High Antibacterial Activity and Low Cytotoxicity and Inhibits Bacterial Adhesion. *Mater. Sci. Eng. C* **2010**, *30*, 175–180. [CrossRef]
46. Zhang, H.; Oyanedel-Craver, V. Comparison of the Bacterial Removal Performance of Silver Nanoparticles and a Polymer Based Quaternary Amine Functiaonalized Silsesquioxane Coated Point-of-Use Ceramic Water Filters. *J. Hazard. Mater.* **2013**, *260*, 272–277. [CrossRef] [PubMed]
47. Puckett, S.D.; Taylor, E.; Raimondo, T.; Webster, T.J. The Relationship between the Nanostructure of Titanium Surfaces and Bacterial Attachment. *Biomaterials* **2010**, *31*, 706–713. [CrossRef]
48. Nieto-Juarez, J.I.; Pierzchła, K.; Sienkiewicz, A.; Kohn, T. Inactivation of MS2 Coliphage in Fenton and Fenton-like Systems: Role of Transition Metals, Hydrogen Peroxide and Sunlight. *Environ. Sci. Technol.* **2010**, *44*, 3351–3356. [CrossRef] [PubMed]
49. Hans, M.; Erbe, A.; Mathews, S.; Chen, Y.; Solioz, M.; Mücklich, F. Role of Copper Oxides in Contact Killing of Bacteria. *Langmuir* **2013**, *29*, 16160–16166. [CrossRef]
50. Otmačić, H.; Telegdi, J.; Papp, K.; Stupnišek-Lisac, E. Protective Properties of an Inhibitor Layer Formed on Copper in Neutral Chloride Solution. *J. Appl. Electrochem.* **2004**, *34*, 545–550. [CrossRef]

51. Sherif, E.M.; Park, S.-M. 2-Amino-5-Ethyl-1, 3, 4-Thiadiazole as a Corrosion Inhibitor for Copper in 3.0% NaCl Solutions. *Corros. Sci.* **2006**, *48*, 4065–4079. [CrossRef]
52. Rui, D.; Li, X.; Jia, W.; Li, W.; Xiao, W.; Gui, T. Releasing Kinetics of Dissolved Copper and Antifouling Mechanism of Cold Sprayed Copper Composite Coatings for Submarine Screen Doors of Ships. *J. Alloys Compd.* **2018**, *763*, 525–537. [CrossRef]
53. Van De Guchte, M.; Serror, P.; Chervaux, C.; Smokvina, T.; Ehrlich, S.D.; Maguin, E. Stress Responses in Lactic Acid Bacteria. *Antonie Van Leeuwenhoek* **2002**, *82*, 187–216. [CrossRef]

Lithium-Ion-Conducting Ceramics-Coated Separator for Stable Operation of Lithium Metal-Based Rechargeable Batteries

Ryo Shomura [1,2], Ryota Tamate [2,3,*] and Shoichi Matsuda [2,3,*]

1. Department of Creative Engineering, National Institute of Technology, Tsuruoka College, 104 Sawada, Inooka, Tsuruoka 997-8511, Japan; SHOMURA.Ryo@nims.go.jp
2. Center for Green Research on Energy and Environmental Materials, National Institute for Material Science, 1-1 Namiki, Tsukuba 305-0044, Japan
3. NIMS-SoftBank Advanced Technologies Development Center, National Institute for Material Science, 1-1 Namiki, Tsukuba 305-0044, Japan
* Correspondence: TAMATE.Ryota@nims.go.jp (R.T.); MATSUDA.Shoichi@nims.go.jp (S.M.)

Abstract: Lithium metal anode is regarded as the ultimate negative electrode material due to its high theoretical capacity and low electrochemical potential. However, the significantly high reactivity of Li metal limits the practical application of Li metal batteries. To improve the stability of the interface between Li metal and an electrolyte, a facile and scalable blade coating method was used to cover the commercial polyethylene membrane separator with an inorganic/organic composite solid electrolyte layer containing lithium-ion-conducting ceramic fillers. The coated separator suppressed the interfacial resistance between the Li metal and the electrolyte and consequently prolonged the cycling stability of deposition/dissolution processes in Li/Li symmetric cells. Furthermore, the effect of the coating layer on the discharge/charge cycling performance of lithium-oxygen batteries was investigated.

Keywords: lithium metal; separator; lithium-ion-conducting ceramics; lithium battery; lithium-oxygen battery

1. Introduction

With the growing demands for electric vehicles and renewable energy, there is a strong need to further increase the energy density of current lithium-ion batteries. Lithium metal is often called an ultimate anode material due to its high theoretical capacity (3860 mAh g^{-1}) and low electrochemical potential (−3.04 V vs. the standard hydrogen electrode) [1–3]. However, the highly reactive lithium metal also leads to continuous electrolyte decomposition, which results in the formation of a thick solid electrolyte interface (SEI) and the accumulation of dead lithium during charging and discharging processes. These processes cause significant degradation of cycling performance. In addition, the formation and growth of lithium dendrites could induce short-circuit and thermal runaway of cells, posing a serious safety concern for the commercialization of lithium metal batteries.

For the practical application of lithium metal anode, there have been many recent attempts to suppress the formation and growth of dead lithium and lithium dendrites. Examples include the development of new liquid electrolytes [4–8], electrolyte additives [9,10], functional separators [11–15], organic and inorganic solid-state electrolytes [16–21], artificial SEI layers [22–27], and 3D anode structures [28–31]. Among them, the modification of the commercial separator is a promising and scalable strategy for realizing lithium metal batteries with high energy density, such as Li-sulfur and Li-oxygen batteries. Ceramic particles including Al_2O_3, SiO_2, and TiO_2 have often been coated onto polyethylene (PE) and polypropylene membrane separators to improve the mechanical strength, thermal stability, and wettability [32–34]. However, a coat of insulating ceramic fillers generally increases the interfacial resistance between the electrolyte and the electrode, which sometimes degrades the cell performance.

One promising way to overcome this problem is to replace the insulating ceramic fillers with Li-ion-conducting solid electrolytes [35–37]. In this study, a facile and scalable blade coating method is used to apply an inorganic/organic composite solid electrolyte layer to the commercial PE membrane separator in order to enhance the stability at the electrolyte–Li metal interface. This solid electrolyte layer is composed of a doped lithium aluminum titanium phosphate (LATP) glass ceramic powder (LICGC from Ohara Inc., Kanagawa, Japan), polyethylene oxide (PEO), and a Li salt (LiTFSI). The PEO/LiTFSI layer acts as the binder of the LICGC particles, as well as the lithium conducting polymer electrolyte layer. The electrochemical performance of the LICGC/PEO/LiTFSI-coated PE separator is evaluated using a Li/Li symmetric cell configuration. The coated separator shows an improved interfacial resistance between the electrolyte and the Li metal, and consequently, it prolongs stable cycling during the Li deposition/dissolution processes. Furthermore, the effect of the composite coating layer on the charge/discharge performance of lithium-oxygen batteries is investigated.

2. Materials and Methods

2.1. Materials

The powdered lithium-ion-conducting glass-ceramics (LICGC ^{TM}, average particle diameter 400 nm) were obtained from Ohara Inc. (Kanagawa, Japan) and used as received. PEO (average Mv ~200,000) and lithium bromide (LiBr) were purchased from Sigma-Aldrich (St. Louis, MO, USA). Tetraethylene glycol dimethyl ether (TEGDME, water content < 10 ppm), lithium bis (trifluoromethanesulfonyl) imide (LiTFSI), and lithium nitrate ($LiNO_3$) were purchased from Kishida Chemical Co., Ltd. (Osaka, Japan) $LiNO_3$ and LiBr were dried at 120 °C under vacuum before use. Acetonitrile was purchased from Wako Pure Chemicals (Osaka, Japan). In all experiments, the electrolyte was a solution of 0.5 M LiTFSI, 0.5 M $LiNO_3$, and 0.2 M LiBr in TEGDME.

2.2. Preparation and Characterization of LICGC/PEO/LiTFSI-Coated Separator

First, PEO (2.0 g) and LiTFSI (1.0 g) were dissolved in acetonitrile (7.0 g). Then, 1.0 g of LICGC was added to 1.0 g of the solution and mixed using a conditioning mixer (THINKY AR-100, THINKY, Tokyo, Japan). The viscosity of the slurry was adjusted by further adding acetonitrile. The solid content of the resulting slurry was ca. 60 wt%. This slurry was blade-coated on one side of the PE separator (W-SCOPE Corporation, Tokyo, Japan), dried overnight at room temperature, and then further dried at 50 °C under vacuum overnight. The mass loading of the coating layer is approximately 0.5 mg/cm^2. The morphology of the separators was characterized by scanning electron microscopy (SEM; JSM-7800F, JEOL, Tokyo, Japan) and energy dispersive spectroscopy (EDS; Oxford detector, Oxford Instruments, Abingdon, Oxon, UK). The cross-sectional SEM sample was prepared using the focused ion beam technique (FIB; SMF-200, Hitachi, Tokyo, Japan). The contact angle measurements were performed with a Drop Master DM 300 (Kyowa Interface Science Co., Ltd. Saitama, Japan). The chemical change of the samples was analyzed by XPS (Axis Ultra, Kratos Analytical Co., Trafford Park, Manchester, UK) with monochromated Al Kα X-rays (hν = 1486.6 eV).

2.3. Assembly and Electrochemical Measurements of Li/Li Symmetric Cell

The cells were assembled in an argon-filled glovebox (UNICO, Ibaraki, Japan). The Li/Li symmetric cells consisted of two Li metal discs (diameter: 16 mm; thickness: 0.4 mm; Honjo Metal Co., Ltd., Osaka, Japan) separated by two PE separators (diameter: 19.5 mm). The amount of electrolyte in each cell was 40 µL. When using the LICGC/PEO/LiTFSI-coated PE separator, the coated side was in contact with the Li metal. Prior to the cycling tests, electrochemical impedance spectroscopy analysis was carried out using a VMP3 potentio/galvanostat (Bio-Logic Science Instruments, Grenoble, France) at a perturbation amplitude of 15 mV over a frequency range of 10^6–10^0 Hz at room temperature. Galvanostatic cycling tests of the Li/Li symmetric cells were carried out at 30 °C with cut-off voltages of −1.0/1.0 V vs. Li/Li$^+$ using a charge/discharge system (HJ1010SM8C; Hokuto Denko Co. Ltd., Tokyo, Japan). Initially, the cells were conditioned at a current density of 0.1 mA/cm^2 for 1 h with three cycles, 0.1 mA/cm^2 for 10 h with one cycle, and 0.2 mA/cm^2 for 10 h with one cycle. Subsequently, the charge/discharge tests were performed at a current density of 0.4 mA/cm^2. The Li surface after cycling was observed by SEM (VE-9800, Keyence, Osaka, Japan).

2.4. Preparation of Porous Carbon Positive Electrode

A homogeneous slurry was prepared using 75 wt% of Ketjen black (KB; Lion Specialty Chemicals Co., Ltd., Tokyo, Japan, EC600J), 5 wt% of single-walled carbon nanotubes (OCSiAl, TUBALL, Luxembourg, average diameter: 1.6 nm, average length: 5 μm), 5 wt% of carbon fiber (Nippon Polymer Sangyo Co., Ltd., Osaka, Japan, CF-N, average fiber diameter: 6 μm, average length: 3 mm), 15 wt% of PAN, and NMP as a solvent. This slurry was blade-coated, and the prepared sheet sample was immersed in methanol. After drying at 80 °C for 10 h, the sample was treated at 230 °C for 3 h in the atmosphere and 1050 °C for 3 h in N$_2$ with a rate of 10 °C/min and a gas flow rate of 800 mL/min.

2.5. Assembly and Discharge/Charge Performance Test of Lithium-Oxygen Cell

The lithium-oxygen cells were fabricated in a dry room (water content < 10 ppm) by stacking the lithium metal foil (20 mm × 20 mm × 0.1 mm, Honjo Metal Co., Ltd., Osaka, Japan), the separator (22 mm × 22 mm × 0.02 mm), the KB-based carbon electrode (20 mm × 20 mm), and the gas-diffusion layer consisting of an array of carbon fibers (overall thickness 110 μm, fiber diameter ~10 μm, TGP-H-030, Toray, Tokyo, Japan). For the electrolyte injection into carbon electrodes, the vacuum impregnation method was adopted. In addition, the electrolyte of 2.5 μL/cm^2 was dropped into the separator. The confining pressure of the cell was controlled at approximately 100 kPa. Repeated discharge/charge test was performed (TOSCAT, Toyo System Co., Ltd., Fukushima, Japan) with a capacity limitation of 4.0 mAh/cm^2 and cutoff voltage of 2.0 V/4.5 V. The current density during discharge and charge was set to 0.4 and 0.2 mA/cm^2, respectively.

3. Results and Discussion

3.1. Morphology of Separator Surface

The organic/inorganic composite solid electrolyte layer was coated on the surface of a commercial PE membrane separator by a simple blade coating method, using an acetonitrile-based slurry containing LICGC particles, PEO, and LiTFSI (LICGC/PEO/LiTFSI = 10/2/1, w/w/w). SEM images of the as-provided PE separator and the LICGC/PEO/LiTFSI-coated PE separator (LICGC/PEO/LiTFSI-PE) are shown in Figure 1a–c. The surface image of coated separator shows uniform distribution of LICGC particles on the separator surface (Figure 1b). In addition, cross-sectional SEM and EDS images confirmed that the formed LICGC/PEO/LiTFSI composite layer was a few microns in thickness (Figure 1c–i). These results indicate that the scalable blade coating approach was able to fabricate the composite electrolyte layer on the PE separator without the aggregation of the ceramic particles. Furthermore, the contact angle measurements for the PE and the LICGC/PEO/LiTFSI-PE separators indicated that the wettability against the electrolyte was significantly improved by introducing the coating layer (Figure 2).

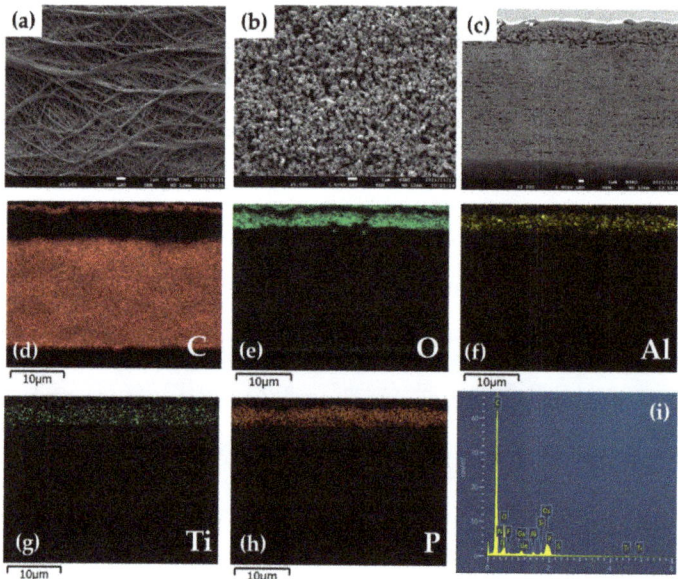

Figure 1. SEM and EDS images of the PE separator with and without coating. Surface SEM images of (**a**) the as-provided PE separator and (**b**) the LICGC/PEO/LiTFSI-coated PE separator. (**c**) Cross-sectional SEM image of the LICGC/PEO/LiTFSI-PE separator. (**d–i**) EDS mapping of the LICGC/PEO/LiTFSI-PE separator.

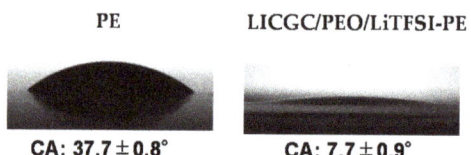

PE   LICGC/PEO/LiTFSI-PE

CA: 37.7 ± 0.8°   **CA: 7.7 ± 0.9°**

Figure 2. Contact angle measurements of the PE and the LICGC/PEO/LiTFSI-PE separators against the electrolyte.

3.2. Impedance Measurements of Li/Li Symmetric Cells

The electrochemical performance of the PE and LICGC/PEO/LiTFSI-PE separators was evaluated using the Li/Li symmetric cell configuration. To ensure a buffer layer between the Li metal electrode and the electrolyte solution, the coated side of the two LICGC/PEO/LiTFSI-PE separators was brought into contact with the Li metal electrode (Figure 3). Figure 4 shows the Nyquist plots of the impedance spectra for the Li/Li cells with PE and LICGC/PEO/LiTFSI-PE separators. Because the coated composite layer was only a few microns thick, the slight difference in thickness between the coated and uncoated PE separators only had a minimal effect on the impedance spectrum. In the impedance spectrum, a high-frequency intercept is related to the bulk resistance of the electrolyte layer, while the depressed semicircular part is considered to be the interfacial resistance between the electrolyte and the Li metal electrode. The cell using the uncoated PE separator showed a bulk resistance of approximately 17 Ω, whereas this value was approximately 25 Ω for the cell using the LICGC/PEO/LiTFSI-PE separator. The slightly higher bulk resistance after coating the separator with LICGC/PEO/LiTFSI could be attributed to the slower diffusion of Li-ion in the composite layer than that in the bulk electrolyte solution. In contrast, the interfacial resistance of the Li/Li cell was significantly suppressed when using the coated separator. This indicates that the inorganic/organic LICGC/PEO/LiTFSI composite electrolyte layer serves as an interfacial buffer to protect the electrolyte solution from directly contacting the Li metal. As a result, the formation of a high-resistance interfacial layer was suppressed, which led to a reduced electrode–electrolyte interfacial resistance.

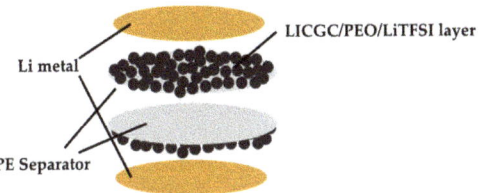

Figure 3. Configuration of the Li/Li symmetric cell using the LICGC/PEO/LiTFSI-PE separator.

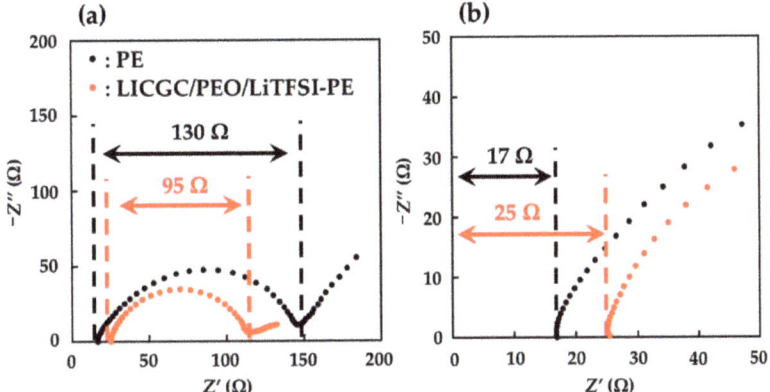

Figure 4. Nyquist plots of the impedance spectra for Li/Li symmetric cells using PE and LICGC/PEO/LiTFSI-PE separators. (**a**) Overview of the impedance spectra and (**b**) expanded spectra in the high-frequency region.

3.3. Li Deposition/Dissolution in Li/Li Symmetric Cells

Figure 5 compares the cycling performance of the Li deposition/dissolution process in the Li/Li symmetric cells using the two separators. The voltage drop that occurred at the first cycle at a current density of 0.4 mA/cm^2 in Figure 5a might be attributed to the irreversible reaction of the native SEI of the lithium metal. For the cell with PE separator, a gradual increase in the cell overpotential was observed after approximately 400 h (20 cycles) at a current density of 0.4 mA/cm^2 and an areal capacity of 4.0 mAh/cm^2 (Figure 5a). This change could be ascribed to the undesired reaction of the electrolyte solution with Li metal and the subsequent formation of a highly resistive SEI. On the other hand, such an increase in cell overpotential was not observed when using the LICGC/PEO/LiTFSI-PE separator even after 800 h (40 cycles), indicating higher stability during the Li deposition/dissolution cycles (Figure 5b). We attribute this to the suppression of electrolyte decomposition at the Li metal surface, as well as the formation of a less resistive SEI due to the presence of the organic/inorganic LICGC/PEO/LiTFSI layer. The latter is consistent with the reduced interfacial resistance of the pristine Li/Li cell using the LICGC/PEO/LiTFSI-PE separator (Figure 4a). In addition, the higher mechanical strength of the composite layer could help suppress the formation and growth of dendritic Li during Li deposition/dissolution [38–40]. Another possible reason for the improved cycling stability is that a large amount of Li$^+$-conducting LICGC particles in the composite layer might regulate Li$^+$ diffusion and homogenize Li$^+$ flux at the Li metal interface [41]. The excellent long-term cycling performance of the Li/Li symmetric cell confirms the effectiveness of the facile coating strategy of an inorganic/organic solid electrolyte layer for stabilizing Li metal electrodes during battery operation. It also should be noted that LICGC/PEO/LiTFSI-PE separator showed no clear chemical change even after repeated Li deposition/dissolution reaction, which was confirmed by XPS analysis (Figure 5c). In particular, there is a concern that Ti^{+4} in LICGC is to be reduced to Ti^{+3} by contacting with metallic lithium, which largely diminishes the Li-ion conductivity of LICGC. However, the results of XPS analysis clearly revealed that the Ti ion in LICGC remained as Ti^{+4}. These results suggest the high stability of the LICGC/PEO/LiTFSI-coated layer.

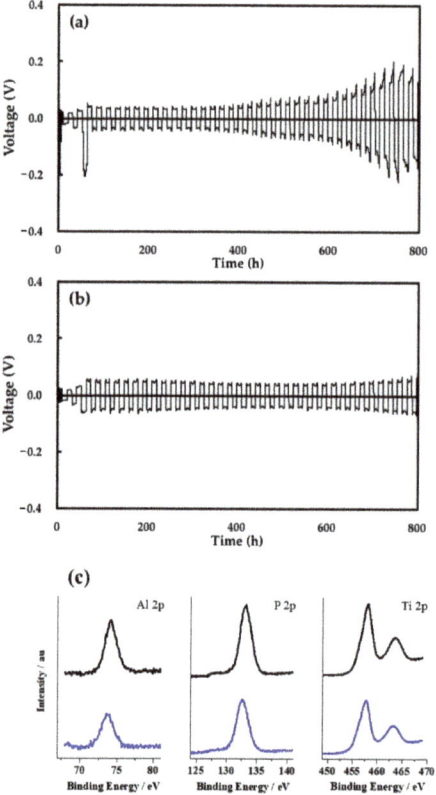

Figure 5. (a,b) Cycling performance of the Li deposition/dissolution process for the Li/Li symmetric cells using (a) the PE separator and (b) the LICGC/PEO/LiTFSI-PE separator. (c) XPS analysis of the LICGC/PEO/LiTFSI-coated PE separator before (black) and after (blue) 10 cycles of Li deposition/dissolution reaction.

3.4. Morphology of Separator Surface after Li Deposition/Dissolution Test

SEM observation was used to investigate the surface morphology of the Li metal electrode after the Li deposition/dissolution cycling test (Figure 6). In the cell using the PE separator, the Li surface after cycling showed a rough and uneven structure (Figure 6a,b). This would be related to unstable Li deposition/dissolution during cycling, owing to the formation of a highly resistive SEI layer. On the other hand, the Li surface in the cell using LICGC/PEO/LiTFSI-PE separator showed a relatively uniform morphology (Figure 6c,d). Although this Li surface still contained pores, the overall structure was denser with fewer pores than the case using the PE separator. Therefore, the LICGC/PEO/LiTFSI composite buffer layer between the Li metal electrode and the electrolyte solution likely suppressed the deposition of heterogeneous and highly porous Li during the Li deposition/dissolution reaction. This result is also consistent with the improved cycling stability of the Li/Li symmetric cell using the coated separator (Figure 5b). Thus, introducing the organic/inorganic solid electrolyte layer to the Li metal surface led to uniform Li deposition and consequently stable long-term cycling performance.

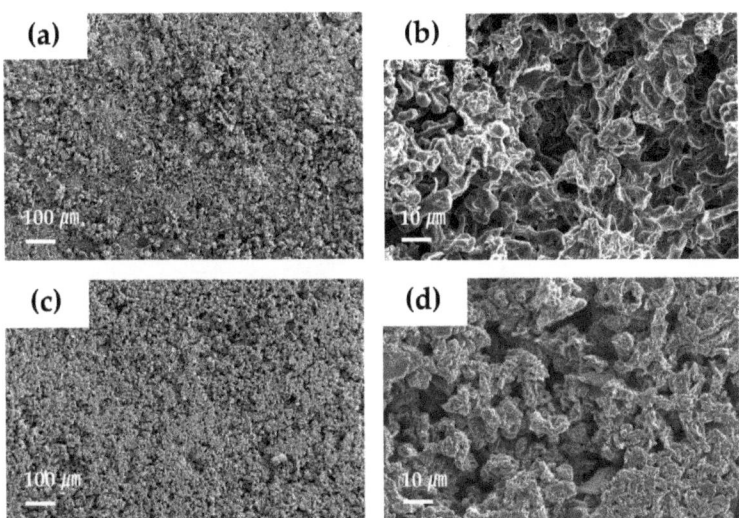

Figure 6. SEM images of the Li metal electrode surface after the Li deposition/dissolution cycling test for the Li/Li symmetric cell (**a,b**) with the PE separator and (**c,d**) with the LICGC/PEO/LiTFSI-coated PE separator.

3.5. Cycling Performance of Li-Oxygen Battery Cells

The results of the repeated Li deposition/dissolution test (Figure 5a,b) and SEM analysis of the electrode (Figure 6) clearly revealed the effectiveness of introducing the LICGC/PEO/LiTFSI coating layer for improving the reversibility of the lithium metal electrode. To demonstrate the effectiveness of the LICGC/PEO/LiTFSI coating layer on the Li metal-based batteries, we performed the discharge/charge cycle test of lithium-oxygen batteries (LOBs), which has the potential to show a higher energy density than conventional lithium-ion batteries. As the mass loading of the coating layer is less than 1 mg/cm^2, the introduction of the coating layer has limited influence on the energy density of LOBs. Here, the LOB cells have a stacked configuration, and their electrolyte contains redox meditators [42,43]. The discharge/charge performance test was conducted at a current density of 0.4 mA/cm^2, a capacity limit of 4.0 mAh/cm^2, and cutoff voltages of 2.0 V/4.5 V. Figure 7a shows the representative voltage profile of the LOB using the uncoated PE separator. During the discharge process, the cell exhibited a voltage plateau at approximately 2.6 V. In the initial and middle parts of the charging process, a stable voltage plateau appeared at approximately 3.5–3.6 V, and the voltage gradually increased to 4.0 V at the end of the charging process. These results are characteristic of the charging profile of LOB cells containing LiNO$_3$ and LiBr as redox mediators [44,45]. However, such stable discharge/charge voltages were only maintained up to the 3rd cycle, and the overpotential increased at the end of charging in the 4th~6th cycles. In contrast, the increase in charging voltage was largely suppressed in the LOB equipped with the LICGC/PEO/LiTFSI-coated separator (Figure 7c). Several mechanisms could cause such an increase in the overpotential, such as the accumulation of lithium carbonate-like solid-state side product on the porous carbon positive electrode [46,47] and deterioration of the lithium metal negative electrode [48,49]. Based on the results in Figure 7a,c, we think that the LICGC/PEO/LiTFSI coating on the separator led to more uniform Li deposition, which helped suppress the elevation of overpotential at the end of the charging process. After seven cycles, the LOB cells with and without the LICGC/PEO/LiTFSI coating both showed a gradual decrease in the discharge voltage, which reached the cutoff voltage of 2.0 V in the 11th cycle (Figure 7b,d).

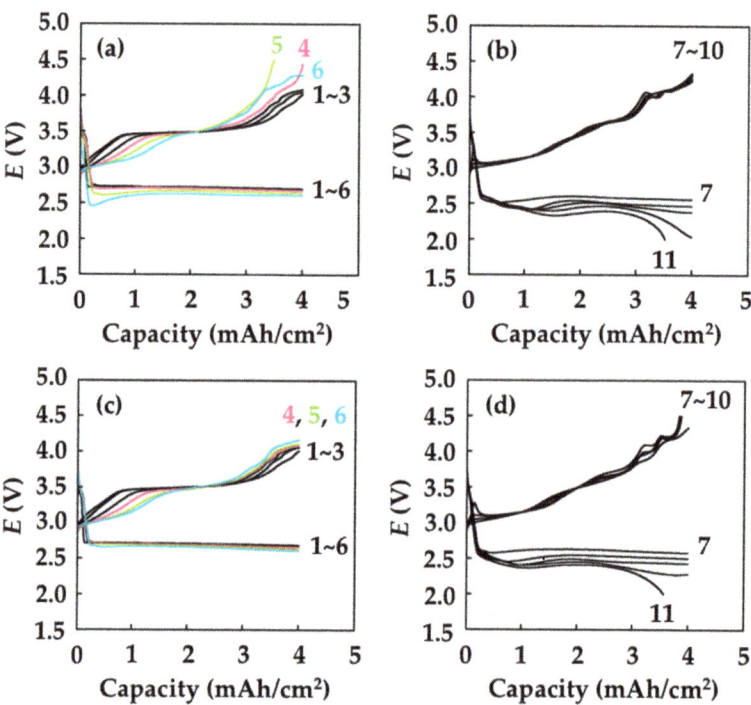

Figure 7. Discharge/charge performance of LOBs. (a) Cycles 1–6 and (b) cycles 7–11 for the LOB equipped with the PE separator. (c) Cycles 1–6 and (d) cycles 7–11 for the LOB equipped with the LICGC/PEO/LiTFSI-PE separator.

4. Conclusions

In this study, a facile and scalable coating method was developed for a commercial polyethylene separator in order to improve the electrochemical performance of lithium metal batteries. An organic/inorganic composite layer containing a lithium-ion-conductive ceramic filler (LICGC) and a solid polymer electrolyte (PEO/LiTFSI) with a thickness of a few microns was coated on the surface of the polyethylene separator via a blade-coating method. We used the commercial PE separator as a substrate separator, but the present coating strategy can be applied to other commercial separators, such as a polypropylene separator. The Li/Li symmetric cell using the LICGC/LEO/LiTFSI-PE separator showed improved interfacial resistance, indicating the formation of a less resistive SEI layer due to the suppression of direct contact between the Li metal and the electrolyte. As a result, long-term stable cycling of Li deposition/dissolution was realized in the cell with the LICGC/PEO/LiTFSI-PE separator. Finally, when the LICGC/PEO/LiTFSI-PE separator was applied in the LOB full cell, the charging overpotential was suppressed. These results indicate that this simple and scalable coating of lithium-ion-conductive fillers composited with solid polymer electrolytes is an effective strategy to improve the electrochemical performance of lithium metal batteries.

Author Contributions: Conceptualization, R.T. and S.M.; methodology and validation, R.S., R.T. and S.M.; writing—original draft preparation, R.S., R.T. and S.M.; writing—review and editing, R.T. and S.M.; supervision, R.T. and S.M. All authors have read and agreed to the published version of the manuscript.

Funding: The present work was partially supported by the ALCA-SPRING (Advanced Low Carbon Technology Research and Development Program—Specially Promoted Research for Innovative Next Generation Batteries) Project of the Japan Science and Technology Agency (JST Grant Number JPMJAL1301). A part of this work was supported by JSPS KAKENHI Grant Number 19K15396.

Institutional Review Board Statement: Not applicable.

Informed Consent Statement: Not applicable.

Data Availability Statement: Not applicable.

Acknowledgments: This work received technical support from the National Institute for Materials Science (NIMS) Battery Research Platform.

Conflicts of Interest: The authors declare no conflict of interest.

References

1. Lin, D.; Liu, Y.; Cui, Y. Reviving the lithium metal anode for high-energy batteries. *Nat. Nanotechnol.* **2017**, *12*, 194–206. [CrossRef]
2. Tikekar, M.D.; Choudhury, S.; Tu, Z.; Archer, L.A. Design principles for electrolytes and interfaces for stable lithium-metal batteries. *Nat. Energy* **2016**, *1*, 16114. [CrossRef]
3. Cheng, X.-B.; Zhang, R.; Zhao, C.-Z.; Zhang, Q. Toward safe lithium metal anode in rechargeable batteries: A review. *Chem. Rev.* **2017**, *117*, 10403–10473. [CrossRef]
4. Suo, L.; Hu, Y.-S.; Li, H.; Armand, M.; Chen, L. A new class of Solvent-in-Salt electrolyte for high-energy rechargeable metallic lithium batteries. *Nat. Commun.* **2013**, *4*, 1481. [CrossRef]
5. Yamada, Y.; Furukawa, K.; Sodeyama, K.; Kikuchi, K.; Yaegashi, M.; Tateyama, Y.; Yamada, A. Unusual stability of acetonitrile-based superconcentrated electrolytes for fast-charging lithium-ion batteries. *J. Am. Chem. Soc.* **2014**, *136*, 5039–5046. [CrossRef]
6. Qian, J.; Henderson, W.A.; Xu, W.; Bhattacharya, P.; Engelhard, M.; Borodin, O.; Zhang, J.-G. High rate and stable cycling of lithium metal anode. *Nat. Commun.* **2015**, *6*, 6362. [CrossRef]
7. Borodin, O.; Self, J.; Persson, K.A.; Wang, C.; Xu, K. Uncharted waters: Super-concentrated electrolytes. *Joule* **2020**, *4*, 69–100. [CrossRef]
8. Chen, J.; Fan, X.; Li, Q.; Yang, H.; Khoshi, M.R.; Xu, Y.; Hwang, S.; Chen, L.; Ji, X.; Yang, C.; et al. Electrolyte design for LiF-rich solid–electrolyte interfaces to enable high-performance microsized alloy anodes for batteries. *Nat. Energy* **2020**, *5*, 386–397. [CrossRef]
9. Zhang, X.-Q.; Cheng, X.-B.; Chen, X.; Yan, C.; Zhang, Q. Fluoroethylene carbonate additives to render uniform li deposits in lithium metal batteries. *Adv. Funct. Mater.* **2017**, *27*, 1605989. [CrossRef]
10. Dai, H.; Xi, K.; Liu, X.; Lai, C.; Zhang, S. Cationic surfactant-based electrolyte additives for uniform lithium deposition via lithiophobic repulsion mechanisms. *J. Am. Chem. Soc.* **2018**, *140*, 17515–17521. [CrossRef] [PubMed]
11. Ryou, M.-H.; Lee, D.J.; Lee, J.-N.; Lee, Y.M.; Park, J.-K.; Choi, J.W. Excellent cycle life of lithium-metal anodes in lithium-ion batteries with mussel-inspired polydopamine-coated separators. *Adv. Energy Mater.* **2012**, *2*, 645–650. [CrossRef]
12. Zhang, W.; Tu, Z.; Qian, J.; Choudhury, S.; Archer, L.A.; Lu, Y. Design principles of functional polymer separators for high-energy, metal-based batteries. *Small* **2018**, *14*, 1703001. [CrossRef] [PubMed]
13. Maeyoshi, Y.; Ding, D.; Kubota, M.; Ueda, H.; Abe, K.; Kanamura, K.; Abe, H. Long-term stable lithium metal anode in highly concentrated sulfolane-based electrolytes with ultrafine porous polyimide separator. *ACS Appl. Mater. Interfaces* **2019**, *11*, 25833–25843. [CrossRef] [PubMed]
14. Poungsripong, P.; Tamate, R.; Ono, M.; Sakaushi, K.; Ue, M. Fabrication of single-ion conducting polymer-coated separators and their application in nonaqueous Li-O2 batteries. *Polym. J.* **2021**, *53*, 549–556. [CrossRef]
15. Hao, Z.; Zhao, Q.; Tang, J.; Zhang, Q.; Liu, J.; Jin, Y.; Wang, H. Functional separators towards the suppression of lithium dendrites for rechargeable high-energy batteries. *Mater. Horiz.* **2021**, *8*, 12–32. [CrossRef] [PubMed]
16. Bouchet, R.; Maria, S.; Meziane, R.; Aboulaich, A.; Lienafa, L.; Bonnet, J.-P.; Phan, T.N.T.; Bertin, D.; Gigmes, D.; Devaux, D.; et al. Single-ion BAB triblock copolymers as highly efficient electrolytes for lithium-metal batteries. *Nat. Mater.* **2013**, *12*, 452–457. [CrossRef]
17. Zhao, Y.; Huang, Z.; Chen, S.; Chen, B.; Yang, J.; Zhang, Q.; Ding, F.; Chen, Y.; Xu, X. A promising PEO/LAGP hybrid electrolyte prepared by a simple method for all-solid-state lithium batteries. *Solid State Ion.* **2016**, *295*, 65–71. [CrossRef]
18. Zhou, W.; Wang, S.; Li, Y.; Xin, S.; Manthiram, A.; Goodenough, J.B. Plating a dendrite-free lithium anode with a polymer/ceramic/polymer sandwich electrolyte. *J. Am. Chem. Soc.* **2016**, *138*, 9385–9388. [CrossRef] [PubMed]
19. Zhao, Q.; Stalin, S.; Zhao, C.Z.; Archer, L.A. Designing solid-state electrolytes for safe, energy-dense batteries. *Nat. Rev. Mater.* **2020**, *5*, 229–252. [CrossRef]

20. Lou, S.; Zhang, F.; Fu, C.; Chen, M.; Ma, Y.; Yin, G.; Wang, J. Interface issues and challenges in all-solid-state batteries: Lithium, sodium, and beyond. *Adv. Mater.* **2020**, 2000721. [CrossRef]
21. Zhao, Q.; Liu, X.; Stalin, S.; Khan, K.; Archer, L.A. Solid-state polymer electrolytes with in-built fast interfacial transport for secondary lithium batteries. *Nat. Energy* **2019**, *4*, 365–373. [CrossRef]
22. Liu, K.; Pei, A.; Lee, H.R.; Kong, B.; Liu, N.; Lin, D.; Liu, Y.; Liu, C.; Hsu, P.; Bao, Z.; et al. Lithium metal anodes with an adaptive "solid-liquid" interfacial protective layer. *J. Am. Chem. Soc.* **2017**, *139*, 4815–4820. [CrossRef]
23. Tu, Z.; Choudhury, S.; Zachman, M.J.; Wei, S.; Zhang, K.; Kourkoutis, L.F.; Archer, L.A. Designing artificial solid-electrolyte interphases for single-ion and high-efficiency transport in batteries. *Joule* **2017**, *1*, 394–406. [CrossRef]
24. Xu, R.; Cheng, X.-B.; Yan, C.; Zhang, X.-Q.; Xiao, Y.; Zhao, C.-Z.; Huang, J.-Q.; Zhang, Q. Artificial interphases for highly stable lithium metal anode. *Matter* **2019**, *1*, 317–344. [CrossRef]
25. Yu, Z.; Cui, Y.; Bao, Z. Design principles of artificial solid electrolyte interphases for lithium-metal anodes. *Cell Rep. Phys. Sci.* **2020**, *1*, 100119. [CrossRef]
26. Zhou, H.; Yu, S.; Liu, H.; Liu, P. Protective coatings for lithium metal anodes: Recent progress and future perspectives. *J. Power Sources* **2020**, *450*, 227632. [CrossRef]
27. Gao, Y.; Yan, Z.; Gray, J.L.; He, X.; Wang, D.; Chen, T.; Huang, Q.; Li, Y.C.; Wang, H.; Kim, S.H.; et al. Polymer–inorganic solid–electrolyte interphase for stable lithium metal batteries under lean electrolyte conditions. *Nat. Mater.* **2019**, *18*, 384–389. [CrossRef]
28. Zhang, C.; Huang, Z.; Lv, W.; Yun, Q.; Kang, F.; Yang, Q.H. Carbon enables the practical use of lithium metal in a battery. *Carbon N. Y.* **2017**, *123*, 744–755. [CrossRef]
29. Zhang, C.; Liu, S.; Li, G.; Zhang, C.; Liu, X.; Luo, J. Incorporating ionic paths into 3D conducting scaffolds for high volumetric and areal capacity, high rate lithium-metal anodes. *Adv. Mater.* **2018**, *30*, 1801328. [CrossRef] [PubMed]
30. Zheng, J.; Zhao, Q.; Liu, X.; Tang, T.; Bock, D.C.; Bruck, A.M.; Tallman, K.R.; Housel, L.M.; Kiss, A.M.; Marschilok, A.C.; et al. Nonplanar electrode architectures for ultrahigh areal capacity batteries. *ACS Energy Lett.* **2019**, *4*, 271–275. [CrossRef]
31. Liu, Y.; Wu, X.; Niu, C.; Xu, W.; Cao, X.; Zhang, J.-G.; Jiang, X.; Xiao, J.; Yang, J.; Whittingham, M.S.; et al. Systematic evaluation of carbon hosts for high-energy rechargeable lithium-metal batteries. *ACS Energy Lett.* **2021**, *11*, 1550–1559. [CrossRef]
32. Zhang, S.S. A review on the separators of liquid electrolyte Li-ion batteries. *J. Power Sources* **2007**, *164*, 351–364. [CrossRef]
33. Jeong, H.-S.; Lee, S.-Y. Closely packed SiO_2 nanoparticles/poly(vinylidene fluoride-hexafluoropropylene) layers-coated polyethylene separators for lithium-ion batteries. *J. Power Sources* **2011**, *196*, 6716–6722. [CrossRef]
34. Zhang, Z.; Lai, Y.; Zhang, Z.; Zhang, K.; Li, J. Al_2O_3-coated porous separator for enhanced electrochemical performance of lithium sulfur batteries. *Electrochim. Acta* **2014**, *129*, 55–61. [CrossRef]
35. Shi, J.; Xia, Y.; Han, S.; Fang, L.; Pan, M.; Xu, X.; Liu, Z. Lithium ion conductive $Li_{1.5}Al_{0.5}Ge_{1.5}(PO_4)_3$ based inorganic–organic composite separator with enhanced thermal stability and excellent electrochemical performances in 5 V lithium ion batteries. *J. Power Sources* **2015**, *273*, 389–395. [CrossRef]
36. Jung, Y.-C.; Kim, S.-K.; Kim, M.-S.; Lee, J.-H.; Han, M.-S.; Kim, D.-H.; Shin, W.-C.; Ue, M.; Kim, D.-W. Ceramic separators based on Li^+-conducting inorganic electrolyte for high-performance lithium-ion batteries with enhanced safety. *J. Power Sources* **2015**, *293*, 675–683. [CrossRef]
37. Liang, T.; Cao, J.-H.; Liang, W.-H.; Li, Q.; He, L.; Wu, D.-Y. Asymmetrically coated LAGP/PP/PVDF–HFP composite separator film and its effect on the improvement of NCM battery performance. *RSC Adv.* **2019**, *9*, 41151–41160. [CrossRef]
38. Stone, G.M.; Mullin, S.A.; Teran, A.A.; Hallinan, D.T.; Minor, A.M.; Hexemer, A.; Balsara, N.P. Resolution of the modulus versus adhesion dilemma in solid polymer electrolytes for rechargeable lithium metal batteries. *J. Electrochem. Soc.* **2012**, *159*, A222–A227. [CrossRef]
39. Harry, K.J.; Higa, K.; Srinivasan, V.; Balsara, N.P. Influence of electrolyte modulus on the local current density at a dendrite tip on a lithium metal electrode. *J. Electrochem. Soc.* **2016**, *163*, A2216–A2224. [CrossRef]
40. Tung, S.-O.; Ho, S.; Yang, M.; Zhang, R.; Kotov, N.A. A dendrite-suppressing composite ion conductor from aramid nanofibres. *Nat. Commun.* **2015**, *6*, 6152. [CrossRef] [PubMed]
41. Zhao, C.Z.; Chen, P.Y.; Zhang, R.; Chen, X.; Li, B.Q.; Zhang, X.Q.; Cheng, X.B.; Zhang, Q. An ion redistributor for dendrite-free lithium metal anodes. *Sci. Adv.* **2018**, *4*, eaat3446. [CrossRef]
42. Matsuda, S.; Yamaguchi, S.; Yasukawa, E.; Asahina, H.; Kakuta, H.; Otani, H.; Kimura, S.; Kameda, T.; Takayanagi, Y.; Tajika, A.; et al. Effect of electrolyte filling technology on the performance of porous carbon electrode-based lithium-oxygen batteries. *ACS Appl. Energy Mater.* **2021**, *4*, 2563–2569. [CrossRef]
43. Matsuda, S.; Yasukawa, E.; Kameda, T.; Kimura, S.; Yamaguchi, S.; Kubo, Y.; Uosaki, K. Carbon-black-based self-standing porous electrode for 500 Wh/kg rechargeable lithium-oxygen batteries. *Cell Rep. Phys. Sci.* **2021**, *2*, 100506. [CrossRef]
44. Xin, X.; Ito, K.; Kubo, Y. Highly Efficient Br^-/NO_3^- dual-anion electrolyte for suppressing charging instabilities of $Li–O_2$ batteries. *ACS Appl. Mater. Interfaces* **2017**, *9*, 25976–25984. [CrossRef]
45. Ue, M.; Asahina, H.; Matsuda, S.; Uosaki, K. Material balance in the O_2 electrode of $Li–O_2$ cells with a porous carbon electrode and TEGDME-based electrolytes. *RSC Adv.* **2020**, *10*, 42971–42982. [CrossRef]
46. Zhao, Z.; Huang, J.; Peng, Z. Achilles' heel of lithium-air batteries: Lithium carbonate. *Angew. Chemie Int. Ed.* **2018**, *57*, 3874–3886. [CrossRef] [PubMed]

47. McCloskey, B.D.; Speidel, A.; Scheffler, R.; Miller, D.C.; Viswanathan, V.; Hummelshøj, J.S.; Nørskov, J.K.; Luntz, A.C. Twin problems of interfacial carbonate formation in nonaqueous Li-O_2 batteries. *J. Phys. Chem. Lett.* **2012**, *3*, 997–1001. [CrossRef]
48. Shui, J.-L.; Okasinski, J.S.; Kenesei, P.; Dobbs, H.A.; Zhao, D.; Almer, J.D.; Liu, D.-J. Reversibility of anodic lithium in rechargeable lithium–oxygen batteries. *Nat. Commun.* **2013**, *4*, 2255. [CrossRef]
49. Sun, F.; Gao, R.; Zhou, D.; Osenberg, M.; Dong, K.; Kardjilov, N.; Hilger, A.; Markötter, H.; Bieker, P.M.; Liu, X.; et al. Revealing hidden facts of li anode in cycled lithium–oxygen batteries through X-ray and neutron tomography. *ACS Energy Lett.* **2019**, *4*, 306–316. [CrossRef]

Article

Tool for Designing Breakthrough Discovery in Materials Science

Michiko Yoshitake

Research Center for Functional Materials, National Institute for Materials Science, Tsukuba 305-40044, Japan; yoshitake.michiko@nims.go.jp; Tel.: +81-298-863-5696

Abstract: A database of material property relationships, which serves as a scientific principles database, and a database search system are proposed and developed. The use of this database can support a broader research perspective, which is increasingly important in the era of automated computer-aided experimentation and machine learning of experimental and calculated data. Examples of the wider use of scientific principles in materials research are presented. The database and its advantages are described. An implementation of the proposed database and search system as a prototype software is reported. The usefulness of the database and search system is demonstrated by an example of a surprising but reasonable discovery.

Keywords: knowledge database; scientific principles; material property relationship; network-type database; interdisciplinary; multidisciplinary; graph search; wide perspective

Citation: Yoshitake, M. Tool for Designing Breakthrough Discovery in Materials Science. *Materials* **2021**, *14*, 6946. https://doi.org/10.3390/ma14226946

Academic Editor: Teofil Jesionowski

Received: 31 October 2021
Accepted: 15 November 2021
Published: 17 November 2021

Publisher's Note: MDPI stays neutral with regard to jurisdictional claims in published maps and institutional affiliations.

Copyright: © 2021 by the author. Licensee MDPI, Basel, Switzerland. This article is an open access article distributed under the terms and conditions of the Creative Commons Attribution (CC BY) license (https://creativecommons.org/licenses/by/4.0/).

1. Introduction

In conventional materials research and development (R&D), researchers explore materials or synthesis processes based on known materials or processes by modifying one or two conditions in the composition or process (conventional search area). The entire search area is very large, for example, the number of five-element systems composed of any combination of 76 practical elements (excluding inert gases and radioactive elements from the periodic table) can be briefly estimated as follows: the number of permutations of five elements from 76 elements (choosing from the largest content) is $76!/(76-5)!$, where $!$ means factorial. If the compounds containing the same five elements but with different compositions of 1 at% are regarded as different compounds, the total number of possible compounds of the five-element system is approximated by $76!/(76-5)! * (100-4)^5 * (1/2)^4$, which is larger than 10^{17}. Here, $(100-4)^5$ (96 at% is the possible maximum concentration) is possible variation of compositions without considering the order in composition and $(1/2)^4$ is for taking the order of five elements in consideration. To increase the search speed, high-throughput experimental techniques [1–3] and automated experimental systems using robotics techniques [4–6] have been developed recently. Machine learning techniques using accumulated data or output data from high-throughput experiments have been introduced in materials R&D [7–11]. Machine learning is a powerful tool for optimizing compositions or process parameters within systems (for example, to find a local minimum) where data are given (that is, the search area consists of various numerical input data). However, because machine learning requires numerical input data, its applications are limited to systems where numerical data for learning exist. By contrast, innovative materials or processes have often been discovered in systems far from existing or explored systems. For example, carbon alloy catalysts for fuel cells [12,13] have no metallic components but contain only carbon and nitrogen, whereas most researchers have tried to decrease the Pt or precious metal content of catalysts. Carbon alloy catalysts could not have been discovered by machine learning using existing data on catalysts containing Pt and/or other metals. To develop these catalysts, it appears that the inventor considered basic scientific principles without being limited by commonly used approaches. The scientific principles and functional mechanism are essentially the same as those of known systems. Here, the knowledge of the inventor appears to have contributed to the discovery. Figure 1 schematically illustrates the automated experiment and machine learning loop (blue lines) and human contribution (red lines) in computer-aided materials R&D. The blue loop in Figure 1 is still under development; however, it is gradually becoming apparent that the red path will become increasingly important in the future. Here, the problem is that individuals acquire knowledge mainly by reading books and papers, which limits the broadness of a field and often results in a narrow outlook on possible approaches. For breakthrough discovery, it is important to support a broader perspective.

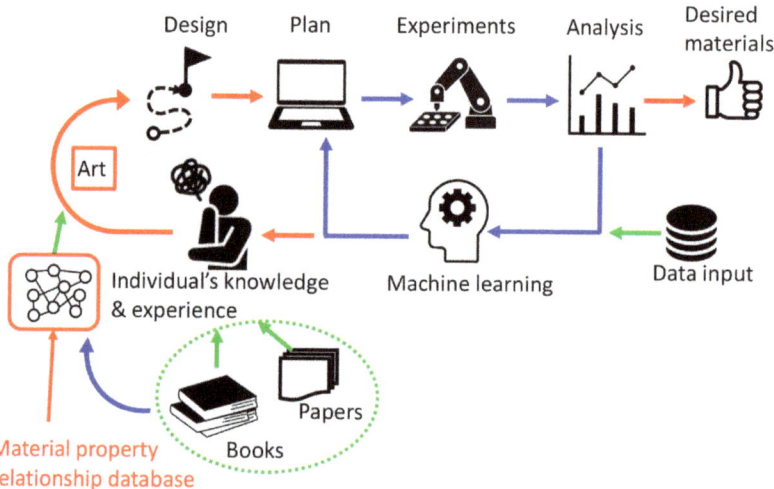

Figure 1. Schematic representation of research process consisting of automated loop with computer aid (**blue**) and human involvement in the process (**red**). Green arrows indicate information inputs.

The author has tried to obtain a broader perspective and has made discoveries, which will be described later. On the basis of these experiences, the author proposed "materials curation", a method of interdisciplinary utilization of scientific principles to solve problems or search for materials from this wider perspective [14–18]. In this method, searches for materials or solutions are conducted beyond the search space in which numerical data are available, as shown schematically in Figure 2, where red indicates more desirable values of target material properties and green indicates less desirable values. To make this method available to many researchers, the author made the concept of a database of scientific principles in materials science [16–18]. The database of scientific principles is used in the third and fourth stages of "materials curation", where the stages are divided into (1) detach from common approaches, (2) consider what the user wants (not needs), (3) describe conditions that satisfy the wants from viewpoint of scientific principles, (4) list methods that can satisfy the conditions in principle, (5) test the method one by one using numerical data, and (6) get new solutions for the wants [16]. On the red path in Figure 1, where the knowledge of an individual human is required, knowledge of scientific principles is acquired mainly from books. The interdisciplinary utilization of scientific principles requires knowledge from multiple fields. However, it is somewhat difficult for individuals to read many books from a broad range of fields. Developing and sharing a database of material property relationships to serve as a database of scientific principles (Figure 1, bottom left) would at least partially solve this problem.

Interdisciplinary support is realized by associating material properties not with material types or material usage but with academic fields, as shown in Figure 3. For example, the electrical conductivity is determined by the same principle described in solid-state physics regardless of the value. Metals, semiconductors, and ceramics (which are typically insulators) have different conductivity values, but those values are determined mainly by carrier density, which depends primarily on band gap energy. Here, the electrical conductivity, carrier density, and band gap energy (each of which is a material property) are connected through solid-state physics (blue lines in Figure 3). Because associations among material properties are made based on published electronic textbooks, the names of the academic fields are mostly based on titles or categories of textbooks from publishers. This article describes the database of material property relationships and the system for searching these relationships.

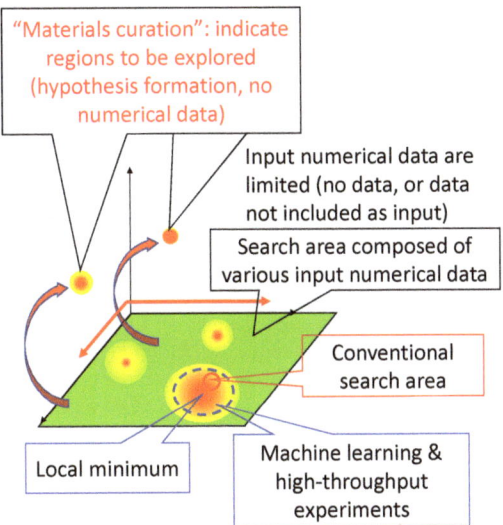

Figure 2. Schematic representation of search space with numerical input data (conventional or machine learning) and without numerical input data (materials curation).

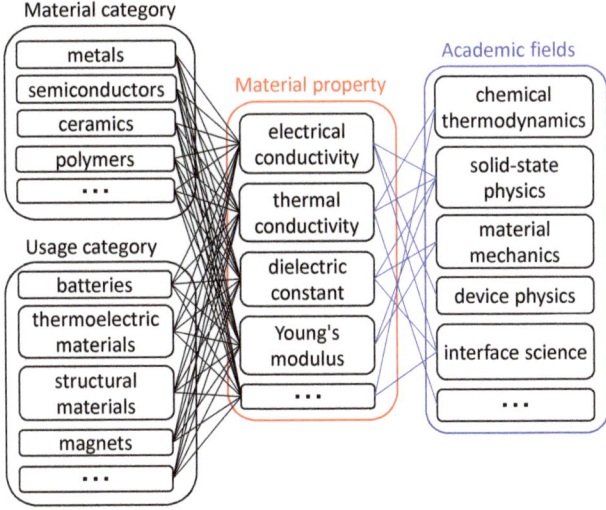

Figure 3. Schematic relationships among material properties (usually categorized by material type or usage; black lines) and scientific principles (usually categorized by academic fields; blue lines).

2. Examples of Knowledge Utilization

Here, examples of knowledge utilization by the author are presented to explain the process of perspective broadening.

2.1. Substrate for the Growth of Ultra-Thin Atomically Flat Epitaxial Alumina Film

Thin epitaxial alumina films have been grown for the study of electron tunneling, model catalysts and so forth. The most popular substrate used for model catalysts is NiAl(110), where the growth of atomically flat, 0.5 nm thick epitaxial alumina is well known [19]. However, it has been found that a thickness of 0.5 nm is not sufficient to avoid the effects of the metallic underlayer (in this case, NiAl). Therefore, many attempts have been made to use other (metallic) substrates. Figure 4 briefly summarizes the results of these attempts. Two types of substrates have been investigated: the (110) plane of pure body-centered cubic (bcc) metals with high melting temperature such as Ta(110) [20] and Mo(110) [21], and the (110) plane of Al-containing intermetallic compounds such as NiAl(110) and FeAl(110) [22]. On the former type of substrate, aluminum is deposited and then oxidized at high temperatures so that it crystallizes. Alumina is known to grow epitaxially but does not form flat films. The reason is that aluminum–oxygen bonds are so strong that in the first step of oxidation, aluminum atoms agglutinate and become islands. This kind of growth is well known to occur in molecular beam epitaxy (MBE) [23]. For Al-containing intermetallic compounds, preferential oxidation produces flat epitaxial alumina films, but the thickness is less than 1 nm, which is insufficient to avoid the effects of the substrate. In the preferential oxidation of Al-containing intermetallic compounds, O atoms react individually with Al atoms on the upper surface because there is no Al–Al bonding at the surface, and agglutination of Al atoms does not occur. If the Al atomic content is less than stoichiometric, Al atoms below the surface diffuse to the surface and bind with O atoms. Because O atoms do not agglutinate, the diffusion of Al atoms is the rate-determining process. Therefore, the agglutination of Al atoms does not occur, and atomically flat epitaxial films are produced. This mechanism is used in MBE, although the supply of metallic atoms is not controlled by diffusion from a substrate but by beam flux, for example, in the growth of GaAs [23]. Thicker alumina epitaxial layers (slightly thicker than 0.5 nm) can be grown by alternately suppling Al and O under controlled conditions [24]. The thickness is limited to less than 1 nm because of the symmetry mismatch of the crystal planes. In ultra-thin (nanometer-order) epitaxial alumina films, oxygen atoms typically align in sixfold symmetry on the plane parallel to the surface. The crystal structure of NiAl and FeAl is bcc-like, where atoms are aligned quasi-hexagonally but do not have sixfold symmetry on the (110) plane. The symmetry mismatch between the substrate and alumina film causes strain, which is thought to prevent further growth of epitaxial alumina. This hypothesis is supported by the fact that when a thicker layer of alumina was grown on NiAl(110) by further deposition of Al and O, the structure changed at a thickness of 0.84 nm, and the alumina became amorphous when the thickness reached 1.62 nm [24,25].

The above findings suggest the possibility of using Al-containing alloys that have a crystal plane with sixfold symmetry. The author was successful in finding such alloys that fulfill the conditions and demonstrated the growth of 1–4 nm thick atomically flat alumina films using Cu-9Al(111) as a substrate [26–28]. The key was to expand the search space beyond intermetallic compounds, which rarely have a plane with sixfold symmetry, and consider alloys as candidate materials.

(a) Al deposition and oxidation

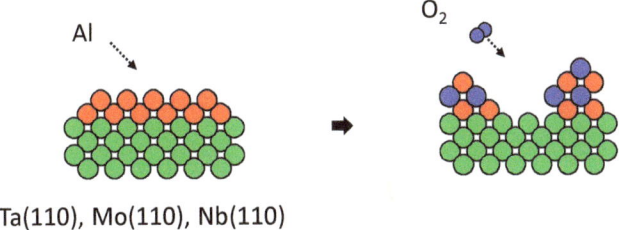

Ta(110), Mo(110), Nb(110)

(b) Preferential oxidation of Al-containing intermetallics

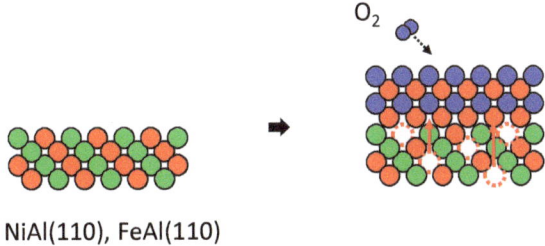

NiAl(110), FeAl(110)

Figure 4. Schematic representation of alumina growth by (**a**) Al deposition and oxidation on high-melting-temperature bcc metal (110) surfaces, and (**b**) preferential oxidation of (110) surface of Al-containing intermetallics having bcc-like structure.

2.2. Thermoelectric Materials

In thermoelectric materials, a voltage is generated between two edges of a material, which are kept at different temperatures. When the two edges are electrically connected via a load, current flows, which can be used as electric power. The efficiency of power generation is expressed as $Z = S^2\sigma/\kappa$, where S is the Seebeck coefficient, σ is the electrical conductivity, and κ is the thermal conductivity. In the early stage of intense research on thermoelectric materials around the beginning of the 2010s, the Seebeck coefficient and electrical conductivity were thought to have a trade-off relationship, and therefore most research focused on controlling the thermal conductivity by fabricating nano structures. However, the author demonstrated that the trade-off can be partially avoided [29,30]. By considering the scientific principles of voltage generation by placing samples of the same material at different temperatures in contact (temperature difference causes difference in electron distribution, accordingly the Fermi level difference, but the shape of density of states (DOS) is the same), and of voltage decrease due to current flow, we can draw a diagram of the relationship between S, σ, κ, and the quantities that determine S, σ, and κ, as shown in Figure 5. [31,32]. One reason for the trade-off relationship is doping, which does not change the main DOS but increases the impurity states (and thus increases σ); consequently, the Fermi level changes, decreasing the generated voltage thus S. However, this explanation between S and σ applies only for doping. A comparison of materials with differently shaped DOSs reveals that there is no trade-off relationship [31]. The reason is that the shape of the DOS depends on the carrier mobility, which is determined by the effective mass of electrons. Therefore, a search for materials considering not the DOS but the shape of the DOS would identify materials that have both large Seebeck coefficients and high electrical conductivity.

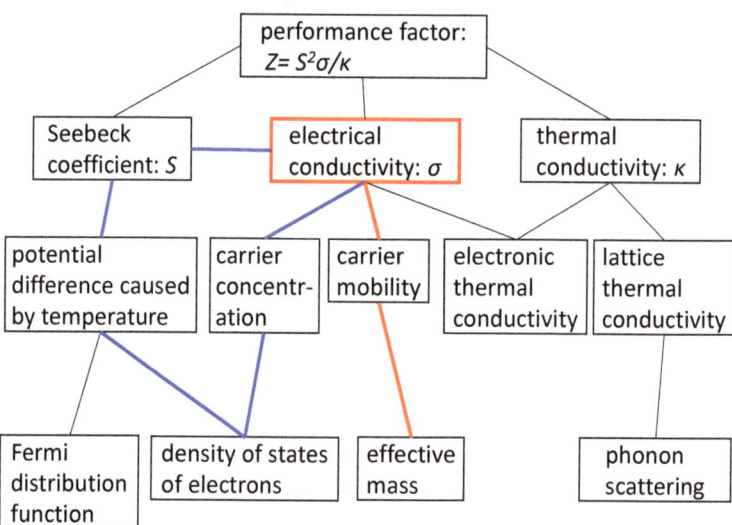

Figure 5. Relationships among various properties affecting the performance factor of thermoelectric materials.

2.3. Prediction of Work Function from Vickers Hardness

The work function is a material property that determines the energy barrier to electron transfer in many devices such as transistors, batteries, and solar cells. Although it is a material property, the value is determined not only by the bulk term (the bulk composition and bulk structure) but also by the surface term (the surface composition, which is not necessarily the same as the bulk composition, and surface atomic arrangement and structures, including the arrangement of steps). Figure 6 shows various material properties that contribute to the work function. In the devices mentioned above, the main functional material is sandwiched between two metallic electrodes, one with low work function and the other with high work function. Most materials with low work function, such as alkali metals, are very reactive. Among low-work-function materials, transition metal carbides (TMCs) and nitrides (TMNs) are less reactive and relatively easy to handle in device processing. Carbides such as TiC and TaC are in practical use.

TMCs are non-stoichiometric compounds, and carbon atoms often deviate from a 1:1 ratio, resulting in the formula TMC_x (x < 1). The work function is affected by the stoichiometry, but only two experimental results on the effects for well-defined surfaces have been reported [33]. First-principles calculations of these two systems have also been reported [34]; they show that carbon deficiency does not affect surface term of the work function. In addition, first-principles calculations have shown that the surface term of the work function of other TMCs remains constant under a carbon deficiency. Therefore, the carbon deficiency affects only the bulk term of the work function. Thus, the question is how to estimate the bulk term of the work function. From the origin of the work function [35], the author found that the Vickers hardness can be used as one measure of the bulk term of the work function of TMCs and TMNs in general [36]. Figure 6 was compiled on the basis of the above consideration. When this diagram is created and published, other researchers who are not familiar with the work function but need to control it for their devices can use it as a reference without following the author's entire thought process as described in [36].

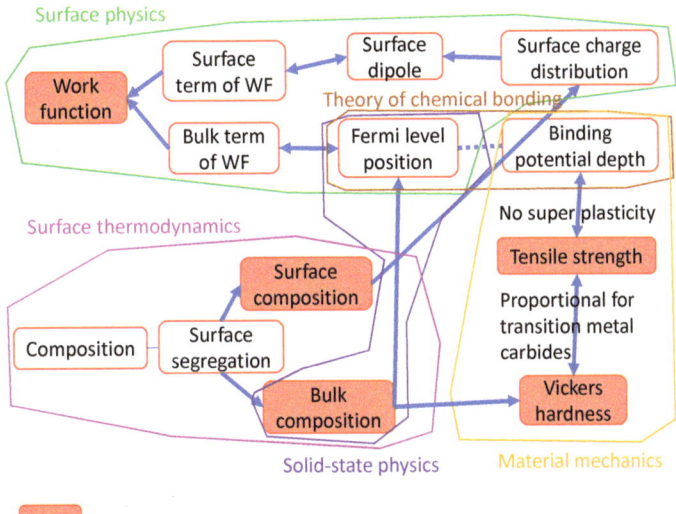

Figure 6. Relationships among factors that contributing to the work function, compiled from descriptions in books and review articles.

3. Relationship between Material Properties

If a diagram of the relationships between various material properties such as Figure 5 is stored as a database and shared among many material scientists, material development is expected to be greatly accelerated. Consequently, the author proposed a system composed of a database of relationships between various material properties and a search tool for the database [16,17,37,38] as shown schematically in Figure 7. Many relationships on material properties, which are given literally, are extracted as pairs of two material properties from texts either by (a) manually, where a person reads textbooks and learns the relationships like Figures 5 and 6, or by (b) automatically using natural language processing techniques and a computer. Extracted pairs of two material properties are input into a database (<Input of relations> in Figure 7). The database of sets of material property pairs is represented as a graph. Users search relations from the database represented as a graph (<Search of relations (users)> in Figure 7). The characteristic feature of the relationship database is its graph-type (network-type) structure, which consists of nodes (material properties) and edges (relations between material properties). This database is completely different from conventional material databases, which contain material names or compositions and the values of material properties such as melting point, density, and dielectric constant. There are no numerical values in the database. Like a train map, this database describes connections. The contents are not numerical data but words such as density. The sources of scientific principles are mainly literal (including mathematical formula), not numerical. Literal information describes essentially universal relationships independent of specific material compositions. Numerical data are useful for specific material systems.

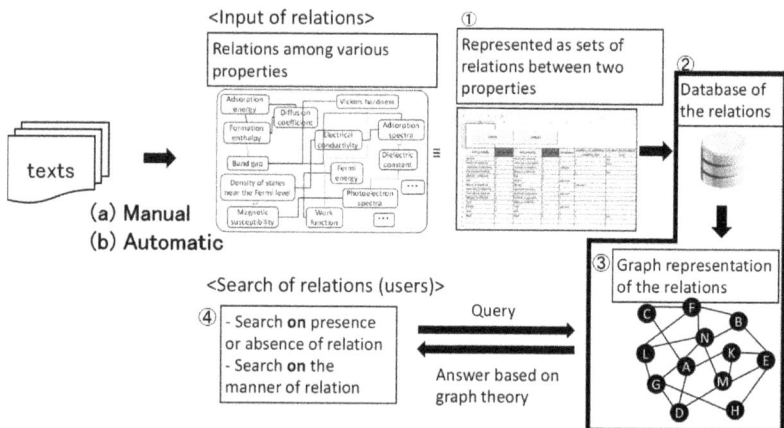

Figure 7. Schematic structure of the proposed system, which enables searches for relationships among material properties.

The advantage of a graph-type database is that it is easy to add or subtract data on connections as shown in Figure 8a. Consequently, it is easy to expand the area of scienfitic principles in the relationship database by connecting a material property mentioned in two textbooks in different academic fields (Figure 8b). Basic techniques for searching for relationships (connections) have been established in the framework of graph theory in mathematics [39] and are widely used in society, for example, in route searches of a train map. Graph-type databases are searched mainly by network searches and path searches, as shown in Figure 9. Here, each node (A, B, C, etc.) represents a material property such as density, thermal conductivity, or Vickers hardness, and each edge shows the relationship between two connected properties. Using a network search, one can, for example, find the material properties that affect the target property M. One example in which a path search is useful is when a material modification that increases material property A causes an unexpected decrease in material property B, which is undesirable. By searching the paths from A to B, one can find relationships that might cause the decrease in B with increasing A on these paths. It is also possible to search for possible ways of avoiding trade-off relationships (Figure 9c) by combining a path search and a network search, for example, by finding nodes that do not have a path to A without passing through H (J in Figure 9c) or finding nodes that connect directly to H but have a long path from A (H in Figure 9c). A node with a long path is usually expected to have less effect on a target node (=property), because there are many other nodes that affect the target node, which are used to avoid a trade-off relationship between A and H.

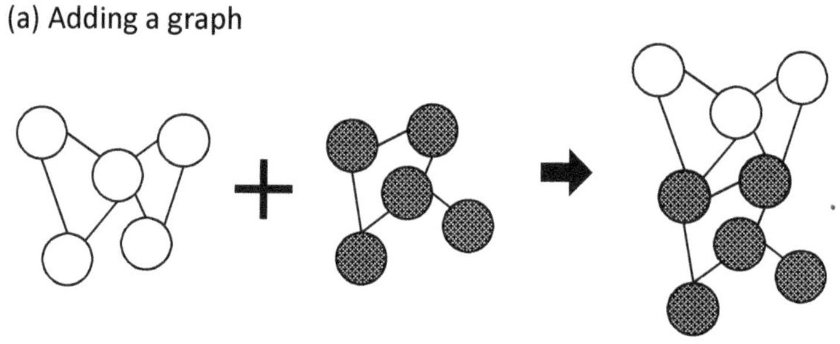

(a) Adding a graph

Figure 8. *Cont.*

(b) Connecting properties through different academic fields

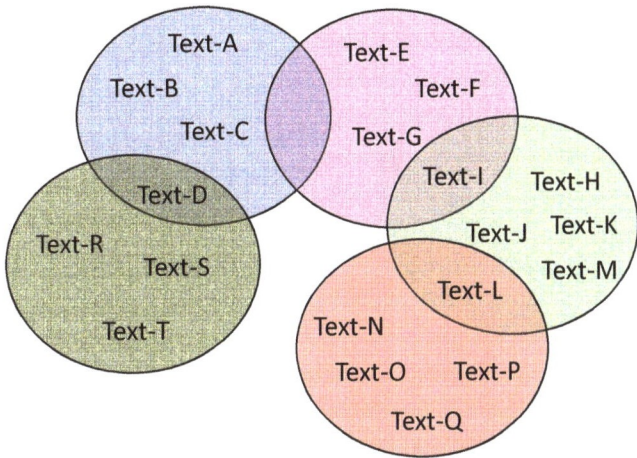

Figure 8. Graph-type database that enables the easy addition of a graph (**a**) and easy expansion of academic fields (**b**), where different colors indicate different academic fields such as materials mechanics, solid-state physics, and chemical thermodynamics.

(a) Network search around M

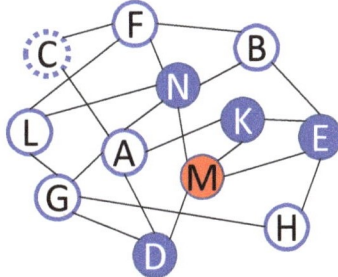

(b) Path search between A and B

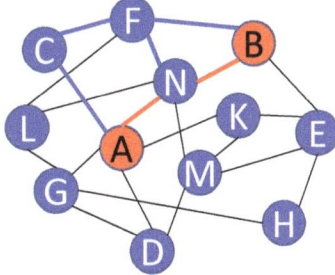

Figure 9. *Cont.*

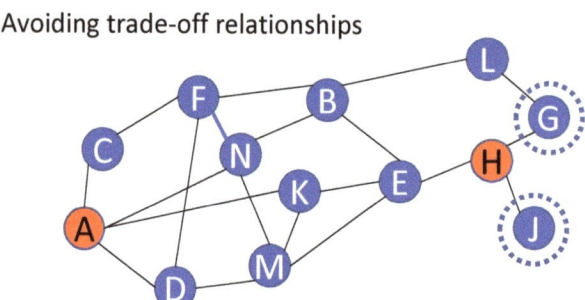

Figure 9. Two basic searches (**a**) network search and (**b**) path search, and search for avoiding trade-offs (**c**), which can be realized by combining path search and network search.

4. Computer Systems

Although the small system shown in Figure 7 has been developed and demonstrated [16,17,40], the number of material properties and relationships stored in the system is quite limited because the relationships between material properties were extracted manually. Computer technology for automated relationship extraction is essential for practical use. The author has collaborated with a company to realize automated relationship extraction from several textbooks on materials science, and a prototype system has been developed as a result of this collaborative project [41]. The relationships between material properties automatically extracted from the 12 textbooks listed in Table 1 are currently included in the web-based system. Figure 10 shows an example of the system output for a path search (Figure 9b) between work function and Vickers hardness, whose relationship was explained in Section 2.3. The descriptions in the textbook are not the same as those the author read, but the system also suggests the possibility of estimating work function values from Vickers hardness (there is a connection), and the properties shown in Figures 6 and 10 (path with red dotted lines) show considerable overlap. In the computer system, a path with nodes (material properties) appearing in the largest number of academic fields (represented by the colored circles around the material properties) is shown with thick edges, indicating the most multidisciplinary path. An example of the system output for a network search is shown in Figure 11. Because it is not commonly known that the work function is related to the Vickers hardness, a network search would be useful for finding properties that can be used to estimate the work function. In this case, a network search beginning with a target property (here, the work function) is used.

Table 1. List of textbooks used for the prototype system.

Book Title	Author(s)	Publisher	Year
Fundamentals of Materials Science	Eric J. Mittemeijer	Springer	2011
Understanding Materials Science	Rolf E. Hummel	Springer	2004
Materials Handbook	François Cardarelli	Springer	2018
The Chemical Bond I–III	D. Michael P. Mingos, ed.	Springer	2016
Ceramic Materials: Science and Engineering	C.Barry Carter, M. Grant Norton	Springer	2013
Electrochemistry for Materials Science	Walfried Plieth	Elsevier	2008
Solid State Electrochemistry I: Fundamentals, Materials and their Applications	Vladislav V. Kharton	WILEY	2009
Electronic Properties of Materials	E Hummel	Springer	2011
Physics of Semiconductor Devices	Simon M. Sze, Kwok K. Ng	WILEY	2006
Principles of Surface Physics	Friedhelm Bechstedt	Springer	2003
Physics of Surfaces and Interfaces	Harald Ibach	Springer	2006
Solid Surface Physics	Heribert Wagner	Springer	1979

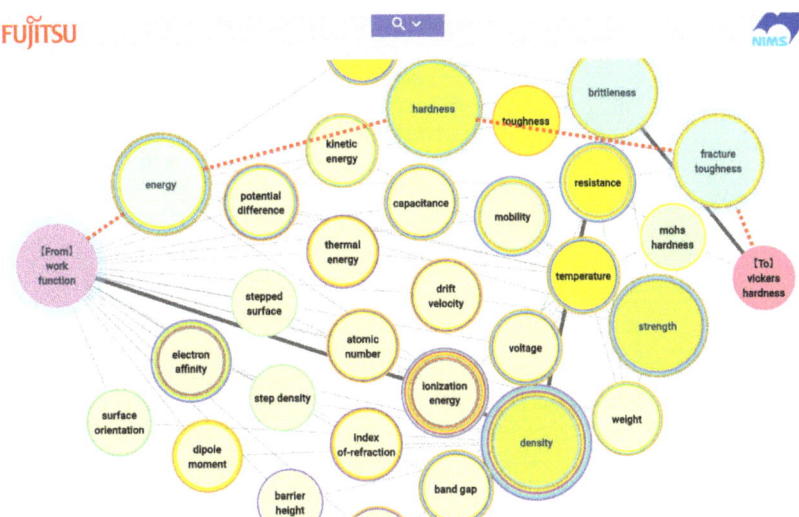

Figure 10. Computer system results screen showing a path search between the material properties of work function and Vickers hardness. Red dotted lines are shown for comparison with the manually compiled relationship in Figure 6.

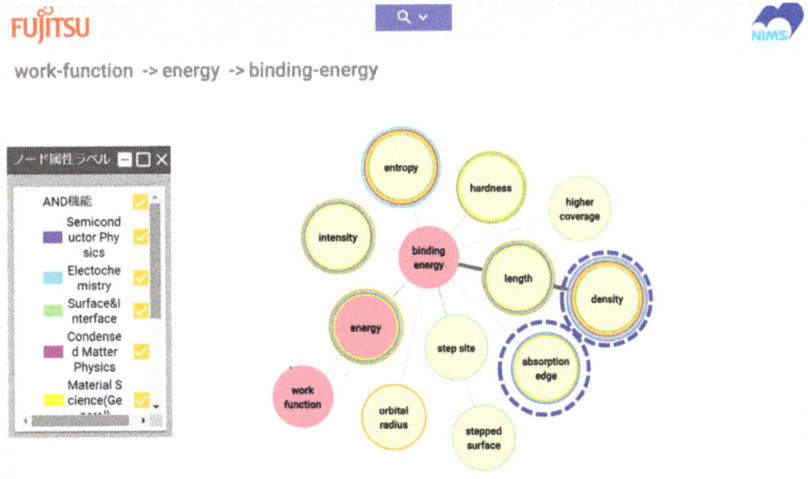

Figure 11. Results screen for sequential network search starting from work function.

The search result in Figure 11 uses the trace function (sequential network search, Figure 9a while retaining the previous network search results); the search begins at work function and reaches binding energy. This result suggests that properties such as density and absorption edge might be used in addition to hardness to estimate the work function. For TMCs, it is expected that experimental results on the effect of carbon deficiency on density may exist, but not results on absorption edges. It is reasonable to consider that density is a measure of binding potential depth in Figure 6, because the density would increase if the bonds in the carbides become stronger (that is, the binding potential is deeper) when both molar mass and lattice constant decrease because of carbon deficiency.

The author checked references on the density of TMCs with carbon deficiency. The effect of carbon deficiency on the density for TiC_x [42] and ZrC_x [43] (group IV TMCs) and VC_x [44] and TaC_x [45] (group V TMCs) is shown in Figure 12a, where the density is calculated from lattice constants obtained by X-ray diffraction measurements and the molar mass in the stoichiometry given in the references. In Figure 12b, the effects of carbon deficiency on hardness, which were previously used as a measure of the bulk term of the work function, are also shown for comparison. The absolute values of the density clearly depend on the atomic radius of transition metals. Therefore, the density is plotted as a relative value, and only the qualitative dependence of density on the stoichiometry is considered. For TiC_x and ZrC_x, whose phase diagrams show a wide region of one carbon-deficient phase, the density decreases monotonously with increasing carbon deficiency (decreasing x), as demonstrated in Figure 12a, in agreement with the trend of hardness in Figure 12b. For VC_x and TaC_x, the density is expected to increase with increasing carbon deficiency near stoichiometry ($0.9 < x < 1.0$) from hardness change with carbon deficiency. Although TaC_x shows the expected dependence on carbon deficiency, density values for $0.9 < x < 1.0$ are missing for VC_x. The density of VC_x decreases with carbon deficiency for $x < 0.87$, which is consistent with the hardness trend. In the phase diagram of the binary system of V and C [46], VC_x exists in the range $0.66 < x < 0.89$ at 1650 °C, where the concentration of C dissolved in metallic V is the maximum. The above range is in agreement with the data range for the density in Figure 12a. Therefore, it is considered that the density, like the Vickers hardness, is also useful as a measure of the bulk term of the work function for VC_x. TaC_x exists in the range $0.68 < x < 0.99$ at 2843 °C, where the concentration of C dissolved in metallic Ta is the maximum. Because the composition at which the hardness is maximum is somewhat unclear, it is difficult to discuss the behavior of TaC_x near the lower limit of x. In summary, it appears that the density can be used as an indicator of the effect of carbon deficiency on the bulk term of the work function in TMCs, at least in the composition range in which the carbon deficiency is smaller and the TMC_x phase exists in the phase diagram.

In the above example, the density of carbon-deficient TMCs was checked manually because there is no retrievable database. However, automated data collection and data presentation, as shown in Figure 12b, should be possible in principle, which would assist an individual researcher in the design process illustrated in Figure 1.

The system presented here is still a prototype. The development of a product and commercialization of the product is necessary in future. In addition, many additional functions such as quantitative relationships, arranging tie-ups with numerical database and machine learning are desired. Finding a new relations based on the structure of the graph database could be also explored, because there are considerable numbers of scientific principles represented in a similar form such in particle mechanics and geostatics and electric field and magnetic field in electromagnetics.

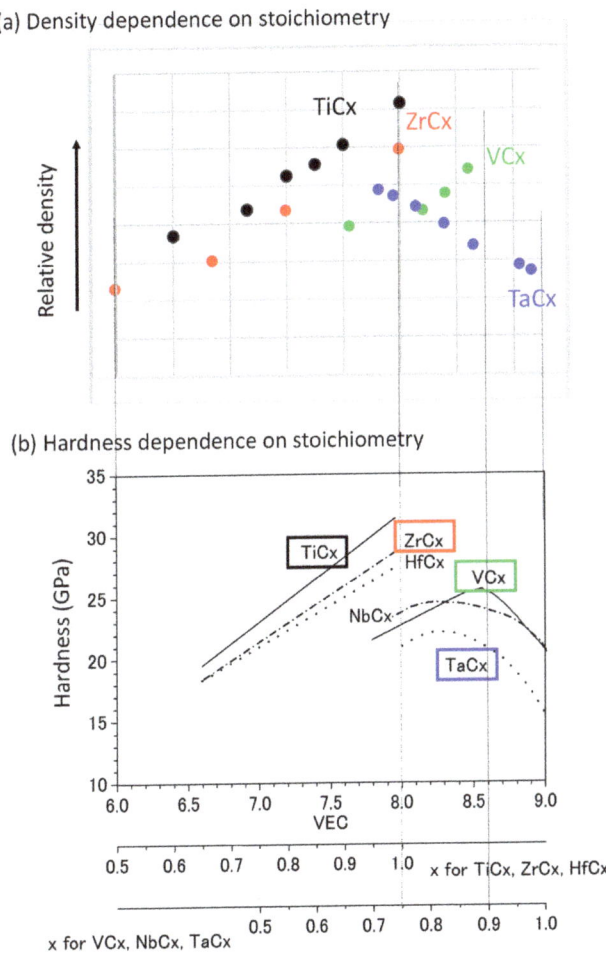

Figure 12. Composition dependence of density of TMCs (**a**) and with that of Vickers hardness [36] (**b**) for comparison. VEC is the abbreviation of "valence electron concentration" [36].

5. Conclusions

A materials informatics method that uses knowledge of scientific principles as well as numerical data was proposed. The use of systematic knowledge of scientific principles enables a broader perspective that is less limited by commonly used approaches. Some examples of material search and prediction using very little experimental data were shown to demonstrate the advantage of using scientific principles. Then, a system consisting of a database of knowledge on the relationships between material properties and a relationship search function, which is being developed by the author and collaborators, was presented. Finally, the author's discovery that work function values can be estimated from the density of materials when the effect of carbon deficiency in TMCs is considered is presented to demonstrate the usefulness of the system.

6. Patents

In the article, the following five patents, (1) property relationship database and search system, (2) those with options on priority, (3) those with modified search, (4) those with user information including search history, (5) those with combined search used for avoiding trade-offs, for example, are related.

(1) Search System, Search Method, and Physical Property Database Management Device, Japanese Patent #6719748, US Patent allowed (publication # 2019/0139279).
(2) Search System, Search Device, and Search Method, Japanese Patent # 6876344, US Patent # 11163829
(3) Search System, and Search Method, PCT/JP2019/028188.
(4) Search System, and Search Method, PCT/JP2019/030108.
(5) Search System, and Search Method, Japanese Patent publication #2021-012502.

Funding: This research was partly funded by Grants-in-Aid from the Ministry of Education and Science of Japan JSPS, KAKENHI Grant Number JP16K06283.

Data Availability Statement: The data presented in this study are available on request from the author.

Acknowledgments: This study is partly supported by Grants-in-Aid from the Ministry of Education and Science of Japan JSPS, KAKENHI Grant Number JP16K06283.

Conflicts of Interest: The author declares no conflict of interest.

References

1. Cawse, J.N. (Ed.) *Experimental Design for Combinatorial and High Throughput Materials Development*; Wiley: Hoboken, NJ, USA, 2002.
2. Potyrailo, R.A.; Maier, W.F. (Eds.) *Combinatorial and High-Throughput Discovery and Optimization of Catalysts and Materials*; CRC Press: Boca Raton, FL, USA, 2006. [CrossRef]
3. Maier, W.F.; Stwe, K.; Sieg, S. Combinatorial and High-Throughput Materials Science. *Angew. Chem. Int. Ed.* **2007**, *46*, 6016–6067. [CrossRef]
4. Laboratory Robotics. Available online: https://en.wikipedia.org/wiki/Laboratory_robotics (accessed on 12 October 2021).
5. Researchers Build Robot Scientist That Has Already Discovered a New Catalyst. Available online: https://phys.org/news/2020-07-robot-scientist-catalyst.html (accessed on 29 October 2021).
6. Burger, B.; Maffettone, P.M.; Gusev, V.V.; Aitchison, C.M.; Bai, Y.; Wang, X.; Li, X.; Alston, B.M.; Li, B.; Clowes, R.; et al. A mobile robotic chemist. *Nature* **2020**, *583*, 237–241. [CrossRef]
7. Isayev, O.; Tropsha, A.; Curtarolo, S. (Eds.) *Materials Informatics: Methods, Tools, and Applications*, 1st ed.; Wiley: Weinheim, Germany, 2019.
8. Rajan, K. (Ed.) *Informatics for Materials Science and Engineering: Data-Driven Discovery for Accelerated Experimentation and Application*; Elsevier: Amsterdam, The Netherlands, 2013.
9. Lookman, T.; Alexander, F.J.; Rajan, K. (Eds.) *Information Science for Materials Discovery and Design*, 1st ed.; Springer: Berlin/Heidelberg, Germany, 2016.
10. Lopez-Bezanilla, A.; Littlewood, P.B. Growing field of materials informatics: Databases and artificial intelligence. *MRS Commun.* **2020**, *10*, 1–10. [CrossRef]
11. Rajan, K. Materials Informatics: The Materials "Gene" and Big Data. *Annu. Rev. Mater. Res.* **2015**, *45*, 153–169. [CrossRef]
12. Maldonado, S.; Stevenson, K.J. Influence of Nitrogen Doping on Oxygen Reduction Electrocatalysis at Carbon Nanofiber Electrodes. *J. Phys. Chem. B* **2005**, *109*, 4707–4716. [CrossRef]
13. Ozaki, J.; Tanifuji, S.; Kimura, N.; Furuichi, A.; Oya, A. Enhancement of oxygen reduction activity by carbonization of furan resin in the presence of phthalocyanines. *Carbon* **2006**, *44*, 1324–1326. [CrossRef]
14. Yoshitake, M. Materials Curation: An Innovative Method for Material Search by Exploring Material Information from a Comprehensive Viewpoint. *Kinou Zair.* **2013**, *33*, 48–55. (In Japanese)
15. Yoshitake, M. Materials Curation: A Tool for Drawing Material Design Principles. *Int. J. Sci. Res.* **2015**, *4*, 571–579.
16. Yoshitake, M.; Kuwajima, I.; Yagyu, S.; Chikyow, T. System for Searching Relationship among Physical Properties for Materials CurationTM. *Vac. Surf. Sci.* **2018**, *61*, 200–205. (In Japanese) [CrossRef]
17. Yoshitake, M. Utilizing Knowledge on Scientific Principles on Material Properties for Materials R&D. *J. Surf. Anal.* **2019**, *26*, 134–135. (In Japanese) [CrossRef]
18. Yoshitake, M. Materials Curation®: Material Search Using Scientific Principles from a Comprehensive Viewpoint. *J. Comput. Chem. Jpn.* **2020**, *19*, 36–42. (In Japanese) [CrossRef]
19. Jaeger, R.M.; Kuhlenbeck, H.; Freund, H.-J.; Wuttig, M.; Hoffmann, W.; Franchy, R.; Ibach, H. Formation of a well-ordered aluminium oxide overlayer by oxidation of NiAl (110). *Surf. Sci.* **1991**, *259*, 235–252. [CrossRef]
20. Chen, P.J.; Goodman, D.W. Epitaxial growth of ultrathin Al_2O_3 films on Ta (110). *Surf. Sci.* **1994**, *312*, L767–L773. [CrossRef]
21. Wu, M.C.; Goodman, D.W. Particulate Cu on Ordered Al_2O_3: Reactions with Nitric Oxide and Carbon Monoxide. *J. Phys. Chem.* **1994**, *98*, 9874–9881. [CrossRef]
22. Graupner, H.; Hammer, L.; Heinz, K.; Zehner, D.M. Oxidation of low-index FeAl surfaces. *Surf. Sci.* **1997**, *380*, 335–351. [CrossRef]

23. Herman, M.A.; Sitter, H. *Molecular Beam Epitaxy*, 2nd rev. and updated ed.; Springer: Berlin/Heidelberg, Germany, 1996; p. 231.
24. Lykhach, Y.; Moroz, V.; Yoshitake, M. Formation of epitaxial Al_2O_3/NiAl (1 1 0) films: Aluminium deposition. *Appl. Surf. Sci.* **2005**, *241*, 250–255. [CrossRef]
25. Yoshitake, M.; Lykhach, Y. Flexible structure of alumina at the interface observed by RHEED. In Proceedings of the 9th International Conference on Atomically Controlled Surfaces, Interfaces and Nanostructure, Tokyo, Japan, 11–15 November 2007; p. 335, Presented at ACSIN-9, PS2-23.
26. Yamauchi, Y.; Yoshitake, M.; Song, W. Morphology and Thickness of Ultra-Thin Epitaxial Al_2O_3 Film on Cu-9%Al (1 1 1). *Jpn. J. Appl. Phys.* **2003**, *42*, 4721–4724. [CrossRef]
27. Nemsak, S.; Yoshitake, M.; Masek, K. Ultra-thin oxide layer formation on Cu–9% Al (1 1 1) surface and Pd growth studied using reflection high energy electron diffraction and Auger electron spectroscopy. *Surf. Sci.* **2006**, *600*, 4357–4360. [CrossRef]
28. Yoshitake, M.; Nagata, T.; Song, W. Electrical properties and stability of an epitaxial alumina film formed on Cu-9 at. % Al (111). *J. Vac. Sci. Technol. A* **2012**, *30*, 021509. [CrossRef]
29. Yoshitake, M. Materials Curation: Case study #2. In Proceedings of the JSAP Autumn Meeting, Hokkaido University, Sapporo, Hokkaido, Japan, 17–20 September 2014.
30. Yoshitake, M. Application of Materials Curation to Thermoelectric Materials. In *Thermoelectric Materials*; NIMS Research Report; NIMS: Tsukuba, Japan, 2015; pp. 181–200. ISBN 978-4-990056360.
31. Nolas, G.S.; Sharp, J.; Goldsmid, H.J. *Thermoelectrics*; Springer: New York, NY, USA, 2001; p. 56.
32. Thermoelectric Properties of Materials. Available online: http://thermoelectrics.matsci.northwestern.edu/thermoelectrics/index.html (accessed on 29 October 2021).
33. Gruzalski, G.R.; Lui, S.-C.; Zehner, D.M. Work-function changes accompanying changes in composition of (100) surfaces of HfC_x and TaC_x. *Surf. Sci. Lett.* **1990**, *239*, L517–L520. [CrossRef]
34. Price, D.L.; Cooper, B.R.; Wills, J.M. Effect of carbon vacancies on carbide work functions. *Phys. Rev. B* **1993**, *48*, 15311–15315. [CrossRef]
35. Yoshitake, M. *Work Function and Band Alignment of Electrode Materials*; Springer: Tokyo, Japan, 2021. [CrossRef]
36. Yoshitake, M. Generic trend of work functions in transition-metal carbides and nitrides, J. *Vac. Sci. Technol. A* **2014**, *32*, 061403. [CrossRef]
37. Search System, Search Method, and Physical Property Database Management Device. Japanese Patent #6,719,748, 19 June 2020.
38. Search System, Search Device, and Search Method. Japanese Patent # 6,876,344, 28 April 2021.
39. Bondy, J.A.; Murty, U.S.R. *Graph Theory with Applications*; Elsevier Science Ltd.: North-Holland, Netherlands, 1976.
40. Yoshitake, M.; Kuwajima, I.; Yagyu, S.; Chikyow, T. Development of Search System of Material Properties Relation Diagram. In Proceedings of the JSAP Spring Meeting, Pacifico Yokohama, Japan, 14–17 March 2017.
41. Yoshitake, M.; Sato, F.; Kawano, H. Developing a Materials Curation® Support System. *J. Surf. Anal.* **2020**, *27*, 22–33. [CrossRef]
42. Zarrinfar, N.; Shipway, P.H.; Kennedy, A.R.; Saidi, A. Carbide stoichiometry in TiC_x and Cu–TiC_x produced by self-propagating high-temperature synthesis. *Scr. Mater.* **2002**, *46*, 121–126. [CrossRef]
43. Nachiappan, C.; Rangaraj, L.; Divakar, C.; Jayaram, V. Synthesis and densification of monolithic zirconium carbide by reactive hot pressing. *J. Am. Ceram. Soc.* **2010**, *93*, 1341–1346. [CrossRef]
44. Lipatnikov, V.N.; Gusev, A.I.; Ettmayer, P.; Lengauer, W. Phase transformations in non- stoichiometric vanadium carbide. *J. Phys. Condens. Matter.* **1999**, *11*, 163–184. [CrossRef]
45. Bowman, A.L. The variation of lattice parameter with carbon content of Tantalum carbide. *J. Phys. Chem.* **1961**, *65*, 1596–1598. [CrossRef]
46. Massalski, T.B. (Ed.) *Binary Alloy Phase Diagrams*; ASM International: Almere, Netherlands, 1990.

Article

High-Resolution Mapping of Local Photoluminescence Properties in CuO/Cu$_2$O Semiconductor Bi-Layers by Using Synchrotron Radiation

Masakazu Kobayashi [1,*], Masanobu Izaki [1], Pei Loon Khoo [1], Tsutomu Shinagawa [2], Akihisa Takeuchi [3] and Kentaro Uesugi [3]

1. Department of Mechanical Engineering, Toyohashi University of Technology, Toyohashi 441-8580, Japan; m-izaki@me.tut.ac.jp (M.I.); khoo@tf.me.tut.ac.jp (P.L.K.)
2. Osaka Research Institute of Industrial Science and Technology, Osaka 536-8553, Japan; tshina@omtri.or.jp
3. Japan Synchrotron Radiation Research Institute, Sayo 679-5198, Japan; take@spring8.or.jp (A.T.); ueken@spring8.or.jp (K.U.)
* Correspondence: m-kobayashi@me.tut.ac.jp

Abstract: The quality of a semiconductor, which strongly affects its performance, can be estimated by its photoluminescence, which closely relates to the defect and impurity energy levels. In light of this, it is necessary to have a measurement method for photoluminescence properties with spatial resolution at the sub-micron or nanoscale. In this study, a mapping method for local photoluminescence properties was developed using a focused synchrotron radiation X-ray beam to evaluate localized photoluminescence in bi-layered semiconductors. CuO/Cu$_2$O/ZnO semiconductors were prepared on F:SnO$_2$/soda-lime glass substrates by means of electrodeposition. The synchrotron radiation experiment was conducted at the beamline 20XU in the Japanese synchrotron radiation facility, SPring-8. By mounting the high-sensitivity spectrum analyzer near the edge of the CuO/Cu$_2$O/ZnO devices, luminescence maps of the semiconductor were obtained with unit sizes of 0.3 μm × 0.3 μm. The devices were scanned in 2D. Light emission 2D maps were created by classifying the obtained spectra based on emission energy already reported by M. Izaki, et al. Band-like structures corresponding to the stacking layers of CuO/Cu$_2$O/ZnO were visualized. The intensities of emissions at different energies at each position can be associated with localized photovoltaic properties. This result suggests the validity of the method for investigation of localized photoluminescence related to the semiconductor quality.

Keywords: oxide semiconductor; electrodeposition; photoluminescence; focused X-ray; imaging

1. Introduction

Multi-layered solar cell devices have been proposed and designed to improve photovoltaics performance (e.g., [1]). For obtaining good photovoltaics performance, it is necessary to understand the influences of layered interface mismatches and boundary segregated impurities on the photovoltaic properties of multi-layered film semiconductor devices (e.g., [2,3]). To investigate local physical properties, such as photoluminescence (PL), at complex heterogeneous interfaces and boundaries, an inspection technique that can be associated with local structure is necessary. However, conventional PL measurements, which ordinarily cover a wide inspection area, are not suitable for the investigation of individual film layers and their interfaces. Frazer et al. reported the relationships between luminescence imaging and lattice defects in Cu$_2$O crystals fabricated by the floating zone method [4]. Here, luminescence imaging by the excitation of a laser beam was utilized and mapped onto a region of a few hundred micrometers. However, such resolution is still insufficient to investigate the local PL within a multi-layered film device with the size of several tens of micrometers, although the resolution has been improved year after year [5].

It has been pointed out that photoluminescence and photovoltaic properties are affected by the presence of lattice defects, such as vacancies and impurities in semiconductor materials and devices [4,6]. Moreover, in multi-layered film semiconductor devices that possess complex heterogeneous structures, local variations of photovoltaic properties are expected to accompany heterogeneity and lack of lattice defects. As such, a method to locally characterize photoluminescence and photovoltaic properties should be developed. Research to directly link the local physical properties to the local structures of bulk devices (e.g., vacancies, impurities, crystal boundaries, interfaces and so on) is likely necessary for finding the best solution of layer structure. Additionally, if the investigation can be conducted non-destructively, the study of property changes during use and after a long period of use would also be possible.

A luminescence spectrograph utilizing synchrotron radiation, named SUPERLUMI, was developed at HASYLAB in the 1990s [7,8] and can measure luminescence properties at high precision [9,10]. Here, high-brilliance synchrotron radiation improved time resolution [7,8]. However, spatial resolution was limited and deemed insufficient. Currently, scanning X-ray microscopy at a synchrotron radiation facility is available, with a 65 nm-size focusing beam [11]. We proposed and applied an imaging technique for local photoluminescence mapping by means of a high-intensity focused X-ray at the undulator beam line in the Japanese synchrotron radiation facility, SPring-8, in our previous research [12]. The spatial distribution of localized photoluminescence in a CuO/Cu_2O semiconductor was measured and demonstrated with a grid size of 0.3 μm × 5 μm.

It is noteworthy that Cu_2O films have recently gained increased attention in the field of photoactive devices, such as photovoltaics (e.g., [13–15]), photonic crystals (e.g., [16]) and photocatalysts (e.g., [17]), due to their optical and electrical characteristics (e.g., [18,19]). The CuO/Cu_2O bi-layer is a potential candidate material for high-performance photoactive material for solar cells, as well as for photocathodes to generate hydrogen by photoelectrochemical water splitting. The bi-layer includes two p-type semiconductors with different bandgap energies, which is a strategy to realize a high-performance photovoltaic layer by extending the photovoltaic wavelength range and improving the quantum efficiency [6]. Since the photovoltaic performance is highly dependent on its semiconductor quality, which affects carrier transportation and recombination loss [6], the importance of local photoluminescence that allows the investigation of local semiconductor quality is evident.

The Cu_2O/CuO bi-layered film utilized in the previous research [12] was prepared form the electrodeposited Cu_2O films by annealing at 673 K for 3.6 ks (1 h) in air [20]. It has been reported that the luminescence properties of the electrodeposited Cu_2O and CuO formed by annealing are different [21]. Currently, hybrid composites such as the bi-layered film of CuO/Cu_2O attract the interest of many, and have been investigated as high-efficiency photocathodes for photoelectrochemical hydrogen evolution reaction [22,23] and electrode materials for batteries [24,25]. The Cu_2O/CuO bi-layered film, the luminescence properties of which were reported and established, is very suitable for sample evaluation in this study to test and develop the improved local mapping method. The FTO film functions as a transparent conductive layer to make electrodeposition possible, while ZnO plays the roles of a conductive film and an n-type semiconductor that forms an n-p junction with a p-type semiconductor of Cu_2O.

In this study, the further improvement of spatial resolution was attempted by using a fine focused X-ray beam set-up (beamline 20XU, Japan Synchrotron Radiation Research Institute (SPring-8), Sayo, Hyogo, Japan) [11]. Furthermore, the energy detection range was extended by installing a high-sensitivity PL-detector (Otsuka Electronics Co., Ltd., Hirakata-shi, Osaka, Japan). It was noticed that a trade-off relationship between the resolution and detection exists, because a smaller beam makes PL detection difficult. The problems that came to light while conducting this experiment and the possibility of an improved method are discussed.

2. Materials and Methods

2.1. Samples

The experimental sample for this study is a bi-layer film constituted of cupric oxide (CuO) and cuprous oxide (Cu_2O). The film was prepared on F:SnO_2 (FTO)/soda-lime glass (SLG) substrates (AGC Fabritech Co. Ltd., Minato, Tokyo, Japan) by electrodeposition in an aqueous solution [26,27]. First, the ZnO layer was prepared by electrodeposition on the substrate in an aqueous solution containing an 80 mmol/L zinc nitrate hydrate (Nacalai Tesque Inc., Nakagyo, Kyoto, Japan) at -0.8 V referenced to an Ag/AgCl electrode and 335 K for an electric charge of 0.5 C cm^{-2} using a potentiostat (Hokuto Denko, HAL 3000, Megro, Tokyo, Japan) connected to a coulomb meter (Hokuto Denko, HF 301, Megro, Tokyo, Japan). The solution was prepared using reagent grade chemicals and deionized water (purified with Milli Pore Ellix-UV-Advantage) (Merck KGaA, Darmstadt, Germany). Next, an aqueous solution with a pH of 13.0 and containing a 0.3 mol/L copper (II) sulfate hydrate, 0.3 mol/L tartaric acid, and 1.5 mol/L sodium hydroxide was used for the electrodeposition of the CuO/Cu_2O bi-layer. The CuO/Cu_2O bi-layer was fabricated by automatically switching the potential at 0.4 V for the CuO layer and at -0.4 V for the Cu_2O layer for a total absolute electric charge of 1 C cm^{-2} at 323 K with a polarization system (Hokuto Denko, HSV-110, Megro, Tokyo, Japan) under light-irradiation by a high-pressure mercury lamp (USHIO, OPTICAL-MODULEX, 500W) (Ushio, Inc., Chiyoda, Tokyo, Japan) [28]. Ag/AgCl and Pt electrodes were used as the reference and counter electrodes. Subsequently, the samples were cut into smaller specimens with the dimension of 5 mm square using a glass-cutter for the synchrotron experiment.

Figure 1 shows the SEM (JEOL Ltd., JSM6700F, Akishima, Tokyo, Japan) image of the cross section of the prepared sample. The stacked layers of ZnO, Cu_2O and CuO can be viewed in this cross-sectional image. The thickness of the upper CuO layer observed here is thinner than the lower Cu_2O layer. The ZnO layer is observed near the FTO substrate. The thicknesses of ZnO, Cu_2O, and CuO were approximated at 0.3 μm, 1.3 μm, and 0.8 μm, respectively. The total thickness of the film was approximately 3–4 μm. The structures of the prepared Cu_2O and CuO films by a similar process have also been confirmed by means of XRD (Rigaku Corp., RINT 2500, Akishima, Tokyo, Japan) inspection, as reported by Izaki et al. [28].

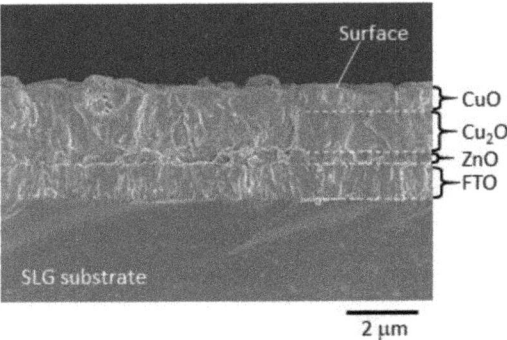

Figure 1. The cross-sectional SEM image for the CuO/Cu_2O bi-layered film formed on a ZnO/FTO-coated substrate.

2.2. Synchrotron Experiment

PL mapping was carried out at the first experimental hutch of BL20XU in the Japanese synchrotron facility, SPring-8. A schematic illustration of the experimental set-up used in this study is shown in Figure 2. The monochromatic X-ray energy of 10 keV was chosen by using an (111) Si double crystal monochromator (standard type of SPring-8, Japan). A probe beam was generated by using a Fresnel zone plate (FZP) as a focal beam [11]. The sample was set at the focal plane of the FZP. The FZP's zone material was made of tantalum with 1 μm thickness and a diameter of 310 μm, as well as an outermost zone width of 50 nm and a focal length of 625 mm at 10 keV. Vertical and horizontal slits were installed to cut off scattering beams. The size of the focused beam can be estimated at 0.3 μm in width and 0.3 μm in height of the sample position. The width of the beam used in this study was approximately 16 times smaller than the beam used in the previous research [12].

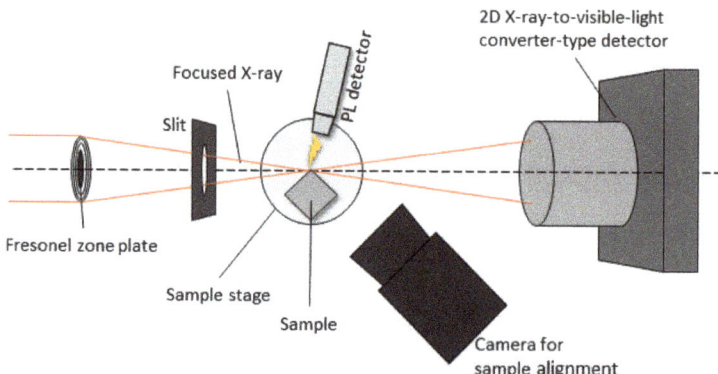

Figure 2. The experimental set-up used in this study.

A small piece of the deposited CuO/Cu$_2$O/ZnO substrate was mounted horizontally on a stage, and a focused X-ray beam was irradiated at the square corner of the sample, as shown in Figure 2. The sample stage had high-precision drive mechanisms for the horizontal, vertical, and rotational movement. Initially, the sample position was roughly adjusted based on the camera image for sample alignment. The detailed position was further calibrated using a 2D X-ray-to-visible light converter-type detector, which consisted of a scintillator, an optical lens, and a CMOS camera (Hamamatsu Photonics K.K., ORCA-Flash4.0, Hamamatsu, Shizuoka, Japan), which was placed 150 mm behind the sample (see Figure 3).

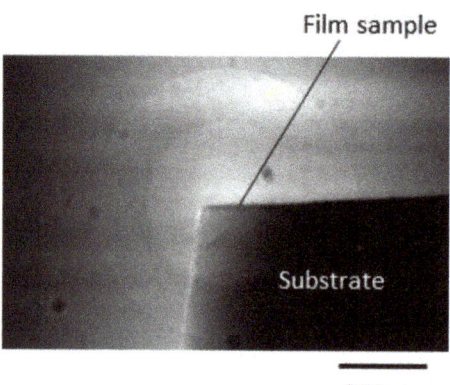

Figure 3. Radiograph obtained by the 2D X-ray detector.

A light-receiving fiber for the PL emission and an array spectrometer MCPD-9800 (Otsuka Electronics Co., Ltd., Hirakata-shi, Osaka, Japan) were placed roughly perpendicular to the X-ray beam. The spectrometer consisted of flexible optical fiber, slits, grating, and array-detecting elements. It can measure light with wavelengths of 360–1100 nm with high sensitivity. The light-receiving fiber was installed at the position where the intensity of light emission became the maximum. The X-ray beam intensity was sufficient to observe PL emissions in the sample of this study. Two-dimensional scans of the PL (i.e., 2D mapping) were performed on the CuO/Cu_2O bi-layered film. The emission spectra were collected by the spectrometer for an exposure time of 10 s at each X-ray irradiation position.

3. Results and Discussion

The photoluminescence spectra obtained by the spectrometer are shown in Figure 4. In this case, the focused synchrotron radiation beam scanned the CuO/Cu_2O film in the depth direction by steps of 0.3 µm. Peaks around 1.4 eV, 2.4 eV, and 3.3 eV were found in the spectra, depending on the scanned depth. Very strong peaks around the vicinity of 3.6 eV are found in almost all of the scanned positions.

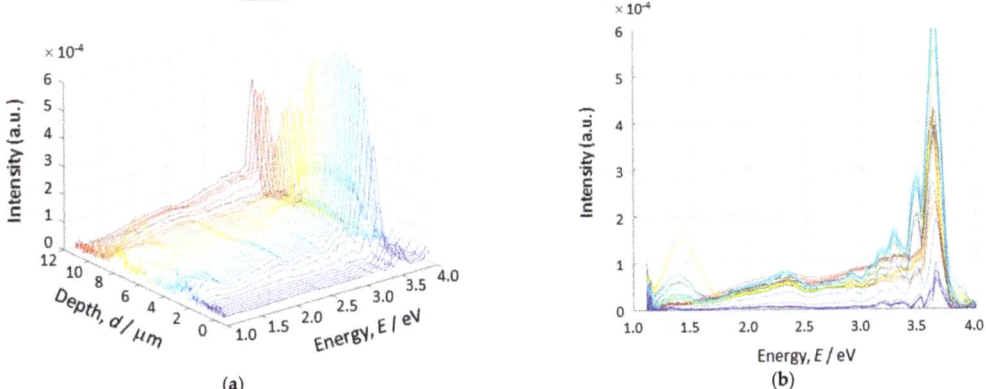

Figure 4. The photoluminescence spectra obtained by the spectrometer during a depth scan. (**a**) Three-axis plot of the emission energy, position (depth), and emission intensity, and (**b**) intensity vs. energy, in which several peaks can be found.

According to the reported references [20,26,29], 2.0 eV–2.1 eV visible light emitted by Cu_2O can be attributed to the direct recombination of the photon-assisted excitons. Additionally, Cu_2O emitted 1.52 eV-light as a defect-related emission [26,30,31]). With regard to CuO, it was reported that the light emission at 1.32 eV [32], 1.4 eV [33], and 1.38–1.56 eV [34] were due to bandgap energies. It was also reported that the (0001)-oriented ZnO layer emits not only near-band emissions at 3.25 eV –3.3 eV by recombination [35], but also visible light emissions at 2.28 eV and 2.8 eV [36]. In addition, the substrates possess emission peaks at 1.9 eV and 2.7 eV [12]. The obtained peaks in Figure 4 are related to the emission light energies reported for each material constituting the film sample. It was noted that the intensity of light emission was weak compared to the previous work [12], due to a smaller-sized X-ray beam. Nonetheless, despite reducing the X-ray beam's size, the local measurement of photoluminescence spectra was successful, although a delicate set-up for the spectrometer was necessary. The light emission spectra obtained at each position were then converted into intensity maps with individual energies, in order to understand the detailed relationship between localized luminescence and film structures.

Figure 5 shows the intensity maps of the measured emissions at (**a**) 1.4 eV, (**b**) 2.0 eV, (**c**) 3.3 eV, and (**d**) 3.6 eV. The vertical and horizontal axes of the figure correspond to the directions parallel to the film's depth and the substrate's surface plane, respectively. In the absence of light emission, we found the film surface placed at the depth position of about 4 µm. Horizontal bands of high intensities are clearly visible in (a), (c), and (d). An extremely weak emission was observed for the whole area of the sample in (b). The high-intensity bands observed in (a), (c), and (d) are not flat. This seems to reflect the film structures at an improved spatial resolution by using the finer focus beam. However, the initial curves of intensity bands observed in the vicinity of zero-position of the horizontal x-axis may have been due to the initial X-ray beam drift.

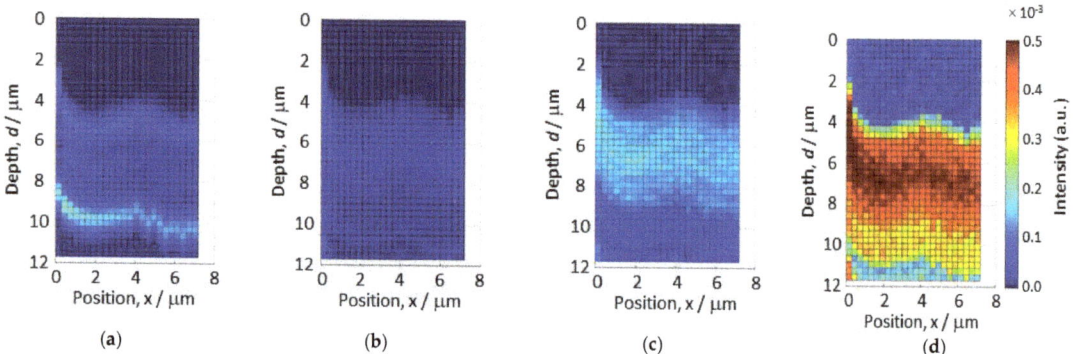

Figure 5. Light emission maps at different energies: (**a**) 1.4 eV, (**b**) 2.0 eV, (**c**) 3.3 eV, and (**d**) 3.6 eV, in CuO/Cu$_2$O bi-layered film prepared on the ZnO/FTO/SLG substrate.

The emissions observed at 1.4 eV can be associated with Cu$_2$O and CuO, as mentioned earlier in this section. Two separated narrow bands are recognized in (a), although CuO and Cu$_2$O layers are stacked in the prepared film. The intensity of the band nearer to the surface is weaker than that of the lower band. Emissions of 2.0 eV shown in (b) are expected in Cu$_2$O. However, no clear light emission was obtained in this study; only weak light was observed. Light emissions at 3.3 eV could not be obtained in the previous work [12] because 3.3 eV was out of range for the spectrometer used in the last experiment. The emission at 3.3 eV can be related to the ZnO layer. As shown in the map (c), light emissions were obtained in a depth range from 4 µm to 8 µm. In this study, it was found that not only spatial resolution but also energy range was extended. When one observes the 3.6 eV map (d), which would almost reflect the whole sample structure in detail, it is revealed that the film thickness observed from the light emission map of the X-ray beam is thicker than that observed by SEM (see Figure 1). The SEM observation of the samples was carried out again after the synchrotron experiment, to confirm the experimental situation of the X-ray scan.

Figures 6 and 7 show the SEM images of the X-ray scanned sample. Top views are shown in Figure 6. A missing part of the film is recognized at the corner of the substrate, as shown in (a). This was due to the pushing and cracking of the substrates with a glass cutter while the small samples were being prepared to be mounted onto the stage. The shape of the missing area was arc-shaped. Magnified images at the end of the arc can be seen in (b) and (c). The bi-layered CuO/Cu$_2$O film seemed to be undamaged by the X-ray irradiation. The non-flat surface in the targeted area observed in the SEM images corresponds to the surface characteristic obtained in Figure 5.

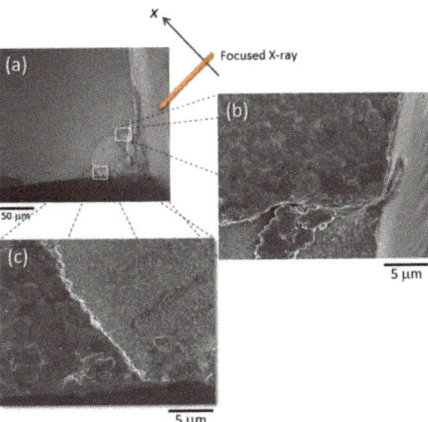

Figure 6. Top views of X-ray scanned sample observed by SEM. Image of the sample corner, which was scanned by X-ray (indicated by an arrow), in low magnification are shown in (**a**). The magnified images within (**a**) are shown in (**b**,**c**).

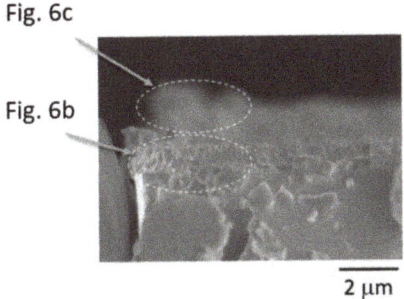

Figure 7. Side view of the X-ray-scanned sample by SEM.

The sample image observed from the side of X-ray irradiation is shown in Figure 7. The two arc edges are shown in Figure 6b,c. Since the mounting of the substrate on sample stages was reproducible for both the X-ray experiment and SEM observation, this shows that there was a probable difference in film height at different film positions. The difference of film thickness in the X-ray scan (Figure 5) and SEM image (Figure 1) can also be explained. Two narrow bands were observed at 1.4 eV separately, as shown in Figure 5a. These emissions came from the CuO and Cu_2O layers. The origin of the two observed bands is understandable by referring to the SEM image shown in Figure 7; that is, emissions occurred not only for the front side but also the rear side exposed to the front view. As the light emission was obtained at 3.3 eV, which corresponds to ZnO emissions distributed through a depth range from 4 μm to 8 μm, the mounted substrate might be slightly tilted. The results obtained from the comparison of X-ray scans and SEM observations strongly point to the importance of sample alignment in order to improve the spatial resolution of X-ray scanning. The preparation method of the sample pieces should also be improved. Two-dimensional PL mapping was achieved in this study, but the utilization of a tomographic technique can be considered to obtain a three-dimensional PL map. The tomographic method, which reconstructs cross-sections of the sample, could solve the issues of sample condition and sample alignment, as revealed in this study.

4. Conclusions

In this study, the scanning of local photoluminescence was attempted on a CuO/Cu$_2$O bi-layered film formed on a ZnO/FTO substrate by a size-reduced, focused X-ray beam in comparison to the previous study. Although there is a trade-off, with the increase in spectrum collection time due to the decrease in beam intensity, the mapping of photoluminescence influenced by microstructures was possible by utilizing a high-sensitivity spectrometer. Acquisition of the two-dimensional PL maps was successfully accomplished through 0.3 × 0.3 step scanning by applying a similar set-up, which recorded a 65 nm-size focusing beam, as reported in [11], although such fine step had only been available for vertical scan in the previous study [12]. The wavelength range measurable by localized photoluminescence was also extended to over 3.1 eV by the spectrometer. However, it was revealed that the utilization of a small beam makes sample alignment difficult. Comparison of PL maps and SEM images of the scanned sample indicated the importance of sample alignment to measure localized PL correctly. Proper care should also be taken to prevent the film from being damaged. These results gave insight for future trials and improvisations towards a high-resolution localized PL mapping, which are expected in the near future.

Author Contributions: Sample preparation, M.I. and P.L.K.; synchrotron experiment, M.K., M.I., P.L.K. and T.S.; synchrotron methodology, A.T. and K.U.; data validation, T.S., A.T. and K.U.; PL data analysis, M.K.; microstructure investigation, T.S.; writing—original draft preparation, M.K.; writing—review and editing, M.I., P.L.K. and A.T.; project administration, M.K. and M.I. All authors have read and agreed to the published version of the manuscript.

Funding: This work was supported in part by a JSPS Grant-in-Aid for Scientific Research (KAKENHI), grant number JP19H02810.

Institutional Review Board Statement: Not applicable.

Informed Consent Statement: Not applicable.

Data Availability Statement: Data will be made available upon reasonable request.

Acknowledgments: The synchrotron radiation experiment in this study was performed with the approval of JASRI (SPring-8) through proposal no. 2017B1404 and 2018A1329.

Conflicts of Interest: The authors declare no conflict of interest.

References

1. Lee, T.D.; Ebong, A.U. A review of thin film solar cell technologies and challenges. *Renew Sustain. Energy Rev.* **2017**, *70*, 1286. [CrossRef]
2. Musselman, K.P.; Marin, A.; Schmidt-Mende, L.; MacManus-Driscoll, J.L. Incompatible Length Scales in Nanostructured Cu$_2$O Solar Cells. *Adv. Fanct. Mater.* **2012**, *22*, 2202. [CrossRef]
3. Wang, Y.; Steigert, A.; Yin, G.; Parvan, V.; Klenk, R.; Schlatmann, R.; Lauermann, I. Cu$_2$O as a Potential Intermediate Transparent Conducting Oxide Layer for Monolithic Perovskite-CIGSe Tandem Solar Cells. *Phys. State Solidi C* **2017**, *14*, 1700164.
4. Frazer, L.; Lenferink, E.J.; Chang, K.B.; Poeppelmeier, K.R.; Stern, N.P.; Ketterson, J.B. Evaluation of defects in cuprous oxide through exciton luminescence imaging. *J. Lumin.* **2015**, *159*, 294–302. [CrossRef]
5. Rodenbücher, C.; Gensch, T.; Speier, W.; Breuer, U.; Pilch, M.; Hardtdegen, H.; Mikulics, M.; Zych, E.; Waser, R.; Szot, K. Inhomogeneity of donor doping in SrTiO$_3$ substrates studied by fluorescence-lifetime imaging microscopy. *Appl. Phys. Lett.* **2013**, *103*, 162904. [CrossRef]
6. Izaki, M.; Fukazawa, K.; Sato, K.; Khoo, P.L.; Kobayashi, M.; Takeuchi, A.; Uesugi, K. Defect Structure and Photovoltaic Characteristics of Internally Stacked CuO/Cu$_2$O Photoactive Layer Prepared by Electrodeposition and Heating. *ACS Appl. Energy Mater.* **2019**, *2*, 4833. [CrossRef]
7. Zimmerer, G. Luminescence spectroscopy with synchrotron radiation: History, highlights, future. *J. Lumin.* **2006**, *119*, 1–7. [CrossRef]
8. Zimmerer, G. SUPERLUMI: A unique setup for luminescence spectroscopy with synchrotron radiation. *Radiat. Meas.* **2007**, *42*, 859–864. [CrossRef]
9. Pankratov, V.; Popov, A.I.; Kotlov, A.; Feldmann, C. Luminescence of nano-and macrosized LaPO$_4$:Ce,Tb excited by synchrotron radiation. *Opt. Mater.* **2011**, *33*, 1102–1105. [CrossRef]

10. Zorenko, T.; Grbenko, V.; Safronova, N.; Matveevskaya, N.; Yavetskiy, R.; Babayevska, N.; Zorenko, Y. Comparative study of the luminescent properties of oxide compounds under synchrotron radiation excitation: Lu_2O_3:Eu nanopowders, ceramics and films. *J. Lumin.* **2018**, *199*, 461–464. [CrossRef]
11. Takeuchi, A.; Uesugi, K.; Suzuki, Y.; Itabashi, S.; Oda, M. Fresnel zone plate with apodized aperture for hard X-ray Gaussian beam optics. *J. Synchrotron Rad.* **2017**, *24*, 586–594. [CrossRef]
12. Kobayashi, M.; Izaki, M.; Shinagawa, T.; Takeuchi, A.; Uesugi, K. Localized Photoluminescence Imaging of Bi-Layered Cuprous/Cupric Oxide Semiconductor Films by Synchrotron Radiation. *Phys. Status Solidi B* **2018**, *256*, 1800119. [CrossRef]
13. Izaki, M.; Shinagawa, T.; Mizuno, K.-T.; Ida, Y.; Inaba, M.; Tasaka, A. Electrochemically constructed p-Cu_2O/n-ZnO heterojunction diode for photovoltaic device. *J. Phys. D Appl. Phys.* **2007**, *40*, 3326–3329. [CrossRef]
14. Musselman, K.P.; Wisnet, A.; Iza, D.C.; Hasse, H.C.; Scheu, C.; MacManus-Discoll, J.L.; Schmidt-Mende, L. Strong Efficiency Improvements in Ultra-low-Cost Inorganic Nanowire Solar Cells. *Adv. Mater.* **2010**, *22*, E254. [CrossRef]
15. Zuo, C.; Ding, L. Solution-Processed Cu_2O and CuO as Hole Transport Materials for Efficient Perovskite Solar Cells. *Small* **2015**, *11*, 5528. [CrossRef] [PubMed]
16. Park, S.-G.; Miyake, M.; Yang, S.-M.; Braun, P.V.; Wiltzius, P. Cu_2O Inverse Woodpile Photonic Crystals by Prism Holographic Lithography and Electrodeposition. *Adv. Mater.* **2011**, *23*, 2749. [CrossRef] [PubMed]
17. Geng, Z.; Zhang, Y.; Yuan, X.; Huo, M.; Zhao, Y.; Lu, Y.; Qiu, Y. Incorporation of Cu_2O nanocrystals into TiO_2 photonic crystal for enhanced UV–visible light driven photocatalysis. *J. Alloy. Comp.* **2015**, *644*, 734. [CrossRef]
18. Grez, P.; Herrera, F.; Riveros, G.; Ramírez, A.; Henríquez, R.; Dalchiele, E.; Schrebler, R. Morphological, structural, and photoelectrochemical characterization of n-type Cu_2O thin films obtained by electrodeposition. *Phys. Status Solidi A* **2012**, *209*, 2470. [CrossRef]
19. Benz, J.; Hering, K.P.; Kramm, B.; Polity, A.; Klar, P.J.; Siah, S.C.; Buonassis, T. The influence of nitrogen doping on the electrical and vibrational properties of Cu_2O. *Phys. Status Solidi B* **2017**, *254*, 1600421. [CrossRef]
20. Meyer, B.K.; Polity, A.; Rappin, D.; Becker, M.; Hering, P.; Klar, P.J.; Sander, T.; Reindl, C.; Benz, J.; Eickhoff, M.; et al. Binary copper oxide semiconductors: From materials towards devices. *Phys. Status Solidi B* **2012**, *249*, 1487. [CrossRef]
21. Chang, K.B.; Frazer, L.; Schwartz, J.J.; Ketterson, J.B.; Poeppelmeier, K.R. Removal of Copper Vacancies in Cuprous Oxide Single Crystals Grown by the Floating Zone Method. *Cryst. Growth Des.* **2013**, *13*, 4914–4922. [CrossRef]
22. Yang, Y.; Xu, D.; Wu, Q.; Diao, P. Cu_2O/CuO Bilayered Composite as a High-Efficiency Photocathode for Photoelectrochemical Hydrogen Evolution Reaction. *Sci. Rep.* **2016**, *6*, 35158. [CrossRef]
23. Jamali, S.; Moshaii, A.; Mohammadian, N. Improvement of Photoelectrochemical and Stability Properties of Electrodeposited Cu_2O Thin Films by Annealing Processes. *Phys. Status Solidi A* **2017**, *214*, 1700380. [CrossRef]
24. Kim, A.-Y.; Kim, M.K.; Cho, K.; Woo, J.-Y.; Lee, Y.; Han, S.-H.; Byun, D.; Choi, W.; Lee, J.K. One-Step Catalytic Synthesis of CuO/Cu_2O in a Graphitized Porous C Matrix Derived from the Cu-Based Metal-Organic Framework for Li- and Na-Ion Batteries. *ACS Appl. Mater. Interfaces* **2016**, *8*, 19514. [CrossRef]
25. Wu, S.; Fu, G.; Lv, W.; Wei, J.; Chen, W.; Yi, H.; Gu, M.; Bai, X.; Zhu, L.; Tan, C.; et al. A Single-Step Hydrothermal Route to 3D Hierarchical Cu_2O/CuO/rGO Nanosheets as High-Performance Anode of Lithium-Ion Batteries. *Small* **2018**, *14*, 1702667.
26. Izaki, M.; Sasaki, S.; Mohamad, F.B.; Shinagawa, T.; Ohta, T.; Watase, S.; Sasano, J. Effects of preparation temperature on optical and electrical characteristics of (111)-oriented Cu_2O films electrodeposited on (111)-Au film. *Thin Solid Film.* **2012**, *520*, 1779–1783. [CrossRef]
27. Shinagawa, T.; Onoda, M.; Fariza, B.M.; Sasano, J.; Izaki, M. Annealing effects and photoelectric properties of single-oriented Cu_2O films electrodeposited on Au(111)/Si(100) substrates. *J. Mater. Chem. A* **2013**, *1*, 9182–9188. [CrossRef]
28. Izaki, M.; Koyama, T.; Khoo, P.L.; Shinagawa, T. Light-Irradiated Electrochemical Direct Construction of Cu_2O/CuO Bilayers by Switching Cathodic/Anodic Polarization in Copper(II)-Tartrate Complex Aqueous Solution. *ACS Omega* **2020**, *5*, 683–691. [CrossRef]
29. Ray, S.C. Preparation of copper oxide thin film by the sol-gel-like dip technique and study of their structural and optical properties. *Sol. Energy Mater. Sol. Cells* **2001**, *68*, 307–312. [CrossRef]
30. Terui, Y.; Fujita, M.; Miyakita, Y.; Sogoshi, N.; Nakabayashi, S. Photoluminescence of Electrochemically-Deposited Granular Cu_2O Films. *Trans. Mater. Res. Soc. Jpn.* **2005**, *30*, 1049–1052.
31. Scanlon, D.O.; Morgan, B.J.; Watson, G.W. Modeling the polaronic nature of p-type defects in Cu_2O: The failure of GGA and GGA+U. *J. Chem. Phys.* **2009**, *131*, 124703. [CrossRef] [PubMed]
32. Wang, L.; Han, K.; Tao, M. Effect of Substrate Etching on Electrical Properties of Electrochemically Deposited CuO. *J. Electrochem. Soc.* **2007**, *154*, D91–D94. [CrossRef]
33. Izaki, M.; Nagai, M.; Maeda, K.; Farina, F.B.; Motomura, K.; Sasano, J.; Shinagawa, T.; Watase, S. Electrodeposition of 1.4-eV-Bandgap p-Copper (II) Oxide Film with Excellent Photoactivity. *J. Electrochem. Soc.* **2011**, *158*, D578–D584. [CrossRef]
34. Nakaoka, K.; Ueyama, J.; Ogura, K. Photoelectrochemical Behavior of Electrodeposited CuO and Cu_2O Thin Films on Conducting Substrates. *J. Electrochem. Soc.* **2004**, *151*, C661–C665. [CrossRef]
35. Yamamoto, A.; Miyajima, K.; Goto, T.; Ko, H.J.; Yao, T. Biexciton luminescence in high-quality ZnO epitaxial thin films. *J. Appl. Phys.* **2001**, *90*, 4973. [CrossRef]
36. Izaki, M.; Watase, S.; Takahashi, H. Low-Temperature Electrodeposition of Room-Temperature Ultraviolet-Light-Emitting Zinc Oxide. *Adv. Mater.* **2003**, *15*, 2000–2002. [CrossRef]

MDPI
St. Alban-Anlage 66
4052 Basel
Switzerland
Tel. +41 61 683 77 34
Fax +41 61 302 89 18
www.mdpi.com

Materials Editorial Office
E-mail: materials@mdpi.com
www.mdpi.com/journal/materials

www.ingramcontent.com/pod-product-compliance
Lightning Source LLC
LaVergne TN
LVHW070154100526
838202LV00015B/1944